Process and Pattern
in Evolution

Oxford University Press

Oxford   New York   Toronto
Delhi   Bombay   Calcutta   Madras   Karachi
Petaling Jaya   Singapore   Hong Kong   Tokyo
Nairobi   Dar es Salaam   Cape Town
Melbourne   Auckland

and associated companies in
Berlin   Ibadan

Library of Congress Cataloging-in-Publication Data
Avers, Charlotte J.
  Process and pattern in evolution.
  Includes bibliographies and index.
  1. Evolution. I. Title.
QH366.2.A933 1989   575   88-5368
SBN 0-19-505275-7

Process and Pattern
in Evolution

Charlotte J. Avers

Rutgers University

New York   Oxford
Oxford University Press
1989

9 8 7 6 5 4 3 2

Printed in the U

on acid-free pape

To Norman

Preface

Evolution provides a broad umbrella that covers and integrates virtually all of biology and is made even more sweeping by contributions from astronomers, chemists, geologists, philosophers, physicists, and others. The scope of biological input to the study of evolution includes empirical and experimental data provided by anthropologists, behaviorists, biochemists, cell and developmental biologists, ecologists, geneticists, molecular biologists, paleontologists, physiologists, systematists, and others. Although it was not possible to devote an appropriate amount of space to each of these sets of contributors, I tried in this book to include highlights from most of them. Because of my own interests and perspective, some of these inputs to evolutionary biology have been given greater emphasis than others. Particular emphasis is placed on the genetic contexts in evolutionary studies. For this reason, students should have taken a course in genetics and should know the fundamentals of organic chemistry.

The text is divided into three parts, of four chapters each. Part I sets the historical, physical, and chemical framework for the origin and evolution of primeval life, and proceeds to discussions of the evolution in early life forms of the genetic code, the genome, gene organization, metabolic pathways, and cellular organization. Part II concentrates on evolutionary processes that underlie the diversity of gene organization, function, and expression; natural selection and neutral evolution; changes in gene frequencies in populations; and tempos and modes of speciation. Part III deals with patterns and trends in evolution that have been deduced from morphological and molecular phylogenetic analysis and from geologic and paleontological studies. The last two chapters present major evolutionary concepts and approaches exemplified in studies of the evolution of the human family, from its ancestral vertebrate beginnings to *Homo sapiens*.

More than 400 illustrations and tables, many from the original research literature, serve to enhance the discussions of evolutionary concepts and theories and the data from which they have been inferred. Each chapter concludes with a summary of the highlights of the topics discussed and an extensive list of references plus additional readings that provide students with the opportunity to explore further the ongoing intellectual challenge in appreciating and understanding the panorama of biological evolution on Earth.

I am grateful to my colleagues who graciously provided photographs, gave thoughtful comments and suggestions, and reviewed the prospectus and the manuscript of this book. In particular, I wish to thank the four scientists who provided detailed and helpful reviews of substantial portions of the final drafts of the manuscript.

Special thanks are due my editor, William Curtis, for his encouragement and good cheer throughout the preparation and production of this book, and Irene Pavitt, who cleaned up many awkward and redundant phrases in the manuscript. I know the book is better for the efforts made by the staff of Oxford University Press.

C.J.A.

Abbreviated Contents

Contents

II Evolutionary Processes

Process and Pattern
in Evolution

PART I

Foundations of Life on Earth

CHAPTER 1

The Origins and History of Evolutionary Ideas

The origins of evolutionary theory can be traced to theological and philosophical traditions rooted in Judeo-Christian and Greco-Roman thinking dating back more than 2000 years. During this time, many changes in perspective have occurred in response to new and modified philosophies and information. The gains of the past few hundred years in particular, have been due to new insights, methods, and questions, and to the willingness to change perspective in response to a broadened base for gathering and integrating knowledge from many disciplines. Modern evolutionary theory faces its own challenges today not only from scholars, but also from religious fundamentalists. We can better appreciate and understand the issues involved in modern evolutionary study and theory if we review some historical developments that influenced and shaped various patterns of thinking about the subject.

The Growth of Natural History

The concept of a static world emerged from Western theological explanations of Creation as a unique event that included the formation of the universe and all forms of life, as described in Scripture. The Greeks and Romans had held to the precept that nature and time were cyclical and that change reflected the guidance of a cosmic intelligence. Orderliness was the expected outcome of supernatural causes imposed on the world, and was beyond the physical limits of nature. Through the centuries, philosophers and theologians sought the divine plan of nature. The development of the physical sciences during the Renaissance led to broadened views of the world, which eventually challenged religious doctrine. It was not until the nineteenth century, however, that the biological implications of these new concepts were recognized and incorporated into the prevailing framework of scholarly thinking.

Early Concepts of Time and the Universe

Western Scripture provided the basis for the view that plants and animals were descended in unchanged form from the creatures that had appeared during the week of Creation. Explanations of the creation of the universe and of all its features were to be found particularly in the first eleven chapters of Genesis. This view of a single event giving rise to all of existence was at odds with the Greco-Roman tradition of a succession of worlds arising and waning in cyclic fashion but saved from doom and destruction toward the end of each cycle by the intervention of the gods. The early Christian theologians were faced with the problem of integrating the intellectually valued pagan literature with Scripture and preserving those concepts that were compatible with orthodox Christian thinking.

The notion of cycles, which contradicted the belief in the one Creation, the Atonement through Christ, and the Day of Judgment, was largely discarded in the writings of Saint Augustine in the fifth century. The cyclic elements of degradation and divine intervention were retained, but catastrophes were limited to the Flood and Day of Judgment. Saint Augustine and Saint Thomas Aquinas, in the thirteenth century, were largely responsible for laying the theological groundwork for evolutionary ideas. The biblical Creation story was interpreted to mean that at the time of creation, God infused the Earth with the necessary potency to produce living things by a natural process that represented the unfolding of divine plan. The world was subject to constant divine intervention in the form of wrath or approval, which manifested themselves as natural catastrophes or miracles. The orderliness of the world was imposed by a supernatural cause, or divine intelligence, and was not subject to the physical laws of nature. The Earth was only a few thousand years old and was destined for destruction on the Day of Judgment. Both the universe and its living creatures were evolving in the period after Creation, but only in accordance with the divine plan stipulated by Augustine, Aquinas, and other early Christian theologians.

The fixed and limited view of the universe held in the Middle Ages was replaced by broader horizons, based on geographical and astronomical observations and study, in the sixteenth century. The new *natural philosophers* were confronted with the Copernican universe, in which the Earth no longer held the central position and the planets and stars were not affixed to perfect transparent spheres, as in the Ptolemaic universe. They looked to astronomy, mathematics, and physics for natural laws, set in place and maintained by the Creator, which could be comprehended by the human mind. In their search for details behind the plan of Creation, these scientists were not in conflict with religion because a study of nature permitted a better comprehension of divine intervention. This enlightened view did not extend to the grim, authoritarian orthodoxy of the Roman Catholic church, which confronted people labeled as heretics and blasphemers with the sure and certain fate of destruction by fire on Earth and in hell. In the seventeenth century, the offices of the Inquisition caused the astronomer Giordano Bruno to be burned at the stake and Galileo to recant his belief in the Copernican universe.

By the late seventeenth century, the studies of natural philosophers had led to unorthodox concepts of the creation and development of the Earth that were greater threats to religious tradition than were the new views of the universe or of its age. Speculation concerning the origin of the Earth and its antiquity could not be divorced from theories growing out of astronomical studies. The founding of the science of geology in the eighteenth century was an outgrowth of the changing perspectives about the Earth in the late seventeenth century. However, the new concepts of geologic change and terrestrial antiquity were not recognized as having biological implications until the nineteenth century. The religious philosophers, who were intent on demonstrating a grand design in nature, were resistant to the natural philosophy of science and remained apart from the intellectual ferment of the eighteenth century. By the nineteenth century, however, biololgy no longer served as safe refuge from perceived unorthodoxy.

Geologic Foundations for the Concept of Evolution

Fossils were known in the ancient Greco-Roman world and were regarded as the remains of organisms either from a former cycle or from an earlier time in the present cycle of worldwide changes. Marine fossils found far from the sea were presumed to have been stranded when the oceans receded, having been deeper and greater in an earlier time or cycle. Human bones were viewed as the remains of legendary heroes, and massive reptilian and mammalian bones were regarded as traces of a race of titans. Aristotle proposed the existence of a graded sequence of organisms composed of an indefinite number of links in a **chain of being** that ranged from the lowest forms to the gods. He did recognize natural groupings somewhat similar to the species identified by later naturalists, but he regarded them as immutable entities. Human beings stood between the known infrahuman primates and the epic heroes and demigods, maintaining this position in every cycle.

Although some early Christian scholars believed that fossils were the remains of plants and animals, they argued that they had been transported to the tops of mountains and other inaccessible places by the waters of the Flood. Others did not believe that fossils were remnants of former living things. They considered them to have originated through magical or mysterious causes and valued them for their medicinal or occult properties.

In the sixteenth century, Leonardo da Vinci and a handful of learned scholars considered fossils to be the remains of organisms from different times and places, and did not associate them with the Flood. The world view of the time, however, was that living things were immutable and that divine benevolence precluded the extinction of creatures fashioned during Creation. It was therefore presumed that living examples of these fossil creatures would be discovered in regions of the world that had not yet been explored. As naturalists explored remote areas, they found many unfamiliar organisms and relatively few that could be considered descendants of European fossil forms. Indeed, these lands yielded their own fossil forms, which were distinct from the known living or fossil European organisms. These problems were compounded by the failure to find living or fossil organisms that could bridge the gaps between the recognized forms in the chain of being, as well as by strong resistance to the idea of species extinction. Despite the puzzling variety and uniqueness of many living and fossil forms, the naturalist credo remained staunchly in favor of the unbroken chain of being, the immutability of species, and the impossibility of extinction.

In the eighteenth century, the study of the physical universe became separated from its earlier theological focus; such study was now referred to more often as *science*. Natural history, however, remained closely tied to theology because of many mysterious features and was thus more readily amenable to supernatural explanations, unlike astronomy, chemistry, and Newtonian physics. The earlier concepts of natural philosophy were replaced by a broader view, called *natural theology*, which sought to integrate data from physical science and natural history with Scriptural interpretations. Although celestial change had to be accepted, natural theologians held to the belief that the Earth and its life forms had not changed since Creation and that God guided historical and natural processes.

Geologic studies of the Earth progressed during the eighteenth century, as different schools of thought presented either theological or scientific explanations for the origin and history of the planet. By the early nineteenth century, two major geologic concepts were embraced by contending groups. James Hutton and his followers supported **uniformitarianism,** while Georges Cuvier and his supporters adhered to **catastrophism.** The uniformitarian view postulated that, given adequate time, the processes of deposition, erosion, and uplift working gradually and continuously, could account for past physical changes on the Earth, just as they accounted for present changes. The catastrophists maintained that the cataclysmic events of fire and flood had molded the planet over time and that the Flood had been the last of these catastrophes to have occurred. Both concepts assumed great age for the Earth, but neither was evolutionary because species were held to be immutable.

As the geologic sciences moved away from catastrophism and support for uniformitarianism grew, the links between geology and theology were broken, just as the links between the physical sciences and religion had been severed a century earlier. Theological premises were still very much a part of the biological sciences during the eighteenth century and began to give way only in the nineteenth. Although uniformitarians accepted a single creation event, based on the appearance of both simple and complex fossil forms in the same deposits, whereas catastrophists proposed multiple creation events be means of divine intervention following catastrophes, both concepts lent support to the occurrence of biological evolution for those naturalists who favored such a view.

The writings of Georges Buffon, Erasmus Darwin, Jean-Baptiste Lamarck, and others led to the idea that organisms were subject to change after their special creation, particularly in response to changing environmental pressures. By the mid-nineteenth century, a theory of organic **evolution** had fully developed. It included basic premises taken from geologic theory and biological mutability, as seen in the progression of forms from simple to complex, in the relationships between fossil and living forms, and in analogies between changes in species through artificial selection and through natural events. All these ideas culminated in the work of Charles Darwin in the mid-nineteenth century. The reality of evolutionary change, now widely accepted by most educated people, and the mechanisms of species mutability and extinction were dimly understood to be somehow related to hereditary properties acquired by species over time. Darwin's great contribution was to provide a plausible mechanism for the accepted fact of biological evolution. The controversies generated by Darwin's writings had to do with the postulated mechanism—natural selection—not with the reality of species mutability leading to evolutionary changes.

The Darwinian Revolution

Charles Darwin's *On the Origin of Species* had a profound impact on biology from the moment of its publication in 1859, and its central theory of **natural selection** has provided a framework for most of biology to the present day. The theory both provoked controversy and encouraged the study of evolutionary phenomena, and it continues to be controversial in some respects while giving direction to many different avenues of scientific inquiry. The concept of natural selection sparked a revolution in thinking about social issues, politics, and other areas of human concern.

The Integrating Principle of Natural Selection

Darwin began his career as the naturalist on the H.M.S. *Beagle*, whose crew had as its primary mission the detailed mapping of the coastlines of South America. During the five-year voyage (December 27, 1831, to October 2, 1836), Darwin collected many specimens and amassed a wealth of geologic and biological data that later caused him to question the theological interpretation of evolution. According to his autobiography, Darwin began to crystallize his ideas on evolution in 1838 after reading Thomas Malthus's *Essay on the Principle of Population* (1789). He referred to the Malthusian principle

a   b

Figure 1.1. (a) Charles Darwin and (b) Alfred Russel Wallace jointly proposed the theory of evolution by means of natural selection at a meeting in London of the Linnaean Society in 1858. (Photograph [a] courtesy of the American Museum of Natural History, New York)

of unchecked population increase in a world of finite resources, which leads to famine, and said he realized that in the struggle for existence among plants and animals, the favorable variations would be preserved and the unfavorable ones would be destroyed. It was not until 20 years later, however, that Darwin presented his ideas formally at a meeting of the Linnaean Society in London. Even then, he was virtually forced to reveal his ideas after he learned that Alfred Russel Wallace independently hit on the evolutionary mechanism of natural selection. Darwin and Wallace (Fig. 1.1) proposed the theory jointly at the London meeting in 1858.

The intellectual climate of Victorian England had long favored the idea of progress as part of the social philosophy that had developed in the eighteenth and early nineteenth centuries. Biological evolution was easier to accept philosophically at this time, and it began to be incorporated into the world view espoused by leading European theologians and social theorists. The British philosopher Herbert Spencer noted a connection between Darwin's theory of natural selection and his own theory of progressive social change. Spencer proposed that society functions as an organism and is responsive to a deterministic mechanism analogous to natural selection, leading progressively to the natural corollaries of a perfect social order and a superior biological condition. This thesis, which became known as **Social Darwinism,** appealed to those Europeans who believed that their policies of economic and social exploitation in Asia and Africa could now be justified biologically as well as culturally. Social Darwinists were convinced that Europeans and their descendants in various parts of the world had achieved political dominance because of natural selection, which had rendered the populations they controlled biologically and culturally less fit.

The disdain for Spencerian evolution influenced some nineteenth-century intellectuals to reject Darwinian evolution, just as the appeal of Social Darwinism influenced others to accept Darwinism. Although many people in the nineteenth and early twentieth centuries viewed Darwinian and Spencerian evolution as interchangeable, most contemporary biologists have little use for Social Darwinian philosophy and reject any notion of a relationship between evolutionary theory and social doctrine. In spite of various abusive uses of Darwinian evolutionary theory, it became the central focus of and a major influence on practically every aspect of biological science, as well as many other fields of science. Evolution is a guiding motif of molecular, cellular, organismic, and population studies in biology. Darwin can truly be said to have initiated a revolution in science.

The Inclusion of Humans in the Evolutionary Perspective

Darwin carefully avoided any controversy about the place of human beings in the drama of evolution, stating only briefly that his theory of natural selection might throw some light on human origins and history. Although the sciences were substituting empirical observation for the earlier underpinnings of theological doctrine, mid-nineteenth-century views were not amenable to the idea that humans might be modified by natural selection, even though they did accept this premise for other life forms. The rise of Social Darwinism, which is associated today with the bigotry, greed, and arrogance of the nineteenth-century Western power elites, helped pave the way for the acceptance of the idea of evolutionary modification of humans themselves.

At the time that the universe, the Earth, and nature were being examined inductively, the idea that human beings were a part of the dynamic evolutionary saga gradually became established. The fossil hominids discovered in Europe, beginning in 1856 with Neanderthal remains in Germany, were not particularly convincing evidence for human evolution. This was due in large part to the similarities between Neanderthals and modern human beings, both of whom are generally accepted to be members of the species *Homo sapiens*. But the discovery in Java in 1890 of middle Pleistocene fossil hominids provided strong support for a long evolutionary history of humankind, rather than a very special and immutable place in creation. The fossils found in Java and others discovered elsewhere in the first part of the twentieth century are sufficiently different from modern humans to clearly be the remains of creatures that lived a long time ago, and they are classified as a distinct species— *Homo erectus* (Fig. 1.2). It was presumed that *H. erectus* was an ancestor of modern *H. sapiens*.

H. erectus is clearly human in its anatomical features and in its possession of a tool-making tradition, so there was little real controversy over its evolutionary relationship to *H. sapiens*. But when Raymond Dart reported the discovery in South Africa in 1924 of an apelike prehuman skull, referred to

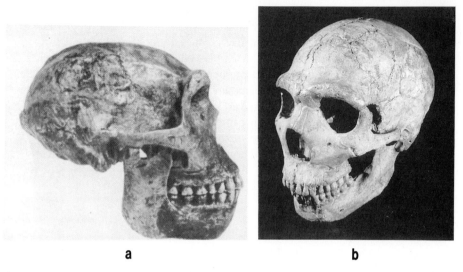

a   b

Figure 1.2. The discovery of (a) Java man in 1890 provided more compelling evidence for a long evolutionary history of humankind than did the earlier discovery in Europe of the remains of (b) Neanderthals, who were sufficiently similar to modern humans to be classified as members of the same species (*Homo sapiens*). Java man and similar fossils are classified as members of the species *H. erectus*. (Photograph courtesy of the Israel Department of Antiquities and Museums, Jerusalem)

as the Taung child (Fig. 1.3), and proposed it to be ancestral to *Homo*, there was almost total rejection of the presumed evolutionary connection. The inclusion of humans in the scheme of evolution became a foregone conclusion early on, but the particular nature of evolutionary change and of the family tree remained controversial for many years. The Taung skull was too apelike to be accepted as a human ancestor by early-twentieth-century paleontologists. In fact, the scandalous Piltdown fossil forgery was accepted as authentic for many years, despite its obvious doctored features, because it fit the belief—held until the middle of this century—that the brain increased in size very early in human evolution (Box 1.1). Controversies continue to surround human evolutionary history, but the emphasis today is on determining the correct interpretations of valid scientific data and not on deciding whether humans are descended from an apelike ancestral stock.

The current view places the human species in the common framework of organic evolution. We are not the central focus in evolutionary history, and, indeed, our own evolutionary past has been opened to more meaningful analysis through the utilization of empirical and theoretical evolutionary models and studies based on other living and fossil forms. Our special feature is our intelligent ability to exert considerable influence over the Earth and its life forms. It is a very different view from the pre-Darwinian vision of human beings as the products of special creation proceeding steadily toward perfection and as divorced from and superior to all other life forms by divine design.

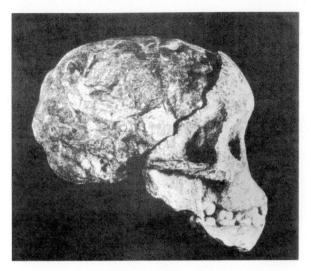

Figure 1.3. The skull of a 3-year-old child, which was found at Taung in South Africa in 1924, was proposed by Raymond Dart to represent an ancient ancestor of modern humans and was classified as a species of a new genus *(Australopithecus africanus)*.

The Molding of Evolutionary Theory Since Darwin

Darwin's book spurred broadly based investigations of evolutionary history by well-established methods of descriptive and comparative embryology and morphology, natural history, and paleontology. Classification schemes emphasized evolutionary relationships among organisms to reflect **descent with modification,** as Darwin had proposed (Fig. 1.4). The study of evolutionary mechanisms, however, lagged far behind. Not only was the nature of heredity poorly understood, but biologists were advocates of Lamarck's theory that organisms changed in response to their felt needs in particular environments. The great advances in understanding evolutionary mechanisms came in the twentieth century, with the discovery in 1900 of Gregor Mendel's earlier genetic studies and, over the following decades, the emergence of the concepts of population genetics and modes of speciation.

The Impact of Genetics on Evolutionary Theory

The discovery of Mendel's work demonstrating particulate inheritance had a negative effect on the acceptance of natural selection, contrary to what one might expect. The early geneticists emphasized discontinuous traits that clearly show Mendelian ratios and dismissed quantitative incremental variation as non-Mendelian and inconsequential in evolution. In addition, Hugo

Box 1.1 The Piltdown Hoax

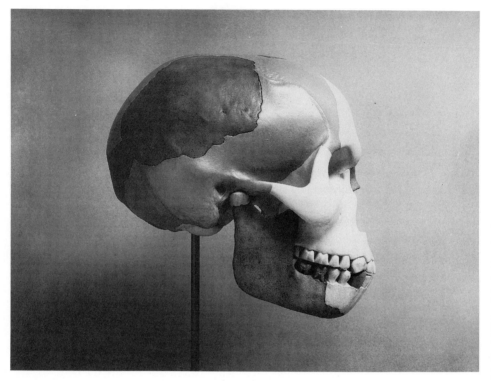

The reconstruction of the Piltdown skull. (Photograph courtesy of the American Museum of Natural History, New York)

On December 19, 1912, Charles Dawson and Arthur Smith Woodward announced the discovery of a Paleolithic human skull in a gravel pit at Piltdown in Sussex, England. Dawson, an amateur paleontologist, requested and received the help of Smith Woodward, a professional paleontologist, in identifying and interpreting the skull and many other fossil materials that he found in the gravel pit. The Piltdown skull was accepted enthusiastically by the British paleontological establishment, perhaps partly because it was the first evidence of early humans to be found in England. Until this time, the British had been treated to a profusion of fossils and artifacts discovered in France and Germany, and they now had fossil evidence that the first man was English. More to the point, however, the nature of the skull, with its humanlike cranium and distinctly apelike jaw and teeth, fit the prevailing ideas about human evolution. Human evolution was believed to have proceeded by the enlargement of the brain in an apelike stock and subsequent humanization of the rest of the skull and skeleton once the "bright" apes descended from the trees and adapted to life on the ground. Piltdown man, which was named *Eoanthropus dawsoni*, was shown in the 1950s to be a fake; it was a human cranium attached to an ape's jaw (recently shown to be an orangutan) and stained with potassium bichromate to make it appear fossilized. All the other fossils and artifacts in this gravel pit also were fraudulent.

Although Dawson, or even the young priest Teilhard de Chardin, who was allowed to work with the material in the gravel pit, was suspected to be the perpetrator of the hoax, it is most likely that both were innocent. The actual forger remains unknown, but strong suspicion has been directed against W. J. Sollas, who had access to all the materials associated with the hoax. Sollas, a distinguished geologist at Oxford, was a bitter opponent of Smith Woodward. It is possible the he engineered the fraud to embarrass Smith Woodward, but was deterred from revealing it by the unexpectedly enthusiastic reception of Piltdown man. We may never know for sure.

de Vries's *Mutation Theory* postulated the emergence of species in one or a few steps by means of discrete mutations that cause major morphological changes, which makes natural selection and the gradual accumulation of small changes irrelevant to the evolution of species. As genetic studies proceeded into the 1920s, it seemed that mutations were generally harmful, and only a few mutations appeared to affect important features of an organism. Biologists thus found it difficult to relate such laboratory observations to the variations actually seen in wild species.

During the 1920s and 1930s, genetics and evolution were reconciled, and the genetic foundation for evolutionary change was firmly established. Genetic studies of natural populations revealed the existence of a wealth of variation and included continuous as well as discontinuous traits as the hereditary raw material for evolution. The emergence of a framework of theoretical population genetics, based largely on the work of Sewall Wright, John B. S. Haldane, Sir Ronald Fisher, Theodosius Dobzhansky, and others, provided an essential perspective for understanding the relative significance of different influences on evolutionary changes in different populations at different times

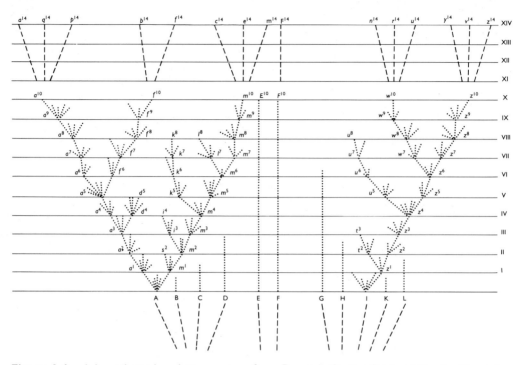

Figure 1.4. A hypothetical evolutionary tree from Darwin's *On the Origin of Species* illustrating descent with modification through natural selection. Eleven different species (A–L) of a genus, shown at the bottom of the illustration, diverge over time, shown by horizontal lines labeled with Roman numerals and each representing 1000 or more generations. Some of the original species, such as A and I, diverge more than others and produce new varieties, such as a^1, m^1, and z^1. These new varieties, in turn, diverge and perhaps give rise to new species, such as p^{14}, b^{14}, and, n^{14}, after thousands of generations. Darwin believed that some species diverge enough to produce new species. Others, such as E and F, remain relatively unchanged or, more likely, go extinct (B, C, D, G, H, K, L).

and places. The basis was thus provided for Darwin's thesis that variations in individual organisms are translated into variations in populations, leading to new species.

The nagging questions about species discontinuities and the maintenance of species distinctness were addressed between the mid-1930s and the mid-1940s by geneticists, systematists, and paleontologists. Their fruitful collaboration led to the **Modern Synthesis,** which laid the foundations for neo-Darwinian evolutionary theory. This broad-based approach to questions about species and speciation produced the inclusive concept of population *divergence in geographical isolation* and the subsequent acquisition of *reproductive isolation.* As reproductively isolated units, species develop genetic distinctness, which they maintain even if they later occupy the same space. The evolution of species is one step toward the evolution of genera (sing. genus) and higher taxonomic categories, or **taxa** (sing. **taxon**).

Three seminal books that greatly influenced the development of the conceptual basis of the Modern Synthesis and emergent evolutionary theory were Theodosius Dobzhansky's *Genetics and the Origin of Species* (1937), Ernst Mayr's *Systematics and the Origin of Species* (1942), and the paleontologist George Gaylord Simpson's *Tempo and Mode in Evolution* (1944). The relevance of the Modern Synthesis for plants was made very clear in G. Ledyard Stebbins's classic *Variation and Evolution in Plants* (1950), which established a common framework for evolutionary mechanisms guiding change in both plants and animals, albeit with a few differences reflective of their different life styles.

The neo-Darwinian phase of evolutionary theory was crystallized by the Modern Synthesis until the mid-1950s to include five basic principles:

1. Ample genetic variation arises and is contained in populations by the random processes of mutation, recombination, and chromosomal changes, not by directed responses according to the needs of organisms.
2. Evolution in populations is influenced by natural selection in particular, as well as by gene flow and random genetic drift, and is characterized by *changes in gene frequencies.*
3. Adaptive genetic variation produces small, stepwise changes in phenotype, which accumulate gradually in evolutionary lineages over long periods of time.
4. The divergence of geographically isolated populations is unimpeded and gradually leads to speciation of reproductively isolated groups.
5. The continued accumulation of genetic differences, under the principal guiding force of natural selection, results in new taxa above the species level by the same processes that produce new species.

Darwin's thesis of the origin of species (and higher taxa) by the step-by-step accumulation of heritable differences that are selectively favored was now firmly grounded in the wide scope of twentieth-century biology, especially in genetics. All this took place before 1953, when James Watson and Francis Crick initiated the age of molecular biology with their model of the DNA molecule as the genetic material.

The profound effects of molecular biology on evolutionary theory and on studies of evolutionary processes are evident in every current aspect of the subject. The gene can be studied even if its molecular product is unknown or its mutant forms have no detectable effect on the phenotype. Genetic relationships underlying descent with modification can be investigated directly at the DNA or protein level, rather than relying mainly on comparative morphology and embryology. The field of molecular evolution has emerged, based on strong neo-Darwinian premises but greatly expanding this perspective and its conceptual foundations. Molecular biology has become an essential component of evolutionary study and, like the synergistic effects of physical, chemical, and geological sciences on biological studies in the past, has energized the whole field of biology and opened areas for exploration.

Some Current Challenges to Neo-Darwinian Theory

Like any complex idea, one part or another of neo-Darwinian theory is subject to question, even if its central notions are accepted. As in the past, Darwin's theory of natural selection as the overriding force guiding evolutionary change is under fire from very different quarters. One difference between past and present challenges, however, is that selectionists no longer have to defend the validity of the concept. Instead, it is incumbent on the challengers to demonstrate the worthiness of their alternative ideas. Natural selection has become the accepted dogma.

Two of the important challenges to the idea that natural selection is among the most important evolutionary mechanisms by which particular genetic variation is established in populations come from molecular biology and from population genetics. Neither molecular nor population biologists deny the significance or operation of natural selection in evolution, but they question the degree of its effectiveness in swamping the influence of random processes of change. Both sets of investigators provide theoretical models, mathematical formulations, and empirical data in support of their proposals, which can therefore be tested and analyzed independently by others.

In 1968, Motoo Kimura proposed that mutation alone, exclusive of selection, is responsible for a substantial fraction of molecular evolutionary change. From a very large molecular data base, Kimura demonstrated the existence of considerable genetic variation that appears to be selectively neutral and, therefore, is established by the process of random mutation in the absence of natural selection. Changes in codons from one synonym to another specifying the same amino acid and substitutions of amino acids that do not alter the properties of the protein products of the genes pointed to mutation rather than selection as the major agent for molecular evolution. Kimura's *neutral theory* of evolution is based on these and other seminal lines of molecular evidence. It has won support from some evolutionists, has faced fierce opposition from others, and remains highly controversial.

Since the 1920s, Sewall Wright has proposed that *random genetic drift* may lead to the fixation or loss of genetic variation in populations, regardless of the adaptive or inadaptive consequences involved. The idea was grudgingly accepted to apply to very small populations, but to be of little consequence in the overall evolutionary picture. From an expanded mathematical theory of population genetics and from molecular studies by Kimura and others showing the pervasive occurrence of neutral mutations apparently fixed in populations by chance, biologists now acknowledge random genetic drift, along with natural selection, as a major agent for evolutionary change. The challenges from population genetics and molecular biology have helped to strengthen basic evolutionary theory by expanding the premises of neo-Darwinism.

In addition to the relative influence of natural selection on evolution, other features of neo-Darwinism have been challenged. Proponents of *macroevolution,* such as Steven Stanley, have suggested that the conventional picture of gradual evolutionary divergence leading to new species and higher taxa is not supported by the fossil record or by comparative morphology. The magnitude of differences among species and among higher taxa is significantly greater than the variations within populations of a species, and thus indicates a distinction between genetic processes responsible for macroevolutionary and for microevolutionary change. The nature of these genetic processes remains undetermined, but clearly involves controls over developmental pathways rather than modifications in minor phenotypic traits. The supporters of macroevolution as well as of conventional microevolutionary transition in speciation look forward to information only now coming from developmental genetics and developmental biology, which is needed to resolve these evolutionary questions.

These are only a few of the controversial ideas in current evolutionary theory. Some of these alternative views have led to new avenues of study and thus to broader perspectives of evolutionary theory. Evolutionary biologists are engaged in dynamic and wide-ranging research, bringing together and relating different kinds of information from developmental biology, ecology, molecular biology, genetics, and other disciplines. For a long time to come, we can look forward to gaining new information about and new insights into the history of evolutionary change and the mechanisms that account for the splendid pageant of life on Earth.

Fact and Theory in Evolutionary Studies

It is important to distinguish between facts, which are undeniable realities, and theories, which are plausible or scientifically acceptable general principles offered to explain phenomena. The goals of evolutionary biology are to learn more about the fact of evolutionary history and about the theoretical mechanisms and processes that cause evolution. The confusion between the established fact that evolution has occurred and the search for the principles and processes that explain evolutionary phenomena has misled a large segment of the public into believing that all of evolution is based only on speculation. This unfortunately general misunderstanding has encouraged certain fundamentalist religious groups to seek public acceptance of their stipulation that evolutionary theory is mere speculation. Their ultimate goal is to erase all mention of evolution from school curricula and textbooks, and substitute special creation for it. Their targets include astronomy, geology, biology, and any other sciences that are at odds with a literal interpretation of Genesis and other biblical books.

Methods for the Study of Evolution

For most of human history, the existence of diverse organisms was believed to be the result of special creation, but is now known to be due to the long history of evolutionary change from simple ancestral unicells to complex life forms. An article of faith, by definition, is accepted as truth without supporting evidence. Scientific or other scholarly truths are not accepted without supporting evidence, which means that questions must be asked and a search for answers undertaken. The reality of evolutionary history was established by the collection of specific data that had been present all along. When all these data were found to be contradictory to theological interpretations of geologic and natural history, they became the focus for a theory of evolution—that is, an explanation of change that relied on natural rather than divine causes. Evolution is no longer a theory; it is fact, just like the fact that the Earth revolves around the sun rather than the other way around, as was believed before Copernicus proved the geocentric view of the universe to be wrong.

The existence of evolutionary history is undeniable, but many details of this history remain unknown. The historical record is incomplete, and even if every fossil now in the rocks were to be discovered and studied, the record of past life and events would still be fragmentary. The large-scale features of evolutionary history are reasonably well understood, but studies of living organisms as well as of fossil forms are needed to determine the patterns of change, the temporal succession of ancient and modern life, and the conditions affecting evolutionary change. The geologic and fossil records provide a timetable of past events and data about world conditions that influenced those events, but relatively little about such important biological properties as behavior, physiology, development, and metabolism.

Two major goals of evolutionary studies are to reconstruct evolutionary history and to determine the mechanisms and processes responsible for the patterns of evolutionary events. Piecing together the events of evolutionary history is approached primarily by empirical (descriptive) studies. By comparing the morphology, embryology, metabolism, DNA, or some other aspect of different modern organisms, it is often possible to correlate events that have characterized patterns of evolution from ancient ancestors to their various living descendants. These comparative studies are based on the fact that evolutionary history is a record of descent with modification. Data from paleontology, geology, and anthropology, among other sciences, also provide crucial information on biological evolutionary history.

The second major goal of evolutionary study is to determine the mechanisms and processes responsible for the patterns of evolutionary events. This research also requires empirical approaches, but to a considerable extent can be approached by experimental methods and mathematical analysis. Evolutionary episodes cannot be recreated in the laboratory because thousands or millions of years cannot be condensed into a few years or even a few lifetimes. However, a model or prediction of a theory can be tested by experimental analysis. Biologists accept the hypothesis of the synthesis of organic molecules on the prebiotic Earth because of the experimental verification that it indeed is possible, not because we have found a world without life and wait to see if organic syntheses take place.

Each theory or mathematical model pertaining to evolutionary mechanisms and processes is useful only if it is open to tests of specific predictions based on the theory or model. If the predictions are not fulfilled, the theory is falsified and must be rejected or, as often happens, be modified in some acceptable way.

Even when data are obtained, it is required that the information be verified by other investigators. If different groups obtain the same results, it is more likely that the results are valid and the theory or model is thereby supported. The free and open publication of scientific studies provides the secure base for the continuing search to learn as much as possible about a subject, to challenge and counter-challenge perceived truths, and to move steadily toward a more complete understanding of our world today and in the past.

The Current Creationist Confrontation

The teaching of evolution is under attack by certain Christian fundamentalist sects, whose members believe in the literal interpretation of the Bible. They know that the United States Constitution forbids the state from sponsoring religious education, so tactics have been devised to get around this barrier. The most militant and broad-scale program has been set in motion by a group that calls itself **creation scientists.** They have proposed that their "theory" of special creation, as written in the Bible, is fully equivalent to the "theory" of evolution and that scientific evidence exists for each theory. Their evidence for special creation is totally spurious, with no scientific basis or validity. Indeed, they rarely present evidence for special creation, but they consistently denigrate and dismiss evidence for evolution on theological grounds.

The creationist confrontation has no bearing on the existence of a deity, which is accepted by the large majority of evolutionists as well as by creationists. The particular orthodox or fundamentalist sects that adhere to the literal interpretation of the Bible, however, believe that no change has occurred since special creation, which happened about 6000 years ago. Thus they deny all the evidence from astronomy, astrophysics, and geology, as well as biology. Indeed, by tracing back through Genesis, Archbishop James Ussher calculated that Creation began at 9:00 A.M. on October 26, 4004 B.C. Modern creationists may or may not accept the specific time and date, but they do accept a 6000-year timetable—or perhaps up to 10,000 years, but no more.

The confrontation between creationists and evolutionists has been based on the presentation by scientists of evidence that supports evolution and the rejection of this evidence by the creationists. No amount of evidence will sway the creationists from their religious belief; faith does not require proof.

The creation scientists have little interest in convincing evolutionists of the truth of biblical dogma. They are concerned primarily with changing the contents of textbooks and school curricula and have therefore devised legal, political, and social strategies to achieve their goal. They rely heavily on the public at large being uninformed, misinformed, or disinterested in science and scientific matters. Their legal maneuvers include obtaining the passage of legislation that will mandate "equal time" for creationism and evolution in school curricula, on the spurious premise that both are based on theories and on evidence for these theories. Neither the creationists nor legislators and their constituents appreciate the distinction between the fact of evolutionary history and the theories dealing with the mechanisms that cause evolution. Such pro-creation, antievolution legislation has been passed in at least two southern states, but has been challenged as unconstitutional by the scientific community and its legal counsel (Box 1.2).

Political pressures have been applied to school boards, which choose public-school textbooks. When an important school district caves in to creationist pressure and selects texts that include the creationist position or that do not discuss evolution, textbook publishers capitulate. The Texas school district is the second largest purchaser of school books in the nation. When Texas demanded books that reduced or deleted the discussion of evolution, publishers agreed to these demands in order to maintain maximum sales. They claim that they cannot afford to publish two editions, one including evolution and the other excluding it, so the effect of this censorship is felt nationwide. Every school district in every state is offered fewer and fewer texts that mention evolution.

The creationists also have prepared and distributed education kits and films to the parents of schoolchildren, exhorting the public to evaluate the curriculum and to insist that creationism be included in classroom teaching. Parents are voters, and they are encouraged to vote for legislators who favor creationism and to influence the passage of laws to enforce the teaching of creationism. There is no equivalent information network for the public to learn more about evolution or to properly evaluate its central position in biology teaching.

How can we place this confrontation in proper perspective for the nation as a whole? Scientific arguments have little effect outside the scientific community, but they do provide the scholarly basis for educational decisions. A program is needed to educate the public about the philosophical, religious, and political issues that are really involved, and to make these issues clearly understandable to those whom the creationists are confusing. Nonscientists must recognize the difference between science and religion, and the basis for accepting a theory as a valid statement that can be tested and whose assumptions are also valid. With proper clarification, it will become obvious to the public that the creationist strategy is to attack evolution, not to defend creationism. By undermining the acknowledgment of evolution as a valid scientific statement and by confusing the fact of evolutionary history with the theory of evolutionary mechanisms, creationism is made to appear to be a reasonable alternative.

In general, the creationist–evolutionist controversy is not a scientific one. It is a religious, political, and educational confrontation that stems from several basic problems, including the failure of the educational system to teach

Box 1.2   Creation Science Law Ruled Unconstitutional

On June 19, 1987, the United States Supreme Court ruled, by a majority of 7 to 2, that states may not require public schools to teach "creation science" if they teach evolution. The decision in *Edwards* v. *Aguillard* struck down a law passed in 1981 by the Louisiana state legislature requiring "balanced treatment" of evolution and creation science in the state's public schools and stipulating state support of in-service training for teachers to prepare them to teach creation science. The law stated that "creation science consists of scientific evidence and not religious concepts" and therefore should be taught in science classes, but that both evolution and creation science should be presented as theories rather than as proved scientific facts. The declared purpose of the law was to protect "academic freedom." No mention was made of a religious motive.

The establishment clause of the First Amendment to the Constitution requires the separation of church and state. The Supreme Court ruled that the Louisiana law violated the establishment clause "because it seeks to employ the symbolic and financial support of government to achieve a religious purpose." Following are additional excerpts from the opinion written by Justice William Brennan, Jr. for the majority:

> It is . . . clear that requiring schools to teach creation science with evolution does not advance academic freedom. The Act does not grant teachers a flexibility that they did not already possess to supplant the present science curriculum with the presentation of theories, besides evolution, about the origin of life. . . . Furthermore, the goal of basic "fairness" is hardly furthered by the Act's discriminatory preference for the teaching of creation science and against the teaching of evolution. While requiring that curriculum guides be developed for creation science, the Act says nothing of comparable guides for evolution. Similarly, research services are supplied for creation science but not for evolution. Only "creation scientists" can serve on the panel that supplies the resource services. The Act forbids school boards to discriminate against anyone who "chooses to be a creation-scientist" or to teach "creationism," but fails to protect those who choose to teach evolution or any other non-creation science theory, or who refuse to teach creation science.
>
> If the Louisiana legislature's purpose was solely to maximize the comprehensiveness and effectiveness of science instruction, it would have encouraged the teaching of all scientific theories about the origins of humankind. But under the Act's requirements teachers who were once free to teach any and all facets of this subject are now unable to do so. . . . Thus . . . the Act does not serve to protect academic freedom, but has the distinctly different purpose of discrediting evolution by counterbalancing its teaching at every turn with the teaching of creation science.
>
> Out of the many possible science subjects taught in the public schools, the legislature chose to affect the teaching of the one scientific theory that historically has been opposed by certain religious sects. . . . Because the primary purpose of the Creationism Act is to advance a particular religious belief, the Act endorses religion in violation of the First Amendment.

the basic information needed by the public to recognize and evaluate the creationists' ideology and their deliberate obfuscation of the difference between science and religion. The problem is ignorance; the solution to this problem is education.

Summary

The belief in an unchanging universe and immutable life forms that persisted until the Renaissance was based on unquestioned acceptance of the biblical account of Creation. Beginning in the West in the sixteenth century, knowledge gained in mathematics, astronomy, physics, chemistry, and geology led to the search for natural laws (science) in place of a divine plan (theology) to explain natural phenomena. Biology was included in this perspective only in the nineteenth century, because many phenomena were poorly understood or explained. Charles Darwin revolutionized biology in 1859 by proposing that natural selection is the mechanism for evolutionary change in organisms. He theorized that life forms had undergone descent with modification from common ancestors and that the fittest survive in the struggle for existence in a world of finite resources.

The fact that evolution had occurred was accepted by educated people in Western society, but the acceptance of the theoretical mechanism of natural selection waxed and waned at various times after 1859. Human beings were finally included in evolutionary scenarios when prehuman fossils were acknowledged as ancestral forms. The recognition of genetics as a strong foundation for evolutionary theory was supported by the collaborative efforts of geneticists, systematists, and paleontologists who initiated the Modern Synthesis in the 1930s and 1940s. Evolutionary theory was further broadened with concepts, methods, and data provided by the new disciplines of population genetics and molecular biology. Scholarly issues continue to be debated today, as in the past, and opinions swing one way or another according to questions asked, information obtained, and interpretations provided. Creation scientists insist that we return to the medieval view of biblical Creation, but they offer only spurious denials of modern science based on their own constricted religious fundamentalist views. Their threat to modern education must be recognized and eliminated.

References and Additional Readings

Blinderman, C. 1986. *The Piltdown Inquest*. New York: Prometheus Books.
Darwin, C. 1859. *On the Origin of Species by Means of Natural Selection*. London: John Murray.
Dawkins, R. 1987. *The Blind Watchmaker*. New York: Norton.

Dobzhansky, T. 1937. *Genetics and the Origin of Species*. New York: Columbia University Press.

Eiseley, L. C. 1956. Charles Darwin. *Sci. Amer.* 194(2): 62.

Ghiselin, M. 1969. *The Triumph of the Darwinian Method*. Berkeley: University of California Press.

Gould, S. J. 1986. Evolution and the triumph of homology, or why history matters. *Amer. Sci.* 74:60.

Gould, S. J. 1987. Darwinism defined: the difference between fact and theory. *Discover*. Jan., p. 64.

Halstead, L. B. 1978. New light on the Piltdown hoax? *Nature* 276:11.

Huxley, J. S. 1942. *Evolution, the Modern Synthesis*. London: Allen and Unwin.

Jukes, T. H. 1984. The creationist challenge to science. *Nature* 308:398.

Kennedy, K. A. R. 1985. The dawn of evolutionary theory. In *What Darwin Began*, ed. L. R. Godfrey, p. 3. Boston: Allyn and Bacon.

Kimura, M. 1979. The neutral theory of evolution. *Sci. Amer.* 241(5):98.

Kohn, D., ed. 1987. *The Darwinian Heritage*. Princeton, N.J.: Princeton University Press.

Mayr, E. 1942. *Systematics and the Origin of Species*. New York: Columbia University Press.

Mayr, E. 1978. Evolution. *Sci. Amer.* 239(3):46.

Mayr, E. 1982. *The Growth of Biological Thought: Diversity, Evolution, and Inheritance*. Cambridge, Mass.: Harvard University Press.

Mayr, E., and W. B. Provine, eds. 1980. *The Evolutionary Synthesis: Perspectives on the Unification of Biology*. Cambridge, Mass.: Harvard University Press.

Numbers, R. L. 1982. Creationism in 20th-century America. *Science* 218:538.

Provine, W. B. 1986. *Sewall Wright and Evolutionary Biology*. Chicago: University of Chicago Press.

Ridley, M., ed. 1987. *The Darwin Reader*. New York: Norton.

Root-Bernstein, R. S. 1984. Ignorance versus knowledge in the evolutionist–creationist controversy. In *Evolutionists Confront Creationists*, ed. F. Awbrey and W. Thwaites, p. 8. San Francisco: American Association for the Advancement of Science, Pacific Division.

Simpson, G. G. 1944. *Tempo and Mode in Evolution*. New York: Columbia University Press.

Stebbins, G. L., Jr. 1950. *Variation and Evolution in Plants*. New York: Columbia University Press.

Stein, G. J. 1988. Biological science and the roots of Nazism. *Amer. Sci.* 76:50.

CHAPTER 2

The Physical and Chemical Setting for Life

The consensus of scientific opinion is that life originated on Earth from collections of organic molecules that were produced early in the planet's history. To better understand these momentous events, we must know about the raw materials and processes involved and about the universe in which these materials were produced.

Cosmology

Cosmology is the branch of astronomy that deals with the structure and origin of the universe, or cosmos, which includes all existing matter and energy in observable space. The universe as we know it came into existence between 12 and 20 billion years ago and has been changing ever since. Some aspects of cosmology have a direct bearing on the nature of the tiny bit of space we occupy in the vastness of the universe.

The Nature of the Universe

The chief structured components of the universe are the masses of matter in galaxies and star clusters, and the dust and gases of interstellar space. Astronomers estimate at least 1 billion (1×10^9) galaxies and an average of 100 billion (1×10^{11}) stars per galaxy. From these figures, they calculate that about 10^{20} stars are distributed in space (10^9 galaxies $\times 10^{11}$ stars per galaxy).

Galaxies as well as stars tend to occur in clusters. Our own galaxy, the Milky Way, is 1 of 17 galaxies in the Virgo cluster, which has a diameter of 2 million light-years. The **light-year** is a measure of distance—the distance traveled in 1 year by light moving at 186,000 miles per second, or about 6×10^{12} miles. The Milky Way is a pinwheel-shaped system whose spiral arms extend across a diameter of about 100,000 light-years. Of the estimated 100 billion stars in the galaxy, our nearest neighbor is Alpha Centauri, which is 4 light-years away. Our own star, the sun, is at the center of the solar system and is encircled by nine planets and the asteroid belt between Mars and Jupiter. These planets are a tiny fraction of the 1 million planets estimated to be present in a galaxy, on the average, and of the 10^{15} planets estimated to occur in the universe (10^9 galaxies $\times$ 10^6 planets per galaxy). Figures of these magnitudes clearly place us as specks in the barely comprehensible dimensions of space and its structured components.

About 99 percent of matter in the universe exists as the elements hydrogen and helium, in a ratio of approximately 10 H:1 He, and the remaining 1 percent consists of all the other, heavier elements (Fig. 2.1). From spectral analyses in different parts of the universe, it is clear that all matter is composed of the same elements; no unexpected elements have been found anywhere. The proportions of the heavier elements vary in different parts of the universe as a result of localized aggregations of stars, dust, and gases distributed among the thinly dispersed elements of interstellar space, and of the ongoing processes of element synthesis, star formation and death, and star explosions, among others.

The Origin of the Universe

In the 1920s, the American astronomer, Edwin Hubble, showed that Andromeda, the galaxy nearest to us, is composed of stars and that other galaxies are separate entities in space and not mere satellites of the Milky Way. Furthermore, Hubble and others clearly demonstrated that all the galaxies are receding from one another. From measurements of their distances from the Earth and their velocities of recession, it was apparent that a simple but powerful correlation exists between distance and speed. What is now known as **Hubble's law** states that the more distant a galaxy, the faster it moves. In particular, the relationship is a simple proportion: that is, if one galaxy is twice as far away from an observer as another, it will be moving twice as fast; if three times as far, it will be moving three times as fast; and so forth (Fig. 2.2).

These galactic features indicate that the universe is not static, as Albert Einstein postulated, but is constantly changing. In fact, these observations and calculations clearly show that the universe is expanding and provide considerable support for the mathematically based theory of universe expansion presented in the 1920s by the Russian mathematician Alexander Friedmann and the Belgian cleric Georges Lemaître. The discovery that interstellar space is not a vacuum but is filled with hydrogen and helium, combined with other

New notation / Previous IUPAC form / CAS version

KEY TO CHART

Atomic Number → 50 +2 ← Oxidation States
Symbol → Sn +4
1983 Atomic Weight → 118.71
18 18 4 ← Electron Configuration

1 Group IA	2 IIA	3 IIIA IIIB	4 IVA IVB	5 VA VB	6 VIA VIB	7 VIIA VIIB	8	9 VIIIA VIII	10	11 IB	12 IIB	13 IIIB IIIA	14 IVB IVA	15 VB VA	16 VIB VIA	17 VIIB VIIA	18 VIIIA	Orbit
1 H +1 −1 1.00794 1																	2 He 0 4.00260 2	K
3 Li +1 6.941 2-1	4 Be +2 9.01218 2-2											5 B +3 10.81 2-3	6 C +2 +4 −4 12.011 2-4	7 N +1 +2 +3 +4 +5 −1 −2 −3 14.0067 2-5	8 O −2 15.9994 2-6	9 F −1 18.9984 2-7	10 Ne 0 20.179 2-8	K-L
11 Na +1 22.9898 2-8-1	12 Mg +2 24.305 2-8-2											13 Al +3 26.9815 2-8-3	14 Si +2 +4 −4 28.0855 2-8-4	15 P +3 +5 −3 30.9738 2-8-5	16 S +4 +6 −2 32.06 2-8-6	17 Cl +1 +5 +7 −1 35.453 2-8-7	18 Ar 0 39.948 2-8-8	K-L-M
19 K +1 39.0983 -8-8-1	20 Ca +2 40.08 -8-8-2	21 Sc +3 44.9559 -8-9-2	22 Ti +2 +3 +4 47.88 -8-10-2	23 V +2 +3 +4 +5 50.9415 -8-11-2	24 Cr +2 +3 +6 51.996 -8-13-1	25 Mn +2 +3 +4 +7 54.9380 -8-13-2	26 Fe +2 +3 55.847 -8-14-2	27 Co +2 +3 58.9332 -8-15-2	28 Ni +2 +3 58.69 -8-16-2	29 Cu +1 +2 63.546 -8-18-1	30 Zn +2 65.39 -8-18-2	31 Ga +3 69.72 -8-18-3	32 Ge +2 +4 72.59 -8-18-4	33 As +3 +5 −3 74.9216 -8-18-5	34 Se +4 +6 −2 78.96 -8-18-6	35 Br +1 +5 −1 79.904 -8-18-7	36 Kr 0 83.80 -8-18-8	-L-M-N
37 Rb +1 85.4678 -18-8-1	38 Sr +2 87.62 -18-8-2	39 Y +3 88.9059 -18-9-2	40 Zr +4 91.224 -18-10-2	41 Nb +3 +5 92.9064 -18-12-1	42 Mo +6 95.94 -18-13-1	43 Tc +4 +6 +7 (98) -18-13-2	44 Ru +3 101.07 -18-15-1	45 Rh +3 102.906 -18-16-1	46 Pd +2 +4 106.42 -18-18-0	47 Ag +1 107.868 -18-18-1	48 Cd +2 112.41 -18-18-2	49 In +3 114.82 -18-18-3	50 Sn +2 +4 118.71 -18-18-4	51 Sb +3 +5 −3 121.75 -18-18-5	52 Te +4 +6 −2 127.60 -18-18-6	53 I +1 +5 +7 −1 126.905 -18-18-7	54 Xe 0 131.29 -18-18-8	-M-N-O
55 Cs +1 132.905 -18-8-1	56 Ba +2 137.33 -18-8-2	57* La +3 138.906 -18-9-2	72 Hf +4 178.49 -32-10-2	73 Ta +5 180.948 -32-11-2	74 W +6 183.85 -32-12-2	75 Re +4 +6 +7 186.207 -32-13-2	76 Os +3 +4 190.2 -32-14-2	77 Ir +3 +4 192.22 -32-15-2	78 Pt +2 +4 195.08 -32-16-2	79 Au +1 +3 196.967 -32-18-1	80 Hg +1 +2 200.59 -32-18-2	81 Tl +1 +3 204.383 -32-18-3	82 Pb +2 +4 207.2 -32-18-4	83 Bi +3 +5 208.980 -32-18-5	84 Po +2 +4 (209) -32-18-6	85 At (210) -32-18-7	86 Rn 0 (222) -32-18-8	-N-O-P
87 Fr +1 (223) -18-8-1	88 Ra +2 226.025 -18-8-2	89** Ac +3 227.028 -18-9-2	104 Unq +4 (261) -32-10-2	105 Unp (262) -32-11-2	106 Unh (263) -32-12-2	107 Uns (262) -32-13-2												O P Q

*Lanthanides	58 Ce +3 +4 140.12 -20-8-2	59 Pr +3 140.908 -21-8-2	60 Nd +3 144.24 -22-8-2	61 Pm +3 (145) -23-8-2	62 Sm +2 +3 150.36 -24-8-2	63 Eu +2 +3 151.96 -25-8-2	64 Gd +3 157.25 -25-9-2	65 Tb +3 158.925 -27-8-2	66 Dy +3 162.50 -28-8-2	67 Ho +3 164.930 -29-8-2	68 Er +3 167.26 -30-8-2	69 Tm +3 168.934 -31-8-2	70 Yb +2 +3 173.04 -32-8-2	71 Lu +3 174.967 -32-9-2	N O P
**Actinides	90 Th +4 232.038 -18-10-2	91 Pa +5 +4 231.036 -20-9-2	92 U +3 +4 +5 +6 238.029 -21-9-2	93 Np +3 +4 +5 +6 237.048 -22-9-2	94 Pu +3 +4 +5 +6 (244) -24-8-2	95 Am +3 +4 +5 +6 (243) -25-8-2	96 Cm +3 (247) -25-9-2	97 Bk +3 +4 (247) -27-8-2	98 Cf +3 (251) -28-8-2	99 Es +3 (252) -29-8-2	100 Fm +3 (257) -30-8-2	101 Md +2 +3 (258) -31-8-2	102 No +2 +3 (259) -32-8-2	103 Lr +3 (260) -32-9-2	O P Q

Numbers in parentheses are mass numbers of most stable isotope of that element. (Weast 1986, inner cover).

Figure 2.1. The periodic table of the elements. (From R. C. Weast, ed., 1986, *CRC Handbook of Chemistry and Physics*. Boca Raton, Fla.: CRC Press)

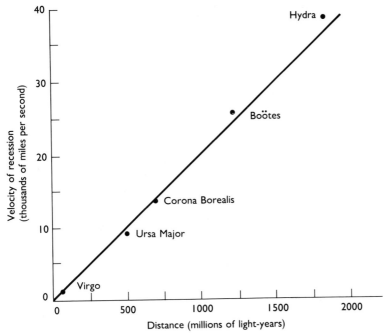

Figure 2.2. Linear correlation between velocity of recession and distance for five galaxies. The correlation, stated in Hubble's law, is that the speed of movement of the galaxy away from the observer is directly proportional to the distance of the galaxy from the observer. The galaxy in Ursa Major, at 500 million light-years distance, is moving away at 10,000 miles per second, whereas that in Hydra is four times farther away (2000 million light-years) and is moving away at a speed of 40,000 miles per second (see Box 2.1).

seminal discoveries by astronomers, promoted the concept of an evolving universe and encouraged the formulation of theories to explain the processes and patterns of change.

The cosmology most favored today is the **Big Bang theory,** which proposes an explosive beginning and an ultimate end of the observable universe, when nothing will remain (Fig. 2.3). The theory was elaborated in the 1940s by George Gamow, an American physicist, and was based on the mathematical expositions of Friedmann and Lemaître that traced the present expanding universe back to a moment in which matter existed in a highly condensed state. Gamow made a number of predictions in support of his theory, some of which have been verified by both observational and experimental evidence. The Big Bang theory makes no attempt to address questions about the nature or conditions of the universe before its beginning in a massive thermonuclear explosion. But given the assumed starting conditions, cosmologists can predict with some confidence the events that occurred after the explosion and will continue to occur in the distant future.

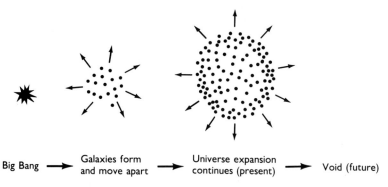

Big Bang $\longrightarrow$ Galaxies form and move apart $\longrightarrow$ Universe expansion continues (present) $\longrightarrow$ Void (future)

Figure 2.3. The Big Bang cosmology postulates the beginning of the universe in the massive thermonuclear explosion of an incrediby dense entity. The universe began to expand immediately afterward. Galaxies formed, and stars condensed within these masses. The universe is still expanding and is predicted to terminate as a void in the future, because the galaxies are moving ever farther apart from one another and ultimately will use up their fuel supply and fade away.

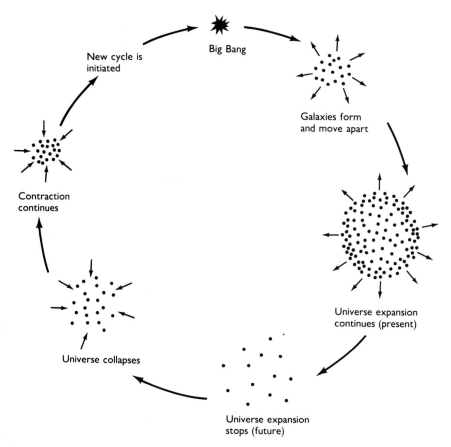

Figure 2.4. The Oscillating Universe cosmology proposes an ageless universe undergoing repeated cycles of explosion, expansion, and contraction to an initial state, from which the next cycle begins.

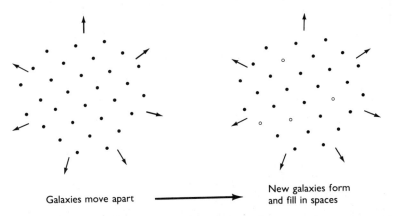

Galaxies move apart ⟶ New galaxies form and fill in spaces

Figure 2.5. The Steady State cosmology predicts a perpetually self-renewing universe, which remains unchanged in its overall large-scale features, as hydrogen atoms are created and new stars and galaxies are condensed. These galaxies, shown as open circles, fill the space left by galaxies moving farther and farther apart in an ever-expanding universe.

The **Oscillating Universe theory** incorporates some premises of the Big Bang theory, but posits no beginning or end of the universe (Fig. 2.4). Instead, the theory suggests that the universe has existed forever in a state of recurring cycles of expansion and contraction. It is expanding at present, but is predicted to reach its maximum extent at some future time and to return to its greatest density after contracting. After the universe attains its maximum density and temperature, with all matter reduced to elemental hydrogen, the inevitable explosion will initiate a new cycle of expansion. This theory may be more satisfying than the Big Bang theory to many because the universe it proposes has existed and will exist interminably. The question of what came before the great explosion is answered simply by reference to the contraction phase. Cosmologists cannot properly evaluate the Oscillating Universe theory, however, because they are uncertain about the density of matter in the universe. The calculations and measurements that have been made show that 10 to 100 times more matter must exist if expansion of the universe is to stop and contraction begin. Given the present estimates of density, physical laws mandate continuing expansion (to emptiness) without subsequent contraction. Perhaps later studies will resolve the question about the "missing" matter that some expect to find.

A third cosmology, the *Steady State theory*, enjoyed some favor in the 1950s and early 1960s but has essentially been abandoned as compelling evidence has been collected in support of an initial Big Bang event leading to expansion of the universe. The Steady State theory proposed an essentially unchanging universe that renews itself continually through the creation of hydrogen to form new stars that replace the old and to fill the spaces left by the separation of galaxies moving away from one another in a perpetually expanding universe (Fig. 2.5). As we shall see, the evidence that supports a Big Bang invalidates a Steady State universe.

The Big Bang Theory

According to the Big Bang theory, the universe began with the explosion of a very hot and very dense entity, between 11 to 12 and 20 billion years ago. At 10^{-43} second after the explosion, which is the earliest time that can be described with current physical theory, the universe is calcualted to have had a density of 10^{90} kg/cm^3 and a diameter of 10^{-33} cm. We think of a rock as a relatively dense substance, but its density is only a few grams per cm^3. The density of the primordial entity is unimaginable and far beyond our experience. The diameter of 10^{-33} cm is even smaller than the nucleus of an atom!

At the instant of the thermonuclear detonation, space-time was disrupted by the force or stresses of gravitation, and the universe began to expand as the fireball spewed its contents in all directions. The temperature must have been in the trillions of degrees at first, and the annihilation of elementary particles (through interaction with antiparticles) produced radiation. A small number of particles persisted, however, and they constitute all matter in the expanding universe. If each particle had interacted with an antiparticle, no matter—only radiation—would have remained. The particles that survived annihilation were the raw materials for a succession of changes initiated within 1 millisecond after the Big Bang. In particular, protons and electrons fused to form neutrons, which was accompanied by the release of neutrinos. By 1 second of time, the temperature is presumed to have dropped to below 10^{10} K, which no longer supported neutron formation. After about 1 minute, it fell to about 10^9 K, which was cool enough for some of the remaining protons to fuse with neutrons and form deuterons (^{2}H, the heavy isotope of hydrogen). Free neutrons are highly unstable and disintegrate in about 15 minutes, but neutrons absorbed by atomic nuclei become parts of stabler elements. Deuterons readily absorb neutrons, which is the major reason for the use of heavy water (^{2}H$_2$O) in nuclear reactors, and further neutron capture would have led to the formation of ^{3}H from ^{2}H and of ^{4}He from ^{3}H.

The synthesis of heavier atomic nuclei from ^{4}He does not take place by neutron capture. It has been shown experimentally that ^{4}He nuclei cannot retain added neutrons and that pairs of ^{4}He nuclei do not fuse to produce a stable element with an atomic mass of 8. Furthermore, the temperatures would have continued to drop as the primordial fireball expanded in space and would have been too low after the first few minutes for continued nucleosynthesis, even of ^{4}He. In the 3-minute-old universe, therefore, all matter probably consisted of hydrogen nuclei (protons), helium nuclei, electrons, and radiation. Atoms could not yet exist because any electrons colliding with hydrogen or helium nuclei would have been ejected instantly as a result of the violence of collisions at extremely high temperatures.

As the universe expanded, its temperature continued to drop. After about 300,000 years, the temperature of the universe may have been about 5000 K and suitable for the formation of atoms from nuclei and electrons. These atoms were mostly hydrogen, and atom formation would have continued until the universe was 1 million years old. It has been calculated that by the time the universe had existed for 100 million years, galaxy formation began; soon afterward, stars condensed within the galactic accumulations of matter. Cosmologists are uncertain whether galaxy formation was restricted to the early years, or took place over a long period of time and may even be occurring today. Current studies of newly discovered galaxies in a part of space that was believed to be strangely devoid of such systems may help to resolve questions about the timing of galaxy formation. These galaxies are very faint and appear to be spherical rather than elliptical or spiral, as are most of the known systems. Perhaps these faint, spherical galaxies represent newly forming systems.

The Big Bang theory predicts that the universe will continue to expand until the galaxies die as their hydrogen supply is used up and that space will become ever emptier as the galaxies move farther and farther apart. In the end, all matter and energy will have disappeared—leaving only a void.

Evaluation of the Cosmological Theories

The Big Bang, Oscillating Universe, and Steady State theories hypothesize expansion of the universe today, but each makes different predictions about the *rate* of expansion now and in the distant past. Both the Big Bang and the Oscillating Universe theories assume a slower pace of expansion now than earlier, whereas the Steady State theory requires a constant rate of expansion throughout time. The first two theories predict the slowing down of expansion due to gravity, which continuously pulls back all the receding components of the universe. Although they agree about the slowdown of expansion, the Big Bang theory maintains that expansion will go on forever, whereas the Oscillating Universe theory predicts slower and slower rates of expansion until it stops, and the contraction phase is initiated. In other words, the pace of universe expansion should become increasingly slower if there are repeated cycles of expansion and contraction than if expansion continues until the universe winks out of existence.

It is theoretically possible to determine the rates of expansion at different times in the history of the universe by the measurement of velocities of recession of the moving galaxies. As mentioned earlier, Hubble showed that the speed of recession of a galaxy is directly proportional to its distance from the observer. This proportion is expressed by the formula

$$v = Hx$$

where v is the velocity of recession, x is the distance from the observer, and H is the constant of proportionality, usually called the Hubble constant. The value of H has units of velocity over distance, and is usually expressed in kilometers or miles per second per million light-years. The Hubble constant tells how fast the universe is expanding (Box 2.1). If H is the same value everywhere in the universe, the rate of expansion has remained unchanged throughout time. If H is smaller today than in earlier times, the universe is expanding at a lower rate today than earlier. The Hubble constant thus per-

Box 2.1　The Red Shift

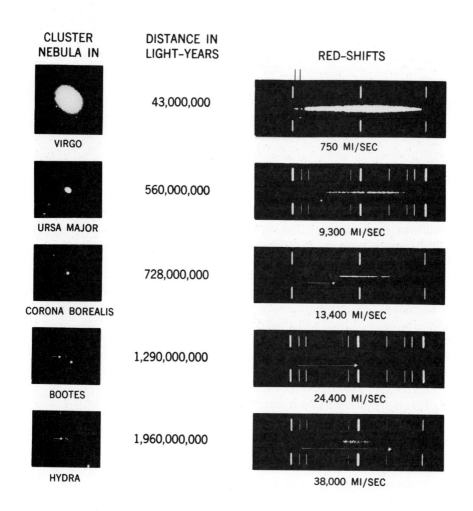

CLUSTER NEBULA IN	DISTANCE IN LIGHT-YEARS	RED-SHIFTS
VIRGO	43,000,000	750 MI/SEC
URSA MAJOR	560,000,000	9,300 MI/SEC
CORONA BOREALIS	728,000,000	13,400 MI/SEC
BOOTES	1,290,000,000	24,400 MI/SEC
HYDRA	1,960,000,000	38,000 MI/SEC

The speed and direction of galaxy movement can be determined from the absorption spectrum of a galaxy. The spectrum of each galaxy shown in the figure is the tapering horizontal band of light in the middle, and comparisons are made by reference to the spectral lines of known wavelength, which are above and below the spectrum. Two absorption lines of calcium, marked by the arrow in the top photograph, cut through the band of light. These calcium lines are shifted slightly toward the right (red end of the spectrum) when compared with their unshifted positions, marked by the two vertical black lines above the top photograph. This small shift in position of the calcium lines corresponds to a speed of 750 miles per second, and the shift toward the red end of the spectrum indicates that the galaxy in Virgo is moving away from the observer. A shift toward the red end of the spectrum occurs in each of the other four cases (horizontal arrows), with a substantially larger shift with increasing distance. These and other data clearly show that all galaxies are receding from us, which indicates that the universe is expanding and that the velocity of recession increases proportionately with distance. By plotting velocity against distance and connecting these points, a straight line is produced and the simple proportionality is thereby displayed. The slope of the line through these points is the Hubble constant H (see Fig. 2.2).

mits comparison of the unchanging Steady State universe (no change in H) with the Big Bang and Oscillating Universe cosmologies (smaller H today than in earlier expansion times). Observations of receding galaxies have clearly shown that H is smaller today than in earlier times; that is, the universe is expanding more slowly now than before, and the Steady State theory must be invalid. (Fig. 2.6). Unfortunately, the extent and accuracy of the measurements of the more distant galaxies contain a relatively large component of uncertainty or possible error. Cosmologists therefore cannot be sure whether the change in H indicates the more rapid slowdown posited by the Oscillating Universe theory or the less rapid pace required by the Big Bang theory.

Another, independent line of evidence in support of an explosive initiation of universe expansion was obtained in the mid-1960s. In 1956, Gamow had predicted that a residue of the original fireball radiation would be detectable by sensitive radioastronomy, albeit greatly diluted by expansion and reduced to the low temperature of about 3 K. There was little reaction to this suggestion at the time, but in 1965 Robert Dicke made the same prediction and stirred considerable interest. The residual *cosmic microwave background radiation* would be distinguishable from all other radiation because it should fill the universe uniformly and thus have the same intensity everywhere a radio antenna was pointed. Before Dicke completed construction of a suitable apparatus, Arno Penzias and Robert Wilson found the predicted radiation. Penzias and Wilson were unaware of the nature of their discovery and were unable to explain the source of the uniform background radiation detected by their radio antenna, which had been set up for a communications-satellite program for Bell Laboratories. Eventually, they and Dicke came to realize that the puz-

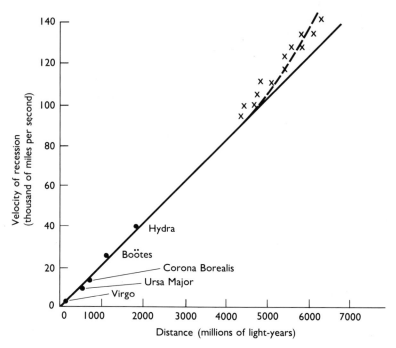

Figure 2.6. The slope of the line, which represents the Hubble constant ($H = v/x$), would be unchanged if all galaxies receded proportionally to their distance from the observer throughout time (solid line). If galaxies farther away are receding faster than nearby galaxies, the H values might be distributed as shown by points marked x, which plot along a different slope (dashed line). Farther galaxies, representative of an earlier time in universe expansion, would provide evidence for a slower rate of universe expansion today than earlier if they indeed can be plotted along the dashed line (showing a larger H value). Such evidence would support the Big Bang and Oscillating Universe theories, but would contradict the predictions of the Steady State theory.

zling uniform radio noise is the residue of the original fireball radiation. Although the fireball residuum does not allow cosmologists to choose between the Big Bang and the Oscillating Universe theories, it is evidence that led to the decline and fall of the Steady State theory.

A third set of observations in support of a Big Bang, whether as a singularity or as the beginning of an expansion cycle in an oscillating universe, is the proportion of hydrogen to helium and their distribution in the universe. As predicted by Gamow, hydrogen and helium, in a ratio of about 10:1, are distributed uniformly throughout interstellar space, whereas nuclei of heavier elements are distributed in various amounts in different parts of the universe. These data are expected on the premise that hydrogen and helium nuclei were produced during the first 3 minutes subsequent to the great explosion, but that heavier elements were synthesized at various, later times in the history of the universe. As we will see in the next section, the heavier elements are made in the stars. Further support for Gamow's theory is the presence of 1

^{2}H nucleus per 30,000 ^{1}H nuclei in the universe today. Deuterons cannot accumulate in any star, but are destroyed or altered as quickly as they are made. The most likely explanation for the observed abundance and distribution of ^{2}H nuclei, therefore, is that they were formed from ^{1}H at temperatures of at least 10^9 K in the primordial fireball, as Gamow postulated.

In order to decide between the once-only Big Bang and the cycles of expansion and contraction in an oscillating universe, cosmologists need unambiguous measurements of the density of matter in the universe. At present, even with the addition of all confirmed amounts of energy and radiation—which are equivalent to matter, according to Einstein's formula $E \doteq mc^2$—there is insufficient calculated density for an oscillating universe. The data indicate that there is too low a density for the gravitational attraction of the different parts of the universe on one another to be strong enough to stop expansion and initiate contraction. Thus the universe will continue to expand into the indefinite future, as predicted by the Big Bang cosmology. New sources of radiation have been found in recent years, however, and they may represent as much as 100 times greater density. Should these preliminary and tenatative calculations be confirmed, they will show that the universe oscillates between phases of expansion and contraction and has done so forever. The Big Bang will then be accommodated within the larger framework as the event that initiates the expansion cycle in an eternity of repeated cycles. Further studies are thus required to choose between the two major cosmological theories.

Stellar Synthesis of the Elements

Hydrogen and helium were the only elements present in the early stages of universe expansion. All the other elements were synthesized later in the hot interiors of stars, where nucleosynthesis continues to occur. Indeed, every atom now in living creatures was made billions of years ago in the stars, some of which may have blinked out long before our solar system formed.

According to the theoretical sequence of the birth, development, and death of an average star, such as the sun, the process of star formation begins with a cold, dilute, turbulent cloud of dust and gases in interstellar space. Condensed pockets of gas develop by the accidental collisions of hydrogen atoms and will become a **protostar** if there are enough atoms to hold the cluster together in the tight grip of its own gravity. The calcualted weight of hydrogen atoms for such a protostar cluster is about 10^{33} g, which turns out to be the density of an average star, such as the sun. Contraction occurs as the protostar collapses due to its own gravity. The contracting protostar gets hotter as its atoms move ever faster as they fall to the center, and within 1 year, the temperature at the star's core has climbed from the numbing cold of interstellar space to about 50,000 K. The protostar will not become a star, however, for another 10 million years, when the temperature at the center reaches about 10 million K and thermonuclear reactions are ignited.

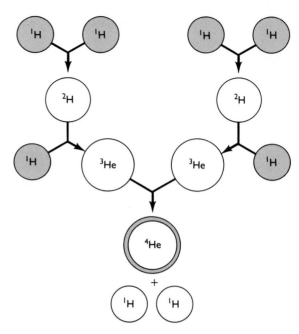

Figure 2.7. The production of helium by means of the proton–proton fusion cycle involves the net consumption of four protons (^{1}H) per helium nucleus (^{4}He) produced, with the formation of deuterons (^{2}H nuclei) as intermediates in the process. This chain reaction is believed to be the major source of energy produced by the sun and similar stars.

At temperatures of at least 10 million K, violent collisions among rapidly moving hydrogen atoms cause the ejection of electrons, leaving naked hydrogen nuclei, or protons. The protons move with enough energy to fuse into deuterons (^{2}H nuclei) on collision, emitting a positron and a neutrino per pair of fused protons. The formation of deuterons eventually leads to the production of helium, by means of the **proton–proton fusion cycle** initiated by the fusion of single protons (Fig. 2.7). Collisions between ^{2}H nuclei and free protons produce ^{3}He, the light isotope of helium. Laboratory studies show that ^{3}He nuclei do not interact successfully with free protons, but that pairs of ^{3}He nuclei fuse to form stable ^{4}He nuclei. Two free protons are released for each ^{4}He nucleus synthesized, so that a total of four protons (of the six involved) are consumed in each nucleosynthesis sequence.

The proton–proton fusion cycle, also called **hydrogen burning**, occupies about 99 percent of the life of a star. During the average 10-billion-year existence of a star, enormous amounts of energy are released in the chain-reaction process of proton fusion. Astrophysicists believe that hydrogen burning is the major source of energy produced in the sun and other average stars, but require more evidence to substantiate the theory.

When much of the hydrogen fuel is used up in helium synthesis, the growing helium core ultimately cools. Gravity predominates at this time, resulting in the collapse of the helium core and a consequent rise in its temperature. This causes the thin hydrogen envelope, or mantle, of the star to expand considerably, which leads to a drop in the surface temperature. The star enters the *red giant* stage of its evolution.

The elements synthesized in a red giant, as well as the fate of the star, vary according to its mass. As expressed in solar masses, with the sun's mass equal to 1, stars of less than about 6 solar masses become *white dwarfs*, which cool gradually until they fade from sight after billions of years. A more massive star may explode as a *supernova*, scattering its elements into space and leaving behind a neutron core, or *neutron star*, which astronomers believe is equivalent to a pulsar. The Crab Nebula is the remains of a supernova that was observed by Chinese astronomers in A.D. 1054, and the radio signals transmitted by the pulsar within the nebula come from precisely where the predicted neutron star would be. A neutron star is therefore an incredibly dense object that is spinning very fast and emitting regular radio-signal pulses with a periodicity indicative of the spin rate. Stars that are even more massive (between 30 and 50 solar masses) explode too, but are believed to form *black holes* in space. Such massive stars may collapse to smaller diameters than the 20-mile limit of a neutron star, perhaps to 4 miles or less. According to Einstein's theory of relativity, a ray of light possesses mass; if this is correct, then the enormous force of gravity at the surface of a 4-mile-wide ball of neutrons would prevent anything from escaping—even light. A black hole is therefore invisible, but theoretically is detectable by its gravitational influence on nearby bodies of matter.

Before its demise as a faded white dwarf or a supernova, the red giant enters a phase of helium burning as its major nucleosynthetic activity. The helium core surrounds a hot center of helium nuclei, which are energized enough at temperatures of about 200 million °C to fuse on collision and form carbon. Gradually, a core of carbon builds up and grows as the helium-burning shell around it continues the nucleosynthetic process (Fig. 2.8). In small stars, the core temperature fails to reach the critical level of 600 million °C, and the carbon core collapses as the star proceeds toward the white dwarf stage. In more massive stars, the heat generated by collapse is much greater, and at 600 million °C the carbon begins to burn and produces magnesium and other elements. The core collapses again as the carbon is used up, raising the temperature in the interior to levels that support other nucleosyntheses, and core collapse stops for a time. In repeated cycles of collapse, heating, and nuclear burning, increasingly heavier elements are created—up to iron (Fig. 2.9).

Until the formation of iron, nucleosynthesis generates nuclear energy and consequent high temperatures that prevent total star collapse. After iron has been produced, nuclear reactions require an input of energy. As nuclear energy is depleted, the iron core collapses until the nuclei are so densely packed that it implodes, releasing a vast amount of energy. During this tre-

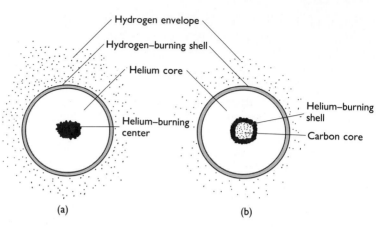

Figure 2.8. Before its demise or explosion into a supernova, a red giant (a) enters a phase of helium burning as the center of its inert helium core heats up enough to initiate the process. As helium burning continues, (b) carbon is produced from the helium, and a carbon core builds up in the center of the large helium core. The helium-burning reactions now take place in a shell surrounding the carbon core. In massive stars, the carbon begins to burn and form heavier elements, which, in turn, undergo nucleosyntheses to produce increasingly heavier elements in a sequence that eventually generates iron in the core of the red giant. The collapse of the iron core results in the release of a vast amount of energy in a supernova explosion.

mendous explosion, producing a supernova, elements heavier than iron are synthesized in a relatively rapid burst of nucleosyntheses and are scattered into interstellar space.

Should the collapsed core, composed of neutrons made from protons and electrons during the violent explosion, stop its contraction on reaching the 20-mile-diameter limit, it remains a neutron star. If the contraction continues, the neutron mass may reach the smaller diameter that leads to a black hole.

The greater abundance of elements lighter than iron is a reflection of the longer time available for nucleosynthesis during the last throes of a star's lifetime. The very brief time for element synthesis during the supernova explosion, similarly, explains the smaller quantity of elements heavier than iron. Such heavier elements are evident in supernova debris, such as the Crab Nebula, a mass that exploded more than 900 years ago. For example, the presence of the unstable heavy element technetium in supernova clouds can best be explained by its formation relatively recently. The element has a very brief half-life and would surely have decayed completely if it had been synthesized early in the star's history. In addition, some heavy elements are found only in supernova debris and at sites of thermonuclear test explosions or as particle-accelerator products on Earth.

The raw materials for element synthesis are the hydrogen and helium atoms produced in the early phase of universe expansion. With the condensation of dust and gases into stars, all the other elements were made within the hot stellar interiors. The observed abundance of the elements in the universe correlates very well with the theoretical abundance derived from calculations based on atomic synthesis processes and the relative stability of the atoms made in stars (Fig. 2.10).

Formation of the Solar System and Earth

We know that life exists on Earth, and we send out messages and radio signals to learn whether intelligent life exists elsewhere in the universe. These efforts at communication are based in large part on current theory, which states that

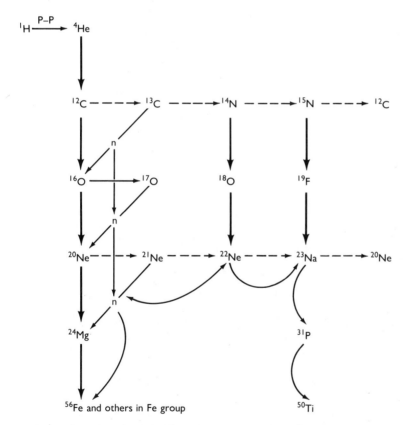

Figure 2.9.   The various nucleosynthetic reactions leading to the formation of iron and of elements lighter than iron in stars. Reactions include fusions involving protons (→), helium nuclei (↓), or neutrons (n), and sequences such as the carbon–nitrogen and neon–sodium catalytic cycles (--→). The synthesis of iron and lighter elements generates high temperatures as nuclear energy is released. The formation of elements heavier than iron, however, requires the input of energy, which ultimately leads to implosion of the iron core and to explosion into a supernova. Bursts of heavy-element synthesis occur at this time.

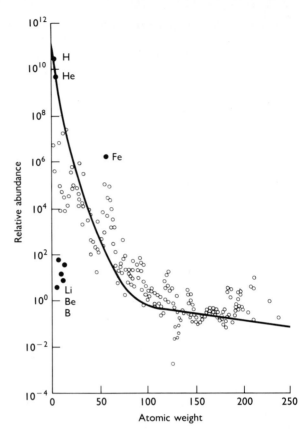

Figure 2.10. Semilogarithmic plot of the observed relative abundance of the elements according to atomic weight. Compared with the theoretically expected abundance curve (solid line), iron is more abundant and the light elements near atomic weight 10 (lithium, beryllium, boron) are far less abundant than expected. These deviations from predicted values can be explained by special features of these elements—for example, the very high stability of iron.

solar system and planet formation occurred many times under similar conditions and, therefore, has probably led to the development of Earth-like planets on which life may have originated and evolved. Some understanding of our own solar system will thus provide a basis for defining conditions suitable to the existence of life on Earth, as well as elsewhere in space.

The Origin of the Solar System from a Dust Cloud

There is general agreement that a solar system, including our own, begins with the condensation of a huge cloud of dust and gases in the intense cold of interstellar space. The central idea of the **Dust Cloud theory** was first proposed in 1644 by René Descartes, was elaborated in the eighteenth century by Pierre Laplace and the philosopher Immanuel Kant, and has been amplified by Fred Whipple and other modern astronomers. Although few astronomers agree on more than the broad outlines of the Dust Cloud theory, enough of a picture can be pieced together to better understand the relationship between the existence of life and the physical surroundings that sponsor and support life on Earth and, perhaps, in other parts of the universe.

The great cloud of dust and gases begins to condense, and it contracts under the force of its own gravity until it becomes a spinning disk of atoms and particles. This solar nebula is very hot at its center and progressively cooler toward its periphery. Circling around the center of the disk are countless atoms and dust grains, which collide with one another and stick together to become larger bodies in orbit around the protosun. The accretion of particles progresses through stages of increasing size, from small bodies to larger ones and, finally, to planets held in elliptical orbit by gravity. (Fig. 2.11). Most, but perhaps not all, planetary satellites probably develop in similar fashion from accreting particles and remain in orbit by gravitational attraction.

The solar system began to form about 5 billion years ago, but there is no agreement on when the protosun became hot enough to ignite its nuclear fuel and begin thermonuclear activity. According to some astronomers, the sun became a shining star before the planets formed; according to others, the planets formed before the sun became a true star. To decide between these alternatives, astrophysicists must learn when the protosun reached the stage of sufficient mass and density for its soaring temperatures to light the nuclear fire. At present, there is not enough data to determine the most probable order of completion of the sun and planets during the tens or hundreds of millions of years in which the solar system eventually emerged from a primordial dust cloud.

The Dust Cloud theory permits astronomers to predict that many solar systems have been formed in the universe, because planet formation is a natural corollary of star formation. An earlier idea, more or less abandoned several decades ago, suggested that the formation of the solar system was a virtually unique event caused by a collision between the sun and a passing star. The force of gravity would have torn streams of flaming gas out of the two stars during the catastrophic event, and some of this material would have been captured by the sun's gravity. The hot gas would have cooled to molten masses, which eventually would have condensed into planets. The idea is unattractive today for several reasons, including the problem of explaining the different structure and features of the Earth and the other planets if they are the products of the same masses of fiery gas that cooled to their present forms. The information available today makes it far more likely that the planets formed along a temperature gradient that was very much like the prevailing temperatures from the center outward to the periphery of the solar system.

It is believed very likely that the processes of solar-system formation are triggered by a nearby supernova event. The turbulent jets of hot gas spewing from the exploding star would stream out through the galactic dust clouds, compressing portions of these clouds to a density appropriate for tight, spi-

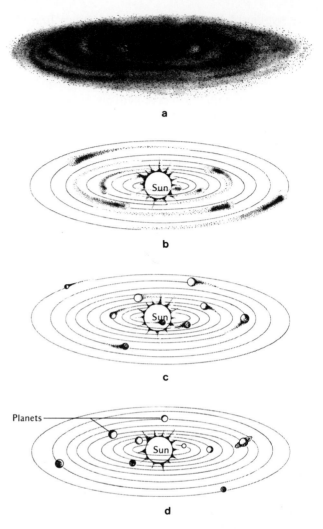

Figure 2.11. Formation of the solar system according to the Dust Cloud theory. (a) The solar system is believed to have begun as a diffuse cloud of dust and gases, which (b) contracted over time to form the central sun (star) and whirling disks of preplanetary material. (c) Whether by accretion of dust grains or by progressive clumping of particles to produce planetesimals, protoplanets eventually formed from these smaller bodies. (d) The protoplanets later developed into the planets of the present-day solar system. (From *Biology: A Human Approach,* 3rd ed., by Irwin Sherman and Vilia Sherman. Copyright © 1983 by Oxford University Press. Reprinted by permission)

raling collapse and turbulence. Each such supernova event would squeeze many clouds, thus leading to the formation of stars in batches.

Evidence in support of the supernova trigger comes largely from studies of the amount of certain isotopes in microscopic dust grains contained in meteorites of the kind called *carbonaceous chondrites.* These meteorites are considered to be primitive, and the tiny dust grains are inferred to be remnants of supernova debris incorporated into the solid meteorite conglomerate that formed early in the history of the solar system. Abnormally high quantities of some isotopes occur in dust grains of carbonaceous chondrites, relative to their number elsewhere in space. For example, in the Orgueil meteorite, which fell in France in the nineteenth century, there is an 80 percent excess of the neon isotope ^{22}Ne in proportion to ^{21}Ne and ^{20}Ne. Similarly, particular isotopes of oxygen, nitrogen, and magnesium are relatively abundant in dust grains of other carbonaceous chondrites. These rare isotopes are made in larger than normal amounts in supernovas, and were injected from supernova jets into dust grains that became part of conglomerates. Such evidence for triggering supernova helps to explain why the internal pressure of a dust cloud may not counteract gravity and oppose contraction. In the vicinity of a supernova, the force of gravity increases enough to allow contraction, leading to solar-system formation.

Formation of the Earth and Other Planets

The solar system includes nine planets that orbit the central sun (Fig. 2.12). The four innermost planets are rocky bodies that are much smaller than the giant planets Jupiter, Saturn, Uranus, and Neptune, but are of the same general size as the outermost planet, Pluto (Table 2.1). Although astronomers lack information drawn from other solar systems as models and must resort to considerable speculation, they can infer a reasonable scenario of major events by reference to the basic features of the Dust Cloud theory and of the solar system as it exists today.

Under the influence of gravity in the spinning solar disk, dust grains aggregated into clumps about the size of asteroids. These clumps, or **planetesimals,**

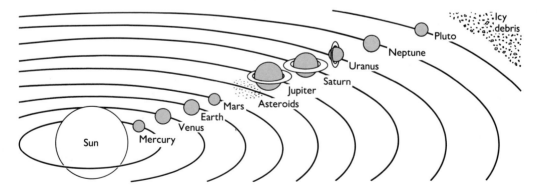

Figure 2.12. In addition to the sun and nine planets, the solar system includes an asteroid belt between Mars and Jupiter and a great cloud of icy debris (and comets) at the edge of the system. Only the four inner rocky planets, within 142 million miles of the sun, are drawn approximately to scale. To have drawn to scale the fiver outer icy planets would have required a width of several feet to place them at their proportionate distances apart (see Table 2.1).

TABLE 2.1
Some Properties of the Sun and Planets

Solar-System Body	Distance from Sun (millions of miles)	Diameter (miles)	Mass (Earth's mass = 1)	Density (water = 1)	Average Surface Temperature (K)	Revolution Period (years)	Number of Moons
Sun	—	867,000	343,000	1.4	5,800	—	—
Mercury	36	3,000	0.1	5.4	600	0.24	0
Venus	67	7,600	0.8	5.2	750	0.62	0
Earth	93	7,900	1.0	5.5	180	1.00	1
Mars	142	4,200	0.1	3.9	140	1.88	2
Jupiter	486	89,000	317.8	1.3	128	11.86	16
Saturn	892	76,000	95.2	0.7	105	29.48	17
Uranus	1,790	32,500	14.5	1.2	70	84.01	15
Neptune	2,810	31,400	17.2	1.5	55	164.79	2
Pluto	3,780	1,900	0.002	0.7	40	248.40	1

continued to collide, accumulate, and compress into solid rocks. The largest planetesimals accreted more matter, because of their greater gravitational attraction, and eventually reached the size of planets. The absence of any significant amount of particulate debris between the planets suggests that most of the primordial clumps were swept up by the growing planetary bodies. Some debris remained, and it now forms the asteroid belt between Mars and Jupiter; the rings that encircle Jupiter, Saturn, and Uranus; and the clouds of icy debris at the farthest reaches of the solar system. Part of the great interest in the appearance in 1985 and 1986 of Halley's comet was the opportunity to examine a body of icy matter that possibly represents a remnant of the primordial solar disk.

As the young sun became more luminous and radiated more heat, the planets developed along the steep temperature gradient outward from the sun. The four inner planets, within 150 million miles from the sun's heat, lost their volatile ices by evaporation but retained rocky and metallic particles. The five outer planets remained largely icy, and because of their greater mass also kept their envelopes of hydrogen and helium gases. The relatively low mass of the four rocky planets prevented their retaining these gases.

The solar system is considered to be 4.6 billion years old, according to the dating of meteorites recovered on Earth. Meteorites are pieces of stone or metal that formed at the same time as the planets, according to the Dust Cloud theory. They therefore provide the same information scientists would be able to obtain from other primary solids in the solar system, including rocks on Earth, but that are not accessible or no longer exist. The oldest dated rocks on Earth, in Greenland, are only 3.8 billion years old. Apparently, the preceding 0.8 billion years of Earth history have been erased by the violent events and processes that shaped and reshaped our planet in its youth. Among these cataclysmic events must have been impacts by meteorites, leaving craters and other large scars on the Earth's surface such as those on the moon, Mars, and other planets and their satellites (Fig. 2.13). Massive and sustained volcanic activity, changes in the planetary interior, and melting and remelting of the Earth must have contributed to the modeling of the primeval planet, including the erasure of almost all the meteorite craters.

The solid phase of the Earth, the **lithosphere,** is stratified into three zones of different temperature, composition, and physical condition (Fig. 2.14). The *core* boundary is about 2000 miles beneath the Earth's surface, measures about 4000 miles across, is very hot and dense, and consists of molten metal— mostly iron and some nickel. Recent studies using measurements of earthquake-wave speeds through molten and solid rock have yielded topographic maps of the core boundary. Instead of the smooth margin it was assumed to have, there appear to be mountains and valleys that may reach heights and depths of up to 6 miles. The friction from liquids sloshing across these features may explain why the planet rotates with a slight jerkiness that makes a day 0.005 of a second longer or shorter than 24 hours every decade.

From its boundary with the core, the semimolten rocky *mantle* extends to within about 60 miles of the Earth's surface. The temperature and nature of the mantle are plainly evident during volcanic eruptions, when hot magma bursts forth and rushes or crawls as semimolten basaltic lava down the mountainside. Much of the Earth's 60-mile-thick *crust,* the solid rocky skin of the planet, may have been deposited as lava early in the planet's history. Indeed, volcanic lava still adds to the Earth's crust in such places as Iceland and the Hawaiian Islands. The crust is composed of lightweight rocks that drift as tectonic plates on the denser, plastic, semimolten mantle underneath. The movement of these plates has molded and remolded the surface features of the continents and ocean basins for billions of years and had a profound influence on biological evolution.

The **atmosphere** is a gaseous envelope surrounding the Earth (Fig. 2.15) and is quite different, in its present composition, from the atmospheres of our nearest neighbors, Mars and Venus. None of the planets is believed to have inherited its atmosphere from the planetesimals that aggregated into the larger bodies. Instead, atmospheric gases probably came from the planet's interior during or shortly after the Earth was formed. A glimpse of the early atmosphere can be obtained today from observations of the *degassing* of the interior by way of volcanic emissions. Modern volcanoes release primarily water vapor, hydrogen, carbon monoxide, carbon dioxide, and nitrogen. Little or no oxygen was present in the primordial atmosphere, even if degassed, because small quantities of oxygen probably were quickly removed by the oxidation of iron and other minerals at the Earth's surface. As we will see in our discussion of chemical evolution on the primeval Earth, the existence of a *nonoxidizing atmosphere* devoid of oxygen was a crucial feature in the eventual origin and evolution of life. The high level of oxygen in today's atmosphere results primarily from photosynthesis.

Hot clouds of water vapor released by the degassing of the planet's interior

a

b

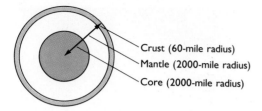

Crust (60-mile radius)
Mantle (2000-mile radius)
Core (2000-mile radius)

Figure 2.14. The dimensions of the three zones of the Earth's lithosphere, shown across the planetary radius of approximately 4000 miles (8000-mile diameter).

remained in a gaseous state, until the Earth's surface had cooled to temperatures below 100°C. At this time, water could fall as rain and remain on the surface as oceans, lakes, and rivers. The fresh water became salty wherever it reacted with carbon dioxide dissolved in it and with carbonates and other minerals. Calculations show that the present level of salinity in the oceans must have been achieved fairly early in the Earth's history and remained more or less the same ever since. The Earth's liquid phase, or **hydrosphere,** is thus considered to have formed later than its atmosphere, under conditions favorable to fluid formation and retention.

Unlike the young Earth, neither Mars nor Venus is hospitable to life. The Venusian atmosphere consists mainly of carbon dioxide and acidic compounds, and surface temperatures on the planet may reach 450°C, conditions distinctly hostile to life in any form we can conceive. Martian temperatures may be moderate near the equator, but the thin atmosphere of carbon dioxide is at such low pressure that any liquid water that might be produced would quickly evaporate. Some water is locked up in ice caps at the Martian poles and is buried under a blanket of frozen carbon dioxide (dry ice). The existence of some frozen water, plus the presence of channels and other surface features reminiscent of waterways, suggest that life exists or did exist on Mars. These and other considerations led the United States to send two *Viking* landing craft to Mars in 1976. The experimental apparatus included in these spacecraft tested samples of the Martian soil for evidence of life activities, but none was found. These negative results have not discouraged future planned attempts to search for life on the red planet. Mars retains many primeval features, unlike the modern Earth, and it may well yield important information that would allow biologists to better understand primeval con-

←——————————————————————————

Figure 2.13. (a) Craters produced by the impact of meteorites and other solid bodies are particularly evident on the lunar surface and range in diameter from a few feet to hundreds of feet. These features have changed very little over billions of years, because little erosion occurs on the moon's surface. (b) Craters on Earth, however, such as Barringer Meteor Crater in central Arizona, which is about 1 mile in diameter, may be eroded within about 10 million years by the forces of wind and water on the Earth's surface. (Photographs courtesy of NASA)

Figure 2.15. Earth photographed from the *Apollo 11* spacecraft at a distance of about 10,000 miles. The presence of an atmosphere is evident in the swirls of clouds over the Earth's surface. (Photograph courtesy of NASA)

ditions that fostered the beginnings of life on our own planet. These prospects provide encouragement for further explorations of the solar system by unmanned spacecraft similar to *Viking 1* and *2; Voyager 1* and *2,* which sent back many spectacular pictures of the outer planets in the 1980s; and *Vega 1* and *2,* the Soviet probes that examined the icy head of Halley's comet in 1985.

Chemical Evolution

Life as we know it is based on water and on organic chemistry. The primeval Earth contained liquid water in its hydrosphere and an atmosphere filled with gaseous molecules that were available raw materials for the synthesis of organic molecules. The first step toward life was therefore completed, and the next two steps could begin. These steps were the synthesis of biologically important **monomers**—such as sugars, amino acids, and nucleotides—and the polymerization of monomers into proteins, nucleic acids, and other significant **polymer** organic compounds on which living systems depend (Fig. 2.16). Ambient temperatures between 0°C and 100°C characterized most of the Earth in primeval times, much as they do today.

Abiotic Synthesis of Organic Monomer Molecules

Of all the requirements for the origin of life, biochemists best understand the reactions and processes most likely to have produced the basic monomer molecules from which the essential organic polymers can be formed. Their

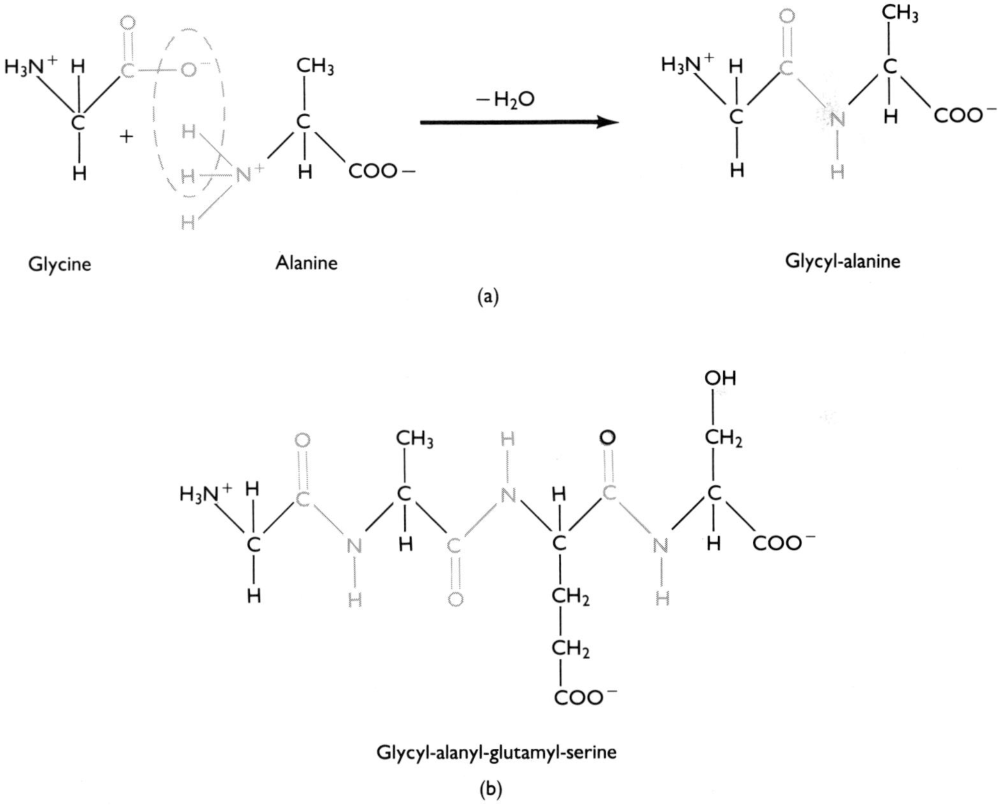

Glycine   Alanine   Glycyl-alanine

(a)

Glycyl-alanyl-glutamyl-serine

(b)

Figure 2.16. The organic synthesis of protein involves covalent joining of amino acid monomers in a dehydration reaction, which may produce (a) a dipeptide from two monomers, or (b) a larger peptide by the addition of amino acid monomers, one at a time, to a growing peptidyl chain. In proteins, the same dehydration reaction joins an amino acid to another amino acid or to a peptidyl chain, and the same peptide bond (gray) connects each amino acid to its neighbors in a polymer molecule.

greater confidence regarding this step toward life is based on an abundance of data from conventional laboratory chemistry and from simulation experiments using artificial systems that mimic the nature and conditions of the primeval atmosphere and hydrosphere. Since the first successful simulation experiments, reported in 1953, many other studies have repeated the **abiotic synthesis** of every essential kind of monomer needed by life as we know it.

In the 1920s, both the Soviet biochemist Alexander Oparin and the British biochemical geneticist John B. S. Haldane wrote of a **nonoxidizing atmosphere** as one precondition for the accumulation of organic molecules on the primeval Earth. Oxygen is a reactive and corrosive gas, and any organic matter made in an oxidizing atmosphere would have been unstable, so organic raw materials could not accumulate and life could not get started. Aerobic life today is protected by enzymes such as catalase and peroxidases from toxic reaction products of molecular oxygen, such as peroxides. Anaerobic life lacks such enzymes, so peroxides and similar molecules are lethal to these organisms.

In addition to an atmosphere lacking molecular oxygen, Oparin and Haldane hypothesized that bodies of liquid water were present, ample amounts of small precursor molecules were available, and various energy sources could drive endergonic (energy-requiring) reactions of organic synthesis. Among the probable energy sources were solar radiation, electricity from lightning discharges during storms, heat in volcanic regions, and energy released by radioactive decay of various elements in the Earth's crust. Once formed, organic molecules accumulated in water on the planet, because no life existed to consume or destroy these compounds and because spontaneous reactions of organic synthesis and degradation are extremely slow in the absence of enzyme catalysts.

In 1953, Stanley Miller, a graduate student working with the prominent astronomer Harold Urey, showed that abiotic synthesis is open to experimental analysis. Miller assembled an apparatus (Fig. 2.17) that simulated conditions for the synthesis of organic molecules in the absence of life during primeval times, as stated in the Oparin–Haldane hypothesis. The four major elements of organic molecules were provided in a gaseous phase containing reduced sources of carbon (CH_4, methane), nitrogen (NH_3, ammonia), oxygen (H_2O), and hydrogen (from these molecules and from hydrogen gas). Electric energy was supplied by spark discharge from electrodes wired to an electric current. The ambient temperature ranged between 0°C and 100°C in the simulated atmosphere and hydrosphere, and any contaminating modern life was destroyed by prior sterilization of the apparatus.

Significant results were obtained in independent experiments allowed to run for about 1 week. A few simple precursors accounted for a large number of major organic products, including amino acids that were produced at a constant rate throughout the experiments (Table 2.2). By withdrawing samples of the brew at intervals during the experiments, Miller deduced that these amino acids were the products of simple condensation reactions involving intermediates such as hydrogen cyanide (HCN), cyanogen (C_2H_2), and alde-

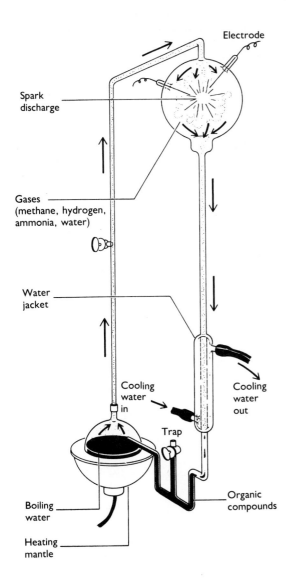

Electrode

Spark
discharge

Gases
(methane, hydrogen,
ammonia, water)

Water
jacket

Cooling
water
in

Cooling
water
out

Trap

Boiling
water

Organic
compounds

Heating
mantle

Figure 2.17. The apparatus used by Stanley Miller in experiments on the abiotic synthesis of organic compounds. Water is added to the flask, and the entire system is evacuated of air. Ammonia, methane, and hydrogen are then introduced through a valve, and these gases circulate clockwise in a stream of steam produced when the water in the flask is heated to boiling. The steam (water vapor) and gases enter the sparking chamber and are subjected to the energy of spark discharges. As the molecules move through a water-cooled condenser, condensation occurs. Nongaseous substances accumulate in the trap, while gaseous materials continue to circulate through the apparatus and past the spark discharge, until the experiment is concluded. Molecules collected from the trap or circulating gases removed through the input valve can be analyzed during and at the end of the experiment. (From *Biology: A Human Approach*, 3rd ed., by Irwin Sherman and Vilia Sherman. Copyright © 1983 by Oxford University Press. Reprinted by permission)

TABLE 2.2
Biologically Significant Organic Molecules Produced in Two of the Experiments on Abiotic Synthesis Conducted by Stanley Miller and Reported in 1953

Organic Compound		Yield (micromoles)	
Name	Formula	Expt. 1	Expt. 3
Glycine	$H_2N—CH_2—COOH$	630	800
Alanine	$H_2N—CH(CH_3)—COOH$	340	90
Aspartic acid	$H_2N—CH(CH_2COOH)—COOH$	4	2
Glutamic acid	$H_2N—CH(C_2H_4COOH)—COOH$	6	5
β-alanine	$H_2N—CH_2—CH_2—COOH$	150	40
α-aminobutyric acid	$H_2N—CH(C_2H_5)—COOH$	50	10
α-aminoisobutyric acid	$H_2N—C(CH_3)_2—COOH$	1	0
Sarcosine	$HN(CH_3)—CH_2—COOH$	50	860
N-methylalanine	$HN(CH_3)—CH(CH_3)—COOH$	10	125
Formic acid	$H—COOH$	2330	1490
Acetic acid	$CH_3—COOH$	152	135
Propionic acid	$C_2H_5—COOH$	126	19
Glycolic acid	$HO—CH_2—COOH$	560	280
Lactic acid	$HO—CH(CH_3)—COOH$	310	43
α-hydroxybutyric acid	$HO—CH(C_2H_5)—COOH$	50	10
Succinic acid	$HOOC—CH_2—CH_2—COOH$	38	0
Iminodiacetic acid	$HOOC—CH_2—NH—CH_2—COOH$	55	3
Iminoacetic-propionic acid	$HOOC—CH_2—NH—C_2H_4—COOH$	15	0
Urea	$H_2N—CO—NH_2$	20	0
N-methylurea	$H_2N—CO—NH—CH_3$	15	0
Total yield of compounds listed*		15%	3%

*Percent yield based on the amount of carbon placed in the apparatus as methane.

hydes (R—CHO). Early in the experiment, the amount of ammonia steadily decreased, and its nitrogen atoms appeared first in hydrogen cyanide, cyanogen, and aldehydes. Amino acids were synthesized more slowly, at the expense of cyanide and aldehydes. The progression of precursors, intermediates, and end products was suggestive of several known reaction sequences, some of which take place in water and some in the air phase (Fig. 2.18). An insignificant amount of organic molecules formed in the control apparatus, which lacked an energy source, thus verifying that abiotic synthesis had indeed occurred in the experimental system. By analogy, these organic syntheses could have occurred on the primeval Earth.

In the years since Miller and Urey conducted their pioneering studies, a large body of experimental evidence has been obtained by Cyril Ponnamperuma, Leslie Orgel, Sidney Fox, and others, showing that virtually all the kinds of biologically important molecules can be synthesized abiotically. Among these molecules are an assortment of nitrogenous bases, such as adenine, and

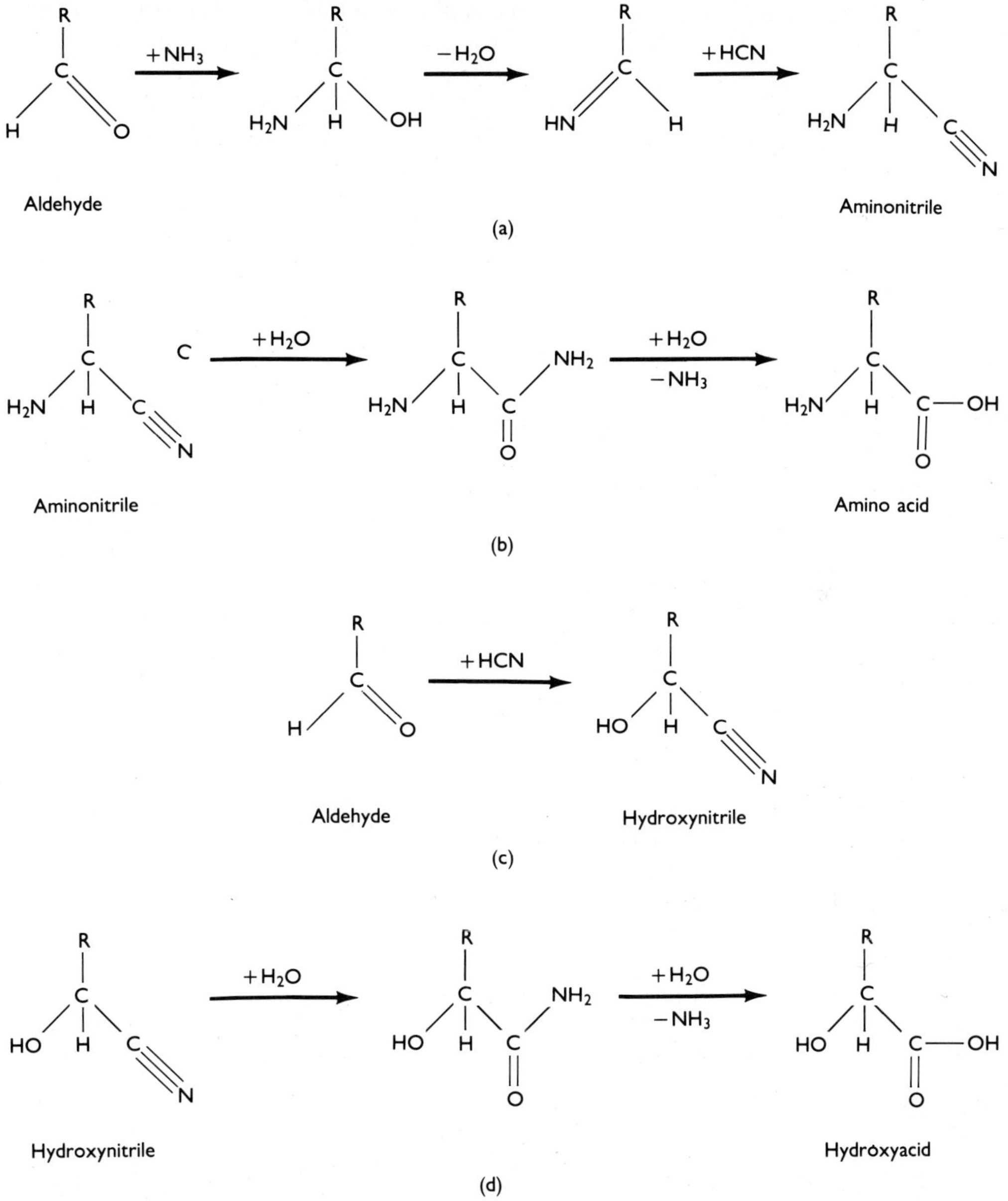

Figure 2.18. Reactions known from laboratory studies to yield organic acids include (a) the formation in air of an aminonitrile from an aldehyde and hydrogen cyanide. On entering a water phase, (b) the aminonitrile is hydrolyzed to form an amino acid. The synthesis of a hydroxyacid proceeds by similar reactions, in which (c) aldehyde and hydrogen cyanide in air yield a hydroxynitrile, which, on entering a water phase, (d) is hydrolyzed to produce a hydroxyacid. Similar reactions presumably took place in the air and water phases in Stanley Miller's apparatus (see Table 2.2), and in the atmosphere and hydrosphere of the primeval Earth.

nucleoside phosphates, such as energy-storing adenosine triphosphate (ATP). Important monomer compounds can be made in experimental systems like Miller's, as well as in systems that utilize energy sources and gas mixtures different from those used in the first experiments. The abiotic synthesis of organic monomers would thus appear to be an undeniable step in chemical evolution during primeval times on Earth. Indeed, organic syntheses must take place in the absence of life elsewhere in the solar system, because amino acids have been found in carbonaceous chondrites. One such carbon-containing meteorite fell in 1969 on the Australian town of Murchison. The Murchison meteorite contains some of the familiar amino acids, as well as amino acids very unlike those known on Earth. The most conceivable explanation for this organic matter is that it was synthesized abiotically.

Polymerization of Organic Monomers

Two major difficulties must be dispensed with in order to understand how polypeptides, polynucleotides, and other polymers were assembled from their monomer precursors. An adequate source of energy is required for covalent linkages between monomer units in a polymer, and water must be removed so that reactants will be concentrated and their polymerized products will not be hydrolyzed in an aqueous environment that thermodynamically favors depolymerization.

Several effective means for the removal of water have been demonstrated experimentally. When water evaporates, reactant molecules become adsorbed to sheets of silicate clays. Interactions between the concentrated molecules and the water that thinly coats these clays leads to the formation of polymers and their preservation on the clays. These polymers might equally well have been synthesized in tidal pools from which water evaporated, and local clays might have served as centers for polymerization in primeval situations, just as they do in the laboratory. The preserved polymers could then have leached back into solution.

When dry amino acids are heated at 60°C for a few hours in the presence of high-energy polyphosphates, proteins containing as many as 200 amino acids are formed. On the primeval Earth, amino acids in solution could have been washed up on solid surfaces, evaporated to dryness, and been polymerized by heat. Monomers in aqueous solution might also be concentrated by freezing, and later heating of the concentrates in ice crystals would favor polymerization. Cycles of dry and wet, or of cold and heat, can therefore lead to the removal of water and the polymerization of the concentrated monomer precursors.

An input of energy is essential if synthetic reactions are to proceed in the thermodynamically "uphill" direction. On the primordial Earth, energy would have been available in various forms, including electrical discharge and solar radiation, as well as in energy-rich molecules such as ATP and polyphosphates. These molecules can readily be made abiotically, and they could

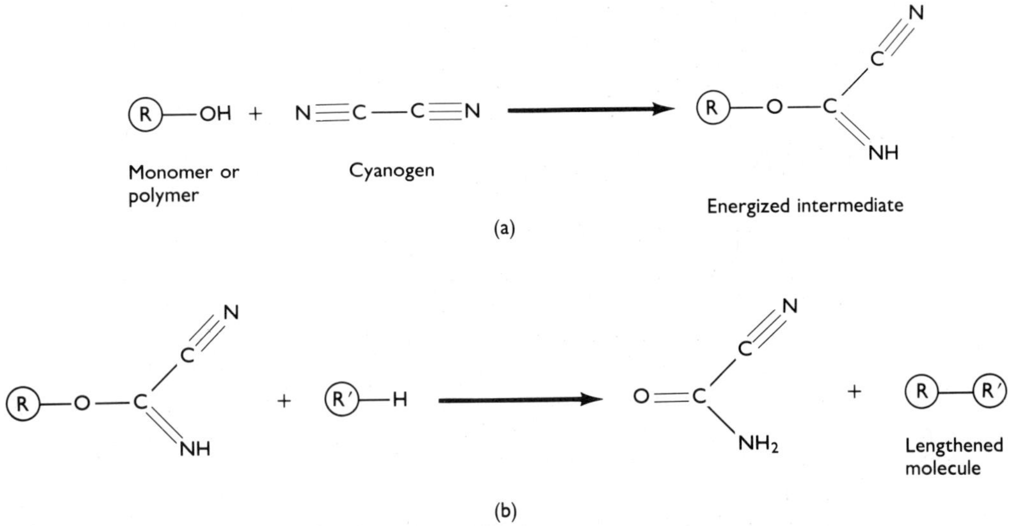

Figure 2.19. As a coupling agent, cyanogen (a) energizes a monomer or polymer molecule to produce an energy-rich intermediate, which may engage in (b) a thermodynamically favorable organic
synthesis reaction, such as lengthening an organic molecule.

have served as energy sources in primeval syntheses, just as they do in living
cells. In living cells, energy released in exergonic reactions can be transferred
for use in an endergonic reaction through the mediation of *coupling agents*.
Cyanogen ($N\equiv C-C\equiv N$) and cyanamide ($N\equiv C-C\equiv C-NH_2$) are two
well-known abiotic coupling agents (Fig. 2.19). Energy released by the
hydrolysis of ATP or polyphosphates can be conserved by coupling the phosphate residues to monomer or polymer molecules to form organophosphates,
which compete well with water for the energetic bonds of cyanogen and other
agents in reactions leading to larger molecules.

By these and other means, a population of diverse polymers accumulated in
the primeval seas, which led Haldane to call them "organic soups." The
absence of life and the ponderous slowness of organic reactions would have
allowed increasingly greater amounts of these important molecules to collect
in the oceans. Nonetheless, the organic soups would have been quite dilute
and an absolute chaos of molecules and reactions. How could such a jumble
have progressed toward life? Taking a cue from living systems, biochemists
presume that combinations of molecules and reactions would have been separated by a boundary from their surroundings, which would accomplish at
least two critical functions. First, a bounded system would preserve a concentrated set of molecules and prevent dilution into the aqueous surroundings.
Second, some bounded systems might have included, by chance, interacting
components that could participate in reactions quite different from hydrolysis
or degradation reactions favored outside the bounded units. We next examine
proposed prelife chemical systems, or *protobionts,* and consider the possible

events leading to the origin of true life forms, or *eubionts,* from these inanimate forerunners of living systems.

Summary

The universe, which includes all the matter and energy in observable space, began to expand after the thermonuclear detonation of an exceedingly small, dense entity. Expansion has continued since the Big Bang, but at a slower rate now than earlier. The Big Bang theory stipulates that the initial explosion was a singular event and that expansion will continue to the point of total emptiness when the hydrogen fuel is used up. The Oscillating Universe theory posits repeating cycles of expansion and contraction, so Big Bangs are recurrent events in endless universe history. Cosmologists cannot choose between these theories until they know the true density of matter and energy in the universe, and whether it is adequate to pull an expanding universe back to an initial state of high density and another Big Bang. According to both theories, all matter—in the form of hydrogen nuclei, helium nuclei, and electrons— formed within the first 3 minutes after the Big Bang, and a residue of radiation persists from the fireball. The elements lighter than iron are synthesized in the stars, but elements heavier than iron are synthesized when stars explode into supernovas.

Solar systems, including our own, form from a cloud of dust and gases in the numbing cold of interstellar space. Gravity leads to the development from this dust cloud of a flattened spinning disk, whose central region heats up and becomes a star (sun). Atoms and dust grains orbiting the center of the solar nebula accrete to planetesimals, which finally sweep up all particles as they grow to planetary size. The solar system is 4.6 billion years old, according to radiometric dating of meteorites. The solid phase of the Earth, or lithosphere, is enveloped in a gaseous atmosphere, which was nonoxidizing at first. The liquid hydrosphere formed later, when temperatures below 100°C led to the fluid state of hot water vapor degassed from the young Earth's interior.

Water in the hydrosphere and various gaseous molecules in the nonoxidizing atmosphere provided the raw materials for the abiotic synthesis of organic monomer molecules, such as sugars, amino acids, and nucleotides. Monomers would have accumulated in the primeval seas, but would not have polymerized unless energy sources were available and water was removed so that depolymerization was not favored by the thermodynamics of an aqueous environment. Laboratory chemistry and many simulation experiments have clearly shown the ease of abiotic synthesis of organic monomers, but biochemists have more speculation and less data on systems favoring polymer formation on the primeval Earth. By whatever means polymers may have formed, they would have assembled into some sort of bounded prelife chemical system that could have been the forerunner of true life forms at a later time in the Earth's history.

References and Additional Readings

Abell, G. O. 1985. The origin of the cosmos and the Earth. In *What Darwin Began*, ed. L. R. Godfrey, p. 223. Boston: Allyn and Bacon.

Bethe, H. A. 1968. Energy production in stars [Nobel lecture]. *Science* 161:541.

Bethe, H. A., and G. Brown. 1985. How a supernova explodes. *Sci. Amer.* 252(5):62.

Butcher, H. R. 1987. Thorium in G-dwarf stars as a chronometer for the galaxy. *Nature* 328:127.

Dicke, R. H. et al. 1965. Cosmic black-body radiation. *Astrophys. Jour.* 142:414.

Dickerson, R. E. 1978. Chemical evolution and the origin of life. *Sci Amer.* 239(3):70.

Gamow, G. 1956. The evolutionary universe. *Sci. Amer.* 195(3):136.

Guth, A. H., and P. J. Steinhardt. 1984. The inflationary universe. *Sci. Amer.* 250(5):116.

Habing, H. J., and G. Neugebauer. 1984. The infrared sky. *Sci. Amer.* 251(5):48.

Miller, S. L. 1953. A production of amino acids under possible primitive earth conditions. *Science* 117:528.

Miller, S. L., and L. E. Orgel. 1974. *The Origins of Life on the Earth*. Englewood Cliffs, N.J.: Prentice-Hall.

Penzias, A. A. 1979. The origin of the elements [Nobel lecture]. *Science* 205:549.

Penzias, A. A., and R. W. Wilson. 1965. A measurement of excess antenna temperature at 4080 Mc/s. *Astrophys. Jour.* 142:419.

Shaham, J. 1987. The oldest pulsars in the universe. *Sci. Amer.* 256(2):50.

Silk, J. 1980. *The Big Bang: The Creation and Evolution of the Universe*. San Francisco: Freeman.

Steward, F. D., P. Gorenstein, and W. H. Tucker. 1985. Young supernova remnants. *Sci. Amer.* 253(2):88.

Vilenkin, A. 1987. Cosmic strings. *Sci. Amer.* 257(6):94.

Viola, V. E., and G. J. Matthews. 1987. The cosmic synthesis of lithium, beryllium, and boron. *Sci. Amer.* 256(5):38.

Waldrop, M. M. 1987. Supernova neutrinos at IMB. *Science* 235:1461.

Weinberg, S. 1977. *The First Three Minutes*. New York: Basic Books.

Wetherill, G. W. 1981. The formation of the Earth from planetesimals. *Sci. Amer.* 244(6):162.

Wheeler, J. C., and K. Nomoto. 1985. How stars explode. *Amer. Sci.* 73:240.

CHAPTER 3

The Origin and Evolution
of Primeval Life

The origin of life remains a deep mystery. Any prelife forms or simple, primeval life disappeared billions of years ago, and no vestige has been preserved in the fossil record. Biologists can, however, apply their knowledge of modern life to trace its origins back to primeval times. Modern organisms are wonderfully diverse, but they share a number of basic features. It is to these basic features that we refer in reconstructing possible scenarios for the appearance and nature of early life. The evolution of biological diversity can also be explored within the context of the changes wrought on the planet by organisms themselves and through the application of the fundamental premise of *descent with modification*. Ancestors and descendants have a common heritage, which has been embellished, modified, and amplified over eons of time. By sorting out the many variations on this heritage, scientists can assemble at least part of the evolutionary history of life during earlier times.

The Origin of Life on Earth

Once the Earth had formed and chemical evolution was under way, two more steps were needed before life could come into existence. Droplets of the organic soup had to be segregated and delimited as bounded systems with their own chemistry and identity, and some of these prelife forms had to acquire characteristics of living entities, particularly a genetic system capable of the replication and transmission of traits from one generation to the next. Reproduction is the hallmark of life.

Basic Features of Living Cells

The protoplasm of a cell is enclosed within the **plasma membrane,** an envelope that is constructed from two layers of phospholipid molecules peppered
with a mosaic of proteins on the surfaces and within the bimolecular structure
(Fig. 3.1). The kinds and arrangements of phospholipids and proteins may
vary from one membrane to another, but every plasma membrane performs
the same basic functions. It separates internal reactions from external disorder, acting as a *selectively permeable* barrier to the entrance and exit of diverse
molecules. Within this enclosure are concentrated reactants quite different in
amount and variety from those in the dilute surroundings, and capable of specific chemical interactions of organic synthesis and degradation. The cell can
exchange both matter and energy with its surroundings, and as such an *open
system,* it may exist in different *steady states* in which the rates of reactions
fluctuate from one moment to another. By these means, the cell never reaches
the state of thermodynamic equilibrium in which work cannot be done,
because the net change in energy does not reach zero. In living cells, the
change in free energy, or $\Delta G°$, fluctuates between energy gain in synthesis
reactions and energy loss in degradation reactions. The living cell, therefore,
maintains its orderliness in the face of encroaching entropy, or disorder, man-
dated by the second law of thermodynamics. Only in death does the cell proceed   inevitably   toward   minimum   free-energy   and   thermodynamic
equilibrium.

    Cells carry on sets of coordinated chemical reactions that constitute **metab-
olism.** The spontaneous rates of organic reactions are notoriously sluggish
because of the **activation energy** requirement, a barrier that must be hurdled
before a reaction can proceed and reactants are converted into end products

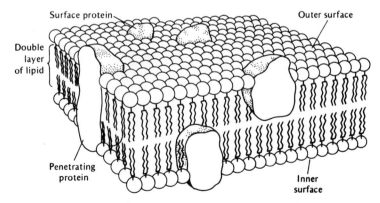

Figure 3.1. The membranes of living cells are constructed from a relatively fluid bilayer of phospholipids, with a mosaic of proteins bound to either surface of the bilayer and penetrating proteins
extending partly or completely through the bilayer. Other lipids, such as cholesterol, may also be part
of the bilayer. (From *Biology: A Human Approach,* 3rd ed., by Irwin Sherman and Vilia Sherman.
Copyright © 1983 by Oxford University Press. Reprinted by permission)

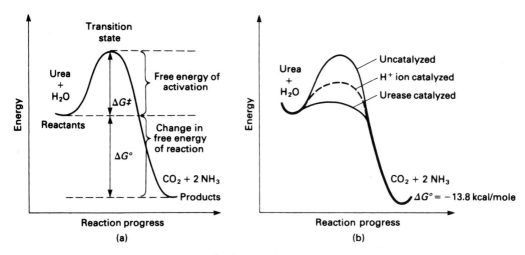

Figure 3.2. The activation-energy barrier. (a) Energy relations are shown for the reaction in which urea is hydrolyzed to carbon dioxide and ammonia, and for the reverse reaction of synthesis. The difference in free energy ($\Delta G°$) between reactants and end products is -13.8 kcal/mole, but the addition of activation energy is necessary to raise the urea molecule to a higher-energy transition state if the reaction is to occur. (b) The activation-energy barrier to reaction progress is reduced by catalysts, which speed up the reaction rate over the spontaneous uncatalyzed rate driven by kinetic energy of molecular movements. The enzyme urease is a more effective catalyst than hydrogen ions, and is far more specific for the particular reaction. Whether or not the reaction is catalyzed, the free-energy difference remains -13.8 kcal/mole. (From C. J. Avers, 1986, *Molecular Cell Biology*. Menlo Park, Calif.: Benjamin-Cummings)

(Fig. 3.2). Just consider that a pinch of sugar in solution may take years to be rendered into its ultimate end products of carbon dioxide and water, whereas this change may take place in seconds or minutes in living cells. We could heat the sugar solution and cause its constituent molecules to move faster, collide more frequently and with greater kinetic energy, and thereby be converted to CO_2 and H_2O more quickly. The same resolution of the activation-energy problem cannot be carried out in living cells because elevated temperatures denature their molecules and destroy their systems. Instead, cells possess **enzymes,** which act as efficient and selective catalysts, governing rates of reactions as well as kinds of possible reactions. By binding with the substrate molecule in a temporary association, the enzyme lowers the activation energy needed to initiate the reaction and greatly speeds up the rate of reaction.

The great preponderance of cellular enzymes are proteins, and they serve as catalysts uniquely in living cells. In the nonliving world, ions of hydrogen and of zinc, magnesium, iron, and other metals can act as catalysts. They are relatively nonspecific, unlike the highly specific enzymes, but they do lower the activation-energy barrier of organic reactions, although less efficiently than enzymes. In addition to their efficiency and specificity, enzymes work in the conditions of mild temperature and low pressure that characterize life.

Another universal feature of cellular metabolism is *coupled reactions,* in which energy released in exergonic reactions may be conserved and made available for energy-requiring endergonic reactions (Fig. 3.3). Energy transfers are accomplished in cells by various routes and are mediated by components of enzymes. Energy may be passed from one reaction to another by *phosphate-group transfers* in conjunction with the ATP–ADP cycle or other phosphorylation–dephosphorylation systems, by *electron transfers* mediated by metal ions or by organic residues in enzymes or their coenzyme components, and by *transfers of organic moieties* assisted by specific coenzymes. Many energy-transfer agents act as common intermediates between pairs of exergonic and endergonic reactions.

Metabolic pathways usually consist of a number of reactions in which the original molecules are modified step by step into particular end products. Each step is catalyzed by a specific enzyme and usually involves the consumption or release of relatively small amounts of free energy (Fig. 3.4). The enormous advantage of the slow, stepwise release of energy is that small packets of free energy can be handled by the metabolic machinery of a cell, and not be dissipated as heat or in an explosion. Cells are not totally efficient, of course, and some energy does escape as heat, but never in explosive amounts that would damage or destroy the living fabric.

In addition to a selective boundary and an organized metabolic system for its organic chemistry, every cell has a **genetic system** whose deoxyribonucleic acid (DNA) is transcribed and translated into products specified in the encoded instructions of its set of genes (Fig. 3.5). The replication and transmission of these genes underlie the continuity of life in successive generations, and alterations in DNA provide the diversity that characterizes biological evolution.

Many of the thousands or tens of thousands of genes of a species are encoded for molecules that form part of the genetic system itself. From about 50 to 100 genes specify the individual molecules of the ribosome, on which the transcribed genetic information is translated. Hundreds of other genes are

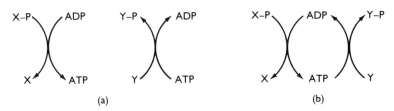

(a)   (b)

Figure 3.3. Coupled exergonic and endergonic reactions. (a) The formation of compound X from phosphorylated X (X–P) releases energy, which is used to produce ATP from ADP in a coupled endergonic reaction. The energy-requiring synthesis of phosphorylated Y (Y–P) from compound Y can occur when coupled to the energy-releasing reaction that produces ADP from ATP. (b) All four reactions may be coupled to in a metabolic pathway, such that energy released in the production of X from X–P is transferred by means of the ATP–ADP cycle to subsidize the energy-requiring production of Y–P from Y.

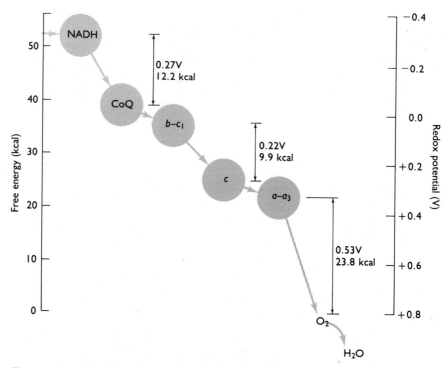

Figure 3.4. The transfer of electrons from oxidized fuel molecules to molecular oxygen takes place in a sequence of respiratory reactions, each catalyzed by a specific enzyme. The more than 50 kcal/mole of free energy that is released (or reduction in redox potential as measured in volts) is distributed among the reaction steps in the sequence. At three of these steps, enough free energy is released to sponsor ATP synthesis in oxidative phosphorylation reactions coupled to electron transport toward oxygen (see Fig. 3.34).

blueprints for the ribonucleic acids (RNA) involved in transcription and translation and for the catalytic, regulatory, and structural proteins that are essential for gene expression.

Primeval life was certainly not as metabolically and genetically complex as even the simplest modern organism. But any primeval life would have possessed the rough beginnings of metabolism and genes enclosed within a selectively permeable boundary that isolated it from the rude surroundings of the dilute primeval seas.

The Search for the Origin of Life

Various approaches have been and are being taken to demonstrate how the appropriate kinds of molecules and reactions could have become organized into the first cells. Three very different **protobiont,** or protocell, systems—

coacervates, proteinoid microspheres, and liposomes—have been proposed as models for the reenactment of primeval events that ultimately led to life. They differ in significant ways, but each has a boundary that separates and concentrates an internal set of molecules from its dilute and disorderly aqueous surroundings.

The first important experimental studies were reported in 1924 by Alexander Oparin, who for more than 60 years pursued the prospect of coacervation as the process of protobiont formation. A **coacervate** basically is a droplet of water in which electrically charged particles are suspended. Coacervates form when colloidal particles in an aqueous medium are joined in a droplet as some of the surrounding water molecules are excluded (Fig. 3.6). The coacervate is sharply delimited from its surroundings by a boundary consisting of a film of bound water molecules, which are attracted to one another and to the internal collidal particles by positive and negative charges. Many

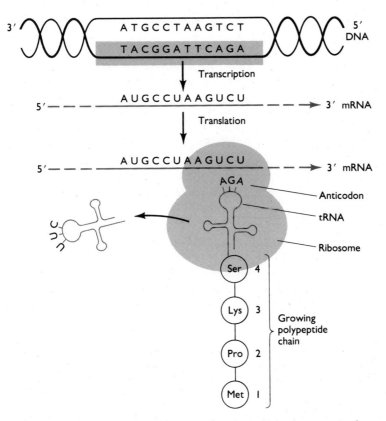

Figure 3.5. Coded information specifying the kinds and sequence of amino acids in polypeptides is transcribed from DNA to messenger RNA (mRNA), and the copied instructions in mRNA are translated into the specified polypeptide at ribosomes in the cytoplasm. Transfer RNAs (tRNA) carry amino acids to sites of synthesis at the ribosomes, and provide recognition (by their anticodon) and interaction signals to mRNA and ribosomes, which permit a high level of translational accuracy.

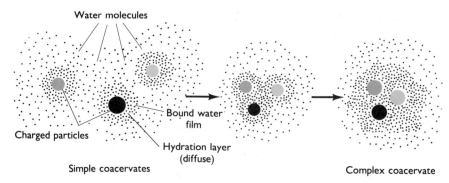

Figure 3.6. Simple coacervate droplets are formed when some of the electrically charged dipolar water molecules become tightly bound to an electrically charged molecule or particle. Water molecules at a greater distance from the charged particle remain part of a free (unbound) water phase. Attraction between electrically charged systems may lead to the fusion of two or more simple coacervates, forming a complex coacervate held in a film of bound water molecules. The film acts as a boundary between the system and its surroundings.

simple coacervates are usually formed in a common medium, and they often merge to produce complex coacervates containing the diverse components of the fused droplets.

Coacervates possess several features that make them a promising protobiont system. The film of bound water molecules serves as a boundary that separates the localized chemical system from the dilute surroundings. Coacervation sweeps up molecules from the aqueous medium and concentrates them in the droplet's interior. In a typical experiment, Oparin showed that 95 percent or more of the protein from a dilute solution of 1 part gelatin to 100,000 parts water ended up in the coacervates that were formed. In media containing mixes of compounds, diverse coacervates are produced, containing different kinds and combinations of molecules. These salient features notwithstanding, there is an important drawback to the coacervate model. Coacervates tend to be rather transient entities that disintegrate fairly quickly, especially under inhospitable conditions of acidity or alkalinity in the surrounding medium. But coacervates have been known to persist under optimum conditions for as long as several years.

The **proteinoid microsphere** model of the protobiont was proposed about 40 years ago by Sidney Fox, whose ongoing studies have produced some very interesting and provocative insights into primeval events leading to life. When mixtures of dry amino acids are heated to 130°C to 180°C, they polymerize into proteinlike molecules called proteinoids. As proteinoids are cooled in water, a large number of small spheres separate out of solution (Fig. 3.7). These structures, called proteinoid microspheres, have certain properties reminiscent of life forms.

Proteinoid microspheres are about 1 to 2 μm in diameter, which makes them similar in size and shape to coccid bacterial cells. The microspheres are remarkably stable and can survive the forces generated in a laboratory centrifuge. They have an osmotically active boundary that can be seen under the microscope as a double layer. When treated with the Gram stain conventionally used by bacteriologists, proteinoid microspheres made from acidic amino acids show a Gram-negative reaction; when made from at least 35 percent basic amino acids, the microspheres prove to be Gram positive. Finally, these structures can be induced to constrict in a process that superficially resembles budding in bacteria and fungi. Whether or not these features are significant, this interesting protocell model deserves the careful consideration it has received over the years, particularly for the insights it gives into conditions for polymerization and the concentration of polymers in bounded structures. It remains to be seen, however, whether the somewhat limited diversity of such chemical systems can have included the spark required to instill life into lifeless structures.

The most recent protobiont model has come from studies by David Deamer, Joan Oró, and others, who have made **liposomes** with a phospholipid boundary that is more cell-like than the boundary of coacervates or proteinoid microspheres. Phospholipids are *amphipathic* molecules; that is, each molecule

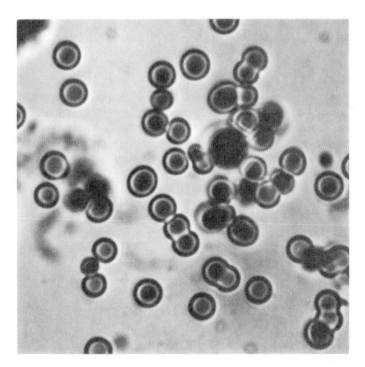

Figure 3.7. When a hot solution of proteinlike polymers, called proteinoids, is cooled in water to 0°C to 25°C, spherical structures form very quickly. These structures are called proteinoid microspheres because of their contents, form, and size. ×1000. (Courtesy of S. W. Fox, from S. W. Fox and K. Dose, *Molecular Evolution and the Origin of Life*, 2nd ed. New York: Dekker)

includes both hydrophilic and hydrophobic regions. The hydrophilic "head" consists of a negatively charged phosphoric-acid group bonded to a positively charged organic residue, such as ethanolamine, and the hydrophobic "tail" is composed of two fatty acids or related compounds. In water, phospholipids aggregate spontaneously to form *micelles,* with the hydrophilic, polar "heads" in contact with the water at the micelle periphery and the hydrophobic, nonpolar "tails" tucked inside the micelle (Fig. 3.8). In systems with an air phase as well as a water phase, the phospholipids may aggregate into bimolecular layers, with polar "heads" at each surface making contact with water and nonpolar "tails" sequestered in the interior of the aggregates. This conformation is characteristic of cell membranes.

When liposomes, or phospholipid vesicles, are washed into solutions containing one or more kinds of organic molecules, the molecules are taken into the vesicles. Many types of vesicles can arise, each with osmotic properties that allow increases in mass and complexity through the selective entry and exit of substances in the surrounding solution. If proteins are incorporated into the phospholipid boundary, as occurs in cell membranes, it behaves even more like a selective plasma membrane. Whether or not proteins are part of the phospholipid layer, the liposome provides a means by which molecules can be concentrated from solution and be spatially isolated from dilute and disorderly aqueous surroundings.

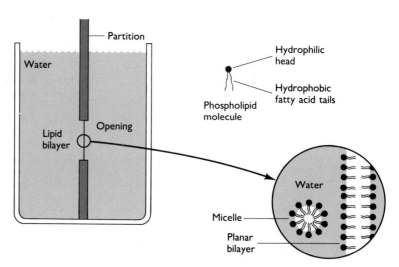

Figure 3.8. When two water-filled compartments are separated by a perforated partition, phospholipids added to the system aggregate spontaneously to form a planar bilayer across the opening. The hydrophilic heads interact with water on each side of the opening, but water is excluded from the hydrophobic region of fatty acid tails in the bilayer interior. When completely surrounded by water, phospholipids may aggregate spontaneously to form micelles. In a micelle, the hydrophilic polar head groups are arranged around the periphery, and the hydrophobic fatty acid chains are sequestered in the interior, away from water.

Each of the three protobiont models provides a reasonable basis for the development of organized structures that can gather into themselves a selected concentration of organic polymers, including proteins and nucleic acids, distinct from the surroundings. Whether any one of these prototypes can become self-sustaining for a long enough time to evolve into a reproducing system is still open to question. The next two sections deal with some of the theoretical premises that underlie the transition to life and the experimental evidence that bears on the development of a suitably dynamic and coordinated chemistry that may have characterized the first life forms.

Protobiont Structures Must Harbor a Dynamic Chemistry

There may be little agreement about the particular structural form of the protocell, but biochemists can agree that such an entity would possess the minimum chemical dynamics for *self-renewal,* or maintenance of existence. Protocells that could repair and replace their chemical components would enjoy the advantage of persisting for longer intervals. Persistence is theoretically advantageous because self-renewing entities have a better chance of accumulating more and different kinds of molecules, some of which can be coordinated in a dynamic display of reactions that mimic simple metabolism.

Various experiments have produced analogies of simple growing or metabolizing systems. For example, Oparin provided coacervates with selected compounds and precursors and showed that a simple flow of metabolites can be established, perhaps leading to an increase in mass. In a typical experiment, he demonstrated that coacervates containing the enzyme *starch phosphorylase* can make and store starch molecules when glucose 1-phosphate precursors are in the surrounding medium. If such coacervates also contain the enzyme *amylase,* starch is hydrolyzed to maltose, which is released into the medium (Fig. 3.9). This system grows as starch accumulates and metabolizes as starch is synthesized and hydrolyzed in coordinated reactions known to occur in living cells.

Starch phosphorylase and amylase are complex proteins made in living cells according to specific genetic instructions that are transcribed and translated by means of an intricate genetic machinery. Such catalysts are unlikely to have arisen by chance in the primeval seas. Much simpler and probably less specific catalysts almost certainly were synthesized abiotically and thus were available to protocells floating in the organic soup. As mentioned earlier, an assortment of metal ions must have been present then, as now, and may have served as catalysts. It is very likely that even more effective catalysts were present in the form of *metal–porphyrin complexes.* Porphyrin is an organic compound composed of four pyrrole units arranged in a ring and can readily be synthesized abiotically from four pyrrole rings and four formaldehyde molecules in a series of condensation and oxidation reactions (Fig. 3.10). Porphyrins tend to coordinate with metal ions, particularly Fe^{2+} and Mg^{2+}, and as metal–porphyrin complexes they are far more efficient catalysts than are metal ions

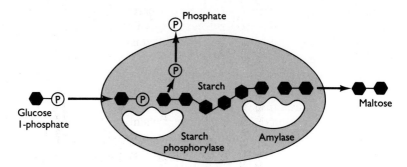

Figure 3.9. In experiments with coacervate systems, A. I. Oparin showed that under appropriate conditions, such systems might increase in mass and biochemical complexity. Coacervates containing the enzyme starch phosphorylase can synthesize and store starch if glucose 1-phosphate is provided as a substrate. If the enzyme amylase is added to the coacervate, the starch is hydroloyzed to produce maltose, which is released into the surrounding medium. Thus the coacervate can increase in mass (due to starch storage) or in biochemical complexity (a one-step reaction expands to become part of a two-step sequence).

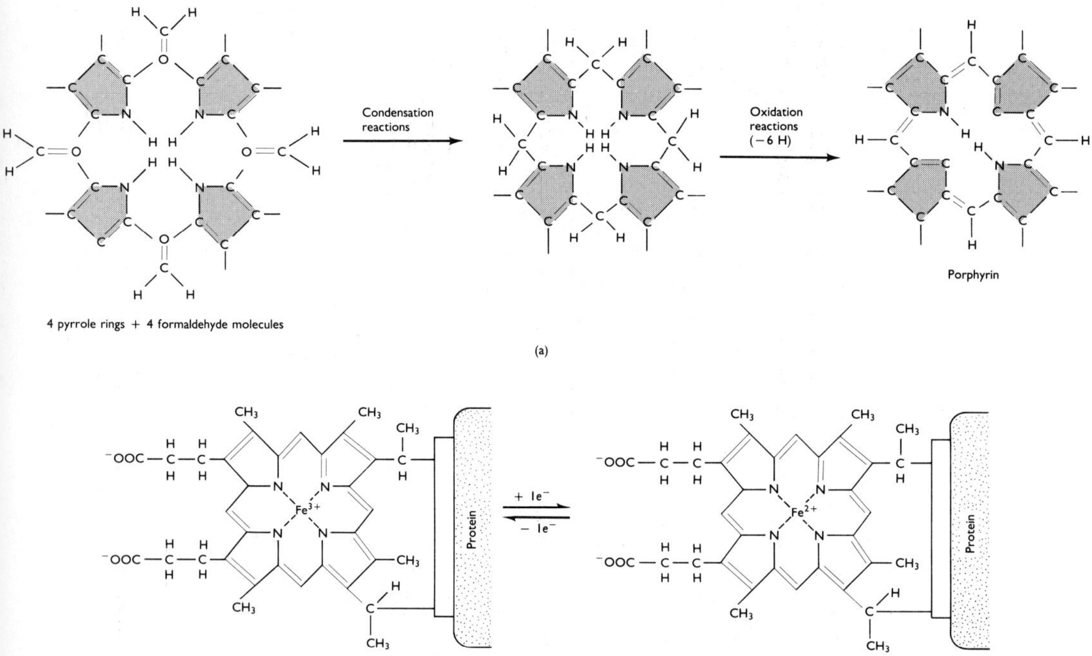

Figure 3.10. (a) Porphyrin, which may be synthesized abiotically from formaldehyde and pyrrole units in a series of condensation and oxidation reactions, often coordinates with metal ions, such as those of iron and magnesium. (b) Cytochrome *c,* a respiratory enzyme, consists of a heme iron–porphyrin complex bonded to a protein moiety. Electron transport to and from cytochrome *c* involves valency changes in the iron ion.

alone. In addition, these complexes are parts of many important cellular enzymes, some of which are believed to be of ancient origin. Catalase, cytochromes, and a number of other enzymes consist of a **heme** iron–porphyrin group linked to a protein. By comparison, a heme molecule is 1000 times more efficient than Fe^{2+} alone in catalyzing the breakdown of hydrogen peroxide, and heme-containing catalase degrades peroxide 10 million times faster than does heme alone.

These and similar considerations help chemists understand how a simple chemical system might become more efficient through improved catalysis and coordinated reactions. But how does such a system move from being a unique experiment in metabolism to being a living entity? Even if some abiotically synthesized proteins were to conjugate with metal–porphyrins and thereby become better catalysts, how are other protocells to receive or develop these advantageous properties in the absence of genes? The variety of porphyrins is relatively small, but there is no limit to the diversity of proteins. Proteins are made from any or all of 20 kinds of amino acids, which are assembled into polypeptide chains of varying lengths. Let us suppose for the sake of argument, however, that as few as 4 kinds of amino acids enter into all possible combinations to form polypeptides only 10 amino acids long. More than 1 million (4^{10}) unique and different polypeptide molecules can be made under these conditions, but the probability is very low that any one polypeptide will be part of any one catalyst in more than a few out of many millions of protobionts in the seas. The problem is compounded by the absence of any known mechanism for the replication of proteins or for their equal partition among cells.

Arguments have raged back and forth over the years between the proponents of a metabolizing protobiont that happened to engulf nucleic acids that perhaps specified at least one essential polypeptide needed for that metabolism, and the supporters of a protobiont that captured a metabolism that matched its established genetic contents. Neither aspect of this chicken-and-egg paradox is particularly attractive, because either alternative calls for an uncomfortably large element of chance. Biologists are looking instead for some protobiont system in which genetic and metabolic systems developed coordinately toward the living condition.

Onward from Protobionts to Living Cells

Because DNA is the genetic material in all cells and many viruses, experiments over the years have been focused on attempts to prod DNA in test tubes to replicate. The results of these experiments have been disappointing because replication apparently does not proceed beyond the point of complementary base pairing. An intrinsic property of nucleic acids is the specific pairing, by hydrogen-bond formation, of their nitrogenous bases. Adenine (A) pairs with uracil (U) or thymine (T), and cytosine (C) pairs with guanine (G), whether the bases are parts of nucleotide monomers or of polynucleotide chains (Fig. 3.11). In a typical experiment, Leslie Orgel showed that poly-uracil, or poly(U), assembles along a polyadenine, or poly(A), template through the complementary base pairing of A and U residues. But in the absence of proteins, the poly(U) strand does not dissociate from its poly(A) partner strand in the duplex and does not sponsor poly(A) formation from adenine-containing monomers in the aqueous medium. DNA replication fails to occur unless the in vitro system has suitable proteins and an energy source.

A fresh perspective on the origin of life has been provided by recent experimental studies of RNA, which in specific instances may have catalytic prop-

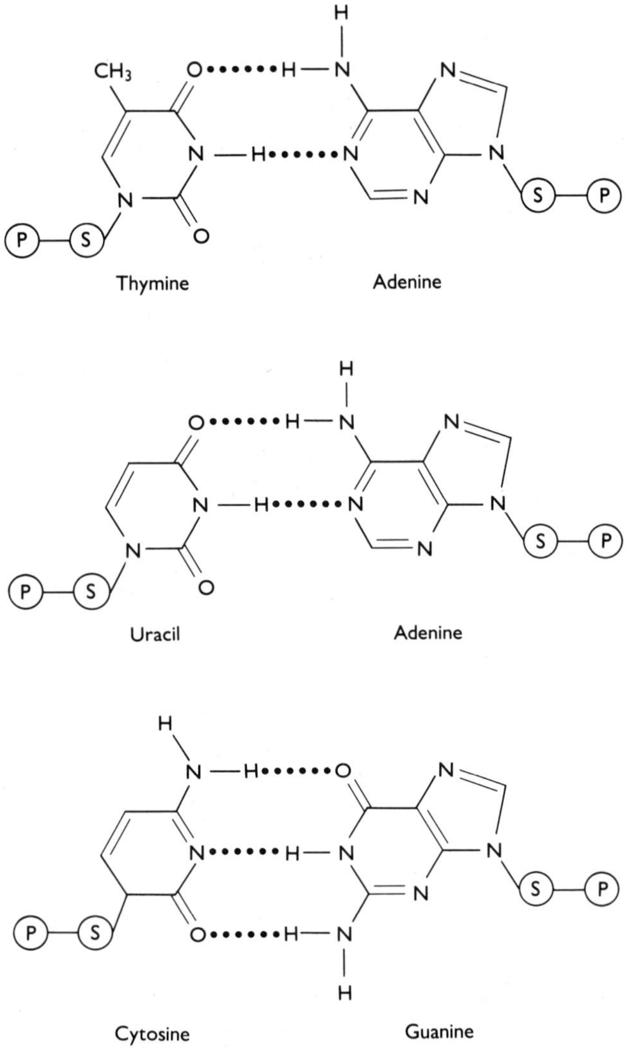

Figure 3.11. Complementary base pairing by means of hydrogen bonds occurs between thymine or uracil and adenine, and between cytosine and guanine. The bases are shown in their nucleotide forms (base–sugar–phosphate).

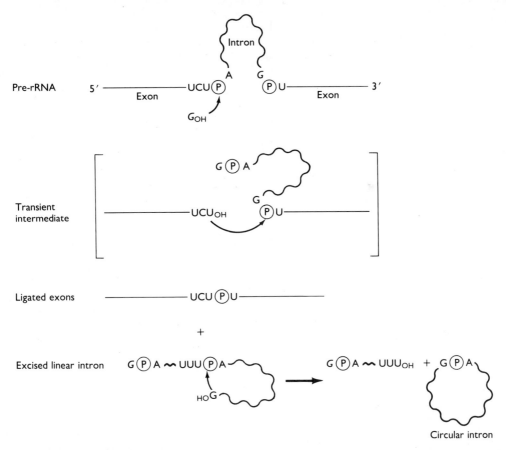

Figure 3.12. Self-splicing RNAs act catalytically in excising an intron and ligating the flanking exons of the RNA copy of the gene. in *Tetrahymena*, the precursor of ribosomal RNA (pre-rRNA) self-splices by means of two consecutive transesterification reactions, which require an external source of guanosine (G_{OH}) but not a protein or an external supply of energy. The excised intron, containing 414 nucleotides, undergoes cyclization to become a covalently closed circle, which is a trademark product of the process in this and other self-splicing RNAs. Phosphate residues are shown by ⓟ; the intron, by a wavy line; and flanking exons, by straight lines.

erties as well as its known capacity for replication. Thomas Cech and his colleagues showed convincingly that a noncoding sequence, or intron, in a ribosomal RNA (rRNA) precursor from the ciliated protozoan *Tetrahymena* can snip itself out of the precursor and splice the two free ends of the molecule to form the mature rRNA (Fig. 3.12). The two-step process took place in the absence of proteins and an external energy source, but did require a molecule of guanosine to initiate the process in vitro. Even more important, the excised intron catalyzed the synthesis of poly(C) molecules up to 30 nucleotides long, from short segments of poly(C) added to the medium. A short sequence in the intron acts as a template for the polymerization reaction. Kinetic analysis of the system clearly showed that the excised intron behaved

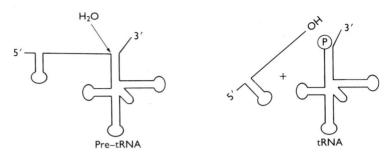

Figure 3.13. The processing of precursor tRNA (pre-tRNA) involves a site-specific cleavage to pro-duce mature tRNA with the correct 5' terminus. The cleavage reaction is catalyzed by RNase P, an enzyme whose catalytic component is its RNA subunit rather than its protein subunit.

like an enzyme, but enzymes were believed to be exclusively proteins. The RNA intron functioned as a polymerase, making copies of itself and other RNA molecules, without the help of protein enzymes.

Subsequent studies by Cech and others have shown that this particular RNA is not the only self-splicing RNA species. Self-splicing appears to be characteristic of all the known classes of introns in both mitochondrial RNA and nuclear RNA. The real goal, however, is to learn whether any or all of these introns have catalytic ability. If they do, then a diverse and widespread group of RNAs, rather than the single (and perhaps unique) rRNA intron from *Tetrahymena,* have both template and catalytic properties.

Quite a different catalytic RNA has been described by Sidney Altman and his colleagues in studies of the enzyme *ribonuclease P,* or RNase P, in *E. coli* and some other bacteria. This enzyme is made of both RNA and protein sub-units, and its function is to cleave about 60 different transfer RNA (tRNA) precursors at a particular bond in order to produce the correct 5' terminals of the mature tRNAs (Fig. 3.13). In various experimental approaches, including studies using transcripts of the enzyme's RNA made in vitro from the gene or isolated from whole cells, it was shown that the RNA subunit alone is able to catalyze the cleavage reaction, in the absence of protein. The products of the reaction are the same as those obtained after cleavage by the whole enzyme. Michaelis-Menten analysis of reaction kinetics shows that the RNA subunit behaves like a conventional enzyme.

Considerable progress has been made in defining the distribution of base pairing in the recognition process preliminary to the binding of the enzymatic RNA to its precursor tRNA substrates. Unlike the "guide sequences" of nucleotides, which are believed to play a role in substrate recognition by intron RNA, the RNA of RNase P apparently binds to scattered nucleotides in its precursor tRNA substrates. These data indicate an important role for the secondary or tertiary structure of the catalytic RNA; that is, the folded conformation of the RNA may be very important for its catalytic function. In addition, it was discovered that the efficiency of catalysis by the RNA subunit is enhanced in the presence of the protein subunit of the enzyme. Much remains to be discovered, including the precise mechanisms of recognition and cleavage by this enzyme, before RNase P can serve as a model for understanding the role of other ribonucleoproteins. In particular, biochemists and biologists are very much interested in the implications for the evolution of the first cells from protobionts that could replicate their nucleic acids without the help of complex proteins, however inefficiently.

The evidence provided in studies of catalytic RNAs raises the enticing possibility that RNA, rather than DNA, was the first genetic material. The genetic code applies equally to RNA and DNA, and both kinds of nucleic acid can be copied by processes that involve the pairing of complementary bases and the construction of polynucleotides of varying lengths. If RNAs in one or more protobionts also possessed catalytic properties, replication could have taken place and produced RNA copies with these properties. By whatever means the replicated molecules might have been partitioned into other protobiont droplets, whether by chance fragmentation or in some other way, there may eventually have developed a protobiont whose RNA evolved into coded instructions that specified some simple polypeptide whose presence enhanced the catalytic replication process.

Engulfing precursors from the primeval seas, protobionts with replicating RNA could be self-sustaining systems as well as producers of entities like themselves. In effect, these entities would be alive. Once living, the primeval cells would come under the guidance of natural selection acting on the genetic diversity that arises by chance processes of mutation and other genetic alterations. The cells with greater advantage, such as faster rates of reactions or the ability to use a broader spectrum of precursors in metabolism, would flourish, while their competitors would be diminished in number or become extinct (Fig. 3.14). By **differential reproduction,** cells with adaptive features

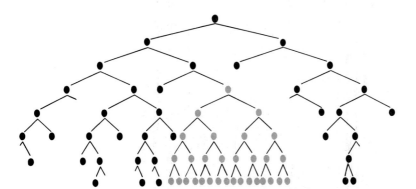

Figure 3.14. By natural selection, inherently better fit individuals (gray) are more likely to survive to reproductive age and thus leave more descendants than are less fit individuals (black), which produce fewer or no progeny. Inherited traits that confer selective advantages on the better fit members of a population are thus transmitted to descendant generations, which are as well adapted as were their parents to particular environments and conditions. The accumulation of inherited adaptive traits in populations leads to evolutionary changes that may result in new biologcal capacities or to new kinds of organisms.

would produce more cells with these advantages, and the procession of life would be subject to evolutionary processes that could never influence the unique and unreproducible protobiont system.

The excitement over catalytic RNA comes from the promise of a solution to the chicken-and-egg paradox. Scientists do not have to wonder whether the right proteins or the right nucleic acids were joined by their respective partner molecules in chance encounters of very low probability. They have the outline of a model system in which catalytic, informational RNA could replicate and, ultimately, specify particular polypeptides spelled out in the sequence of nucleotides. Biologists may be on the very brink of solving the mystery of the origin of life.

Genetic Amplification in Primeval Life

Whether or not primeval genes were made of RNA and were replaced by DNA at some later time, the nucleic acids must have had an *informational* function and come to specify the kinds and sequence of amino acids in proteins. The origin and development of the whole genetic apparatus is beyond scientific understanding now, but biologists can speculate productively about selected features of the genes themselves. In particular, they can use the knowledge gained from observations and experiments conducted over decades to inquire into at least two basic problems: (1) the origin and evolution of the genetic code, and (2) the increase in gene number and diversity as life evolved. These problems are focused on the nature and evolution of the informational component of the genetic apparatus.

Basic Features of the Genetic Code

The genetic code was deciphered in landmark experiments conducted between 1961 and 1967 by Marshall Nirenberg, Sydney Brenner, Francis Crick, and others. In 1961, Nirenberg and Henry Matthaei showed that poly(U) guides the incorporation of the amino acid phenylalanine into chains of polyphenylalanine, which indicates that the code word, or **codon,** for this amino acid is UUU (Fig. 3.15). In the next 6 years, a total of 64 codons were identified, the last of which was one of the three punctuation codons that specify the end of a genetic message, or STOP codons.

Five basic features characterize the genetic code:

1. The code is a *triplet,* and each of the 64 codons consists of a unique permutation of 3 nucleotides drawn from the 4 kinds of nucleotides in DNA or RNA (A, C, G, and T or U); $4^3 = 64$.
2. The code includes 61 codons that specify 20 amino acids, and 3 STOP codons, or *punctuations,* that mark the end of a genetic message.

3. Codon *synonyms* exist for 18 of the 20 encoded amino acids, and only methionine and tryptophan are specified by 1 codon each.
4. The code is *consistent* because each of the 61 codons specifies only 1 of the amino acids in the set of 20.
5. The code is essentially *universal* because the same codons specify the same amino acids and punctuations in virtually all organisms, including viruses. A few codons have a different meaning in mitochondrial genes, but this small number of exceptions can be accommodated within the basic framework of a universal code.

It is immediately apparent that the distribution of codons and their amino acid counterparts is not at all random. Within each family, the codons ending in U and C (5'-*xy*U-3' and 5'-*xy*C-3') always encode the same amino acid, and the codons ending in A and G (5'-*xy*A-3' and 5'-*xy*G-3') specify the same amino acid, except for the single codons meaning methionine and tryptophan. In 8 of the 16 families of codons, all 4 codons share the initial doublet and specify the same amino acid. In these *unmixed families,* therefore, codon meaning is independent of the 3' base. Three amino acids (leucine, serine,

	U		C		A		G	
U	UUU	Phe	UCU	Ser	UAU	Tyr	UGU	Cys
	UUC	Phe	UCC	Ser	UAC	Tyr	UGC	Cys
	UUA	Leu	UCA	Ser	UAA	STOP	UGA	STOP
	UUG	Leu	UCG	Ser	UAG	STOP	UGG	Trp
C	CUU	Leu	CCU	Pro	CAU	His	CGU	Arg
	CUC	Leu	CCC	Pro	CAC	His	CGC	Arg
	CUA	Leu	CCA	Pro	CAA	Gin	CGA	Arg
	CUG	Leu	CCG	Pro	CAG	Gin	CGG	Arg
A	AUU	Ile	ACU	Thr	AAU	Asn	AGU	Ser
	AUC	Ile	ACC	Thr	AAC	Asn	AGC	Ser
	AUA	Ile	ACA	Thr	AAA	Lys	AGA	Arg
	AUG	Met	ACG	Thr	AAG	Lys	AGG	Arg
G	GUU	Val	GCU	Ala	GAU	Asp	GGU	Gly
	GUC	Val	GCC	Ala	GAC	Asp	GGC	Gly
	GUA	Val	GCA	Ala	GAA	Glu	GGA	Gly
	GUG	Val	GCG	Ala	GAG	Glu	GGG	Glu

Figure 3.15. The genetic coding dictionary, showing messenger RNA codons and the amino acids or STOP punctuations they specify. Amino acids may be specified by 1, 2, 3, 4, or 6 different codons, so that 18 of the 20 encoded amino acids are represented by 2 to 6 synonymous mRNA codons. The codon AUG specifies Met as the initiating amino acid of the translation product, as well as Met throughout the polypeptide, and STOP codons UAA, UAG, and UGA indicate termination of the genetic message.

arginine) are specified by 6 codons each, but they abide by the common-doublet rule. Finally, acidic amino acids (aspartic acid, glutamic acid) are grouped together; basic amino acids (histidine, arginine, lysine) are clustered; and all codons with a middle U (xUy) specify amino acids with a nonpolar R group (Fig. 3.16).

In addition to these features of the code itself, we must consider the relationship between the genetic information in the sequence of nucleotides and the translation of this information into a sequence of amino acids in the polypeptide product. Genetic analysis has shown that the genetic message transcribed into mature messenger RNA (mRNA) is read linearly along a **reading frame,** such that each amino acid is added in sequence starting from the amino terminus of the polypeptide and ending at the carboxy terminus. The addition or deletion of even one base in the gene sequence can alter the whole reading frame, leading to a defective polypeptide (Fig. 3.17). Such *frameshift mutations* provided the basis for Crick and others to demonstrate the existence of a reading frame for translation, which has important implications for the nature of the code since primeval times.

Expansion of the Genetic Code to Its Present Form

On the basis of the known features of the modern genetic code and of the colinear translation of a sequence of codons into a sequence of amino acids, biologists can make some reasonably educated suggestions about the nature of the primeval code and its subsequent evolution. To begin with, it is very unlikely that there was anything but a triplet code since earliest times. If the original codons were doublets of bases and later became triplets, the entire reading frame of every gene would have been thrown off and would have yielded proteins that were defective or had altered functions (see Fig. 3.17). For this reason, codons of more than three nucleotides can also be ruled out.

It is reasonable to assume that the primeval code specified fewer than the present 20 amino acids. Chemical analysis of many kinds of proteins has shown that few of them are composed of all 20 kinds of encoded units and that proteins with a more ancient heritage have disproportionately greater amounts of aspartic acid, glutamic acid, alanine, serine, and leucine, and little or no methionine and tryptophan, when compared with more recently evolved proteins. Furthermore, abiotic-synthesis experiments yield predominantly glycine, proline, aspartic acid, glutamic acid, alanine, and valine. It is tempting to assume from these observations that in its earlier stages, the code consisted of pairs of unmixed codon families, such that each set of eight synonyms specified a different amino acid. With time, two or all four codons in any one of the family-pairs may have taken on a new meaning, and a new amino acid came to be specified. Are each of the six synonyms for leucine, serine, and arginine a remnant of an original set of eight synonyms? Are half

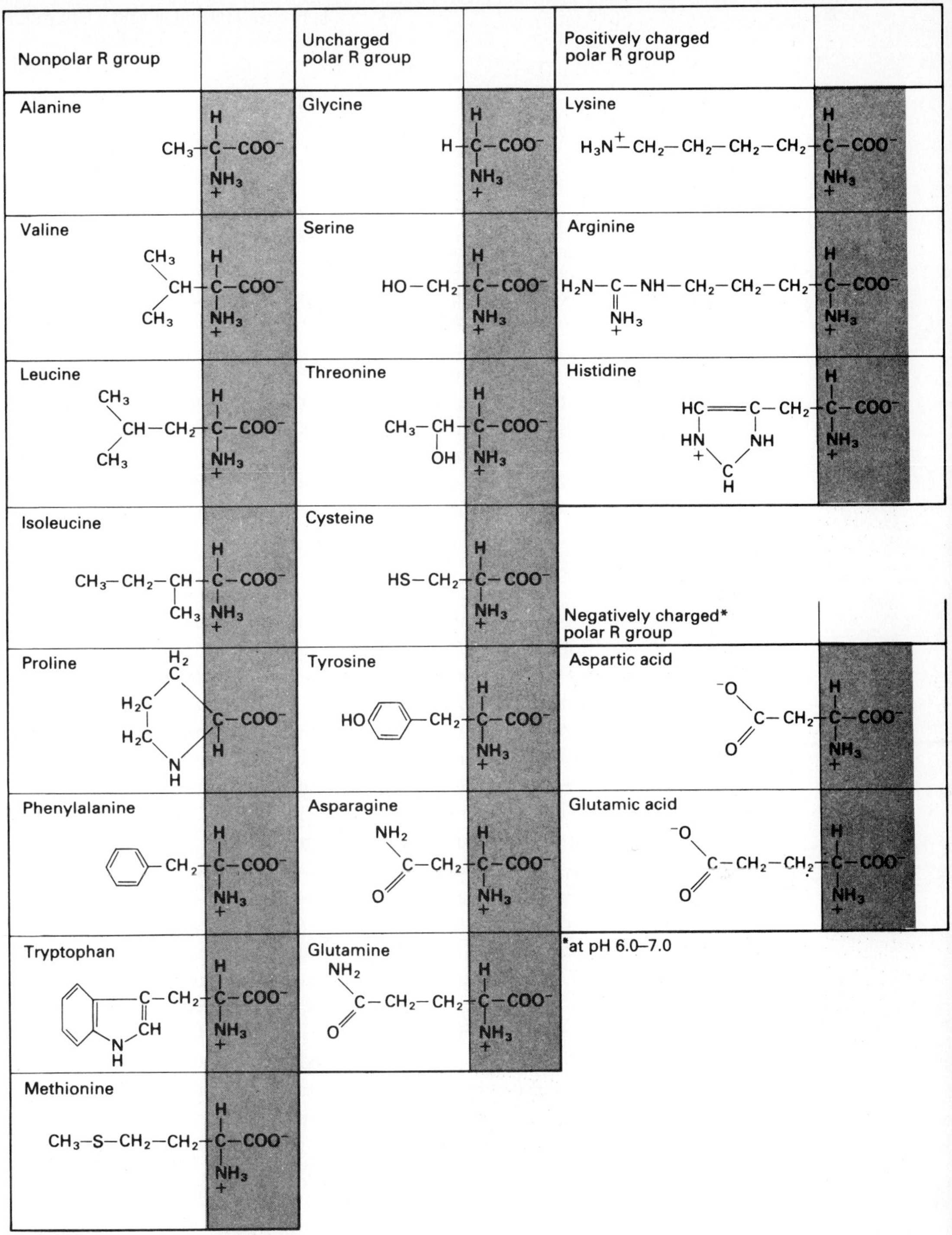

Figure 3.16. In all 20 amino acids specified in the genetic code, the α-carbon atom is bonded to a carboxyl group, an amino group, and a hydrogen atom, but the fourth residue (R) varies from one amino acid to another. The R side chain determines the electrical and polar or nonpolar nature of the amino acid. All the amino acids but glycine may exist in the D- or L-isomeric forms, because the α-carbon is asymmetric. Virtually all natural proteins are composed exclusively or predominantly of L-amino acids.

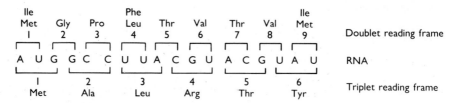

Figure 3.17. If a primeval doublet code had been modified to a triplet code during evolution, the entire reading frame would have become distorted, and every protein encoded in a primeval genome would have been changed. It is unlikely that such a profound modification would have been adaptive. Hence, it is unlikely that the primeval genetic code evolved from a doublet to a triplet reading frame.

the codon families remainders of a reduction from eight to four synonyms per amino acid, allowing more amino acids to be encoded? Are other codon families even more recent products of evolutionary divergence in codon meanings, leading to 1, 2, or 3 codons per amino acid in the modern repertory of 20? Even if they seem plausible, these suggestions remain very theoretical because there are no models of current living systems with fewer than 20 encoded amino acids.

In discussing the expansion of the genetic code and its amino acid or punctuation assignments, no attention was paid to the important role played by transfer RNA. Biochemists know of no stereochemical or chemical basis for direct interactions between template DNA or RNA and the amino acids guided into covalently bonded sequences in polypeptides. In modern systems, amino acids are brought to the correct codon sites by means of a recognition process that involves complementary base pairing between messenger RNA codons and transfer RNA **anticodons.** A number of interesting features of transfer RNAs have helped shed some light on the problem of an expanding genetic code and a divergence of codon assignments.

On the basis of the discovery that only 32 tRNAs are required to handle 61 codons specifying amino acids and that inosine and other modified bases are part of the anticodon triplets in many tRNAs, Crick proposed the **wobble hypothesis.** He suggested that the 5′ base of the anticodon can pair in certain cases with more than one kind of complementary 3′ base in the mRNA codon. This 5′ anticodon residue is the *wobble base,* which means that its pairing properties are not rigidly restricted (Fig. 3.18). In accord with the hydrogen-bonding abilities of the bases, Crick suggested that in the 5′ wobble position of an anticodon, uracil can pair with guanine as well as adenine, guanine can pair with uracil as well as cytosine, and inosine can pair with adenine, uracil, or cytosine. Neither adenine nor cytosine can pair with bases other than their standard partners in the 3′ position of an mRNA codon.

The wobble hypothesis also drew on the observation that codon synonyms have the same initial doublet but differ in their 3′ base. Crick predicted the particular codons that would bind with particular tRNAs, and experiments indeed verified his predictions. For example, three different serine-carrying tRNAs were found to bind to particular pairs of the six different serine-specifying mRNA codons, in accordance with the wobble hypothesis (Fig. 3.19). Evolutionary expansion of the genetic code may therefore have proceeded from relatively few tRNAs for codon synonyms in earlier times, to more tRNAs for newly specified codon pairs.

New information from studies of mitochondrial tRNAs has shown that as few as 22 to 25 tRNAs can cope with all 61 codons in mitochondrial mRNA.

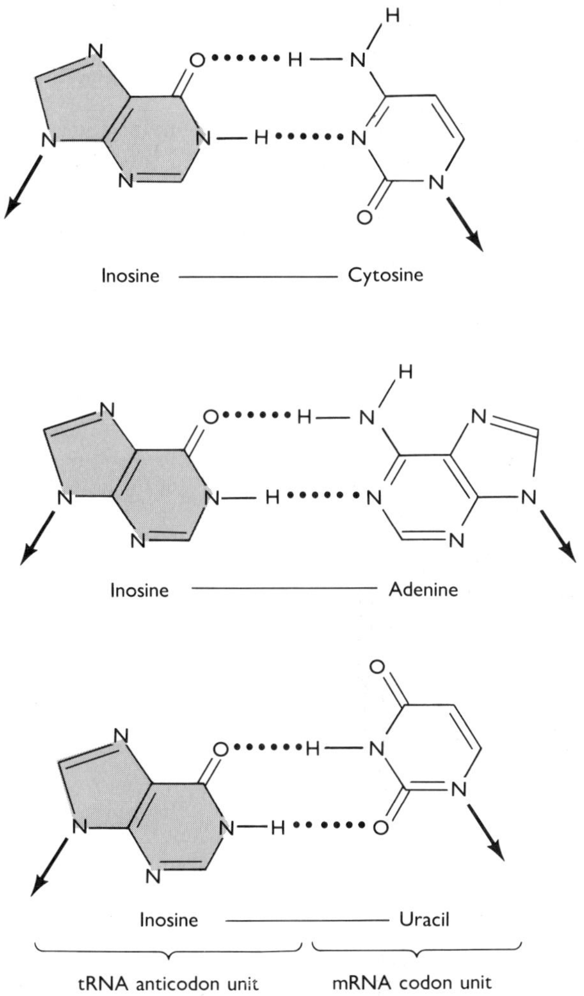

Pairing according to the wobble hypothesis	
5′ base in tRNA anticodon	3′ base in mRNA codon
A	U
U	A or G
C	G
G	U or C
I	A, U, or C

Figure 3.18. Wobble in the tRNA anticodon. (a) Less restricted base pairing, or wobble, is characteristic of certain bases when they occupy the 5′ end of a tRNA anticodon triplet. Inosine (deaminated adenine) can pair with cytosine, adenine, or uracil in the complementary 3′ site of a mRNA codon triplet. (b) Uracil and guanine, as well as inosine, may act as wobble bases, but cytosine and adenine pair only according to the strict C–G and A–U rule, regardless of the position they occupy in the anticodon.

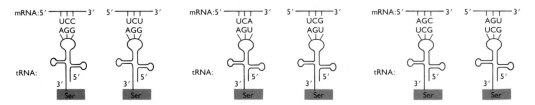

Figure 3.19. Six different serine-specifying mRNA codons can be handled by three different tRNA^Ser anticodons, because of wobble in the 5′ base of the anticodons (AGG, AGU, UCG). Wobble base 5′–G can pair with C or U in the mRNA codons UCC and UCU, and AGC and AGU; wobble base 5′–U can pair with A or G in the mRNA codons UCA and UCG.

The system depends in part on U being the 5′ wobble base in anticodons of tRNAs that can interact with all four codons of an unmixed family, and in part on the heavier use of one or two of the codons in an unmixed family of four (Fig. 3.20). In addition, it is possible that the unusual structure or composition of many mitochondrial tRNAs makes for their capacity to interact with more than the pairs of codons that are the maximum in nucleocytoplasmic translation. Mitochondrial systems show that a smaller contingent of

UUU			UCU			UAU			UGU		
	Phe	AAG	UCC			UAC	Tyr	AUG	UGC	Cys	ACG
UUC				Ser	AGU						
UUA			UCA			UAA	STOP		UGA		
	Leu	AAU					Ter			Trp	ACU
UUG			UCG			UAG			UGG		
CUU			CCU			CAU			CGU		
CUC			CCC			CAC	His	GUG	CGC		
	Thr	GAU		Pro	GGU					Arg	GCA
CUA			CCA			CAA			CGA		
							Gln	GUU			
CUG			CCG			CAG			CGG		
AUU			ACU			AAU			AGU		
AUC	Ile	UAG	ACC			AAC	Asn	UUG	AGC	Ser	UCG
				Thr	UGU						
AUA			ACA			AAA			AGA		
							Lys	UUU		Arg	UCU
AUG	Met	UAC	ACG			AAG			AGG		
GUU			GCU			GAU			GGU		
GUC			GCC			GAC	Asp	CUG	GGC		
	Val	CAU		Ala	CGU					Gly	CCU
GUA			GCA			GAA			GGA		
							Glu	CUU			
GUG			GCG			GAG			GGG		

Figure 3.20. The coding dictionary for yeast mitochondrial mRNA codons (5′ → 3′, left) and tRNA anticodons (3′ → 5′, right). The 5′ wobble base of each anticodon is underlined. The four codons in an unmixed codon family, in which all four codons with the same initial base doublet specify the same amino acid, interact with a single anticodon whose wobble base is U. The unmixed codon family for arginine (CGU, CGC, CGA, CGG) interacts with a single anticodon, but its 5′ base is A. These Arg codons are rarely present in yeast mitochondrial mRNAs. (From S. G. Bonitz et al., 1980. *Proc. Natl. Acad. Sci.* 77:3167, Fig. 2)

tRNAs, as may have been present in primeval cells, can handle all the codons. New tRNAs with greater specificity might have been added during cellular evolution, eventually reaching the present total of 32 tRNAs.

Why are only 20 amino acids included in the genetic code when many others are not? Dozens of amino acids are known to be products of protein degradation or to be modified from encoded amino acids after the latter have been added to the growing polypeptide chain. Only the 20 encoded amino acids are assembled into polypeptides during translation. According to the *"frozen accident" hypothesis,* the original allocation of codons was a matter of chance. As life forms became more complex, natural selection would have favored the maintenance of the present code if it produced the fewest harmful effects. Too much tampering with the code might have altered essential proteins of cell structure and function. Some modifications were tolerated or might even have been adaptive, but large-scale changes probably would have proved nonadaptive in complex systems.

The *"stereochemical" hypothesis* states that the code is the outcome of relationships between specific amino acids and specific nucleic acid codons. Chemists know of no basis, however, for such associations or interactions. Indeed, all the available evidence shows that interactions occur only between mRNA and tRNA, regardless of the amino acid carried by the tRNA (Fig. 3.21). It is true, however, that proteinoid microspheres preferentially absorb different amino acids when different nucleic acids exist in the microspheres. Poly(A) favors the absorption of lysine; poly(U) favors phenylalanine; poly(C) favors proline; and poly(G) favors glycine. These correlations are meaningful because AAA is the codon for lysine, UUU for phenylalanine, CCC for proline, and GGG for glycine. But the absorption of monomers may be a distinct phenomenon unrelated to coding and the accuracy of translation. The "stereochemical" hypothesis is open to experimental tests, however, whereas an ancient "frozen accident" will remain speculative and untestable.

Ulf Lagerkvist has proposed a third possibility to explain the allocation of codons and the limit of 20 encoded amino acids. Two major premises underlie the hypothesis: first, codon allocations are not random; and second, codon families have different probabilities for misreading. According to experimental evidence from studies of codon interactions with transfer RNA anticodons, codons in unmixed families have a high probability for misreading, whereas codons in mixed families have a low probability for misreading. Changes in codon meaning in unmixed families would lower translational fidelity and would therefore be selected against. Changes in codon meaning in mixed families would not lower translational fidelity, because of their low probability for interacting with the wrong transfer RNA and its anticodon, and would not necessarily prove disadvantageous. The allocations and limit of the genetic code are therefore an outcome, in large part, of the high premium on accuracy in translation, which depends on the proper recognition of and interaction between codons and their anticodons. Far from being accidental, the genetic code has evolved under stringent selection pressures that favor maximum translational accuracy.

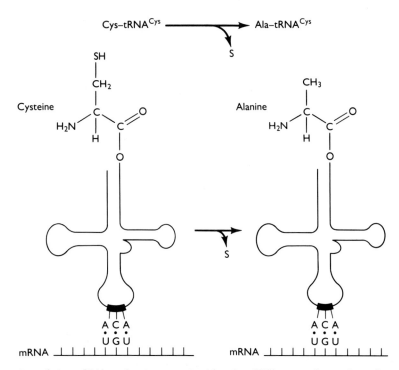

Figure 3.21. The meaning of the mRNA codon is recognized by the tRNA anticodon and not by its bound amino acid, as was shown in studies of tRNA^{Cys} carrying its normal cysteine residue (Cys–tRNA^{Cys}) or carrying an alanine (Ala–tRNA^{Cys}) produced by chemical desulfuration of Cys–tRNA^{Cys}. Because alanine was inserted into sites usually occupied by cysteine in the polypeptide, the tRNA^{Cys} must have recognized the mRNA codon and thus inserted into the polypeptide chain any amino acid that it carried. If the amino acid were part of the recognition system, Ala–tRNA^{Cys} would have put alanines only where alanines belong and not where cysteines belong in the growing polypeptide chain.

New Genes and the Expansion of the Genome

It is a foregone conclusion that primeval life was genetically and metabolically simple. The earliest cells could have functioned very well as scavengers of the myriad abiotically synthesized organic molecules that had accumulated for hundreds of millions of years prior to the origin of life. Primeval consumers would have synthesized very few proteins and would therefore have performed reasonably successfully, although their replicating nucleic acids included few gene sequences. As time passed, however, the organic soup would have become thinner as primeval life consumed molecules that were scarcely replenished in very slow abiotic reactions. The selective edge would have been held by the occasional cells that could make the molecules they needed from precursors absorbed from their surroundings. But each reaction step in a sequence must be catalyzed, or the reaction rate is inefficiently slow, and more specific catalysts are wholly or partly protein in nature. How could life have acquired the capacity to carry out more complex metabolism and the genes encoded for enzymes that catalyzed the increasing number of organic reactions? Part of the answer was suggested in 1945 by Norman Horowitz, who proposed the idea of *evolution backward*.

Horowitz based his idea on the growing body of evidence from studies at the time, particularly from biochemical genetics of the fungus *Neurospora*. It became increasingly clear from these studies that genes specify enzymes, which, in turn, govern each step in a sequence of metabolic reactions. In essence, Horowitz proposed that the organic soup was diluted as life multiplied and used up available organic molecules. Any cell with the genetic capacity to make essential molecules from simpler precursors would have had a selective advantage in depleted environments. If compound F was needed but was unavailable, any cell with a properly encoded gene could make the enzyme that catalyzed the synthesis of F from its precursor, E. Such cells would survive and reproduce to leave descendants with the same inherited capacity. Should E become scarce, chance events leading to the appearance of a new gene might allow some cells to use precursor D to make E and then F. By the expansion of such a sequence, with each step catalyzed by an enzyme under genetic control, more complex metabolism might have evolved.

$$
\begin{array}{ccccc}
\text{Gene} & e & d & c & b & a \\
 & \downarrow & \downarrow & \downarrow & \downarrow & \downarrow \\
\text{Enzyme} & e & d & c & b & a \\
\text{Substrate} & F \leftarrow E \leftarrow D \leftarrow C \leftarrow B \leftarrow A
\end{array}
$$

Metabolic pathways are not dependent exclusively on precursors absorbed from external sources. Reaction products retained within the cell can also be mobilized as raw materials for other pathways (Fig. 3.22). Successful life forms developed intersecting and interacting pathways, as is evident from modern cellular metabolism. Energy conservation by coupled endergonic and exergonic reactions, and energy transfer through common intermediates and other agents, must have been incorporated into metabolism. The universal existence of energy-mobilizing systems in modern organisms indicates an ancient heritage that has been passed down to descendants through countless generations. Metabolic advantages were probably due to chance genetic events, including the appearance of genes that were incorporated into evolving cells under the guidance of natural selection. That is, by differential reproduction, any advantageous traits were transmitted to successive cell generations, which ultimately became predominant over less advantageous cell types in populations or led to their extinction.

How did the genome expand from the relatively few genes in primeval life to the thousands of genes that exist in modern organisms? Based on genetic studies conducted over decades, biologists believe that the enlargement of the genome results initially from processes leading to the **duplication** of nucleotide sequences. Duplicate sequences of any length may be produced by errors during replication, transposition of genes from one lineage into another, retention of extra genome copies that ordinarily are transmitted to progeny

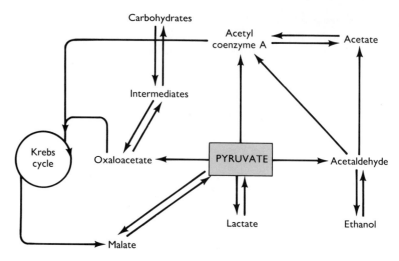

Figure 3.22. Intersecting biochemical pathways involving relatively few intermediate metabolites are a common feature in modern life forms. Reaction products made in these pathways, such as pyruvate, may be routed to different processing systems, depending on internal and external conditions. Precursors for metabolism may be provided from internal sources as well as from the external environment.

cells, unequal crossing over and other errant recombinations, and other small-scale and large-scale events.

Because mutation and other processes of genetic modification are random events, originally identical nucleotide sequences diverged and became different. It is highly unlikely that the same changes occurred in molecules continually subjected to random processes. With time, these sequences might have diverged sufficiently to specify altered products involved in related but distinct metabolic processes, or have been so changed that their products no longer resemble one another in structure or in function.

A classic example of gene duplication and divergence in evolution is provided by the globin genes, which specify globin polypeptides of myoglobin and hemoglobin. When the amino acid sequences of α globin and β globin are aligned in the most favorable way for comparison, 63 residues, among more than 140 amino acids in each polypeptide, are in the same location. The α-globin and β-globin genes are organized identically into coding and noncoding sequences (exons and introns), which are almost the same length, and have splice junctions in identical location (Fig. 3.23). It is too great a coincidence for all these similarities to have evolved independently in polypeptides that participate as parts of the same hemoglobin molecule in the common function of oxygen transport.

Even more persuasive evidence for the duplication and subsequent divergence of the globin genes comes from a broader comparison. For example, the genes for human β globin and δ globin differ by only 10 nucleotides, and

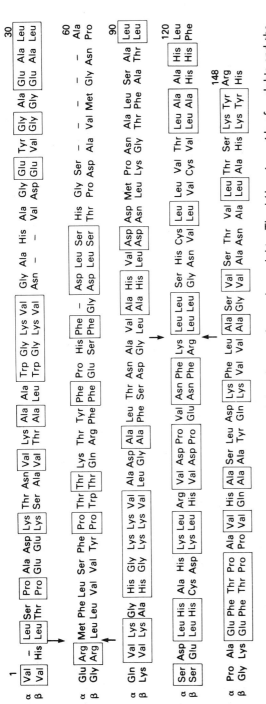

Figure 3.23. Amino acid sequences of α-globin and β-globin chains in human hemoglobin. The 141 amino acids of α globin and the 146 amino acids of β globin have been aligned over 148 positions to enhance the similarities of the globins. The exon–intron splice junctions (at arrows) are essentially the same in the two globins, even though the original gene duplicated and began to diverge about 400 million years ago (see Fig. 3.24). Only 63 amino acids (boxed) occupy the same locations in the two polypeptide chains. (Adapted from *Evolution*, by T. Dobzhansky, F. J. Ayala, G. L. Stebbins, and J. W. Valentine, Fig. 9-17. W. H. Freeman and Company. Copyright © 1977)

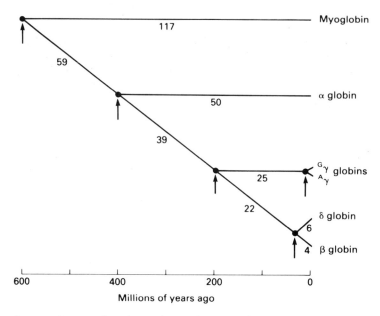

Figure 3.24. A divergence diagram showing the relationships and origins of the human globin genes. Each duplication is indicated by an arrow, and the numbers along the branches indicate the number of base substitutions believed to have occurred over the past 600 million years of evolution from a metazoan ancestor to modern human beings. It is not known if the myoglobin gene duplication led first to an α-globin gene or to a β-globin gene after divergence. Whichever came first, divergence after a second gene duplication, about 400 million years ago, led to both the α-globin and β-globin lineages.

the 2 human γ-globin genes differ in only 1 nucleotide residue. By combining information from the fossil record, comparative studies of living species, and molecular features of genes and polypeptides, biologists can trace an evolutionary history of gene duplication and divergence leading to the present set of human globin genes (Fig. 3.24).

The ancestral gene may have appeared 600 million years ago in the first invertebrate animals, and perhaps encoded a polypeptide similar to the muscle protein *myoglobin*. Myoglobin consists of one polypeptide chain bound to a heme group, whose coordinated iron atom is involved in binding molecular oxygen in the oxygen-storing molecule. The myoglobin gene probably duplicated about 600 million years ago to produce the forerunner of α-globin and β-globin genes. The β-globin gene duplicated several times to produce ϵ globin (in embryonic hemoglobin), γ globin (in fetal hemoglobin), and δ globin (an alternative to β globin in postnatal hemoglobin). The ancestral α-globin gene also underwent duplication and divergence. The human genome contains three copies of the α-globin gene and two copies of the equivalent embryonic ζ-globin gene, which are divergent duplicates of the α-globin gene (Table 3.1). The cluster of α-like genes is on human chromosome 16, and the

TABLE 3.1
Human Hemoglobins

Type of Hemoglobin	Globin Chains Present	Amount Synthesized
Embryonic	$\zeta_2\epsilon_2$ (Gower I)	Uncertain
	$\alpha_2\epsilon_2$ (Gower II)	Uncertain
	$\zeta_2\gamma_2$ (Portland)	Uncertain
Fetal	$\alpha_2\gamma_2$ (Hb F)	Major hemoglobin of fetus
Adult	$\alpha_2\beta_2$ (Hb A)	Major postnatal (adult) hemoglobin (98%)
	$\alpha_2\delta_2$ (Hb A$_2$)	Minor postnatal (adult) hemoglobin (2%)

cluster of β-like genes is on human chromosome 11 (Fig. 3.25). All the globin components of hemoglobin bind a heme group, whose coordinated iron atom readily binds and releases molecular oxygen to all the cells bordering the bloodstream.

The globins are not an isolated example of gene duplication and divergence in evolution. Computer searches of vast data banks of molecular sequence

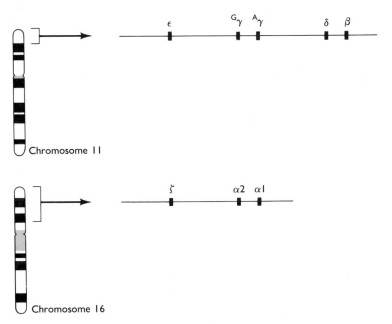

Figure 3.25. The β-globin gene cluster is located near one end of human chromosome 11, in the order shown, and the α-globin gene cluster occupies a region in one arm of human chromosome 16. Recent studies have revealed a second ζ-globin gene and a third α-globin gene in human chromosome 16.

information have yielded more and more similarities of sufficient magnitude to indicate genetic relationship rather than coincidence. These molecular studies will be discussed in Chapter 9. It thus seems that genome amplification certainly occurred, by processes leading to nucleotide sequence duplication and subsequent divergence as the result of continued genetic change. Biologists infer, of course, that natural selection and other evolutionary processes were important in the successes and failures during descent with modification over the billions of years that life has existed on Earth.

Evolutionary Changes in Early Life Forms

From the geologic and fossil records left in the Earth's crust over billions of years and from comparative studies of living organisms, at least some of the evolutionary history of life forms can be traced from ancient times to the present. In this part of our discussion, some broad features of cellular evolution will be examined, with particular emphasis on important metabolic processes that profoundly affected the nature of the planet and its life forms.

Early Life Forms Were Unicellular Anaerobes

The oldest remaining rocks of the Earth's crust are 3.8 billion years old, and the earliest known fossils are about 3.5 billion years old. Although the first 800 million years of geologic history have been lost, the record of past events since then is available for analysis and interpretation (Fig. 3.26).

There is no question that primeval life was **unicellular** and that life continued exclusively in this form for about 3 billion years. All fossils are of single-celled organisms until almost 700 million years ago, when the first evidence appears for multicellular organisms. The consistency of the fossil record is very persuasive evidence for unicellularity as the ancient state, and the fossils are supported by comparative studies of living organisms. The simplest life forms, including all the bacteria and protists, are invariably unicellular.

Primeval life must have been **anaerobic,** according to independent evidence from the geologic record and from comparative biochemical data. The primeval atmosphere is presumed to have been nonoxidizing, as we discussed in Chapter 2. The atmosphere remained largely deficient in molecular oxygen until about 2 billion years ago, when a significant transition to an oxygen-rich atmosphere began to take place. The basis for this conclusion comes from dated deposits of uraninites and banded iron formations and from other sources. *Uraninites* are uranium compounds (UO_2) that can accumulate in stream beds only in the absence of oxygen. When the concentration of atmospheric oxygen is greater than about 1 percent, grains of uraninites are oxidized to U_3O_8, which dissolves in water and, therefore, cannot accumulate. The youngest uraninites are 2 billion years old; none are found after this time.

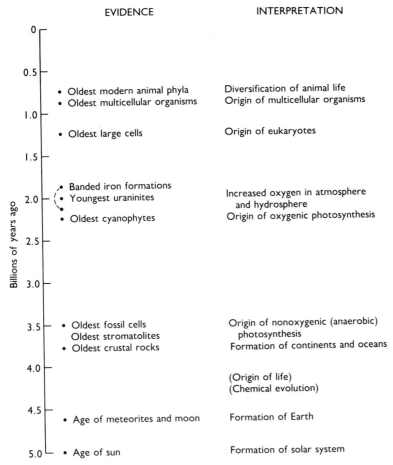

Figure 3.26. Selected highlights in the evolutionary history of Earth over the past 5 billion years, showing interpretations based on major lines of evidence from astrophysical, geologic, and fossil studies.

The curtailment of uraninite deposition indicates the presence of atmospheric oxygen in high enough concentration to cause the oxidation of uraninites to the soluble forms of uranium.

Banded iron formations are sedimentary deposits of iron embedded in silicate matrix. They are the major reserves of iron on which the world economy depends. Much of this material was deposited within the relatively brief period of a few hundred million years, beginning a little over 2 billion years ago, and may have been the result of a transition in oceanic oxygen concentration. Iron was dissolved in its ferrous (Fe^{2+}) form in the ancient anoxic seas. With the introduction of and increase in oxygen, presumably from the activity of oxygenic photosynthesizing organisms, ferrous iron was converted to its ferric (Fe^{3+}) form and, as hydrous ferric oxides (rust), was precipitated (together with silica) as rusty layers on the ocean floors. The oceans were thus swept free of dissolved iron, which was converted to iron-oxide-rich sediments that accumulated in thick deposits at the bottom of the seas. It is significant that the oldest abundant deposits of ancient oxygenic photosynthesizing cyanophytes are found in sediments about 2.3 billion years old, which correlates with the time of banded iron formation.

Comparative studies of metabolic pathways in living organisms indicate an ancient anaerobic past and the later evolution of aerobic life styles. When biosynthetic reaction sequences are compared, it is apparent in numerous cases that the earlier steps in a pathway are oxygen-independent and the later steps are oxygen-dependent reactions. Presumably, the later reactions were added more recently to the pathway, and thus they probably evolved after the atmosphere began to accumulate molecular oxygen. The earlier reactions reflect the origin of the pathway in more ancient life that was anaerobic and flourished in a nonoxidizing atmosphere. In many of these examples, the anaerobic steps may be the only reactions carried out by an organism, whereas the aerobic steps do not occur in the absence of the anaerobic ones. We can invoke a basic principle here, first suggested by Oparin, that reactions and other traits that are present in all organisms are more ancient than features found in only some life forms but not in all. Universal features have been passed along to all descendants from the most ancient common ancestor. Later in evolution, variations and additions arise, and they are transmitted only to generations descended from the more recent founders of divergent lineages. Biochemical pathways that exemplify the anaerobic–aerobic distinction include those that carry out fatty acid synthesis, photosynthesis, and aerobic respiration.

Fatty acids are lengthened during synthesis by the addition of two-carbon units in repeated reactions, the first of which are identical and lead to saturated fatty acid products. A double bond can be introduced into the growing chain when it is 8 units long or 10 units long, but only in bacteria. The completed fatty acid produced by bacteria (prokaryotes) usually has a double bond, but the molecule remains fully saturated on completion of its synthesis in nucleated organisms (eukaryotes). None of these steps in the biosynthetic sequence requires molecular oxygen (Fig. 3.27). Unsaturated fatty acids and more complicated polyunsaturated fatty acids are produced in eukaryotes through oxidative desaturation, a process by which double bonds are formed by removing two hydrogen atoms and combining them with an oxygen atom to form water. Molecular oxygen is required for this process.

Just as in many biochemical pathways, the earliest steps in this sequence occur in all organisms capable of synthesizing fatty acids, and they are the only reactions in the most primitive organisms. Organisms that evolved later, such as the eukaryotes, have longer pathways, including the reaction of oxidative desaturation. Furthermore, the more highly evolved eukaryotic organisms carry out a larger number of oxygen-dependent than oxygen-independent reactions. As we will soon see, in the processes of photosynthesis and cellular respiration, the more primitive organisms perform only the earlier

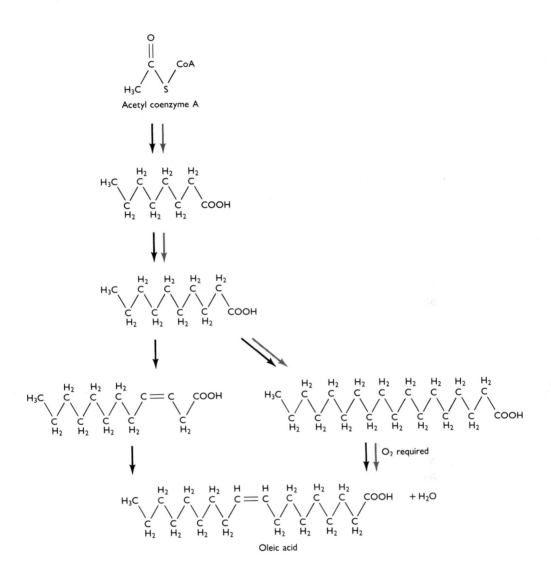

Figure 3.27. Saturated fatty acids are synthesized by repeated additions of two-carbon units in the energized form of acetyl coenzyme A, in both prokaryotes and eukaryotes. In bacteria (black arrows), one or more double bonds may be introduced into the growing molecule as early as the 8- to 10-unit stage, and oxygen is not required. In eukaryotes (gray arrows), however, the synthesis of a saturated fatty acid continues to completion of the chain. Double bonds can then be introduced into the saturated molecule by oxidative desaturation, in which hydrogen atoms are removed, combined with O_2, and released as water molecules. Oleic acid is one kind of unsaturated fatty acid, and is produced by either pathway in bacteria but only by O_2-requiring oxidative desaturation in eukaryotes.

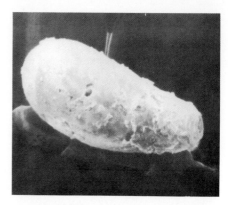

Figure 3.28. Scanning electron micrograph of a fossil cell from 800-million-year-old deposits in the Grand Canyon of Arizona. This unicellular organism is believed to have been eukaryotic, because it is morphologically more complex and much larger (about 100 μm long) than any known fossil or living prokaryotic cell. (From "The Evolution of the Earliest Cells" by J. W. Schopf. Copyright © 1978 by Scientific American, Inc. All rights reserved)

reactions, which are oxygen independent, and the more advanced organisms have longer pathways, which are oxygen dependent.

Although cellular organization will be considered in Chapter 4, it is pertinent at this point to mark the prokaryotic nature of simpler, more primitive life forms and the eukaryotic nature of more advanced organisms. It may be presumed that the prokaryotic cellular plan is more ancient and that eukaryotes evolved from prokaryotic ancestors. This presumption is amply supported by the fossil record, in which cells of the greater size and complexity typical of eukaryotes do not appear until about 1.5 to 1.4 billion years ago (Fig. 3.28). Only prokaryotelike cells occur in the preceding 2 billion years of fossil deposits. Eukaryotes must have evolved after the atmosphere became oxygen rich, according to the time correlates of the geologic record.

Photosynthesis Changed the Earth Forever

Photosynthesis is the process by which cells convert radiant energy into chemical energy stored in organic molecules. In the prokaryotic cyanophytes and prochlorophytes and in all eukaryotic green cells and organisms, O_2 is released as a by-product of reactions that take place in the light, and CO_2 is reduced to carbohydrates in the light-independent "dark reactions." These organisms provide an endless primary source of food and energy that sustains virtually all life on the planet, as long as the sun keeps shining. In addition, the oxygen they release into the atmosphere is the requirement for aerobic life, which includes the vast majority of organisms on Earth at the present time. The trauma of transition from a nonoxidizing atmosphere to one with oxygen must have affected primeval anaerobic life profoundly. Few anaerobic forms could have survived the transition, but some did come through and now occupy the few airless pockets that remain on the Earth. Any life that could tolerate oxygen or detoxify its reaction products enjoyed a great selective advantage. Such life forms were able to flourish and repopulate the Earth.

Photosynthesis originated in the anaerobic conditions of the ancient past, perhaps 3.5 billion years ago. Evidence for this timetable comes from datings of fossil **stromatolites**, which are pillarlike mounds of layered mats of debris left by the activities of photosynthetic bacteria and cyanophytes (Fig. 3.29). Fossil microorganisms of these groups have been identified in stromatolites, and communities of such photosynthetic species are found in stromatolites being built in several locations; especially remarkable structures are in a lagoon at Shark Bay in Western Australia.

The earliest life forms must have been **heterotrophic,** obtaining energy and carbon from organic molecules they scavenged from their rich surroundings of the organic soup. Relatively few genes or proteins were required for their simple activities. While some primitive heterotrophs developed more complex metabolic pathways, others must have evolved genes that encoded proteins that allowed autotrophic activities in addition to their ancient heterotrophic heritage. **Autotrophs,** including photosynthesizers, can derive all their basic energy and carbon from nonorganic sources. In every case, however, they also possess heterotrophic pathways.

From comparative studies of living photosynthetic organisms, in conjunc-

Figure 3.29. Stromatolites in the lagoon at Shark Bay, Western Australia. These pillarlike mounds consist of layered mats of debris and cells, and are produced by the activities of photosynthetic bacteria, including cyanophytes. Fossil structures containing remains of microorganisms are sufficiently similar to these modern stromatolites to permit their identification as ancient stromatolites produced by biological activities as early as 3 to 3.5 billion years ago. (Photograph by M. Walker)

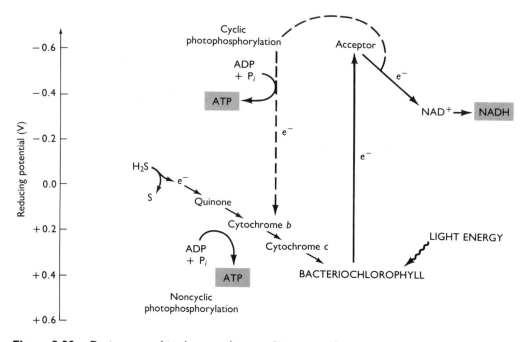

Figure 3.30. During anaerobic photosynthesis in the green sulfur bacteria and purple sulfur bacteria, electrons from H_2S are passed down an electron transport chain of carriers to light-activated bacteriochlorophyll, and from there to NAD^+ to produce NADH. The synthesis of ATP in noncyclic photophosphorylation is coupled to electron transport from H_2S to bacteriochlorophyll, and ATP synthesis in cyclic photophosphorylation is coupled to the transport of electrons back to cytochromes in the chain. These processes are more productive in obtaining and storing energy than are the fermentative pathways on which nonphotosynthetic anaerobes depend for their energy needs.

tion with information from the geologic and fossil records, a reasonable evolutionary sequence of events can be constructed. Such anaerobic photosynthesizers as the green sulfur bacteria, including *Chlorobium*, and purple sulfur bacteria, including *Chromatium*, have an **electron transport chain** of components by which electrons from H_2S or simple organic compounds are passed along to NAD^+ (nicotinamide adenine dinucleotide) in the presence of **bacteriochlorophyll,** a porphyrin-based light-sensitive pigment. During electron transfer from donor compounds to bacteriochlorophyll, ATP is synthesized in a **photophosphorylation** process (Fig. 3.30). These anaerobes have no associated cycle of CO_2 reduction or any other system of "dark reactions," but they have the advantage of a more efficient pathway than fermentation alone by which ATP is synthesized and NAD^+ is reduced to NADH. In effect, these bacteria transduce radiant energy to chemical energy in ATP and NADH, which are energy-transferring agents in many metabolic sequences. Less ATP and NADH is obtained by the fermentative processing of organic molecules, which is the more ancient heritage in these and other organisms.

Photosynthetic species of the purple nonsulfur bacteria, such as *Rhodospi-rillum*, have a **C₃ cycle** of CO_2 reduction to carbohydrates, which takes place at the expense of ATP and NADH made in the light. They may have evolved from a lineage related to the anaerobic purple sulfur bacteria, but have an aerobic metabolism and arose after the atmosphere became oxidizing.

Most of these photosynthetic bacterial groups are nonoxygenic, having only the system of photosynthetic pigments that are part of **photosystem I.** The oxygen-producing cyanophytes, however, have chlorophyll *a* rather than bacteriochlorophyll, and a **photosystem II** linked by an electron transport chain to the photosystem I pigments involved in the reduction of NADP⁺ (nicotinamide adenine dinucleotide phosphate) to NADPH. Photosystem II is at a low enough reducing potential for it to accept electrons from water, rather than from H_2S or organic compounds used by the green and purple sulfur bacteria and purple nonsulfur bacteria (Fig. 3.31). When electrons are extracted from water in the light, oxygen is released as a by-product of the reactions. The electrons are channeled to chlorophyll *a* in photosystem II, passed down the electron transport chain to chlorophyll *a* in photosystem I, and end up in NADPH. The ATP made by photophosphorylation during

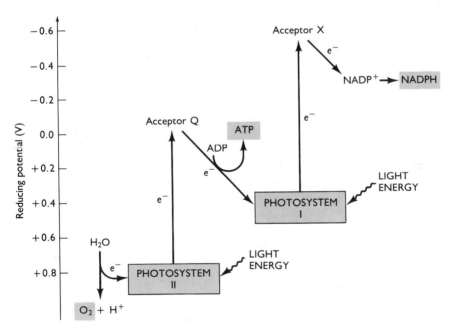

Figure 3.31. The evolution from anaerobic ancestry of oxygen-releasing photosynthetic prokary-otes, such as the cyanophytes and prochlorophytes, must have involved a number of modifications and innovations. One such innovation featured the addition of a photosystem II component linked to photosystem I by the electron transport chain that funnels electrons to photosystem I. Because pho-tosystem II can obtain electrons from water, oxygen is released as a by-product of the reaction in which water molecules are dissociated in the light (compare Fig. 3.30).

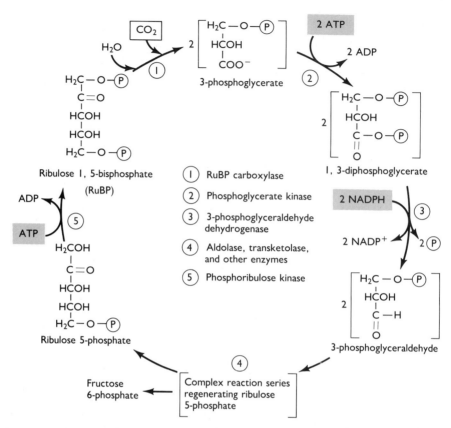

Figure 3.32. The C_3 cycle of CO_2 reduction to carbohydrates in the so-called dark reactions of photosynthesis. The first detectable product of the cycle is the three-carbon compound 3-phospho-glycerate, produced from half-molecules of a transient six-carbon intermediate made when ribulose 1, 5-bisphosphate (RuBP) reduces (fixes) CO_2 from the air. On the addition of ATP, 1, 3-diphos-phoglycerate is produced, and it can accept hydrogens and electrons from NADPH to form 3-phos-phoglyceraldehyde. Various reactions occur in the processing of 3-phosphoglyceraldhyde, including the formation of fructose 6-phosphate and ribulose 5-monophosphate. Fructose 6-phosphate enters a major pathway leading to the synthesis of carbohydrates. Ribulose 5-monophosphate is phospho-rylated by ATP to regenerate RuBP for another turn of the cycle. The enzymes that catalyze the C_3 cycle of reactions are shown in the center of the cycle. These reactions of carbohydrate synthesis are light independent, but they use the chemical energy of ATP and NADPH made in the light reactions of photosynthesis (see Fig. 3.31).

electron transport between the two photosystems, and the NADPH produced at the end of the light reactions, subsidize CO_2 reduction in the C_3 cycle of the "dark reactions" (Fig. 3.32).

During the evolution of oxygenic photosynthesis in the cyanophytes, photosystem II was linked to the more ancient photosystem I. The switch to water as an electron donor became possible, leading to a widely available and nontoxic source of electrons quite different from the sulfur compounds required by bacterial photosynthesizers. The release of oxygen is incidental to photosynthesis, but it led to a profound alteration in the seas and atmosphere. As oxygen levels rose over the millennia, life forms adjusted to the new conditions or retreated to anoxic places if they survived.

Aerobic Life Becomes Predominant

The extraction of energy and carbon from organic molecules takes place by fermentative pathways in virtually all cells. **Fermentations** are the only means by which anaerobic organisms derive carbon and energy, but constitute the earlier oxygen-independent steps in longer pathways that characterize aerobic life. The pivotal molecule in food processing is three-carbon **pyruvic acid,** which is the end product of oxygen-independent glycolysis and other fermentative pathways (Fig. 3.33). In the absence of oxygen or the metabolism to decarboxylate pyruvate and form two-carbon acetate products, pyruvate is converted to lactate or to ethanol, depending on the enzymes in the cells. The bulk of potential free energy in organic foods like glucose remains locked up in the lactate or ethanol products and is not available for cellular biosyntheses.

The enormous advantage of **aerobic respiration** is its role in oxidizing glucose and other organic foods to CO_2 and H_2O. A considerable amount of free energy is conserved in ATP made in **oxidative phosphorylation** reactions coordinated with free-energy release during oxidation-reductions. Only 2 ATP result from the oxidation of glucose to pyruvate in glycolysis, whereas an additional 36 ATP per glucose can be derived from aerobic respiration reactions that fully oxidize the pyruvate intermediates to CO_2 and H_2O (Fig. 3.34). Aerobic life became predominant during the transition to an oxidizing atmosphere because of the significant advantages of free-energy release and conservation in respiration, and not merely because it could tolerate oxygen or detoxify its reaction products.

Aerobic respiration probably evolved independently in several bacterial lineages, as is suggested by a comparison of pathways in living organisms. All the systems share many features in oxidation-reductions, but differ mostly in their terminal oxidases. These and other considerations lead to the proposal that the final steps in electron transport arose later in evolution and that different terminal oxidases passed electrons to different final electron acceptors, one of which was molecular oxygen. All the oxidases receive electrons from a chain of **cytochromes,** one of which is cytochrome c (Fig. 3.35).

In the purple nonsulfur bacteria, the same chain of cytochromes functions in both photosynthesis and aerobic respiration. Richard Dickerson has made the interesting suggestion that aerobic respiration evolved in part by the addition of a terminal oxidase to the cytochrome chain involved in bacterial photosynthesis. In addition to transferring electrons to bacteriochlorophyll, cytochrome c_2 can pass electrons to a cytochrome oxidase complex, which, in turn, gives up the electrons to O_2 in the terminal reaction, by which O_2 is reduced

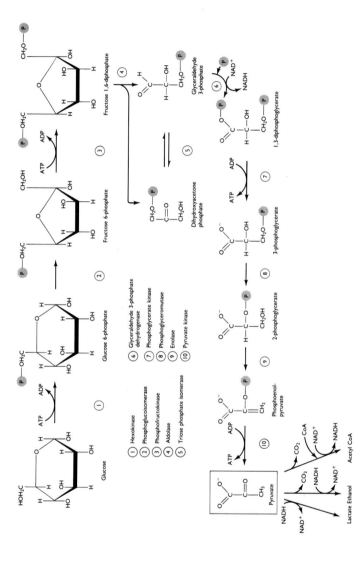

Figure 3.33. The oxidation of glucose to pyruvate in glycolysis occurs in two stages. In stage I, glucose is phosphorylated and then rendered into two three-carbon compounds, at the cost of two ATPs. In stage 2, each half of the original glucose molecule is processed in the same reaction pathway, which is an economic metabolic device. At the end of the pathway, pyruvate has been formed, and its fate depends on the enzymatic capacity of the species and the availability of molecular oxygen. Anaerobic reaction products of pyruvate are lactate or ethanol. Aerobically, pyruvate is decarboxylated, and acetyl CoA is produced. A total of four ATPs are produced in stage-2 reactions, but the net yield is only two ATPs per glucose because two ATPs were used in the stage-I reactions. The enzymes that catalyze each step are listed.

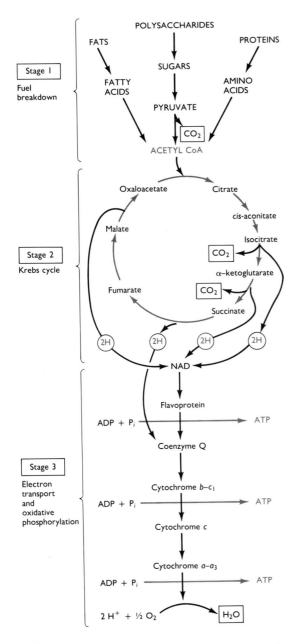

Figure 3.34. Three stages of fuel processing in cells. rats, polysaccharides, and proteins can be broken down to simpler, soluble constitutents in various pathways of stage 1. Soluble intermediates are channeled to the Krebs cycle, by acetyl CoA, where they are fully oxidized to CO_2 and their energy is transferred to NAD and FAD in stage 2. In stage 3, electrons are transferred from reduced NAD and FAD to O_2 along a chain of cytochromes and other electron carriers, with coupled reactions of ATP synthesis taking place at three of the transport steps. Stages 2 and 3 are oxygen dependent and are largely responsible for the efficient conservation of fuel energy in ATP, the general energy currency of cells.

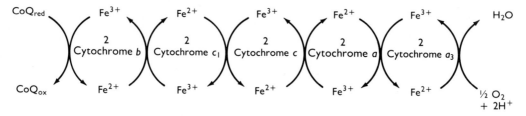

Figure 3.35. The pathway of electron transfer to O_2, emphasizing the valency changes in the iron atom in cytochromes.

to H_2O (Fig. 3.36). If Dickerson's hypothesis were confirmed, it would mean that respiration evolved later than photosynthesis. Although the majority of the scientific community accepts this as the most probable sequence, some biologists insist that respiration evolved before photosynthesis. This controversy remains to be resolved.

Dickerson has also suggested that photosynthesis was lost during bacterial evolution in some lineages whose bacteriochlorophyll was discarded. The deletion of the bacteriochlorophyll "loop" would have removed the capacity for photosynthesis but left unchanged the electron transport chain of cytochromes in the respiratory sequence (see Fig. 3.36). Support for the relationship between bacterial photosynthesis and aerobic respiration has been pro-

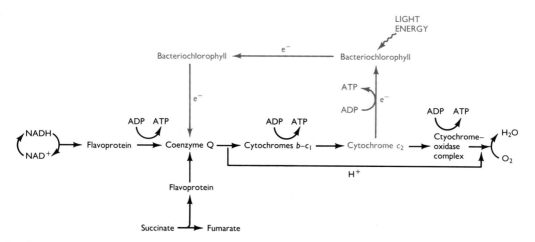

Figure 3.36. *Rhodopseudomonas*, a purple nonsulfur bacterium, uses the same chain of electron carriers in its photosynthetic activities, through the bacteriochlorophyll loop (gray), as in its aerobic respiratory activities, in which electrons are passed along cytochrome carriers to O_2. ATP is synthesized by photophosphorylation during electron flow to bacteriochlorophyll in the light, and by oxidative phosphorylation during respiratory electron flow to cytochrome oxidase and O_2. If the bacteriochlorophyll loop were suppressed or deleted, the respiratory pathway would function very much the same as it does in nonphotosynthetic aerobic bacteria and in eukaryotic mitochondria.

vided by Dickerson's comparative analysis of cytochrome c in anaerobic photosynthesizers and in aerobic purple nonsulfur bacteria and nonphotosynthetic species.

Previous studies of different cytochromes c failed to show any significant homology among these molecules in different bacterial groups. Amino acid composition varied considerably, as did the length of the polypeptide chain— from about 80 residues to chains with more than 130 amino acid units. By comparing three-dimensional tertiary structure, however, Dickerson showed that all these cytochromes c in bacteria are remarkably similar in the conformation of the polypeptide chain and in the organization of their heme-binding region (Fig. 3.37). Such a degree of similarity is strongly suggestive of evolutionary homology—that is, descent with modification from a common ancestral gene and its polypeptide product. Although cytochrome c is the product of only one gene among many genes encoded for photosynthetic and respiratory polypeptides, these studies strengthen the probability of evolutionary relationship and derivation of these processes in cells from simple anaerobic photosynthesizers to organisms that carry out aerobic respiration but may or may not have remained photosynthetic.

Up to this point, we have dealt with photosynthesis and aerobic respiration in prokaryotic organisms. These processes take place in the noncompartmentalized cytoplasm of prokaryotes, but are restricted to organelles in eukaryotes. What is the relationship of chloroplasts and mitochondria to prokaryotic cells? By what means did eukaryotes acquire their photosynthetic and respiratory organelles? Although these systems will be discussed at greater length in Chapter 4, it is appropriate to sketch in the broad outlines of ideas about the origins of chloroplasts and mitochondria.

The major proposal of an endosymbiotic origin of both kinds of organelles was made in 1967 by Lynn Margulis on the basis of new molecular information she incorporated into the original suggestion of **endosymbiosis,** made by August Weismann in the 1880s. Margulis states that chloroplasts are modern descendants of ancient prochlorophytes or cyanophytes and that mitochondria are modern descendants of ancient respiring bacteria. These formerly free-living organisms were engulfed by cells in independent episodes at different times, and a symbiotic association developed (Fig. 3.38). The host cell provided a secure and nourishing environment for its resident organisms, which, in turn, gave their host the very important advantages of photosynthesis or respiration. Ultimately, these free-living cells evolved into resident organelles no longer capable of an independent existence outside the host environment.

The alternatives to endosymbiosis consist of various proposals whereby the cell itself developed internal compartments that held the respiratory apparatus and the photosynthetic systems. These organelles would have evolved in much the same way as other compartments of the eukaryotic cell, such as lysosomes and even the nucleus itself, by the invagination and ultimate *internalization* of pieces of the plasma membrane that later assumed the different

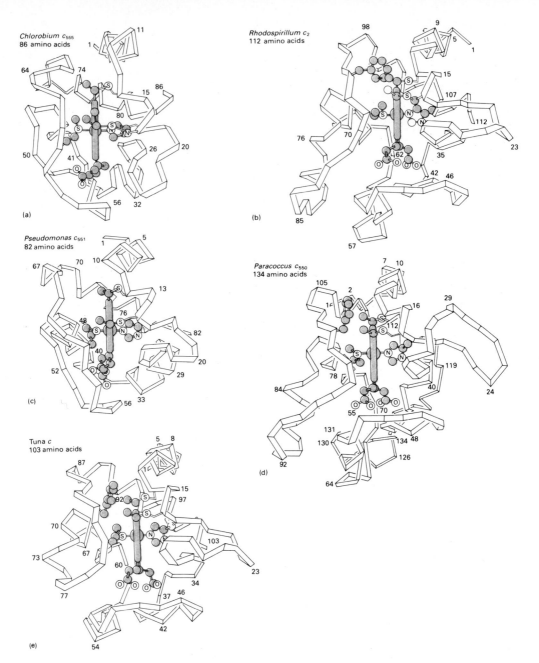

Figure 3.37. Although varying considerably in the number and sequences of amino acids, cytochrome c shows a remarkably similar tertiary structure and heme-binding region in a spectrum of bacterial and eukaryotic species. Cytochrome c participates in electron transport in (a) photosynthesis in *Chlorobium,* an anaerobic green sulfur bacterium; (b) photosynthesis and respiration in *Rhodospirillum,* a purple nonsulfur bacterium; (c) respiration in *Pseudomonas* and (d) *Paracoccus,* both of which are nonphotosynthetic aerobic bacteria; and (e) respiration in mitochondria of tuna and other eukaryotes. These similarities indicate homologies (genetic relatedness) that are not apparent from molecular primary structures. (From C. J. Avers, 1986, *Molecular Cell Biology.* Menlo Park, Calif.: Benjamin-Cummings. Data based on studies by R. E. Dickerson)

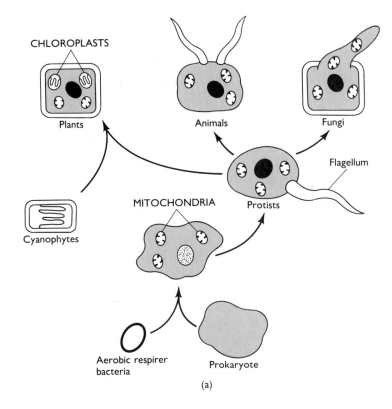

CHLOROPLASTS

Plants

Cyanophytes

Animals

Fungi

Flagellum

MITOCHONDRIA

Protists

Aerobic respirer
bacteria

Prokaryote

(a)

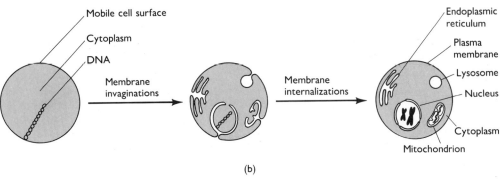

Mobile cell surface

Cytoplasm

DNA

Membrane
invaginations

Membrane
internalizations

Endoplasmic
reticulum

Plasma
membrane

Lysosome

Nucleus

Cytoplasm

Mitochondrion

(b)

Figure 3.38. Two of the hypothetical evolutionary pathways describing the origin and modifications responsible for mitochondria and chloroplasts in eukaryotic cells. (a) The endosymbiosis theory proposes that modern mitochondria are descendants of an ancestral free-living bacterial organism capable of aerobic respiration, and that modern chloroplasts are descendants of ancestral cyanophytes. The ancestral organisms established themselves as endosymbionts in host cells, both deriving benefit from their association. Eventually, the endosymbionts lost their ability for an independent existence and became indispensable organellar parts of the host cells. (b) Mitochondria and chloroplasts may have evolved by the internalization of parts of the plasma membrane that had invaginated. The membranous pieces eventually became established as internal membrane systems independent of the plasma membrane. Prokaryotes with a mobile cell surface might thus have evolved subcellular compartmentalization, without episodes of endosymbiosis.

specificities of modern cellular membranes enclosing a variety of metabolic pathways.

As we will soon see, the case for an endosymbiotic origin of chloroplasts is very strong. The evidence in support of an endosymbiotic origin of mitochondria, however, is somewhat ambiguous. It well may be that the two kinds of organelles evolved by very different pathways, even though they have significant features in common. One important shared feature is their own genetic machinery for the transcription and translation of genes in organelle DNA that encode some of the polypeptides of photosynthetic and respiratory pathways.

Summary

Any prelife or primeval life form must have possessed one or more features known to be present in all modern cells, including a selectively permeable boundary, metabolism, and a genetic apparatus. Each of the three major protobiont models has a boundary, and each serves as a container for concentrating organic molecules from a dilute aqueous environment. Coacervates are usually transient; proteinoid microspheres are relatively unvaried; and liposomes mimic some of the features of living cells. Whatever their dynamic nature might be, none would be able to replicate and transmit its favorable metabolic traits in the absence of nucleic acid informational template molecules. New data showing that some RNAs have catalytic as well as informational properties may help to resolve the chicken-and-egg paradox of metabolism first or genes first, because such RNAs provide both a catalytic and a genetic component for subsequent evolution.

Once reproduction became possible, natural selection would have influenced the future evolution of the first life forms. Life became more complex as a result of the expansion of the genetic coding dictionary to 64 codons specifying 20 amino acids and 3 STOP codons. The code probably was triplet to begin with; otherwise, the reading frame of existing genes would have been destroyed. The expansion of the code may have involved a recognition function for the third base of some codons, rather than the initial doublet of bases alone, changing the specificity of some codons to amino acids that were added to the smaller repertory of ancient life. New genes and an expanding genome may have been brought about by the duplication of nucleotide sequences through copying errors and other genetic events known to occur in modern organisms. New genetic information specifying new RNAs and proteins may have emerged eventually as random mutations sponsored greater degrees of sequence divergence from initially identical duplicates. The globin genes, among others, provide models for gene duplication and divergence.

The earliest life forms must have been unicellular, anaerobic, and heterotrophic, and had a prokaryotic cellular organization. Unicellular organisms are the only life forms found in the fossil record for almost 3 billion years after life originated on Earth. Oxygen was essentially absent from the primeval atmosphere, and appeared in small amounts only about 2 billion years ago, according to geologic and geochemical observations of uraninites and banded iron formations. Early life was genetically and metabolically simple, scavenging abiotically synthesized organic molecules from its aqueous surroundings. New genes specifying new enzymes and functions provided a photosynthetic capacity in ancient anaerobic bacteria, perhaps 3.5 billion years ago. These and subsequent photosynthetic bacteria were nonoxygenic, possessing only photosystem I. They reduced NAD^+ to NADH and produced ATP by photophosphorylation during electron transport from H_2S to bacteriochlorophyll. Later, a C_3 cycle was added whereby NADH and ATP subsidize the production of organic molecules from CO_2. Oxygenic photosynthesis evolved in the prokaryotic cyanophytes and prochlorophytes, which possess chlorophyll and photosystem II linked to photosystem I. Chloroplasts in eukaryotic cells are believed to be endosymbionts derived from cyanophytes or prochlorophytes ingested by cells and retained.

Increasing levels of atmospheric oxygen supplied by oxygenic photosynthesis favored aerobic organisms and led to the extinction or diminution of anaerobic life. Aerobic respiration probably evolved independently in several bacterial lineages. It is a highly efficient means by which energy can be extracted from relatively little organic food and be stored as ATP. It is uncertain whether eukaryotic mitochondria are ancient aerobic bacterial endosymbionts or are indigenous cell organelles.

References and Additional Readings

Altman, S. et al. 1986. Enzymatic cleavage of RNA by RNA. *Trends Biochem. Sci.* 11:515.

Anderson, S. et al. 1981. Sequence and organization of the human mitochondrial genome. *Nature* 290:457.

Avers, C. J. 1986. *Molecular Cell Biology.* Menlo Park, Calif.: Benjamin/Cummings.

Baltscheffsky, H., H. Jornvall, and R. Rigler, eds. 1987. *Molecular Evolution of Life.* New York: Cambridge University Press.

Been, M. D., and T. R. Cech. 1988. RNA as an RNA polymerase: net elongation of an RNA primer catalyzed by the *Tetrahymena* ribozyme. *Science* 239:1412.

Bonitz, S. G. et al. 1980. Codon recognition rules in yeast mitochondria. *Proc. Natl. Acad. Sci. U.S.* 77:3167.

Bretscher, M. S. 1985. The molecules of the cell membrane. *Sci. Amer.* 253(4):100.

Byerley, G. R., D. R. Lower, and M. M. Walsh. 1986. Stromatolites from the 3,300–3,500 Myr Swaziland supergroup, Barberton Mountain Land, South Africa. *Nature* 319:489.

Cairns-Smith, A. G. 1985. The first organisms. *Sci. Amer.* 252(6):90.

Cavalier-Smith, T. 1987. The origin of cells: a symbiosis between genes, catalysts, and membranes. *Cold Spring Harbor Sympos. Quant. Biol.* 52:805.

Cech, T. R. 1987. The chemistry of self-splicing RNA and RNA enzymes. *Science* 236:1532.

Crick, F. H. C. 1966. Codon–anticodon pairing: the wobble hypothesis. *Jour. Mol. Biol.* 19:548.

Crick, F. H. C. et al. 1976. A speculation on the origin of protein synthesis. *Origins of Life* 7:389.

Darnell, J. E. 1985. RNA. *Sci. Amer.* 252(4):68.

Dickerson, R. E. 1978. Chemical evolution and the origin of life. *Sci. Amer.* 239(3): 70.

Dickerson, R. E. 1980. Cytochrome *c* and the evolution of energy metabolism. *Sci. Amer.* 242(3):136.

Dickerson, R. E., and I. Geis. 1983. *Hemoglobin: Structure, Function, Evolution, and Pathology.* Menlo Park, Calif.: Benjamin/Cummings.

Eigen, M. et al. 1981. The origin of genetic information. *Sci. Amer.* 244(4):88.

Fox, S. W., ed. 1965. *The Origins of Prebiological Systems.* New York: Academic Press.

Fox, S. W. 1980. Metabolic microspheres. *Naturwissenschaften* 67:378.

Hardies, S. C., M. H. Edgell, and C. A. Hutchison III. 1984. Evolution of the mammalian β-globin gene cluster. *Jour. Biol. Chem.* 259:3748.

Hartman, H., P. Morrison, and J. G. Lawless, eds. 1987. *Search for the Universal Ancestors: The Origins of Life.* Boston: Blackwell Scientific.

Horowitz, N. H. 1945. On the evolution of biochemical synthesis. *Proc. Natl. Acad. Sci. U.S.* 31:153.

Jukes, T. H. et al. 1987. Evolution of anticodons: variations in the genetic code. *Cold Spring Harbor Sympos. Quant. Biol.* 52:769.

Lagerkvist, U. 1980. Codon misreading: a restriction operative in the evolution of the genetic code. *Amer. Sci.* 68:192.

Leinfelder, W. et al. 1988. Gene for a novel tRNA species that accepts L-serine and cotranslationally inserts selenocysteine. *Nature* 331:723.

Lowe, D. R. 1980. Stromatolites 3,400-Myr old from the Archean of Western Australia. *Nature* 284:441.

McLachlan, A. D. 1987. Gene duplication and the origin of repetitive protein structures. *Cold Spring Harbor Sympos. Quant. Biol.* 52:411.

Margulis, L. 1971. Symbiosis and evolution. *Sci. Amer.* 225(2):48.

Miller, S. L., and L. E. Orgel. 1974. *The Origins of Life on the Earth.* Englewood Cliffs, N.J.: Prentice-Hall.

Nirenberg, M. W., and J. H. Matthaei. 1961. The dependence of cell-free protein synthesis in *E. coli* upon naturally occurring or synthetic polynucleotides. *Proc. Natl. Acad. Sci. U.S.* 47:1588.

Oparin, A. I. 1957. *The Origin of Life on Earth.* New York: Academic Press.

Oparin, A. I. 1971. Routes for the origin of the first forms of life. *Subcell. Biochem.* 1:75.

Sawada, I., and C. W. Schmid. 1986. Repetitive human DNA sequences. I. Evolution of the primate α-globin gene cluster and interspersed *Alu* repeats. *Cold Spring Harbor Sympos. Quant. Biol.* 51:471.

Schopf, J. W. 1978. The evolution of the earliest cells. *Sci. Amer.* 239(3):110.

Schopf, J. W., and B. M. Packer. 1987. Early Archean (3.3-billion to 3.5-billion-year-old) microfossils from Warrawoona Group, Australia. *Science* 237:70.

Trachtman, P. 1984. The search for life's origins—and a first "synthetic cell." *Smithsonian* 15(3):42.

Uzzell, T., and C. Spolsky. 1974. Mitochondria and plastids as endosymbionts: a revival of special creation? *Amer. Sci.* 62:334.

Vidal, G. 1984. The oldest eukaryotic cells. *Sci. Amer.* 250(2):48.

Walter, M. R. 1977. Interpreting stromatolites. *Amer. Sci.* 65:563.

Watson, J. D. et al. 1988. *Molecular Biology of the Gene*, 4th ed. Menlo Park, Calif.: Benjamin/Cummings.

Weinberg, R. A. 1985. The molecules of life. *Sci. Amer.* 253(4):48.

Zaug, A. J., and T. R. Cech. 1986. The intervening sequence RNA of *Tetrahymena* is an enzyme. *Science* 231:470.

CHAPTER 4

Some Evolutionary Themes in the Living World

The existence in all cells of a genetic system based on a common code in informational DNA, which is transcribed into RNA and translated into protein, indicates that all life is descended from an ancient common ancestor. In Chapter 3, some of the episodes of divergence leading to autotrophs, to aerobes, and to more complex life forms in general were discussed. In this chapter, some of the major relationships among the diverse forms of modern life will be sought, which may provide some idea of the basic trunk and larger branches of the evolutionary tree of life. In other words, we will trace major events that led to the fundamentally different kinds of organisms that now populate the Earth.

Prokaryotes and Eukaryotes

With the advent of electron microscopy, which allows observations of subcellular details never before possible, the whole perspective on evolutionary dichotomy changed. All cellular life may be identified as prokaryotes or eukaryotes, rather than as plants or animals—the only or major groups of earlier classifications. It is clear that a fundamental divergence in evolution produced different plans of cell organization, and only later gave rise to the familiar plants and animals, as well as the other *kingdoms*, or major groupings of organisms.

Differences in Subcellular Organization

Except for the viruses, which are acellular and nonprotoplasmic, all life can be divided into organisms whose protoplasm is prokaryotic or eukaryotic in its subcellular organization (Fig. 4.1). The DNA of **prokaryotes** is localized in the **nucleoid** region, which is not separated by membranes from the surrounding **cytoplasm.** The plasma membrane is the only permanent membranous enclosure of the cell. In **eukaryotes,** DNA is segregated within the **nucleus,** which is delimited by a pair of nuclear membranes from the surrounding cytoplasm of the cell. In addition, eukaryotic cells possess membranous compartments that are physically separated from the plasma membrane. Among these compartments are the endoplasmic reticulum, mitochondria, lysosomes, microbodies, and chloroplasts. The hallmark of any eukaryotic cell, regardless of its other features, is its membrane-bounded nucleus. All prokaryotic photosynthesizers do have photosynthetic membranes, called *thylakoids,* distributed in the cytoplasm. Thylakoids in these organisms are believed to be differentiated parts of the plasma membrane, although the matter is somewhat controversial. Whether or not they are part of the plasma membrane, they are not separated by a membrane from the cytoplasm. The thylakoids in photosynthetic eukaryotes are always enclosed in the membrane-bounded chloroplasts.

If the only distinction between prokaryotes and eukaryotes were the presence or absence of a nucleus, the primary nature of the dichotomy might be no more persuasive than the earlier idea of a primary divergence into plants and animals based on their habits or morphological appearance. Prokaryotes and eukaryotes differ in many fundamental traits that are of such significance in evolution that biologists can be quite confident that this dichotomy is the result of a major divergence in the ancient past.

Prokaryotes not only lack membrane-bounded organelles and nucleus, but also are exclusively unicellular, asexually reproducing, and devoid of a mitotic or meiotic mechanism by which replicated DNA is partitioned into progeny cells. Many, but not all, eukaryotes are multicellular and sexually reproducing. Most have a mitotic mechanism of nuclear division, but only the sexual species carry out meiosis in association with this mode of reproduction (Table 4.1). These distinctions are of greater evolutionary significance than is the presence or absence of a rigid cell wall and a means for locomotion, which were two major bases for the now-discarded thesis that the primary dichotomy led to plants and animals. The ambiguity of these earlier criteria was apparent from the many exceptions among all the organisms lumped together as plants and from the exceptions among the unicellular forms grouped with animals into a kingdom (Table 4.2).

Prokaryotes and eukaryotes are alike in certain features, as we would predict from our survey of primeval life. They are composed of one or more *cells,* the basic structural unit of life forms other than the viruses. Cells arise from preexisting cells (apart from the initial origin of life) and transmit a set of genetic instructions that determines the resemblance or identity of progeny

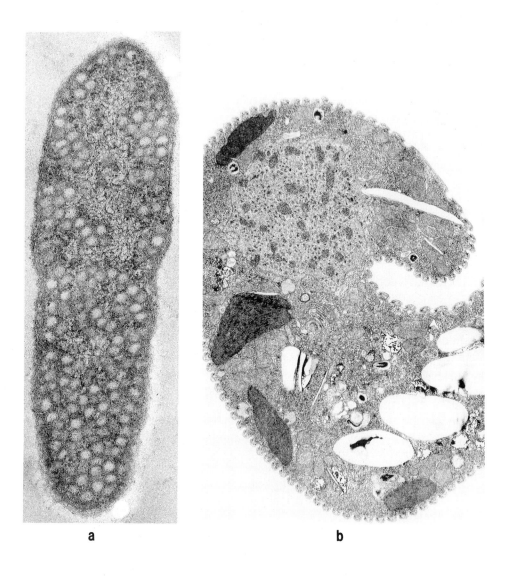

a   b

Figure 4.1. Subcellular organization. (a) Prokaryotic cells, such as the photosynthetic bacterium *Rhodopseudomonas sphaeroides,* have a protoplast consisting of an enveloping plasma membrane, and a ribosome-filled cytoplasm surrounding DNA in a nucleoid region. The cell wall is a secretion of the living protoplast. (Electron micrograph courtesy of G. Cohen-Bazire) (b) Eukaryotic cells, such as the protist *Euglena gracilis,* have their DNA in chromosomes in a membrane-bounded nucleus. The surrounding cytoplasm contains a variety of membranous and nonmembranous structures and organelles. (Electron micrograph by H.-P. Hoffmann)

TABLE 4.1
Some Major Differences Between Prokaryotes and Eukaryotes

Characteristic	Prokaryotes	Eukaryotes
Cell size	Mostly small (1–10 μm)	Mostly larger (10–100 μm)
Genetic system	DNA not associated with proteins in chromosomes	DNA complexed with proteins in chromosomes
	Nucleoid not membrane-bounded	Membrane-bounded nucleus
	One linkage group	Two or more linkage groups
	Little or no repetitive DNA	Repetitive DNA
Internal membranes (organelles)	Transient, if present	Numerous types, e.g., mitochondrion, chloroplast, lysosome, Golgi
Tissue formation	Absent	Present in many groups
Cell division	Binary fission, budding, or other means; no mitosis	Various means; associated with mitosis
Sexual system	Unidirectional transfer of genes from donor to recipient, if present	Complete nuclear fusion of equal gametic genomes; associated with meiosis
Motility organelle	Simple flagella in bacteria, if present	Complex cilia or flagella, if present
Nutrition	Principally absorption, some photosynthesis	Absorption, ingestion, photosynthesis

TABLE 4.2
Kingdom Classification Systems

"Traditional"	Dodson, 1971	Whittaker, 1969	Margulis, 1982
Plantae	Monera	Monera	Monera
bacteria	bacteria	bacteria	bacteria
blue-green algae*	blue-green algae	blue-green algae	blue-green algae
chrysophytes	Plantae	grass-green algae	grass-green algae
green algae	chrysophytes	Protista	Protoctista
red algae	green algae	chrysophytes	chrysophytes
brown algae	red algae	protozoa	green algae
slime molds	brown algae	Fungi	red algae
true fungi	slime molds	slime molds	brown algae
bryophytes	true fungi	true fungi	protozoa
tracheophytes	bryophytes	Plantae	slime molds
Animalia	tracheophytes	green algae	Fungi
protozoa	Animalia	red algae	true fungi
metazoa	protozoa	brown algae	Plantae
	metazoa	bryophytes	bryophytes
		tracheophytes	tracheophytes
		Animalia	Animalia
		metazoa	metazoa

*Cyanophytes.
Source: Data from E. O. Dodson, 1971. The kingdoms of organisms. *Syst. Zool.* 20: 265; R. H. Whittaker, 1969. New concepts of the kingdoms of organisms. *Science* 163:150; L. Margulis and K. V. Schwartz, 1982. *Five Kingdoms.* San Francisco: Freeman.

and parents. Viruses do not reproduce in the cellular mode, but they do have a unique set of genes from which viruses like themselves are assembled in living host cells. The continuity of life depends on the same basic genetic premises in all organisms, whether cellular or viral in nature.

Every cell has a plasma membrane; a mix of substances in solution and suspension that makes up the cytoplasm; a genetic apparatus that includes DNA (of which genes are made), RNA intermediary molecules, and ribosomes (on which polypeptides are synthesized during translation); and metabolic pathways whose reaction steps are catalyzed by enzymes. Of course, many differences in detail have been incorporated in these broad features during evolution.

Theories of the Origin of the Eukaryotic Cell

All the theories can be divided into two categories of explanation for the evolution of compartmentalized eukaryotic cells from noncompartmentalized prokaryotic ancestors. One category stipulates *autogenous origins* of membrane-bound compartments; that is, organelle membranes and contents arose by the internalization of differentiated areas of the cellular plasma membrane that had come to enclose particular parts of the cellular metabolic system. The other category postulates *exogenous origins* of at least some organelles, which are proposed to be endosymbiont descendants of formerly free-living cells that took up permanent residence in a host cell.

The starting point for these theories is that eukaryotes are descended from prokaryotes. The fossil record clearly shows the later arrival of eukaryotes in a world that had been exclusively prokaryotic for more than 2 billion years. The similarities in genetic systems, metabolism, and other basic features of prokaryotes and eukaryotes indicate descent with modification rather than the independent origin of eukaryotes and the development of all these shared features by pure coincidence. In addition, both proposals posit a *mobile cell surface,* by which plasma membrane infoldings might be internalized or whole cells be engulfed in endocytotic events not unlike those known in many modern cells (Fig. 4.2).

A provocative theory of the autogenous origin of organelles by membrane infolding inside the cell was described in the 1970s by Thomas Cavalier-Smith. He suggested that the **lysosome** was the first internal compartment to evolve in prokaryotic cells with a mobile cell surface. The ingestion of solid foods would have been possible, but the nutrients would have been unavailable to the cell unless the solids were digested to soluble forms. Powerful but potentially dangerous digestive enzymes may have been located on the outer cell surface and collected into endocytotic pockets, along with solids from the surroundings. Once digested, soluble nutrients moved into the cell. Later in evolution, adaptive changes may have included the enclosure of digestive enzymes in membranous vesicles at the cell surface (Fig. 4.3). Solid foods taken into these vesicles would be digested, and the soluble nutrients could

Yeast cell   Leucocyte

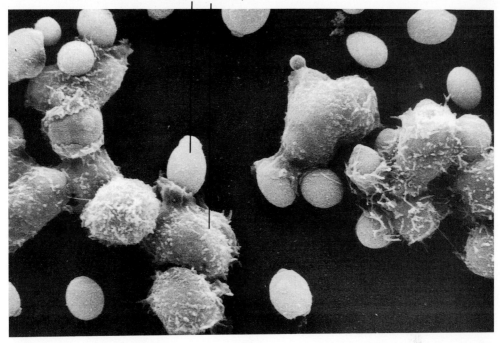

Figure 4.2. Phagocytic polymorphonuclear leucocytes (a type of white blood cell) in this scanning electron micrograph are seen engulfing yeast cells. Yeast cells, though, have a rigid cell wall and are not capable of ingesting solids, whether particles or cells. (Photograph courtesy of D. F. Bainton)

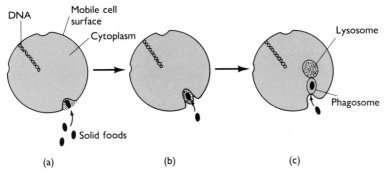

Figure 4.3. According to a proposal by Thomas Cavalier-Smith, the lysosome may have been the earliest membranous organelle to evolve. (a) Ingestion of solid foods would have been possible through membrane infolding and endocytosis, but the powerful digestive enzymes would probably have been located in endocytotic pockets on the outer surface of the cell, where they could not harm (digest) the internal cellular components. (b) Later in evolution, solid foods might have been packaged along with the digestive enzymes in membranous vesicles formed at the cell surface. Foods would thus have been digested outside the cell, but in specialized enclosures. (c) Ultimately, the digestive enzymes came to be packaged separately in subcellular lysosomes. Incoming solids would have been made soluble after the fusion of the entering phagosome and the lysosome inside the cell, as in modern eukaryotes.

enter the cell. Ultimately, vesicles containing digestive enzymes may have been internalized, as are lysosomes in modern cells. Solids could be endocytosed by the cell and be sequestered in vesicles not unlike phagosomes. As in modern cells, digestion would take place after the fusion of lysosome and phagosome inside the cell, and soluble products would be made available in the cytoplasm. The differentiation of endoplasmic reticulum and Golgi membranes later in evolution provided a system by which lysosomes formed internally and exocytosis and endocytosis involved the plasma membrane in membrane recycling (Fig. 4.4). This system exists in modern eukaryotic cells, all these membranes being produced from the primary source of plasma membrane, which accepts and releases chunks of membrane in recycling events. The different properties of these internal membranes and their metabolic specifications are due in large measure to differences in the proteins that make up the organelle systems.

Cavalier-Smith has also suggested an autogenous sequence by which DNA came to be enclosed in the nucleus. In prokaryotes, the DNA molecule is attached to the plasma membrane, which is not a safe place in cells with an active mobile cell surface. The DNA might be expelled by exocytosis or taken into the interior by endocytosis and perhaps be digested there in lysosomes. Genetic changes that led to the detachment of DNA from the cell membrane

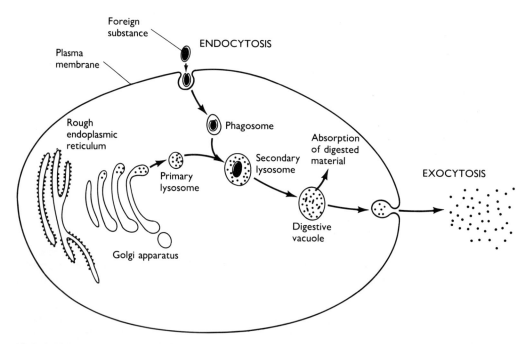

Figure 4.4. Lysosomes in modern eukaryotic cells are formed from the Golgi apparatus, which itself is differentiated from endoplasmic reticulum membranes. Plasma membrane elements taken into the cell during endocytosis and returned to the membrane during exocytosis represent parts of the membrane-recycling system in eukaryotic cells.

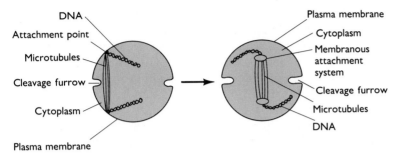

Figure 4.5 According to a proposal by Thomas Cavalier-Smith, detachment of DNA from the active mobile cell surface would have reduced the probability of genes being expelled by exocytosis or being taken into the interior and digested. The orderly segregation of replicated DNA would have been aided by the evolution of a rigid microtubular apparatus between the separating attachment points. Later in evolution, this system would have been dissociated from the mobile plasma membrane. All eukaryotes possess microtubules, which suggests their early appearance in eukaryotic evolution and retention in all descendant lineages.

thus might be adaptive, but the orderly segregation of replicated DNA molecules during cell division would require an alternative rigid attachment system. In accordance with known eukaryotic mechanisms, Cavalier-Smith postulated the emergence of a system of *microtubules* by which replicated DNA molecules are separated and guided toward opposite poles of the cell (Fig. 4.5). Microtubules are fibrous components of all eukaryotic cells, suggesting their appearance very early in eukaryotic evolution and their retention in all descendant lineages. The eventual enclosure of DNA in the nuclear envelope presumably resulted from membrane infolding that evolved into a permanent system separate from the plasma membrane and possessing unique features related to nuclear function and nucleocytoplasmic interactions.

Mitochondria and chloroplasts are presumed to have evolved by similar autogenous events, but with important differences. Both have a genetic apparatus by which some of their polypeptide and all their RNA components are specified and synthesized within the organelles themselves. All autogenous origin theories postulate the enclosure of part of the cellular genetic machinery within these organelles, along with the capacity to make all their molecules from the combined efforts of the nucleocytoplasmic and organelle systems. It is precisely because of the organelles' genetic systems, however, that exogenous origins have been proposed for mitochondria and chloroplasts in eukaryotic cells. Both kinds of organelles are presumed to have a genetic system that is a remnant of the genetic apparatus originally in the free-living cells that ultimately evolved into mitochondria and chloroplasts.

The **endosymbiosis theory**, revived by Lynn Margulis in 1967, is the major proposal today for exogenous origins of the mitochondrion, chloroplast, and uniquely eukaryotic flagellum. She has suggested an ancestral prokaryotic cell that was anaerobic and capable of ingesting solids by means of its mobile cell surface. Such a cell, not unlike the modern archaebacterium *Thermoplasma,* engulfed some respiring prokaryotes that were not digested but were retained in a mutually beneficial (symbiotic) relationship with their host (Fig. 4.6). The host provided nutrients and protection for the *endo*symbionts, which, in turn, provided energy-efficient aerobic respiration for their host. In an anaerobic world undergoing a transition to aerobic conditions, the host would have derived substantial benefits from its endosymbiotic retainers. With time, the endosymbionts were streamlined into mitochondria, which became incapable of existing outside the host. Mitochondria presumably retained part of the original genetic machinery, but became totally dependent on the host for genes and gene products lost to the host during evolution.

Mitochondrial genes today specify all their transfer and ribosomal RNAs,

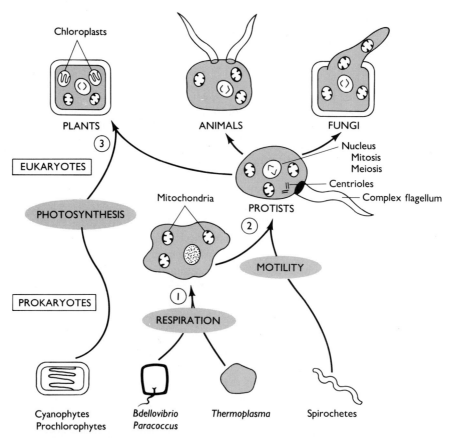

Figure 4.6. The endosymbiosis theory of organelle origin in eukaryotes, as proposed by Lynn Margulis. Ancient free-living prokaryotic organisms similar to particular modern species were engulfed by host cells in a sequence of independent episodes (1, 2, 3). The engulfed organisms and their hosts became established in a mutually beneficial association, or symbiosis. The endosymbionts ultimately evolved into organelles (mitochondria, 1; flagella, 2; and chloroplasts 3,) that were totally dependent on the host, but provided the host with the advantageous properties of aerobic respiration, controlled motility, and photosynthesis.

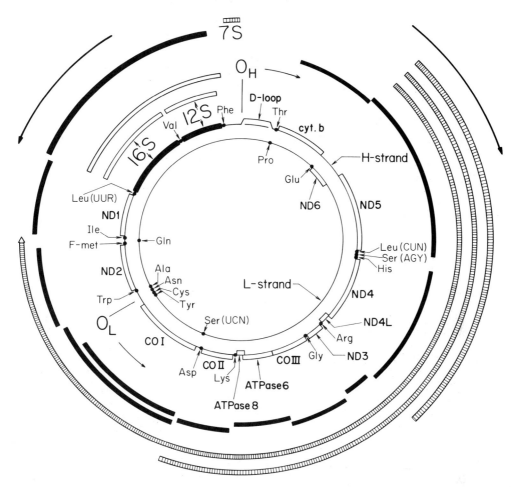

Figure 4.7. The organization of the human mitochondrial genome. Genes for 2 rRNAs (16S, 12S), 22 tRNAs, and 13 polypeptides are distributed economically around almost the entire sequence of 16,569 base pairs of the circular mitochondrial DNA molecule. All 13 polypeptides specified by genes have been identified; these are three subunits of cytochrome oxidase (CO I, CO II, CO III), two subunits of ATP synthetase of oxidative phosphorylation (ATPase 6, ATPase 8), cytochrome *b* (cyt *b*), and seven subunits of NADH dehydrogenase (ND1, ND2, ND3, ND4, ND4L, ND5, ND6). (From G. Attardi et al., 1986, Seven unidentified reading frames of human mitochondrial DNA encode subunits of the respiratory chain NADH dehydrogenase. *Cold Spring Harbor Sympos. Quant. Biol.* 51:103, Fig. 1)

but only one or a few of the polypeptide subunits in some of the cytochrome enzymes and several other enzymes of the electron transport chain and ATP synthesis reactions of aerobic respiration (Fig. 4.7). All the remaining polypeptide subunits are synthesized in the cell cytoplasm from nuclear transcripts and imported into the mitochondria. In addition, the enzymes of the Krebs cycle, such enzymes as DNA and RNA polymerases, and all or almost all of the dozens of ribosomal proteins and membrane proteins are specified by nuclear genes and shipped into the mitochondria from their sites of translation at cytoplasmic ribosomes. In mammalian mitochondria, the genome of over 16,500 base pairs consists of only 37 genes: 13 for polypeptides, 22 for transfer RNAs, and 2 for ribosomal RNAs.

Margulis has proposed two other episodes of endosymbiosis, both of which occurred later in eukaryote evolution than the engulfing of proto-mitochondria. One led to flagella, which are said to have evolved from ingested spirochetelike prokaryotic organisms. Later, and only in lineages leading to plants, prokaryotic cyanophytes and prochlorophytes established residence in eukaryotic hosts and there evolved into modern chloroplasts.

Evidence on the Origins of Mitochondria and Chloroplasts

Any evidence of autogenous or exogenous origins of mitochondria and chloroplasts must be judged according to contrasting predictions made by these theories. If these organelles are descendants of ancient prokaryotes, they should resemble their prokaryotic ancestors more than their eukaryotic hosts. If they arose autogenously, however, there should be homology with eukaryotes rather than with prokaryotes. The crux of any such comparison is to determine whether resemblances are indeed **homologous**—that is, genetically related—or perhaps only **analogous**—that is, superficially similar because of independent pathways of genetic change that led to convergent end products, perhaps as a result of similar selection factors. A prime example of evolutionary **convergence** is the flying habit of insects and birds. Both kinds of animals have wings, but they are produced by totally different genetically programmed pathways of development in insects and birds. Wings arose independently in insect and bird lineages; insects are not the ancestors of birds.

The case for an endosymbiotic origin of chloroplasts from cyanophytes and prochlorophytes, the two groups of prokaryotic oxygenic photosynthesizers, has gained considerable strength in recent years. Symbiotic associations between cyanophytes and eukaryotic cells are well known among modern organisms. Once inside the host, cyanophytes lose their rigid cell wall and continue their photosynthetic activities, to the benefit of the host. In some cases, the symbionts are unable to exist outside the host, and they are then referred to as *cyanelles* (Fig. 4.8). Much less is known about the recently discovered prochlorophytes, but *Prochloron* was found as a symbiont in ascidians, a group of marine nonvertebrate chordates.

Evidence from nucleic acid and protein sequencing and from comparisons of the photosynthetic systems in chloroplasts and the two groups of oxygenic photosynthetic prokaryotes reveals unambiguous homologies. In addition, cyanophytes share many photosynthetic components with chloroplasts found uniquely in the eukaryotic red algae. Both have essentially the same phyco-

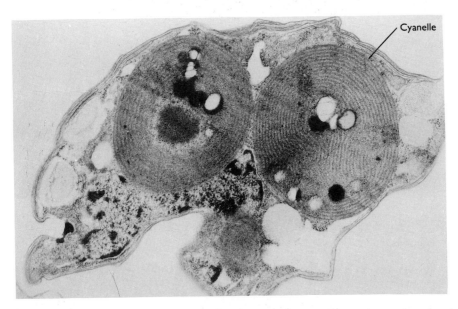

Cyanelle

Figure 4.8. Electron micrograph of a section through the protist *Cyanophora paradoxa,* a unicellular flagellate that contains from two to four cyanelles. Cyanelles are photosynthetic structures that closely resemble free-living cyanophytes in their thylakoid organization and photosynthetic pigments. The organelle cannot be cultured outside the cell, and it has a relatively small DNA genome. All these features indicate that the cyanelle is an endosymbiont that has evolved from a cyanophyte ancestral origin (compare Fig. 4.9). (Photograph courtesy of L. P. Vernon)

biliprotein pigments in phycobilisome particles, which are arranged along unstacked thylakoids, and both lack chlorophyll *b* but have chlorophyll *a* (Fig. 4.9). Chloroplasts in most other eukaryotes resemble *Prochloron* in many respects, including the presence of chlorophyll *b* as well as chlorophyll *a* and the absence of phycobiliproteins.

These homologous features indicate a close genetic relationship between cyanophytes and chloroplasts in red algae, and between prochlorophytes and chloroplasts in eukaryotic green cells. The evolution of chloroplasts from *two* prokaryotic lineages makes the case for endosymbiosis even stronger. If the red algae and green algae had evolved their chloroplasts autogenously, biologists would have to assume that each group of algae had evolved independently into eukaryotes from different prokaryotic ancestors that just happened to develop the same nucleocytoplasmic system but quite different chloroplast systems (Fig. 4.10). It is far more likely that the two groups of algae share a eukaryotic inheritance derived from a common ancestral lineage, and *as eukaryotes* each group incorporated different endosymbionts that became the different kinds of modern chloroplasts. The *polyphyletic* (more than one ancestral lineage) derivation of chloroplasts is thus a powerful argument for the endosymbiotic origin of these organelles in eukaryotic hosts.

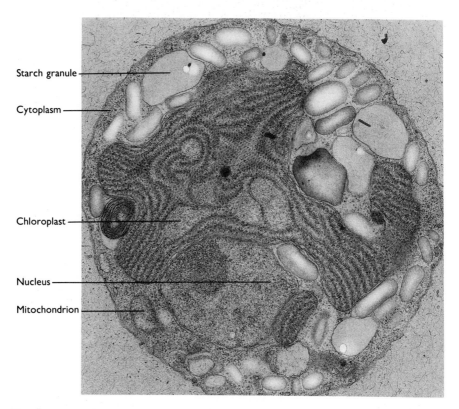

Figure 4.9. Electron micrograph of a section through the unicellular red alga *Porphyridium cruentum*. The chloroplast, with its thylakoids encrusted with granular phycobilisomes, is strikingly similar in appearance to a cyanophyte cell. (Photograph courtesy of E. Gantt, from E. Gantt and S. F. Conti, 1965. *Jour. Cell Biol.* 26:365, Fig. 7)

The origin of mitochondria is unclear because the comparisons point as much to their eukaryote-like features as to their prokaryote-like traits and their considerable variability (Table 4.3). The mitochondrial genome resides in a duplex, circular DNA molecule that is not separated by a membrane from the surrounding matrix, just like the genome in noncompartmentalized prokaryotic cells. The genes themselves, however, may or may not be subdivided into **exons** and **introns**. Mitochondrial genomes vary from one group of organisms to another. In mammals, the more than 16,500 base pairs of mitochondrial DNA include 37 genes immediately adjacent to one another, and every gene lacks exons and introns. In fungi, such as yeast, a similar number of genes, some with introns and exons, are scattered among the 75,000 base pairs of mitochondrial DNA, but lengthy stretches of noncoding DNA intervene between genes, just as they do in eukaryotic genomes (Fig. 4.11). Ribosomes vary from 55S to 80S particles in mitochondria, in contrast with the 70S particles typical of prokaryotes (and of chloroplasts). Mitochondria make

do with as few as 22 transfer RNAs, as compared with the 32 tRNAs in prokaryotes and eukaryotes and the 30 or 31 tRNAs in chloroplasts.

The mix of traits unique to mitochondria with others that are prokaryote-
like and eukaryote-like creates a complex and puzzling situation. Biologists
are hard put to identify mitochondrial characteristics as homologous or analogous to their counterparts in the nucleocytoplasmic systems of the cell. The
physical appearance of mitochondria led August Weismann to suggest their
origin from engulfed bacteria. These observations by light microscopy
appeared to be supported later by electron microscopy, and were made even
more appealing when it was shown in the 1960s that mitochondria have their
own DNA. Studies of mitochondrial profiles in random sections revealed
these organelles to be the approximate size and shape of bacteria. When mitochondria were analyzed from a complete series of sections through a cell,
however, the three-dimensional reconstruction showed a large mitochondrial

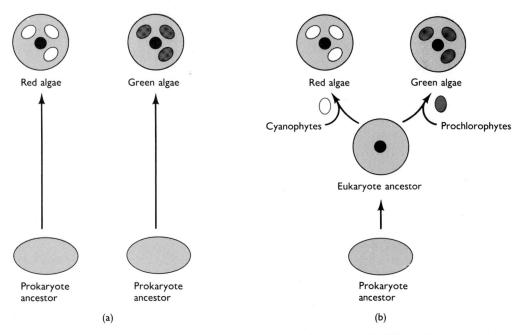

Figure 4.10. Two alternative proposals for the origin and evolution of different chloroplasts in red
algae and green algae. (a) Red algae and green algae may have evolved independently from different
prokaryotic ancestors, which could explain the differences in the chloroplasts of the two groups of
eukaryotic algae. This explanation depends on the parallel evolution of identical nucleocytoplasmic
systems from different ancestors of the two groups of eukaryotic algae, which is unlikely. (b) The
two groups of algae may have diverged from a eukaryotic lineage that had evolved from a prokaryotic
ancestor, and subsequently incorporated different endosymbionts that evolved into the two kinds of
chloroplasts. The endosymbiosis proposal more readily explains why red algae and green algae (and
those plants that evolved from green algae) have identical nucleocytoplasmic systems but different
kinds of chloroplasts. The endosymbiont is thought to have been a cyanophyte in red algae, and a
prochlorophyte in green algae.

TABLE 4.3
Comparison of Selected Features of the Genomes in Human Mitochondria, Yeast Mitochondria, *E. coli* (prokaryote), and *Drosophila melanogaster* (eukaryote)

Feature	Human Mitochondria	Yeast Mitochondria	*E. coli* (prokaryote)	*D. melanogaster* (eukaryote)
Genomic DNA	I duplex; circular	I duplex; circular	I duplex; circular	4 duplexes; linear
contour length	5.5 μm	25–26 μm	1300 μm	>50,000 μm
number of kilobase pairs	16.57	75–78	~4,000	~150,000
intergenic spacers	Absent	Present	Absent	Present
Genes				
polypeptide-specifying genes	13	7 known + URFs	3000–4000	5000–10,000
tRNA genes	22	25	32 + multiple copies	32 + multiple copies
rRNA genes*	2; adjacent	2; far apart	3 (5–10 copies of each)	2 (hundreds of copies)
noncoding leader and trailer segments	Absent	Present	Present	Present
STOP codons	Absent in some genes	Present	Present	Present
exon–intron organization	Absent	Present in some genes	Absent	Present
Transcription	Within mitochondria	Within mitochondria	Within nucleoid	Within nucleus
number of promoters	I	5 or more	Numerous	Numerous
pre-mRNA	Absent	Present	Absent	Present
mRNA	Transcribed directly	Processed from pre-mRNA	Transcribed directly	Processed from pre-mRNA
poly(A) tail	Posttranscriptional	Absent	Absent	Posttranscriptional
Translation	Within mitochondria	Within mitochondria	Within cytoplasm	Within cytoplasm
ribosomes	55–60S	80S	70S	80S
codon usage	UGA = Trp	UGA = Trp	UGA = STOP	UGA = STOP
	AUA = Met	AUA = Ile	AUA = Ile	AUA = Ile
	AG_G^A = not used	$AG_G^A = \frac{Arg}{not}$ used?	AG_G^A = Arg	AG_G^A = Arg
	CUN = Leu	CUN = Thr	CUN = Leu	CUN = Leu

* Human mitochondrial rRNA genes = 12S and 16S; yeast mitochondrial rRNA genes = 15S and 21S; *E. coli* rRNA genes = 16S, 23S, and 5S; *D. melanogaster* rRNA genes = 38S (processed to 18S and 28S rRNAs) and 5S.
Source: Reproduced with permission from C. J. Avers, 1984. *Genetics*, 2nd ed. Boston: Willard Grant Press.

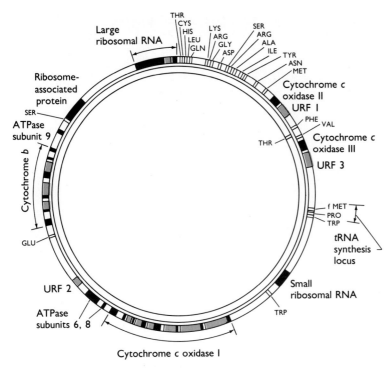

THR
CYS
HIS
LEU
GLN
LYS
ARG
GLY
ASP
SER
ARG
ALA
ILE
TYR
ASN
MET

Large ribosomal RNA

Ribosome-associated protein

SER

ATPase subunit 9

Cytochrome b

GLU

URF 2

ATPase subunits 6, 8

Cytochrome c oxidase I

Cytochrome c oxidase II
URF 1
PHE VAL
THR
Cytochrome c oxidase III
URF 3

f MET
PRO
TRP

tRNA synthesis locus

Small ribosomal RNA

TRP

Figure 4.11. The organization of the mitochondrial genome in the yeast *Saccharomyces cerevisiae*. Distributed around the 76,000 base pairs of the circular DNA molecule are 2 rRNA genes, 25 tRNA genes, 1 gene specifying RNA of an RNase P type of tRNA-processing enzyme, and 8 identified plus several unidentified genes (URF) that encode polypeptides. The 8 known polypeptide-specifying genes are encoded for three subunits of cytochrome oxidase (CO I, CO II, COIII), three subunits of ATP synthetase of oxidative phosphorylation (ATPase 6, 8, and 9), cytochrome *b* (Cyt *b*), and a protein associated with the mitochondrial ribosome (*var* 1). Interestingly, there are no genes for NADH dehydrogenase, like those in the human mitochondrial genome and all others that have been analyzed. Three genes (CO I, Cyt *b*, 21S rRNA) are organized into introns and exons, and considerable amounts of noncoding spacer DNA are present throughout the genome. (Courtesy of L. A. Grivell)

reticulum rather than many small, sausage-shaped profiles (Fig. 4.12). The two-dimensional profiles appeared to be cuts through portions of the larger, irregular reticular structure. Mitochondria thus seem to be quite different in size and shape from known prokaryotes. Furthermore, no examples are known of associations in which an endosymbiont exists inside a prokaryotic host cell; the host is always eukaryotic, and the internalized symbiont residents may be either prokaryotes or eukaryotes. If respiring bacteria were engulfed by ancient anaerobic prokaryotes and if these endosymbionts and their hosts evolved together toward the eukaryotic state, the capacity for ingestion has been lost in all modern bacterial lineages.

The endosymbiosis theory significantly stimulated a variety of analytic approaches to the problem of organelle origins. Support for the endosym-

a

Figure 4.12. The single reticulate mitochondrion in the yeast *Saccharomyces cerevisiae*. (a) Three-dimensional reconstruction of the mitochondrion in a budding yeast cell, showing the growing portion of the organelle in the large bud (left) and the major portion in the mother cell (right). (b) Tracings made of mitochondrial profiles from electron micrographs of a complete consecutive series of 70 sections through the budding cell are shown for sections 37 to 43. The separate mitochondrial profiles in sections 37 to 40 are connected to one another in sections 41 to 43, and this entity is connected in turn to the rest of the organelle reticulum. Continuity between profiles is apparent in the cumulative view of the superimposed tracings from sections 37 to 43, shown in the inset. (From H. -P. Hoffmann and C. J. Avers, 1973, Mitochondrion of yeast: ultrastructural evidence for one giant, branched organelle per cell. *Science* 181:749)

biotic origin of chloroplasts is very strong, but mitochondrial origins remain vague. Margulis included in her theory the proposition that the eukaryotic flagellum evolved from an endosymbiotic spirochetelike prokaryote that became part of its eukaryotic protistan host. The major line of information in this regard has come from her studies of an unusual group of spirochetes that live on and in protists. These spiral bacteria depend for their locomotion on bundles of typical bacterial flagella, but some also have microtubules, which are not found in other prokaryotes but are ubiquitous in eukaryotic cells. It has yet to be determined, however, whether or not these similarities are an outcome of genetic homology. It is also possible that the two genes that encode microtubule proteins may have evolved independently in the spirochetes or may have been transferred from the eukaryotic host nucleus into spirochete DNA. Such *horizontal gene transfers* have been identified in a number of different prokaryotes, and are also known to occur among nuclei, mitochondria, and chloroplasts in eukaryotic cells. Molecular analysis, including sequencing studies, are required to confirm the most likely source of origin for genes that otherwise do not seem to belong in a genome.

b

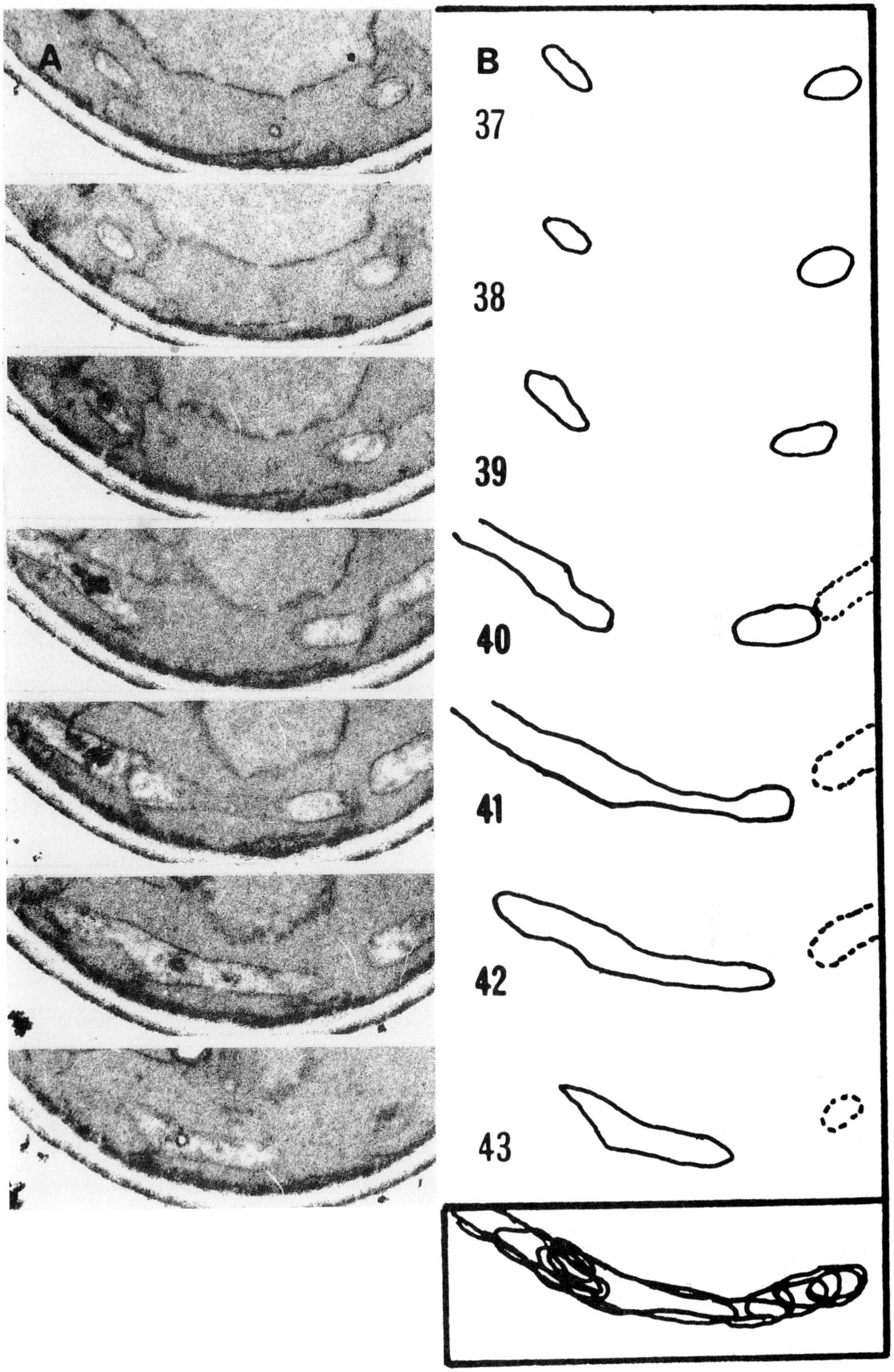

A

B

37

38

39

40

41

42

43

The Evolutionary Tree of Life

The graphic depiction of biological evolution in the form of a tree is familiar to everyone. The organisms making up the roots, trunk, major branches, and smaller branches and twigs, however, are the subject of lively controversies. We now focus on two of the various proposals for major features of the evolutionary tree of life on Earth. We will be concerned, in particular, with relationships among kingdoms of organisms. Discussions of smaller groups within the kingdoms will be included in other chapters.

The Five-Kingdom System of Classification

The world of prokaryotes and eukaryotes has been subdivided into **kingdoms,** the major units of classification, or *taxa.* Each kingdom includes organisms that share fundamental characteristics but vary in many other features, and is divided further into phyla, classes, orders, families, genera, species, and other taxa. The aim is to establish groups that are *phylogenetically* related; that is, groups that are genetically related through evolutionary descent. Because of gaps in the fossil record, different judgments regarding criteria to be applied, and other difficulties, no single classification scheme is universally accepted. Different viewpoints prevail at every level, from identification of kingdoms to identification of individual species.

    In 1969, Robert Whittaker reevaluated and regrouped all cellular life into five kingdoms that represent broad relationships in regard to levels of organization and different modes of nutrition that affect different kinds of organization (Fig. 4.13):

1. **Monera,** consisting of all the prokaryotes, which obtain nutrients by photosynthesis and absorption.
2. **Protista,** consisting of a great variety of unicellular eukaryotes whose affinities are unclear, so they cannot be grouped conveniently into the other three eukaryote kingdoms, that obtain nutrients by photosynthesis, absorption, and ingestion.
3. **Plantae,** consisting of all the eukaryotic algae and the multicellular plants, which obtain nutrients by photosynthesis.
4. **Animalia,** consisting of the multicellular eukaryotes that obtain nutrients by ingestion.
5. **Fungi,** consisting of the unicellular and multicellular spore-forming eukaryotes which obtain nutrients by absorption.

    The separation of all the bacteria, cyanophytes, and prochlorophytes into a single kingdom of prokaryotes takes into account their many basic distinctions from eukaryotes. The subdivision of eukaryotes into four kingdoms is more controversial. Some proposals include only 2 eukaryote kingdoms,

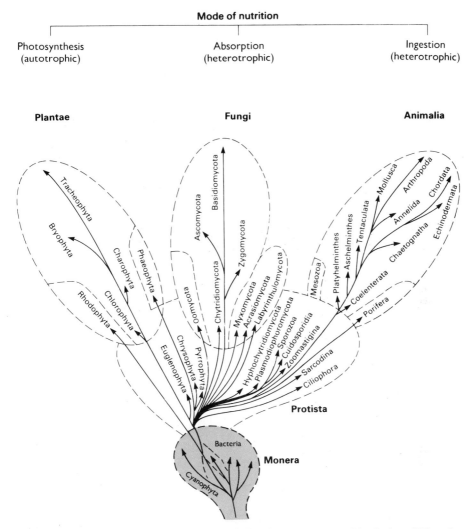

Figure 4.13. The five-kingdom system of classification proposed by Robert Whittaker groups all prokaryotes into the kingdom Monera and separates eukaryotes into the unicellular Protista and the multicellular (or multinucleate) Plantae, Fungi, and Animalia. Heterotrophic monerans absorb organic nutrients in solution from their surroundings, and autotrophic monerans make their own foods by photosynthesis. Protists include species that absorb nutrients, are photosynthetic, or ingest solid foods and digest them internally. Plants, fungi, and animals differ in organization and life style, according to the single mode of nutrition that characterizes each kingdom. The five-kingdom system emphasizes evolutionary relationships and postulates moneran ancestry for the protists, which later gave rise from different lineages to the three kingdoms of multicellular eukaryotes. (From R. H. Whittaker, 1969, New concepts of kingdoms of organisms. *Science* 163:150, Fig. 3. Copyright © 1969 by the American Association for the Advancement of Science)

plants and animals, or as many as 30. Even among proponents of the five-kingdom system, considerable difference of opinion exists concerning the particular groups of organisms to be included in the four kingdoms of eukaryotes. Margulis, for example, believes that major evolutionary trends and pathways are more accurately reflected in limiting plants to only the multicellular photosynthesizing eukaryotes that develop from an embryo, thereby excluding all the algae; fungi are only the species that form spores and lack flagella; thus excluding all the slime molds and water molds; animals remain essentially as proposed by Whittaker, thereby including all the multicellular eukaryotes that ingest food and develop from a blastula, the hollow ball of cells produced by cell divisions from the fertilized egg. The protists, called Protoctista by Margulis, are all the organisms, whether or not they are unicellular, excluded from the other eukaryote kingdoms. She believes that multicellularity arose many times from unicellular forms and that many multicellular organisms are more closely related to particular unicells than they are to any other multicellular organisms; such forms are therefore included in the Proctoctista. The underlying theme in her proposal or any other is to classify organisms in the way that best reflects their evolutionary heritage and relationships. Each proposal and its details are acceptable to some biologists and not to others, depending on their own perceptions and judgments about evolutionary lineages.

The five-kingdom system basically stipulates that eukaryotes evolved from prokaryote ancestors and that protists were the first eukaryotes to appear. The fossil record is fragmentary at best, but undeniable protistan fossils have been dated as about 900 million years old. Based on large cell size and other external features of older fossils, J. William Schopf and others have identified even older fossil cells as eukaryotes; they are about 1.4 billion years old (see Fig. 3.28). Uncertainties remain about specific ancestral protistan lineages that gave rise to fungi, plants, and animals, as defined by Whittaker. The consensus, however, is that all three groups arose independently from different lineages. The roots and base of this evolutionary tree are thus composed of prokaryotes. The trunk expanded later, when eukaryotic protists appeared and diversified into many lineages. The three major branches of the tree consist of the more recently evolved kingdoms: plants, animals, and fungi, as shown in Fig. 4.13.

Prokaryotic cell organization mandates a minimum of diversity in cell size but allows great variety in biochemical pathways of both heterotrophic and autotrophic nutrition. The existence of a rigid cell wall in virtually all prokaryotes does not interfere with autotrophy, but heterotrophs must obtain all their organic nutrients by absorption; ingestion is impossible when the cell membrane is covered by a rigid coat. Multicellularity never evolved in prokaryotes.

The eukaryotic organization of protistan cells apparently opened the way for a great deal of diversity in cell size and shape and for a variety of life styles made possible by the addition of ingestion as a mode of nutrition in some

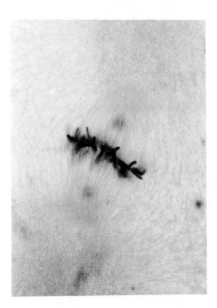

Figure 4.14. Mitotic cell from the plant *Claytonia virginica.* The evolution of a mitotic spindle apparatus ensured a high degree of accuracy in the separation of replicated chromosomes in every cell generation, regardless of the number of chromosomes involved. (Photograph by N. V. Rothwell)

lineages. Accompanying the eukaryotic status of protistan cells was the first appearance of chromosomes, allowing diversification of the genetic programs, and a mitotic mechanism that ensures the accurate distribution to progeny cells of any number of chromosomes in a genome, or a full set of DNA (Fig. 4.14). Meiosis and gamete fusions in sexual reproduction permitted even greater genetic diversity. Eventually, eukaryotic multicellularity emerged and became the predominant body plan in animals, fungi, and plants.

The major mode of nutrition influenced the kinds of body plan and the distinctive life styles of the three most recent kingdoms of organisms. Body plan is restricted in fungi, which must absorb all their nutrients from their immediate surroundings and therefore must present all of most of their cells to these nutrient sources. Body size and plan may vary to a much greater extent in plants and animals, as long as some portion of the organism is differentiated into organs suitable for photosynthesis or ingestion, by which nutrients are obtained and made available to all other parts of the body. Mobility and activity are advantageous features for animals, which must seek the foods to be ingested. Plants, on the contrary, do very well as passive organisms that continually take in radiant energy from the sun and CO_2 from the air to produce organic foods. Very different sets of selection factors would therefore have guided the evolution of the three recent kingdoms of organisms, based in large measure on their fundamentally different modes of nutrition, as well as on other features related to life style.

Archaebacteria, Eubacteria, and Eukaryotes

About 10 years ago, Carl Woese, George Fox, and their colleagues proposed a very different rationale for the division of organisms into kingdoms. On the basis of biochemical and genetic features in life forms, they suggested that the major distinction between organisms is not their prokaryotic or eukaryotic organization. Their studies of bacteria revealed an unusual group, which they named the **archaebacteria,** whose molecular features include unique composition of the cell wall and membrane lipids as well as traits that are as much like those of prokaryotes as of eukaryotes (Table 4.4). Archaebacterial cells resemble bacteria in size and shape, and their subcellular organization is prokaryotic in plan.

The archaebacteria are a heterogeneous collection of species that flourish in extreme habitats. The *methanogens* are anaerobic bacteria that obtain car-

TABLE 4.4
Comparison of Selected Traits in Archaebacteria, Eubacteria, and Eukaryotes

Trait	Archaebacteria	Eubacteria	Eukaryotes
Average cell size	$1-10\ \mu m$	$1-10\ \mu m$	$10-100\ \mu m$
Nuclear membrane	Absent	Absent	Present
Membranous organelles	Absent	Absent	Present
Muramic acid in cell wall	Absent	Present	Absent
Aliphatic chains of membrane lipids	Branched, ether-linked	Unbranched, ester-linked	Unbranched ester-linked
Split genes	Present (few)	Absent	Present (many)
Ribosomes			
subunit sizes	30S, 50S,	30S, 50S	40S, 60S
small rRNA size	16S	16S	18S
large rRNA size	23S	23S	25–28S
sensitivity to			
chloramphenicol	No	Yes	No
kanamycin	No	Yes	No
anisomycin	Yes	No	Yes
CCUCC binding site at 3' end of 16S (18S) rRNA	Present	Present	Absent
Transfer RNAs			
ribothymidine in TψC arm	Absent	Present	Present
dihydrouracil in D arm	Absent	Present	Present
initiating amino acid	Methionine	Formylmethionine	Methionine

Source: Data from C. R. Woese, G. Fox, and other sources.

bon from formate, acetate, and methanol, but not from sugars or proteins, and derive energy by producing methane (marsh gas) from CO_2 and H_2 in air. The *extreme halophiles* are aerobic bacteria that obtain their salt requirement from the NaCl dissolved in brine and salt works; they live in high salt concentrations that would tear apart the membranes of other cells. The *thermoacidophiles* are a mixed lot that share the property of living in hot springs and similar environments, where the temperature may rise to nearly 90°C and the pH may be as low as 1 to 3. In fact, *Sulfolobus* species may die at temperatures below 55°C, whereas most cells would die at 55°C due to the denaturation of their proteins. *Thermoplasma* includes only one species, *T. acidophilum*, found in hot coal-refuse piles and in hot springs in Yellowstone National Park. In addition to its bizarre life style, this species is unusual among the bacteria in having histonelike basic proteins coating its DNA, and it lacks a cell wall as do *Mycoplasma* and a few other bacteria.

Notwithstanding the unclear affinities of the archaebacteria to one another and their similarities to conventional bacterial groups, Woese, Fox, and others have set them apart from all other bacteria, which they refer to as the **eubacteria.** The two "primary kingdoms" of archaebacteria and eubacteria are distinct from the eukaryotes, which are presumed to represent the third major branch that emerged from a common ancestor of all known life forms. This universal ancestor, or *progenote*, was an unknown organism much simpler than present-day prokaryotes, but perhaps resembling archaebacteria in certain features. The progenote is assumed to have diverged into three major stems, two of which are the archaebacterial and eubacterial groups, which evolved prokaryotic cell organization. The third stem emerged from a non-prokaryote, called an *urkaryote*, which achieved final eukaryotic status after endosymbionts took up residence as mitochondria and chloroplasts in the urkaryotic host (Fig. 4.15). According to this evolutionary scenario, the eukaryotes did not diverge from a prokaryotic ancestor, which is the primary premise of the five-kingdom system and its variants.

The approach that has been taken to determine evolutionary genealogies from ancient times to the present is to analyze particular molecules that are presumed to retain features old enough to serve as phylogenetic markers of ancestry and descent. Chief among such molecules are ribosomal and transfer RNAs, whose nucleotide sequences can be determined and catalogued for comparison. From these sequencing studies, the proponents of the "three-stem" theory have concluded that each of the three primary groups is equally distant from the other two. They interpret these data to mean that the groups are equally ancient in origin, which indicates that they evolved independently from a primordial common ancestor and not from one another.

The interpretations are controversial, in part because little is known about the specific functions carried out by different parts of these RNA molecules in the complex processes of translation. If functionally related parts of the molecules, rather than total sequence variance, are compared, biologists are in a better position to see if functions differ coordinately with sequences. For

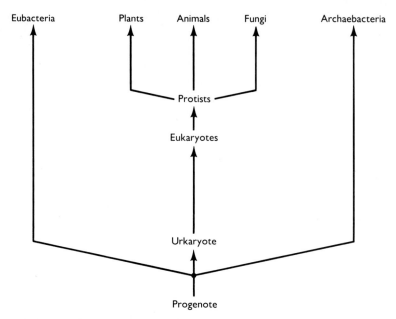

Figure 4.15. The evolutionary tree proposed by Carl Woese and George Fox separates the prokaryotic eubacteria from the prokaryotic archaebacteria on the basis of important molecular differences between the two groups, and deemphasizes their common plan of subcellular organization. All modern organisms are postulated to have had a progenote ancestor, which diverged early in evolution to produce three major lineages. Two of these lineages evolved prokaryotic organization, but the third lineage (urkaryote) evolved into the eukaryotic protists, from which plants, fungi, and animals are descended.

example, the binding of messenger RNA to the prokaryotic ribosome in the initiation of translation is mediated by base pairing between a pyrimidine-rich sequence near the 3′ end of 16S ribosomal RNA and a purine-rich region preceding the AUG initiator codon in messenger RNA. This pyrimidine-rich rRNA sequence is conserved in archaebacteria but is missing entirely in eukaryotic rRNA (Fig. 4.16). The eukaryotic ribosome must have evolved an alternative mechanism for specific recognition of mRNA initiator regions, whereas archaebacteria and eubacteria share the conserved 3′ terminus of their rRNAs. The divergent rRNA sequences throughout the molecules may be less significant in regard to translation than the shorter divergent sequences of known function that sharply distinguish prokaryotic and eukaryotic protein synthesis at the ribosome.

Very different processes are believed to be responsible for divergence in sequences that are more vital to cellular processes and molecular configuration than to less essential features. The ribosomal RNAs differ greatly in nucleotide sequence in different groups of organisms and in the mitochondrion. When the molecules are viewed on the basis of their *primary structure* (sequence), they are quite diverse; but when they are folded into their two-

dimensional *secondary structure*, all the rRNAs are remarkably similar in organization and conformation (Fig. 4.17). The sum total of nucleotide divergence
may be due more to a high level of random substitution than to natural selection. If substitutions occurred that did not significantly alter the functions and
conformation of the molecules, they might well have been incorporated as
tolerable modifications during evolution. Nucleotide substitutions that would
alter conformation or any essential functions of the molecules, however,
would be subject to relentless selection and would be incorporated only if
they were harmonious with the conformation assumed or the function performed by the sequence in that region. Nucleotide substitutions encompassing
substantial regions are well known in many kinds of molecules, but such
regions rarely are involved in the basic conformation or functions of the molecules. You may recall from the discussion of the globin genes and polypeptides that globin conformation and function remained unchanged even when
only 63 amino acid residues (and codons) were conserved out of more than
140 residues in α globin and β globin of human hemoglobin.

The proposal of a common ancestor that diverged into three main stems of
an evolutionary tree must be given careful consideration. Biologists have been
redirected from their preoccupation with the evolution of eukaryotes to
search more seriously into more ancient evolutionary events. Whether or not
the archaebacteria are as ancient a group as has been suggeted, and whether
or not the eukaryotic lineage extends to much earlier times than has been
supposed, are two of many questions that remain to be answered by further
molecular analysis. Such analyses should include more than sequence information alone. At the least, molecular biologists must look at specific functional regions of whole molecules and at the secondary and tertiary structures
of the molecules chosen for study. The evolutionary history of life is written
in its genes and gene products. Biologists can understand this writing only if
they know how to distinguish between more important and less important
differences in clues that relate the past to the present.

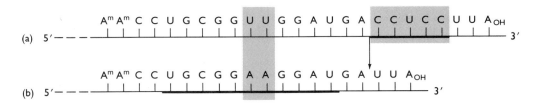

Figure 4.16. (a) The 3′ end of 16S rRNA (small subunit) in eubacteria and archaebacteria includes
a conserved sequence of the pyrimidines CCUCC, which mediates the binding of mRNA to the ribosome in the initiation of translation. (b) This sequence is missing in mammalian and other eukaryotic
small-subunit rRNA, which indicates the existence of a different mRNA–rRNA recognition mechanism
in eukaryotic translation initiation. Another difference is the presence of a pair of uracils (pyrimidines)
in bacteria but a pair of adenines (purines) in eukaryotes, in an otherwise conserved region of this
rRNA.

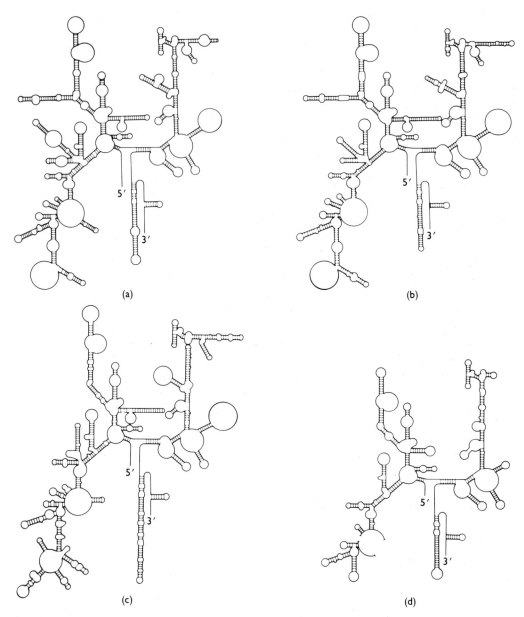

Figure 4.17. The secondary structure of 16S or 18S rRNA, from the small subunit of the ribosome, is strikingly conserved in (a) *E. coli*, a eubacterium; (b) *Halobacterium volcanii,* an archaebacterium; and (c) *Saccharomyces cerevisiae,* a eukaryotic yeast. (d) Features common to all the sequenced 16S and 16S-like rRNAs, including those from mitochondrial and chloroplast ribosomes. (From H. F. Noller, 1984. *Ann. Rev. Biochem.* 53:119, Fig. 4)

Some Eukaryotic Themes in Evolution

Unicellular organisms are quite successful in a variety of environments; many of them can reproduce more than once every hour; and they represent more than half of the total biomass on Earth. Multicellular organisms must possess different features from unicells because they exploit different resources. Among these characteristics we would expect important distinctions between the organisms' genomes and between their individual genes. A few of the more general aspects of eukaryotic life will be discussed here, and further discussions of genetic systems and groups of particular multicellular organisms will be reserved for other chapters.

Requirements for and Advantages of Multicellularity

A basic requirement for the eventual development of multicellular organisms is the association of unicells in colonies, each of which acts as a cooperative enterprise. Examples of such systems can be found in various eukaryotic species and among a few prokaryotes. The prokaryotic *myxobacteria,* or gliding bacteria, exist in colonies composed of daughter cells that remain associated after cell divisions. They secrete digestive enzymes, which are pooled and thereby allow the whole group to more efficiently break down organic solids, whose products are then absorbed by the individual cells. When the food supply is exhausted, the cells aggregate to form a multicellular "fruiting body," within which the bacteria produce spores that are resistant to adverse conditions (Fig. 4.18). With the return of more favorable conditions, the spores germinate to produce cells that associate in new colonies. Such a colony may be formed by binary fission initiated by any one of the bacteria, which clearly indicates the unicellular nature of the species. In contrast to the myxobacteria, with their undifferentiated cells in loose association, many cyanophytes can form long filamentous chains of cells that remain associated after cell divisions, and in which some of the cells can differentiate into thick-walled *heterocysts.* Atmospheric nitrogen can be processed into organic molecules within the heterocysts, and these products of nitrogen fixation are shared by undifferentiated cells in the filaments.

   Eukaryotes have evolved a number of different and more effective means by which their cells are bound together and cooperate as distinct functional parts of an organism. In almost every group of green algae, which are ancestral to the land plants, trends from unicellular to multicellular organisms are evident. In the family Volvocales, for example, species like *Chlamydomonas* exist as flagellated unicells, *Gonium* is a colonial species in which 4 to 32 flagellated cells move as a unit, and *Volvox* species may form colonies consisting of more than 50,000 cells held together by fine strands of cytoplasm (Fig. 4.19). Each cell in a *Gonium* colony is the same as any other, and new colonies can be produced by cell divisions begun by each member of the cluster. *Volvox* can easily be considered multicellular because the whole colony is

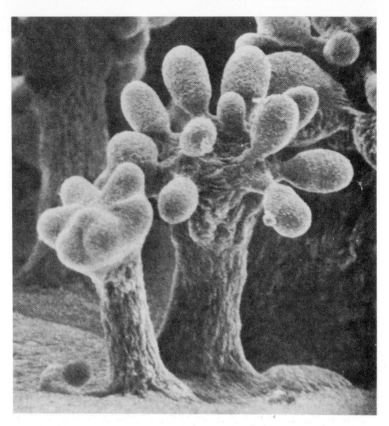

Figure 4.18. Scanning electron micrograph of the fruiting body of aggregated cells of a myxobacterium. ×600 (Photograph by J. Pangborn)

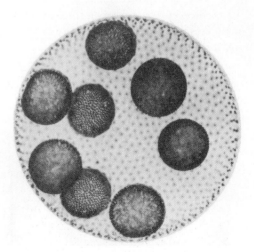

Figure 4.19. A *Volvox* colony is a sphere consisting of a single layer of 500 to more than 50,000 cells embedded in a gelatinous material, and an interior filled with watery mucilage. Some small developing daughter colonies can be seen embedded in the matrix of the parent colony.

structurally and functionally polarized. The synchronous beating of its many flagella permits *Volvox* to move in specific directions rather than spinning aimlessly. The reproductive cells, which usually are located in one region of the colony, produce small new colonies that are held within the parental hollow sphere for a time before being released as independent colonies. The individual cells cannot live independently, and the whole colony dies if it is broken up.

Large seaweeds and land plants achieve and maintain multicellular forms by means of adhesion between adjacent cell walls secreted by the living protoplasm in each cell (Fig. 4.20). Pectins and celluloses knitted together in the cell walls also help bind the individual, differentiated cells in the organism. Multicellular fungi, whose cell walls are made of chitin rather than cellulose, are held together much like plants. Animal cells do not have rigid walls, and interconnections by cytoplasmic bridges exist for only brief periods of time during early stages of embryonic development. The problem of cell adhesion in animals has been handled by a variety of devices, including *cell adhesion molecules* and their surface protein receptors; desmosomes, or specialized *adhesion junctions* between cells; and *extracellular matrix* composed of large molecules secreted by cells that remain together in the loose meshwork of secretions (Fig. 4.21).

Multicellular organisms are not chaotic aggregates of cells; they are made up of specialized communities of cells, or *tissues*, which may exist together as cooperative assemblies of an *organ*. The secrets of development have been probed for over a century, but very important insights are now being revealed

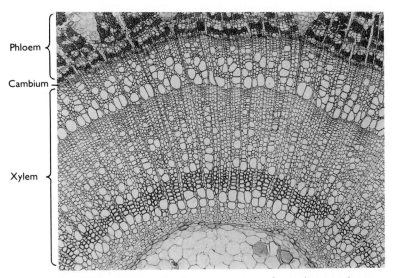

Phloem

Cambium

Xylem

Figure 4.20. Light micrograph of a cross section of woody stem showing a variety of tissues, each composed of differentiated cells bound together by their adhering cell walls. (Photograph courtesy of R. E. Triemer)

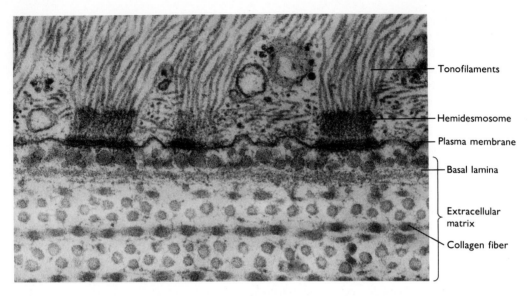

Figure 4.21. Electron micrograph of the basal portion of a larval epidermal cell from the newt *Taricha torosa*. Hemidesmosomes line the intracellular face of the plasma membrane and serve as adhesion junctions between the cell (above) and the extracellular matrix (below). Tonofilaments (intermediate filaments) radiate from each desmosomal plaque into the cell interior. The basal lamina and collagen fibers are evident in the extracellular matrix. x58,000 (Photograph courtesy of D. E. Kelly, from D. E. Kelly, 1966. *Jour. Cell Biol.* 28:51, Fig. 11. Copyright by Rockefeller University Press)

in greater detail by molecular and genetic analyses. Cells appear to differentiate in response to chemical signals that may be responsible for the spatial patterning of tissues and organs in particular locations in the body, as well as for the development of individual cells into one kind of specialized unit rather than another. Every cell in a multicellular organism carries a full set of genes, but some genes are expressed and others are not. The devices that regulate such *differential gene expression* are more complex in eukaryotes than in prokaryotes, and must underlie the programmed developmental pathways in all multicellular organisms.

The evolution of developmental pathways can be seen in its broadest perspective by simple comparisons among living species. The vertebrate animal embryo that will develop into a fish is almost the same in its earliest stages as the embryo that will develop into a human being. Different body plans take shape later in embryo development, however, and give rise to the different adult form of a fish or a human being (Fig. 4.22). From this example and many others, it is reasonable to deduce that the earliest processes of development are conserved in evolution, whereas later steps can be varied or even added to the basic foundations of embryogenesis. Genetic studies have shown that mutations affecting earlier development are not tolerated nearly as much as those involved later in development. This is understandable because almost any change in the foundations will influence all subsequent development. Mutations affecting later events can be accepted, but only if they are compatible with the existing programs and products. For example, early developmental pathways that establish the nervous system are more likely to be conserved in evolution, which is evident from the development of some sort of nervous system in almost all animals. Embellishments on the basic nervous system occur later in development, and they exist in great variety and levels of complexity among different groups of animals. None of these variations interferes with the establishment of a nervous system early in embryonic development.

The recent discovery of conserved nucleotide sequences called *homeo boxes* may provide important clues about the nature of genes responsible for spatial patterning in multicellular organisms (Fig. 4.23). Homeo boxes were identified in molecular studies of *Drosophila* homeotic mutants, in which whole

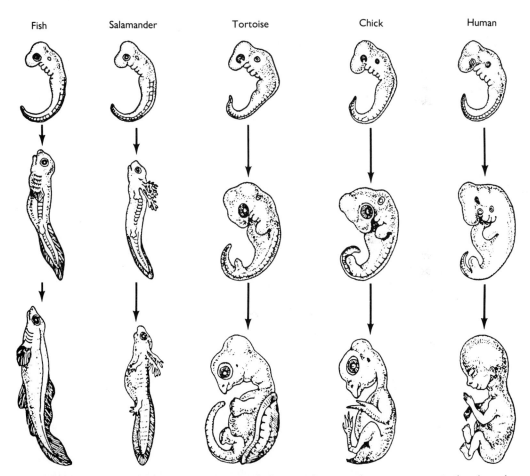

Fish   Salamander   Tortoise   Chick   Human

Figure 4.22. Early stages in development of the vertebrate embryo are very similar, but the embryos become increasingly different with increasing evolutionary distance. Differences appear much earlier in development for more distantly related organisms, but appear relatively later for more closely related organisms. (Drawing by Ernst Haeckel)

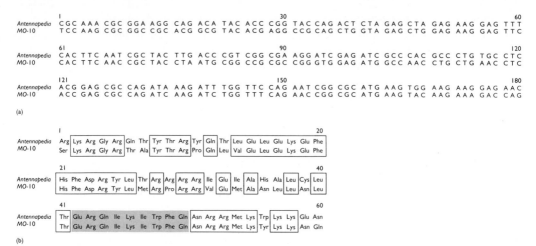

Figure 4.23. Homeo boxes encoding homeo domains in proteins from an invertebrate (*Drosophila melanogaster*) and a vertebrate animal *(Mus musculus)*. (a) The 180 nucleotides of the homeo box region of the *Drosophila* homeotic gene *Antennapedia* and the mouse gene *MO-10* are homologous to each other and to homeo boxes in a wide range of organisms. The *Antennapedia* sequence is used as a standard of comparison for all such sequences. (b) The 60 amino acids of the homeo domain encoded by the homeo box in the protein products of these genes show identity of amino acids in 45 (boxed) of the 60 units, including stretches up to 16 amino acids long (40–55). Such high resemblance indicates a common ancestry and function in all homeo domain-containing proteins. The function may be that of binding of the protein at its homeo domain to DNA. By binding to specific DNA sequences, these proteins could be involved in the regulation of gene expression. The sequence of 9 amino acids (gray tint) from unit 42 to unit 50 is thought to be the region where the protein contacts DNA in a binding interaction.

organs may develop in the wrong places on the flies. Legs rather than antennae may be formed on the head, or a second pair of wings may be produced instead of the normal stabilizer organs on the thorax. Equally interesting are the homeotic segmentation mutants, which may have body segments in addition to the head, thorax, and abdomen. Homeo-box sequences have since been found in many different kinds of animals, from simple nematode worms to human beings. Although the role of homeo-box sequences remains uncertain, it is tempting to speculate that they are conserved sequences in genes that govern spatial patterning in development. The highly conserved nature of these sequences indicates a role in basic processes with a long evolutionary history. That these may be developmental processes is suggested by their presence in genes that are concerned with the formation and location of multicellular body parts.

Multicellular organisms have the enormous advantage of carrying out their activities through the division of labor among organized sets of cells dedicated to specific tasks. A redwood may grow 300 feet tall but receive water through its roots in the soil, transport the water upward through xylem tissue to all its parts, make its organic foods in photosynthetic leaves high in the air, and transport the foods to the rest of the tree through phloem tissue. The trees in California may be 2000 years old, having replaced old cells with new cells for all those centuries. In all that time, they have been changing the soil, enriching the air with oxygen, and providing habitats for bacteria, fungi, insects, and other life forms.

Unlike the passive plants, animals move about in the constantly changing conditions of the world around them. Muscles and other machinery permit movement, but the nervous system allows an animal to respond rapidly and appropriately to ever-changing conditions by elegantly coordinated activities of different parts of the body. Behavioral patterns are genetically wired in lower organisms, but are the outcome of experience and learning as well as genetic programs in higher animals. In human beings, the complex nervous system includes a large and highly developed brain, which gives us the capacity to manage a continual stream of new problems; to communicate complicated ideas; to distinguish among past, present, and future, and thus to plan; and to adapt to a broad variety of living conditions. Through cultural as well as genetic evolution, we have come to dominate other life forms on the planet.

Genome Evolution and the C Value Paradox

Every species has a characteristic amount of DNA in its genome, which is called its **C value** and is expressed in picograms (pg), base pairs, or molecular weight of DNA. The **C value paradox** refers to the great discrepancies between the size of the genome and the morphological complexity of the species—even among some closely related species. We would expect a corresponding increase in genome size and evolutionary ranking of organisms if more genes are involved in specifying increasingly complex systems of structure, development, and life style. Some correspondence between genome size and morphological complexity is evident when C values are compared across the broadest range of organisms, but no correspondence is evident when different groups in the animal kingdom are compared (Fig. 4.24).

It takes more DNA to make an alga or a fungus (2×10^7 to 10^8 base pairs) than to make a bacterium (6×10^5 to nearly 10^7 base pairs), and a minimum of 10^8 base pairs are needed to produce a multicellular plant or animal. The C value paradox becomes evident, however, once the quantum evolutionary jump to multicellularity has occurred. Within the flowering plants alone, genome size varies from about 4×10^8 to over 4×10^{11} base pairs of DNA, which is a 500-fold difference among species that share a recent ancestry and exhibit relatively similar features. Some mollusks and insects have more DNA than most of the vertebrates, and even among the vertebrates, some amphibians have almost 100 times more DNA than any of the mammals. If more genes are needed to make the more highly evolved mammal than to make the simpler amphibian, why do mammals have less DNA than many amphibians? Why do some amphibians need nearly 10^{11} base pairs of DNA, whereas others of similar morphological complexity can manage with the 100-fold smaller genome of about 10^9 base pairs? The answers to these questions lie with another question: Does all this DNA consist of genes that code for proteins?

If all the DNA in a genome consisted of protein-specifying genes, a typical

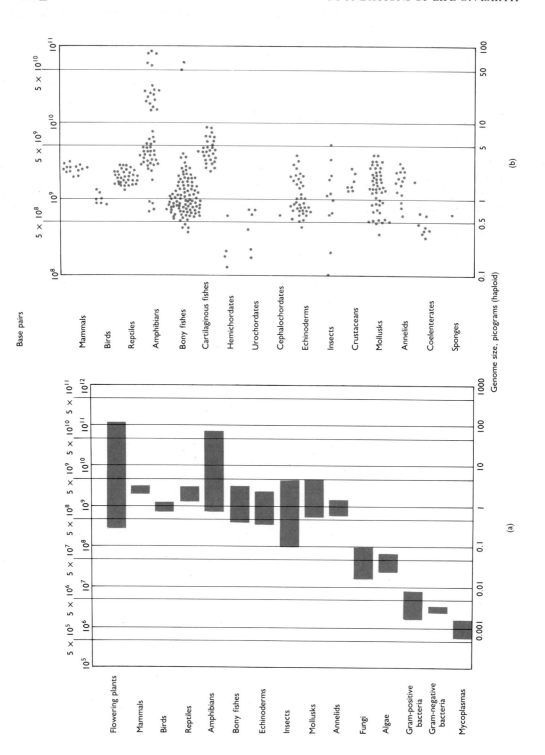

mammalian genome of 3×10^9 base pairs would include 600,000 genes of an average length of 5×10^3 base pairs. At least two lines of evidence indicate the gene number to be in the thousands or tens of thousands, not the hundreds of thousands. Gene number can be estimated from quantitative studies of different messenger RNAs produced by gene expression in different kinds of cells of the organism. In mammals, about 10,000 mRNAs are produced in a given cell type, and most of these mRNAs are the same in most of the cells that have been studied. If allowance is made for some differences in gene expression among the various kinds of cells, the total number of genes expressed by transcription in the organism is between 2 and $5 \times 10,000$.

A second and independent line of evidence comes from genetic studies. In *Drosophila*, for example, an estimated 5000 essential genes are inferred to be present, according to genetic analysis of all the loci that can be mutated in the genome of 10^8 base pairs of DNA. Based on an average gene length of 2×10^3 base pairs in this organism, it would appear that only 10^7 base pairs out of the total of 10^8 are encoded for proteins (5×10^3 genes $\times 2 \times 10^3$ base pairs per gene $= 10^7$ base pairs of DNA). Only 10 percent of *Drosophila* genomic DNA is represented by protein-specifying genes (10^7 base pairs of genes $\div 10^8$ base pairs of total DNA $= 10^{-1}$ of the genome).

If genes constitute a mere fraction of genomic DNA in eukaryotes, what is the nature of the remainder? From molecular analyses, it is known that eukaryotic genomes have varying amounts of **repetitive DNA,** much of which has no coding function. (Fig. 4.25) In general, larger genomes contain more repetitive DNA and smaller genomes contain less. In simple eukaryotes, only 10 to 20 percent of DNA is repetitive; as much as 50 percent is repetitive in many animal groups; and in very large genomes of amphibians and flowering plants, up to 80 percent of the total DNA may be repetitive. The relationship between the size of the genome and the size of the repetitive DNA component is not strict, but it does provide some help in explaining the C value paradox. Additional information comes from sequence analysis of nonrepetitive DNA.

A substantial amount of nonrepetitive DNA is distributed within and between protein-specifying genes in eukaryotes. These nonrepetitive sequences are *intergenic spacers* and *intron segments* within the genes them-

←——————————————————————————————————

Figure 4.24. Genomic DNA. (a) Some correspondence is evident between genome size (C value expressed in base pairs of DNA) and morphological complexity of a broad spectrum of organisms. It takes more DNA to make a eukaryotic alga or fungus (2×10^7 to 10^8 base pairs) than to make a prokaryotic bacterium (6×10^5 to nearly 10^7 base pairs), and at least 10^8 base pairs are required for a multicellular plant or animal. Each horizontal bar represents the range of C values recorded for each group of organisms listed. (b) C values expressed in picograms (lower scale) and in base pairs (upper scale) of DNA for various animal groups. The range of C values is much greater in some groups (such as amphibians) than in others (such as reptiles, birds, or mammals). No correspondence is evident between C values and morphological complexity in these groups, a feature referred to as the C value paradox.

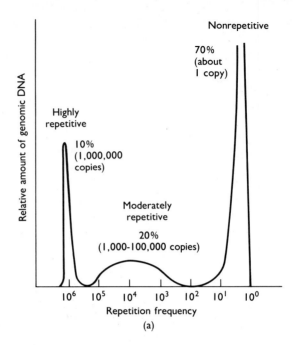

Figure 4.25.   Repetitive DNA in eukaryotic genomes. (a) The major component of genomic DNA is nonrepetitive in most cases, but an average of 30 percent of the genome consists of highly and moderately repetitive sequences. (b) Prokaryotes contain little or no repetitive sequences, whereas eukaryotes vary in their repetitive DNA content and in the proportions of moderately and highly repetitive sequences.

selves; none are known to have a coding function because they either are not transcribed or, if transcribed into messenger RNA, are excised prior to translation at the ribosome. The cluster of β-like globin genes in human chromosome 11 is spread out over 60,000 base pairs of DNA, of which only 5 percent occurs in exon coding segments and the remaining 95 percent is noncoding. The ovalbumin gene in the chicken is 7564 base pairs long, but only 1158 of these base pairs specify the 386 amino acids of this egg protein; some of the remaining DNA has a regulatory function, but most of it consists of intron segments that are excised from mRNA transcripts (Fig. 4.26).

   Biologists have a good idea from basic genetics about the kinds of processes

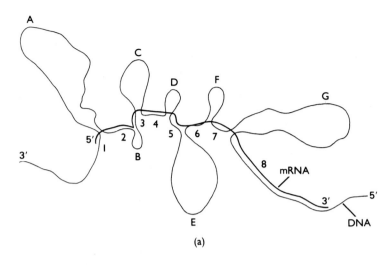

(a)

Coding sequences (exons)	Intervening sequences (introns)	Location in gene (nucleotide numbers)	Sequence size (number of nucleotides)
1		1–47	47
	A	48–1636	1589
2		1637–1821	185
	B	1822–2072	251
3		2073–2123	51
	C	2124–2704	581
4		2705–2833	129
	D	2834–3233	400
5		3234–3351	118
	E	3352–4309	958
6		4310–4452	143
	F	4453–4783	331
7		4784–4939	156
	G	4940–6521	1582
8		6522–7564	1043
			7564

(b)

Figure 4.26. The organization and processing of the gene that encodes ovalbumin protein in the chicken. (a) Tracing of an electron micrograph of a hybrid duplex molecule consisting of ovalbumin DNA and the mature rRNA transcribed and processed from this gene, showing that eight exons of the gene sequence are all that remain in the complementary mature mRNA strand. The seven introns (A–G) of the gene sequence are missing in mRNA, but are evident from the size and locations of the unpaired loops of DNA in the hybrid duplex. (b) The number of nucleotides in different parts of the gene are shown in tabulated form. Exons 1 to 8 contain a total of 1872 nucleotides, and excised introns A to G contain a total of 5692 nucleotides; the overall length of the gene is 7564 nucleotides. The polypeptide of 386 amino acids is specified by only 1161 nucleotides (1158 + 3 for STOP), and the remaining 711 nucleotides of the exon segments are regulatory sequences at the 3′ and 5′ ends of the gene, which are not translated. (Tracing from F. Gannon et al., 1979. *Nature* 278:428, Fig. 3b. Copyright © 1979 by Macmillan Journals Limited)

that can produce repetitive DNA sequences and influence their distribution in the genome. Duplication errors, gene amplifications, and other events associated with DNA replication can produce multiple copies of a nucleotide sequence. Some of the copies remain together in the chromosome, and other copies may be distributed throughout a genome by transpositions and other genetic processes leading to sequence rearrangements in chromosomes. But what are the advantages of repetitive DNA with no known coding capacities? The answers are hard to find because of current uncertainties about repetitive DNA functions. One imaginative speculation was made by Richard Dawkins, who suggested that repetitive DNA consists of sequences that benefit only itself and not the cell in which it is carried along passively by chromosome replication. He called this material "selfish DNA," which is perpetuated during evolution because of its ability to multiply and spread in the eukaryotic genome without damaging the organism that serves as its host.

   Although it is true that some repetitive DNA is dispensable, other repetitive DNA appears to be essential for survival. For example, tens of millions of copies of genetically inert, short nucleotide sequences surround the *centromere* in every chromosome of higher eukaryotic organisms (Fig. 4.27). Disruption or other damage to this repetitive DNA causes severe problems in the directed movement of chromosomes to the poles during nuclear division and may cause cell death. Its genetic inertness makes centromeric repetitive DNA a buffer against changes in this essential component of chromosome structure

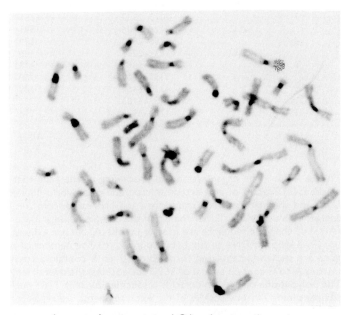

Figure 4.27. Human chromosome complement, showing stained C bands primarily at the centromere region of each chromosome. C bands identify regions of concentrated highly repetitive DNA, and occur at the centromere region of all the chromosomes in almost every eukaryotic species that has been studied.

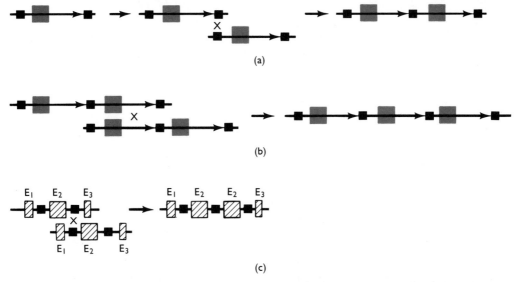

Figure 4.28. The consequences of recombination of coding sequences (gray boxes) as a result of unequal crossing over at homologous short repeated sequences of DNA (black boxes) dispersed within and between genes. The duplication of a gene by unequal crossing over may occur (a) once or (b) repeatedly to produce two or more copies of the original coding sequence, (c) Unequal crossing over involving short DNA repeats in intervening sequences (introns) between exons (E_1–E_3) may also lead to a recombinant gene containing one or more duplicated coding segments. (From A. J. Jeffreys and S. Harris, 1982, Processes of gene duplication. *Nature* 296:9, Fig. 1. Copyright © 1982 by Macmillan Journals Limited)

and activity, and, therefore, it hardly qualifies as "selfish DNA," as defined by Dawkins.

One feature of eukaryotic genome evolution about which more is known today is gene organization into exons and introns, or split genes. Shortly after the existence of split genes became known, from new methods for DNA sequencing introduced in the mid-1970s, Walter Gilbert suggested that crossing over within introns might lead to exon duplication and exchange between genes (Fig. 4.28). By such events, new genes could be created relatively quickly, and new proteins could be produced from the modified sequences. Such proteins might include enzymes, structural proteins, and other molecules of great advantage to the eukaryotic organism, and they would evolve much faster by exon duplication and exchange than by gene mutation alone.

Exon duplication during eukaryotic gene evolution is well documented, and new examples are reported more and more often from computer-based searches for sequence homologies within the gene. For example, collagens are stiff cablelike molecules secreted by various cell types and held together with these cells in connective tissues throughout the body. Individual collagen polypeptide chains contain about 1000 amino acid residues, about half of which are glycines and hydroxylated derivatives of prolines and lysines. In the chicken, the gene for 1 of the collagens is organized into at least 52 exons

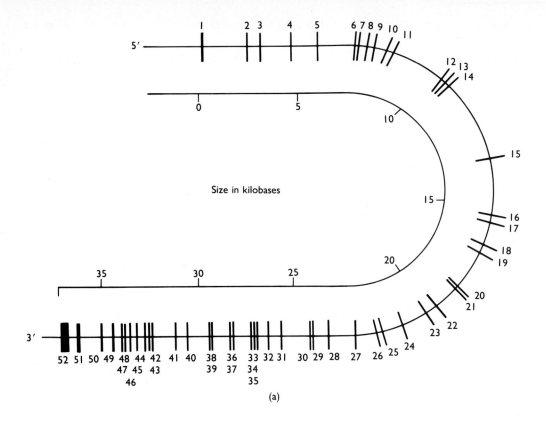

Size in kilobases

(a)

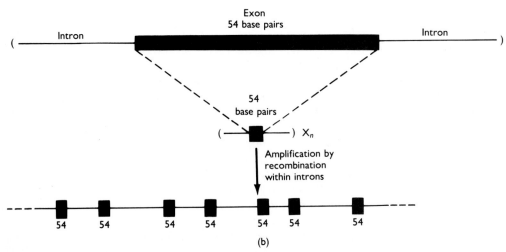

(b)

Figure 4.29. The organization of the chick α-2(type I) collagen gene. (a) The gene consists of 52 exons (black bars) interspersed with 51 introns over a length of about 38,000 base pairs of DNA. (b) Many of these exons are 54 base pairs long, which suggests that it is the length of the ancestral coding segment that duplicated many times during the evolutionary history of the gene. (From B. de Crombrugghe and I. Pastan, 1982, Structure and regulation of a collagen gene. *Trends Biochem. Sci.* 7:11, Figs. 1 and 2)

alternating with 51 introns over a length of 38,000 base pairs (Fig. 4.29). From structural and sequencing analysis of the gene, it appears that many of these exons are exactly 54 base pairs long. It is unlikely that all these exons arose merely by chance, and far more likely that they are duplicates of an ancestral coding sequence that was 54 base pairs long. Sequence divergence during the subsequent evolution of the exons has occurred by base substitutions and other ongoing mutational processes.

The genes for human low density lipoprotein (LDL) receptor and for the precursor of epidermal growth factor (EGF) have a homologous region consisting of 8 contiguous exons that range in size from 105 to 228 base pairs and are separated by introns that vary in length between 130 and 6000 base pairs. The homologous region spans more than 15,000 base pairs in the LDL receptor gene and 23,000 base pairs in the EGF precursor gene. Of the 400 amino acids encoded in this region, fully 33 percent are identical in the two proteins. Did the homologous regions evolve by convergence to become so similar, or did they evolve from a common ancestral gene that duplicated and diverged afterward? Analysis of the organization of exons and introns in the region provided the answer to the question.

Thomas Südhoff, working with Joseph Goldstein and Michael Brown, compared the 400 amino acids in the protein sequences and their corresponding nucleotide sequences. When the two protein sequences were aligned for maximum homology, five of the nine introns interrupted the coding sequences at precisely the same amino acid in the LDL receptor and the EGF precursor (Fig. 4.30). This level of exact correspondence indicates the conservation of

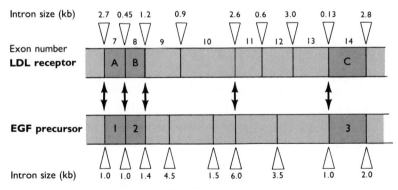

Figure 4.30. A high level of correspondence in the organization of 8 exons and 9 introns in a portion of the genes for the LDL receptor and the EGF precursor proteins, and the identity of many of the 400 amino acids in this region of both proteins, provide strong evidence for homology of the entire region in both genes. Three repeats of 40 amino acids are lettered A, B, and C in the LDL receptor protein and are numbered 1, 2, and 3 in the EGF precursor protein. Intron locations are indicated by large arrowheads, and their sizes (in kilobases) are given. The five introns that interrupt identical residues in the aligned sequences are indicated by two-headed arrows. Note the much greater discrepancy in lengths of the corresponding introns than of the corresponding exons in this portion of the two genes. (From T. C. Südhoff et al., 1985, Cassette of eight exons shared by genes for LDL receptor and EGF precursor. *Science* 228:893, Fig. 1. Copyright © 1985 by the American Association for the Advancement of Science)

exon–intron junctions in divergent descendants of a common ancestral gene, rather than coincidental similarities due to convergence. Conservation of exons and their borders was not matched, however, by the sizes and sequences of the intervening introns. As we will soon see, the great diversity of non-repetitive DNA represented by intron sequences is a predicted feature of eukaryotic split genes. The role of introns in genome evolution appears to be more concerned with exon duplication and exon exchange than with any role in C value increase corresponding to evolutionary rankings among eukaryotes.

On the Nature and Antiquity of Split Genes

In his seminal 1978 note on the possible significance of split gene organization, Walter Gilbert predicted that exons in genes code for repeated and unique amino acid sequences that correspond to useful portions of protein structure. Such portions might be functional regions, elements involved in protein folding, domains or subdomains, or any other kind of segment that can be assorted independently during evolution. Furthermore, the noncoding introns were presumed to be random sequences that by chance can change rapidly in size and order, whereas coding exons are subject to the stricter rules of natural selection. Although Gilbert restricted his discussion to eukaryote evolution, W. Ford Doolittle almost immediately followed with the suggestion that split genes might be of ancient origin and that present-day prokaryotes and simple eukaryotes had lost some of or all their introns in exchange for streamlined genomes more suitable to the advantages of rapid replication and the proliferation of progeny.

Three important questions, therefore, can be asked about split gene organization. First, do exons in genes correspond to domains or other assortable segments in proteins? Second, are size and sequence more varied in introns than in exons, reflecting their different functions and consequent evolutionary directions? Third, are split genes a eukaryotic characteristic, or do they have an earlier origin? Biologists have a variety of data, some of a tentative and preliminary nature, with which to formulate answers to these fundamental evolutionary questions.

The first question, on the correspondence between exons in genes and functional segments in protein, can be addressed by comparisons between genes and their encoded products. Single exons in a gene can often be correlated with separate functional segments in the corresponding regions of the polypeptide translation product. The N-terminal signal sequence in exported proteins is usually encoded in a single exon at the beginning of the genetic message; functionally distinct segments of immunoglobulin chains are encoded in separate exons; the three functional domains in vertebrate globin polypeptides are encoded in the three exons in all the vertebrate globin genes; and many other similar examples are known (Fig. 4.31).

Proteins with repeated structure characteristically are encoded in corresponding repeated exons, such as 3 repeats of 40 amino acids rich in cysteine residues in the LDL receptor and the EGF precursor, which are encoded in 3 repeats of the same exon in the 2 genes for these proteins. The regular turns of the helix of collagen molecules are built up from 40 repeats of an exon that specifies amino acids in a half-turn. These and similar examples of the correspondence between single and repeated exons and the single and repeated functional segments of the polypeptides provide ample support for Gilbert's hypothesis that split genes contain collections of exons that code for useful portions of protein structure. As we will see in Chapter 5, split genes may contain exons recruited from different genes, as well as exon repeats in the same gene. Split gene organization thus provides a system for exon duplication and shuffling through recombinations within intron sequences between exons. Split genes present opportunities for speeded-up protein evolution, as Gilbert proposed.

Sequence analysis helps answer the second question, on differences between intron patterns and exon patterns in split genes. Exon size is distributed across a relatively narrow range, with a maximum of about 600 nucleotides and a peak at 120 to 150 nucleotides, representing 40 to 50 amino acids. Intron sizes tend to scatter randomly, ranging from as few as 10 nucleotides in length to as many as 10,000. The variable and often unexpectedly large size of eukaryotic genes is due more to their introns than to their coding segments. Genes composed of 50,000 base pairs are not uncommon, and genes can range up to 200,000 base pairs.

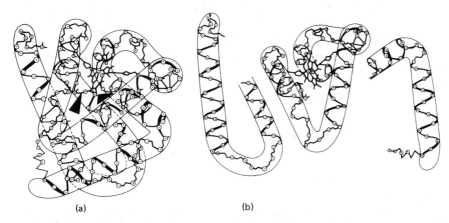

(a) (b)

Figure 4.31. The three domains in a vertebrate globin polypeptide correspond to the three exons in globin gene organization. (a) Two arrowheads show the points at which the structure of a globin chain is interrupted by the two introns in the corresponding gene sequence. (b) The two introns interrupt the polypeptide in the α-helical regions to break the polypeptide into three portions that correspond to the three exons of the globin gene. The product of the central exon surrounds the heme component of the globin chain. (From W. Gilbert, 1981, DNA sequencing and gene structure. *Science* 214:1305, Fig. 11. Copyright by the Nobel Foundation)

In the homologous region of the LDL receptor and EGF precursor genes, the 8 exons range in size from 105 to 228 base pairs, whereas the 9 introns (2 flanking and 7 between exons) vary in size from 130 to 6000 base pairs and have quite different sequences in the 2 genes. The high degree of homology in size and sequence of the three exons in the α-globin and β-globin genes clearly shows conservation over the hundreds of millions of years since the original gene duplication. The two introns in each gene, however, differ greatly in nucleotide sequence when the α-globin and β-globin counterparts are compared, and the second intron in the β-globin gene is seven times longer than that in the α-globin gene.

Such a magnitude of divergence in introns known to be parts of duplicated genes or gene regions, and the conservation of considerable homology in exons of these same genes, clearly show that introns and exons are subject to different evolutionary influences. Exons code for functionally and structurally important features of proteins and incorporate only those genetic changes that are adaptive or tolerated, in accordance with natural selection. Introns provide a space in which recombinations take place, leading to repeated and shuffled collections of exons. This intron space can be modified by chance alone, with little or no effect on recombination processes. A wide range of base substitutions, additions, and deletions can thus be incorporated comfortably into intron sequences, in the virtual absence of direction by natural selection. Another prediction based on Gilbert's hypothesis thus appears to be confirmed by the evidence collected from eukaryote split gene sequences.

It now remains for us to deal with the third question, which concerns the antiquity of exon–intron organization in eukaryotic genes. It is easy to suppose that exon–intron organization appeared during eukaryote evolution, because almost all prokaryotic genes lack introns. Simple eukaryotes have a mix of genes with and without introns, but more advanced eukaryotes have very few genes without introns. According to this view, introns have been inserted into genes that were originally uninterrupted coding sequences. At the other extreme, it has been suggested that introns are vestigial links between useful coding sequences, left over from earlier times when they tied together shorter and simpler reading frames. According to this view, the earliest organisms had split genes, and present-day prokaryotes and simple eukaryotes lost introns in response to selection pressures for rapid multiplication; it takes less time to replicate a streamlined genome than one encumbered with introns.

One approach to the problem is to compare the same gene in organisms of different evolutionary backgrounds. In vertebrate animals, every globin gene—including the myoglobin gene, which has probably existed for at least 700 million years—has three exons separated by two introns. The leghemoglobin gene in leguminous plants (beans, peas, and others) has four exons and three introns, with the central intron splitting the central exon into two discrete structural modules. Does the leghemoglobin gene represent the ancestral gene, which has been retained in the plant kingdom since the time of the ancestral eukaryote common to plants and animals, more than 1 billion years ago? If so, then vertebrate globin genes presumably have lost the "extra" intron (Fig. 4.32). In the insect *Chironomus thummi*, all 12 globin genes lack introns; have these genes been streamlined even more severely in invertebrates during the 700 million years since they shared an ancestor with the vertebrate lineage? These questions cannot be answered unambiguously unless the age and origin of the leghemoglobin gene are known. Unfortunately, they are not at the present time.

A more enlightening test to decide between the two alternative theories is to compare genes whose products existed before the separation of prokaryotes and eukaryotes. Gilbert has studied such systems by comparing genes coded for enzymes of the ancient process of glycolysis. The enzymes have the same function and tertiary structure in all cells, but their genes do not have introns in prokaryotes and simple eukaryotes, such as yeast, and have introns in vertebrate animals and flowering plants.

A particularly informative comparison would be between the structures of the same gene in flowering plants and vertebrate animals. Such a gene presumably resembles the original gene in their unicellular common ancestor more closely than it resembles its counterpart in simple eukaryotes, which

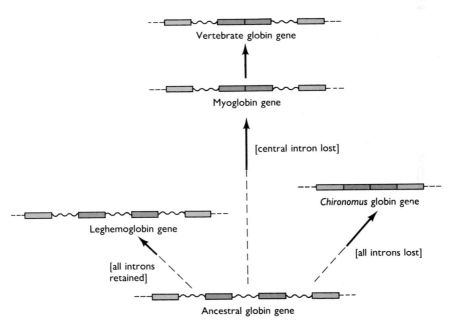

Figure 4.32. A hypothetical sequence based on the assumption of an ancient origin for exon–intron organization and of the loss of introns during the evolution of prokaryotes and simple eukaryotes. On this hypothesis, the leghemoglobin gene in leguminous plants might represent a retained ancestral state, and a more extreme case of "streamlining" the genome occurred through the loss of introns in the globin genes of the insect *Chironomus* than through the loss of introns in vertebrate globin genes during animal evolution subsequent to the divergence of plants and animals from a common ancestor about 1 billion years ago.

have evolved over a much longer time. The gene for the glycolytic enzyme *triose phosphate isomerase* is organized very similarly in maize (corn), a flowering plant, and in the chicken. At least three introns are in identical positions in both genes, and at least one intron is present in the maize gene but absent in the chicken gene. Although the information is preliminary and much more data are needed, it appears likely that a split gene existed in the unicellular ancestor more than 1 billion years ago and that introns in this gene were lost during the evolution of prokaryotes and simple eukaryotes.

In a very different approach to the question of split gene antiquity, Periannan Senapathy conducted a statistical analysis of known gene sequences in prokaryotes and eukaryotes and of random sequences generated by computer programs. The data were interpreted as showing that introns are probably vestigial links between short reading frames that have been retained since ancient times in eukaryotic organisms but were lost from prokaryotic genes. This scenario coincides with Doolittle's suggestion that split genes were present in primordial organisms and evolved to the intronless state only in prokaryotes and simple eukaryotes. The same assumption is a significant component of the "three-stem" theory, advanced by Carl Woese and others, which proposes the existence of a common ancestral progenote that diverged into three lineages in ancient times. The archaebacterial and eubacterial lineages evolved prokaryotic features, including intronless genes, but the eukaryotic lineage evolved from an urkaryote ancestor, presumably with split gene organization that has been retained to varying degrees in present-day eukaryotic organisms.

The major advantage of a streamlined genome is the greater speed of DNA replication and proliferation of progeny. The major advantage of split gene organization would appear to be the ease of exon duplication and shuffling, leading to speedier protein evolution than is possible by mutation alone. If these features indeed represent differences in selection factors guiding the divergent patterns of genome evolution, biologists are on the threshold of some fundamental insights into a very basic theme in biological evolution.

Biologists can now inquire into other features of genome evolution in relation to graduated scales of complexity in different groups of organisms. Experimental studies can be conducted with a wide variety of systems, and perhaps with model genomes in mitochondria and chloroplasts. Chloroplast genomes are relatively manageable in size, consisting of about 1.2 to 1.6 × 10^5 base pairs, whereas eukaryotic nuclear genomes range between 10^8 and 10^{11} base pairs. Furthermore, chloroplast genes consist of both split and unsplit sequences, and comparisons are possible with cyanophyte and prochlorophyte genomes, which are believed to be phylogenetically related to chloroplast genomes. Mitochondrial genomes in most eukaryotes are even smaller than chloroplast genomes, and provide the possibility of comparison between streamlined genomes, as in animal mitochondria, and mitochondrial genomes with intergenic spacers and both split and unsplit genes, as in fungi. Models such as these would also permit molecular biologists to compare nuclear and organelle genomes of totally different organization—for example, vertebrate animals' nuclear genomes, bursting with repetitive DNA and split genes, and mitochondrial genomes, which are the most streamlined ones known. It seems that, once again, the new methods of molecular biology have opened up avenues for experimental studies that were never dreamed of only a decade ago. It is truly an exciting time in all of biology, particularly in evolutionary studies.

Summary

All cellular life shares the features of plasma membrane boundary of the living cytoplasm, enzymatically catalyzed metabolism, and a genetic apparatus by which encoded genetic information is transcribed and translated. Prokaryotes are unicellular, asexual organisms lacking mitosis and meiosis, and harbor their DNA in a nucleoid not separated by a membrane from the surrounding cytoplasm. Eukaryotes may be multicellular or unicellular, sexual or asexual, and they possess a membrane-bounded nucleus containing DNA in choromosomes. Alternative theories exist for the origin of eukaryotic cells from ancestral progenote or prokaryote forms by internalization of infolded cellular membranes (autogenous origins) or by endosymbiosis (exogenous origins). The evidence favoring the exogenous origin of chloroplasts is strong, but is ambiguous or absent for other parts of the eukaryotic cell.

Two major evolutionary trees have been proposed as representations of phylogenetic relationships among kingdoms of organisms. The five-kingdom system is based on the descent of eukaryotic protists from ancestral prokaryotic monerans, and the subsequent divergence of plants, animals, and fungi from different protistan lineages. The "three-stem" system posits a hypothetical progenote as the common ancestor of all existing life. Divergence produced an archaebacterial stem and a eubacterial stem, both of which evolved prokaryotic cellular traits, and a hypothetical urkaryote that evolved into eukaryotic life. Prokaryotes are thus not considered to be the ancestors of eukaryotes; all three stems are presumed to be of equal ancient age, according to interpretations of their molecular features.

Among many evolutionary themes in the history of eukaryotes are the emergence of multicellularity, a significant increase in the amount of genomic DNA (C value) compared with that in prokaryotes, and split gene organization. Multicellularity provides for a division of labor among organized sets of cells (tissues, organs) dedicated to specific tasks in the individual organism. Although present in simplified form in some colonial organisms, multicellularity in more evolutionarily advanced grades is established by means of shared cell wall secretions (plants and fungi) or adhesion molecules and structures that bind cells (animals). Programmed developmental pathways are the outcome of differential gene expression, and involve chemical signals and spatial patterning during development and differentiation. Genome evolution did not proceed by the increase in gene numbers alone, which is evident by the general absence of correspondence between C values and evolutionary rankings (C value paradox). A large fraction of the eukaryotic genome consists of noncoding DNA, in introns within genes and in spacers between genes, whose functions remain unclear for the most part. Split genes are made up of exons, which are coding segments that usually specify functional or structural domains in proteins, and introns, which are noncoding sequences between exons. It is possible that split gene organization is of great antiquity. If this is true, introns have been lost in prokaryotes and from many genes in simple eukaryotes, but have been retained during evolution leading to the most highly evolved plants and animals. Intronless DNA can be replicated rapidly, an advantage in prokaryotes and simple eukaryotes. Split genes take longer to be replicated, but they provide the potential for exon shuffling and thus for more rapid evolution of proteins than by mutation alone.

References and Additional Readings

Alexander, F., P. R. Young, and S. M. Tilghman. 1984. Evolution of the albumin: α-fetoprotein ancestral gene from the amplification of a 27-nucleotide sequence. *Jour. Mol. Biol.* 173:159.

Almassy, R. J., and R. E. Dickerson. 1978. *Pseudomonas* cytochrome c_{551} at 2.0 Å resolution: enlargement of the cytochrome *c* family. *Proc. Natl. Acad. Sci. U.S.* 75:2674.

Antoine, M., and J. Niessing. 1984. Intron-less globin genes in the insect *Chironomus thummi thummi. Nature* 310:795.

Attardi, G. et al. 1986. Seven unidentified reading frames of human mitochondrial DNA encode subunits of the respiratory chain NADH dehydrogenase. *Cold Spring Harbor Sympos. Quant. Biol.* 51:103.

Blake, C. C. F. 1984. Exons and the evolution of proteins. *Trends Biochem. Sci.* 8:11.

Brandon, C.-I. et al. 1984. Correlation of exons with structural domains in alcohol dehydrogenase. *EMBO Jour.* 3:1307.

Brimacombe, R. 1984. Conservation of structure in ribosomal RNA. *Trends Biochem. Sci.* 9:273.

Cavalier-Smith, T. 1975. The origin of nuclei and of eukaryotic cells. *Nature* 256:463.

Cavalier-Smith, T. 1987. The origin of cells: a symbiosis between genes, catalysts, and membranes. *Cold Spring Harbor Sympos. Quant. Biol.* 52:805.

Cavalier-Smith, T., ed. 1985. *The Evolution of Genome Size.* New York: Wiley.

Chambon, P. 1981. Split genes. *Sci. Amer.* 244(5):60.

Darnell, J. E. 1983. The processing of RNA. *Sci. Amer.* 249(4):90.

de Crombrugghe, B., and I. Pastan. 1982. Structure and regulation of a collagen gene. *Trends Biochem. Sci.* 7:11.

Doolittle, R. F. 1985. Proteins. *Sci. Amer.* 253(4):88.

Doolittle, W. F. 1980. Revolutionary concepts in evolutionary cell biology. *Trends Biochem. Sci.* 5:146.

Doolittle, W. F. 1987. What introns have to tell us: hierarchy in genome evolution. *Cold Spring Harbor Sympos. Quant. Biol.* 52:907.

Dover, G. A., and R. B. Flavell, eds. 1982. *Genome Evolution.* New York: Academic Press.

Dyer, B. D., and R. Orr, eds. 1987. *The Origin of Eukaryotic Cells*. New York: Van Nostrand Reinhold.

Fox, G. E. et al. 1980. The phylogeny of prokaryotes. *Science* 209:457.

Frederick, J. F., ed. 1981. *Origins and Evolution of Eukaryotic Intracellular Organelles*. *Ann N.Y. Acad. Sci.* 361.

Gall, J. G. 1981. Chromosome structure and the C value paradox. *Jour. Cell Biol.* 91:3s.

Gehring, W. J. 1987. Homeo boxes in the study of development. *Science* 236:1245.

Gehring, W. J., and Y. Hiromi. 1986. Homeotic genes and the homeobox. *Ann. Rev. Genet.* 20:147.

Gilbert, W. 1978. Why genes in pieces? *Nature* 271:501.

Gilbert, W. 1981. DNA sequencing and gene structure [Nobel lecture]. *Science* 214:1305.

Gilbert, W. 1985. Genes-in-pieces revisited. *Science* 228:823.

Gilbert, W. 1987. The exon theory of genes. *Cold Spring Harbor Sympos. Quant. Biol.* 52:901.

Gilbert, W., M. Marchionni, and G. McKnight. 1986. On the antiquity of introns. *Cell* 46:151.

Gō, M., and M. Nosaka. 1987. Protein architecture and the origin of introns. *Cold Spring Harbor Sympos. Quant. Biol.* 52:915.

Gray, M. W., and W. F. Doolittle. 1982. Has the endosymbiont hypothesis been proven? *Microbiol. Rev.* 46:1.

Grivell, L. A. 1983. Mitochondrial DNA. *Sci. Amer.* 248(3):78.

Heinhorst, S., and J. M. Shively. 1983. Encoding of both subunits of ribulose 1,5-bisphosphate carboxylase by organelle genome of *Cyanophora paradoxa*. *Nature* 304:373.

Hoffmann, H.-P., and C. J. Avers. 1973. Mitochondrion of yeast: ultrastructural evidence for one giant, branched organelle per cell. *Science* 181:749.

Jaynes, J. M., and L. P. Vernon. 1982. The cyanelle of *Cyanophora paradoxa*: almost a cyanobacterial chloroplast. *Trends Biochem. Sci.* 7:22.

Jeffreys, A. J., and S. Harris. 1982. Processes of gene duplication. *Nature* 296:9.

Marchionni, M., and W. Gilbert. 1986. The triosephosphate isomerase gene from maize: introns antedate the plant–animal divergence. *Cell* 46:133.

Margulis, L. 1981. *Symbiosis in Cell Evolution*. San Francisco: Freeman.

Margulis, L., and K. V. Schwartz. 1987. *Five Kingdoms*, 2nd ed. New York: Freeman.

McLachlan, A. D. 1987. Gene duplication and the origin of repetitive protein structures. *Cold Spring Harbor Sympos. Quant. Biol.* 52:411.

Nei, M., and R. K. Koehn, eds. 1983. *Evolution of Genes and Proteins*. Sunderland, Mass.: Sinauer.

Ozeki, H. et al. 1987. Genetic system of chloroplasts. *Cold Spring Harbor Sympos. Quant. Biol.* 52:791.

Ruddle, F. H., C. P. Hart, and W. McGinnis. 1985. Structural and functional aspects of the mammalian homeo-box sequences. *Trends Genet.* 1:48.

Senapathy, P. 1986. Origin of eukaryotic introns: a hypothesis, based on codon distribution statistics in genes, and its implications. *Proc. Natl. Acad. Sci. U.S.* 83:2133.

Stanier, R. Y. 1970. Some aspects of the biology of cells and their possible evolutionary significance. *Sympos. Soc. Gen. Microbiol.* 20:1.

Straus, D., and W. Gilbert. 1985. Genetic engineering in the precambrian: structure of the chicken triosephosphate isomerase gene. *Mol. Cell Biol.* 5:3497.

Südhoff, T. C. et al. 1985. Cassette of eight exons shared by genes for LDL receptor and EGF precursor. *Science* 228:893.

Van Valen, L. M., and V. C. Maiorana. 1980. The archaebacteria and eukaryotic origins. *Nature* 287:248.

Whittaker, R. H. 1969. New concepts of kingdoms of organisms. *Science* 163:150.

Woese, C. R. 1981. Archaebacteria. *Sci. Amer.* 244(6):98.

Woese, C. R., and G. E. Fox. 1977. Phylogenetic structure of the prokaryotic domain: the primary kingdoms. *Proc. Natl. Acad. Sci. U.S.* 74:5088.

Yang, D. et al. 1985. Mitochondrial origins. *Proc. Natl. Acad. Sci. U.S.* 82:4443.

PART II

Evolutionary Processes

PART II

CHAPTER 5

Genetic Diversity

Even a casual observation of life on Earth reveals an astonishingly high level of diversity. Millions of species exist today, and countless species have become extinct. These varieties of life are genetically different from one another, and a number of genetic processes contribute to this great profusion of unique species. In this chapter, we will discuss three categories of diversity: (1) differences in the same or related genes; (2) genotypic diversity, resulting from rearrangements of existing sets of alleles; and (3) phenotypic diversity, arising from regulation of gene expression. By genetic analysis, molecular analysis, and other means, a large body of evidence has been gathered showing that much genetic diversity exists in the world of life and that ongoing processes apply equally well to past, present, and future variety. Indeed, this basic knowledge has been used to genetically engineer new variety.

Mechanisms that Produce Genic Diversity

The modern concept about the origin of inherited variation was first proposed in 1901 by Hugo De Vries, the Dutch botanist who was one of the three rediscoverers of Gregor Mendel's ground-breaking but neglected 1865 report on his studies with garden peas. De Vries presented considerable evidence in support of his theory that sudden heritable changes, which he called **mutations**, are responsible for inherited variations. Curiously, many of the variants described by De Vries in the evening primrose, *Oenothera*, are products of unusual chromosome behavior in these plants, but they do not invalidate the basic concept of mutation. And modern molecular studies have shown that genes may arise by processes other than mutation.

Spontaneous Mutations Produce New Alleles

Changes in one or more nucleotides of a gene sequence produce alternative forms, or **alleles,** of the gene. These modifications are perpetuated by the replication and transmission of DNA in its original and mutated forms to successive generations of cells or organisms. Spontaneous mutations have no known cause, by definition, but *copying errors during DNA replication* may be responsible for a large fraction of allelic variation (Fig. 5.1). It is well known that radiation and some chemicals in the environment are responsible for a substantial fraction of spontaneous mutations. Genetic studies demonstrated that mutations can be induced by such agents in laboratory populations.

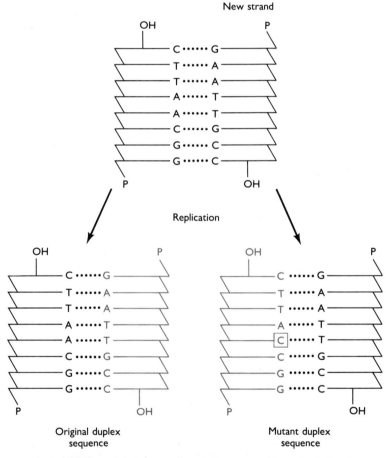

Figure 5.1. Copying errors during DNA replication are thought to cause a large fraction of spontaneous mutations, as could occur if the new strand (gray) incorporated a C instead of an A opposite T in the parental strand of a (mutant) duplex DNA. Error-free replication produces an exact copy of the original duplex.

Hemoglobin type	Residue number				
	143	144	145	146	STOP codon
Normal β–chain nucleotide sequence amino acid sequence	—CAC—AAG—UAU—CAC—UAA —His — Lys — Tyr — His — COOH				
Hb Rainier (missense)	—CAC—AAG—**UGU**—CAC—UAA —His — Lys — **Cys** — His — COOH				
Hb McKees Rocks (nonsense)	A —CAC—AAG—**UAG** —His — Lys —COOH				

Figure 5.2. Human hemoglobin variants illustrating missense and nonsense mutations at the carboxy terminus of the β-globin chain. In Hb Rainier, an amino acid different from the wild-type residue is specified; in Hb McKees Rocks, the β chain is shortened by two amino acids due to a change from a codon for the amino acid tyrosine to a STOP codon (UAA or UAG).

Experimental studies clearly show that mutations are (1) *rare*, occurring at average rates of about once in 10^5 to 10^8 cell generations for most genes; (2) *random*, happening at any time in any gene in any cell without regard to benefit to that cell; (3) *recurrent*, permitting the calculation of mutation rates for individual genes; and (4) *reversible*, producing new alleles from originals by *forward mutations* or restoring the original sequence by *reverse mutations* of alternative sequences. All these characteristics of mutations have been verified repeatedly in studies of mutations induced by physical and chemical agents. Furthermore, these mutagenic agents increase the mutation rates proportionately, so mutants can be analyzed in far greater numbers than arise by only spontaneously occurring changes. Indeed, it is possible to alter selected nucleotides in genes by methods of *site-directed mutagenesis*, and not rely on chance alone to produce useful allelic variants for experimental studies.

Mutations resulting from nucleotide substitution may or may not be evident in the phenotypic expression of the organism. If the substitution creates a codon that specifies an amino acid different from the original, and if the original amino acid is critical to the conformation or function of the protein, the mutation may be harmful to the organism—even lethal. For example, the change from CTC to CAC in the codon for amino acid 6 in human β globin leads to the presence of valine instead of glutamic acid, producing sickle-cell hemoglobin in place of normal hemoglobin. Individuals with two copies of the sickle-cell allele develop the symptoms of sickle-cell anemia, which may lead to premature death. A substitution of 1 base in 1 codon specifying 1 amino acid out of 146 residues in human β globin has an unmistakable and deleterious phenotypic effect in this particular case. Similar *missense mutations* (Fig. 5.2) may have no effect at all, producing either a different amino acid with no apparent effect on the protein function or a synonymous codon specifying the same amino acid in the mutant and the **wild-type**, or predominant, form. If nucleotide substitution involves a change from an amino acid–

specifying codon to a STOP codon, this *nonsense mutation* may cause shortened defective proteins to be produced because of the premature termination of the original coded message.

The whole constellation of mutations and the great range of their effects, from none to profound, are as characteristic of **regulatory sequences** as of sequences for **structural genes** encoding proteins for the construction and functions of the organism. Mutations in regulatory sequences can be distinguished by alterations in gene expression with no change in the structure of the protein coded by the gene. Alterations in the **promoter,** for example, may affect the binding of RNA polymerase and the initiation of transcription of the gene sequence. If the sequence were to be transcribed and then translated, the protein would be perfectly normal. Other effects of regulatory elements on gene expression will be discussed later in the chapter, but the important point to note here is that mutations of importance in evolution are not restricted to structural genes.

Long before biologists knew that DNA is the genetic material and adopted molecular approaches to genetic analysis, the whole foundation for genetics was established by classical transmission analysis of the gene and its behavior. These studies were based on interpretations of the transmission patterns of allele distribution to the progeny from genetically marked parents, either by standard crosses or by pedigree analysis. Many of these studies had a direct bearing on evolutionary questions, including the consequences of mutations on variation in populations of organisms. Luther Stadler, who was the first to show that x-rays induce mutations in plants, and Herman Muller, who was the first to show that x-rays induce mutations in animals, provided data on the relative benefit or harmfulness of mutations.

Stadler assessed the effects of mutations on barley plants and calculated that only about 1 per 1000 mutations is beneficial to the organism. These mutations improve seed yield, drought and pest resistance, and other features. This figure is still used in discussions of evolutionary phenomena. Muller designed an experimental method by which he determined the relative proportions of three operational classes of mutations: (1) those that lead to some *morphological* change; (2) those that are *detrimental,* as shown by reduced life expectancy; and (3) those that are *lethal,* as shown by death before reaching the age of reproduction. Muller's **ClB method,** named for the genetic markers in stocks of *Drosophila melanogaster* he constructed specifically for these experiments, was designed originally to analyze mutations on the X chromosome (Fig. 5.3). The method was extended later to analyze nonsex chromosomes (autosomes), and it can be applied equally well to the study of spontaneous and of induced mutations.

By far the most frequent of these three classes of mutations are those that are detrimental to the organism. Whatever the specific effect of these mutations, the mutants are less successful before and after reaching the adult stage than are individuals with the wild-type alleles. These data provided the first indications of the **genetic load** of harmful hidden variation carried in virtually every diploid sexually reproducing species, including human beings. It raises

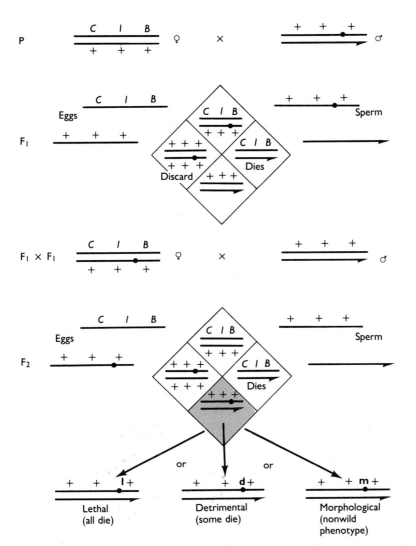

Figure 5.3. The frequency of spontaneous or induced mutations in the X chromosome of Drosophila melanogaster can be analyzed quantitatively and qualitatively by the CIB method. C is a crossover-suppressing element; I is a recessive lethal; and B is a dominant morphological (bar eye) marker for the whole chromosome. In the F_2 progeny, the sex ratio is expected to be 2 females:1 male, because CIB males die. Depending on the ratio actually observed, the X-linked mutation (large dot) may be identified as a lethal, detrimental, or morphological type. Data from such experiments can also be used to calculate the spontaneous mutation rate of a particular X-linked locus or of a category of mutations, such as recessive lethals.

our concern about excesses of radiation and chemical mutagens in the environment and food supplies. These agents increase the rate of spontaneous mutation, and a large fraction of the increased number of mutations will be detrimental, as shown by Muller's and other experimental studies.

Genetic analysis of the nature of and the relative frequency of mutations in populations has also had a strong bearing on the quantitative approach to evolutionary questions. Such quantitative analysis has clearly shown that mutation alone cannot account for the *rate* of evolutionary change. Other mechanisms for producing genetic diversity are far more effective than mutation in influencing the speed of evolution. But mutation is responsible for generating *new* genetic information in the form of new allelic alternatives and, ultimately, of new gene sequences that encode new gene products. A world without mutation is a nonevolving world, and the Earth would still be home to the simplest primeval life rather than the diverse organisms of past and present times if not for mutation.

Exon Shuffling Produces New Genes from Old Sequences

As was discussed in Chapter 4, the duplication of exons by means of recombination within introns appears to be an established evolutionary pattern, along with the duplication of whole genes through intergenic recombination, miscopying during replication, and other genetic processes. Nucleotide substitution and other kinds of mutation lead to sequence divergence of originally identical duplicated genes or exons. Divergence is expected because of the random nature of mutation, but the particular sites of divergence and the extent of evolved differences depend to a large degree on the agency of natural selection. Harmful or lethal changes ordinarily will not be perpetuated, but beneficial or neutral mutations can be incorporated and accumulate over time.

Walter Gilbert's suggestions about the nature and advantages of exon–intron gene organization in eukaryote evolution included the exciting idea that exon exchange between genes would be facilitated by recombination within introns. Such **exon shuffling** might produce new genes relatively quickly, and new proteins from the modified sequences. New proteins would include enzymes, structural proteins, and other molecules of advantage to the eukaryotic organism. These proteins would thereby evolve much faster by exon shuffling than by gene mutation alone. An exon that codes for a receptor binding site in one gene might be introduced into another gene and provide the binding site in one step.

Strong support for Gilbert's suggestion was first provided in 1985 by Joseph Goldstein, Michael Brown, and their colleagues in studies of homologous regions in the human LDL receptor and a number of other gene products with functionally similar parts. The LDL receptor in the plasma membrane binds the cholesterol transport protein LDL (low density lipoprotein) and helps it enter the cell, where it is processed. The LDL receptor gene is more than 45,000 bases long and contains 18 exons, which correlate with functional domains in the protein. One domain is a typical hydrophobic signal sequence in exported proteins; these first 21 amino acids are cleaved from the precursor, leaving 839 amino acids in 5 recognizable domains of the mature protein.

Of the 18 exons in the LDL receptor gene, 13 encode protein sequences that are homologous to sequences in other proteins (Fig. 5.4). Five of these 13 exons include repeats of 40 residues each, and this unit is homologous to a sequence of 40 amino acids in plasma protein C9 (factor IX), a component of the complement system. Three of the 13 exons encode a repeated sequence shared with the precursor of epidermal growth factor (EGF), which is a hormone that stimulates cell division, and with 3 different blood-clotting proteins. The remaining 5 exons of the group of 13 encode nonrepeated sequences that are shared with the EGF precursor. These homologies indicate that the LDL receptor is a mosaic of exons shared with other genes. Such extensive homology is unlikely to have arisen by chance in independently evolving genes. It is far more likely that it is the result of exon shuffling, leading to the incorporation of the same coding segments in genes of different origin. These segments code for regions with similar functions, such as binding, in different proteins.

Mechanisms that Produce Genotypic Diversity

The **genotype** is the actual genetic constitution of an organism, as distinguished from its expressed features, or **phenotype**. The genotype is usually indicated by notation of the particular alleles present for genes being studied; for example, the genotypes *AA* and *Aa* may be expressed as normal pigmentation, in contrast with an albino phenotype expressed in recessives with the *aa* genotype. Given the same set of genes in different combinations of alleles, an enormous variety of genotypes may exist in a single species. Among the 5 billion people in the world today, no two individuals have the same genotype unless they are identical twins or other sibling sets produced from the same fertilized egg. Genotypic diversity in viruses and in cellular life is the consequence of various genetic processes.

Independent Assortment and Genetic Recombination of Alleles

The basic premises of **independent assortment** of members of different pairs of alleles were established by Mendel's experiments with garden peas, and they stand fast to the present day. The familiar 9:3:3:1 ratio of phenotypic classes in the F_2 progeny of a dihybrid cross is the outcome of independent assortment of members of different pairs of alleles carried on different chro-

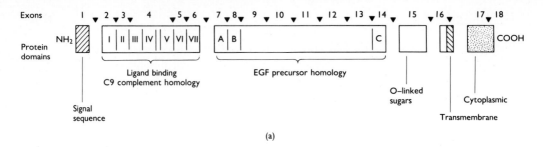

(a)

Protein	Species	Residue	Amino acid sequence
LDL receptor (A)	Human	297–331	C - - - L D N N G G C S H V C . (8) . C L C P D G F Q L V A Q - R R C
LDL receptor (B)	Human	337–371	C - - - Q D P - D T C S Q L C . (8) . C Q C E E G F Q L D P H T K A C
LDL receptor (C)	Human	646–690	C E R T T L S N G G C Q Y L C . (14) . C A C P D G M L L A R D M R S C
EGF precursor (1)	Mouse	366–401	C - - - A T Q N H G C T L G C . (8) . C T C P T G F V L L P D G K Q C
EGF precursor (2)	Mouse	407–442	C - - - P G N V S K C S H G C . (8) . C I C P A G S V L G R D G K T C
EGF precursor (3)	Mouse	444–482	C - - S S P D N G G C S Q I C . (9) . C D C F P G Y D L Q S D R K S C
EGF precursor (4)	Mouse	751–786	C - - - L Y R N G G C E H I C . (8) . C L C R E G F V K A W D G K M C
Factor X	Human	89–124	C - - - S L D N G D C D Q F C . (8) . C S C A R G Y T L A D N G K A C
Factor IX	Human	88–124	C - - - N I K N G R C E Q F C . (9) . C S C T E G Y R L A E N Q K S C
Protein C	Bovine	98–133	C - - - S A E N G G C A H Y C . (8) . C S C A P G Y R L E D D H Q L C

(b)

A	Alanine
C	Cysteine
D	Aspartic acid
E	Glutamic acid
F	Phenylalanine
G	Glycine
H	Histidine
I	Isoleucine
K	Lysine
L	Leucine
M	Methionine
N	Asparagine
P	Proline
Q	Glutamine
R	Arginine
S	Serine
T	Threonine
V	Valine
W	Tryptophan
Y	Tyrosine

(c)

Figure 5.4. Exon organization and protein domains in the human LDL receptor, and its homologies with other proteins. (a) Six protein domains in the LDL receptor protein are identified beneath the heavy-outline boxes. The 7 cysteine-rich 40–amino acid repeats in the LDL binding domain are shown by roman numerals I to VII. The three cysteine-rich repeats in the EGF precursor homology domain are lettered A to C. Introns interrupt the coding regions at locations shown by arrowheads, between exons numbered I to 18. (b) Alignment of the amino acids in segments A, B, and C from the LDL receptor with homologous regions from the EGF precursor and several proteins of the blood clotting system. Amino acids in the boxes are those that occur at a given position in more than 50 percent of the sequences. The number of amino acids composing the variable region in the middle of each sequence is shown in parentheses. (c) The standard single-letter designation of the amino acids. (From T. C. Südhoff et al., 1985, Cassette of eight exons shared by genes for LDL receptor and EGF precursor. *Science* 228:815, Figs. 5 and 8. Copyright © 1985 by the American Association for the Advancement of Science)

mosomes in the genome (Fig. 5.5). Later studies revealed parallels between gene behavior and chromosome behavior, which indicated that genes are located in chromosomes, first theorized by William Sutton in 1902. As chromosomes assort independently during meiosis, so do the alleles they carry (Fig. 5.6).

An astonishingly high level of genotypic diversity can be generated by independent assortment alone, but only in sexual organisms. Meiosis is absent in asexual species, including all the prokaryotes. In sexual species, the extent of independent assortment varies with the amount of allelic diversity in organisms and with the number of chromosomes in the genome. Mendel just happened to study seven inherited traits that are expressions of one gene on each of the seven pairs of chromosomes in the garden pea (Table 5.1). If he had made crosses between parents differing in each of these 7 *unlinked* genes, a total of 2187 genotypic classes would have been produced in the F_2 progeny. His experimental maximum of a difference in only 3 pairs of alleles between parents produced a total of 27 genotypic classes in the F_2. The number of expected genotypic classes is 3^n, where n is the number of different unlinked pairs of alleles involved (Table 5.2).

It is obvious that a species with three pairs of chromosomes will produce far fewer genotypic classes by independent assortment than will a species with a higher chromosome number. Similarly, diversity depends on the presence and number of alleles of the genes on these chromosomes. In these discussions, we have taken the simplest situation of one pair of alleles per gene.

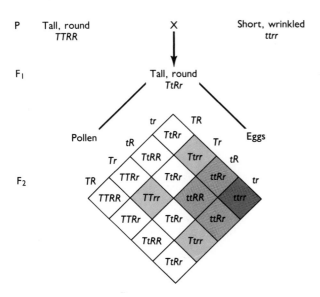

Figure 5.5. The independent assortment of members of pairs of alleles for height *(T, t)* and seed shape *(R, r)* yields a 9:3:3:1 ratio of phenotypic classes in the F_2 progeny of F_1 heterozygous parents; $9/16$ tall, round *(T-R-)*: $3/16$ tall, wrinkled *(T-rr)*: $3/16$ short, round *(ttR-)*: $1/16$ short, wrinkled *(trrr)*.

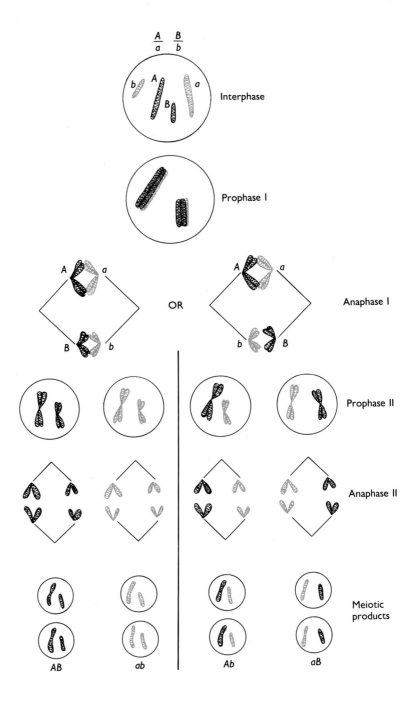

Figure 5.6. During meiosis, members of homologous pairs of chromosomes assort independently during anaphase I reduction in chromosome number. The independent assortment of homologous chromosomes is responsible for the independent assortment of members of pairs of alleles in these chromosomes.

TABLE 5.1
Chromosome Numbers of Various Organisms

Organism	Diploid Number
Human (Homo sapiens)	46
Chimpanzee (Pan troglodytes)	48
Rhesus monkey (Macaca mulatta)	42
Cat (Felis domestica)	38
Dog (Canis familiaris)	78
Horse (Equus caballus)	64
Cow (Bos taurus)	60
Mouse (Mus musculus)	40
Chicken (Gallus gallus)	78
Toad (Xenopus laevis)	36
Leopard frog (Rana pipiens)	26
Housefly (Musca domestica)	12
Fruit fly (Drosophila melanogaster)	8
Australian ant (Myrmecia pilosula)	2
Nematode (Caenorhabditis elegans)	11 ♂, 12 ♀
Garden pea (Pisum sativum)	14
Corn (Zea mays)	20
Onion (Allium cepa)	16
Potato (Solanum tuberosum)	48
Sunflower (Helianthus annuus)	34
Bread wheat (Triticum aestivum)	42
Rice (Oryza sativa)	24
Baker's yeast (Saccharomyces cerevisiae)	34
Red bread mold (Neuospora crassa)	14

TABLE 5.2
Relationship Between the Number of Unlinked (independently assorting) Pairs of Alleles and the Number of Kinds of Gametes, Gamete Combinations at Fertilization, Genotypic Classes, and Phenotypic Classes that Can Be Produced by Heterozygotes

Number of Unlinked Allele Pairs (genes)	Number of Kinds of Gametes Formed by Each Sex	Number of Gamete Combinations Produced by Random Fertilization	Number of Genotypic Classes in Progeny	Number of Phenotypic Classes in Progeny
1	2	4	3	2
2	4	16	9	4
3	8	64	27	8
4	16	256	81	16
5	32	1,024	243	32
6	64	4,096	729	64
7	128	16,384	2,187	128
8	256	65,536	6,561	256
9	512	262,144	19,683	512
10	1,024	1,048,576	59,043	1,024
n	2^n	4^n	3^n	2^n

But dozens of alleles do exist for many known genes. Any base substitution in the hundreds or thousands of bases in the gene sequence produces a different allele.

The great diversity in sexual species not only is made possible by independent assortment, but also is generated *regularly* in every meiotic division during the lifetime of the individual, and in every generation. Truly, the appearance of sexual reproduction in eukaryotes opened an evolutionary reservoir of continual and extensive production of genetic diversity, the raw material for evolutionary change. None of this variety represents new information, however, because it is existing alleles that are reassorted. An *AA* or *aa* genotype could give rise to *Aa,* if one of the alleles were to mutate. But the pace of mutation is slow, and mutation is random. By independent assortment, *Aa* can be produced regularly and in large numbers of individuals in any sexual population in any generation. If this concept is extended to the hundreds of alleles in the average species and to the countless possible genotypes, it is clear that independent assortment alone produces a vast amount of genotypic diversity in sexual species.

The genome contains many more genes than chromosomes, and independent assortment is limited essentially to pairs of alleles that are unlinked—that is, located on different chromosomes. New genotypic combinations of alleles are produced in sexual and asexual eukaryotes, and in viruses and prokaryotes whose genes are *linked* in a single DNA molecule (chromosome), by the more widespread process of **genetic recombination** by means of **crossing over** between homologous segments of DNA or chromosomes (Fig. 5.7). The probability of crossing over between linked genes is directly proportional to the distance between them, which allows geneticists to construct gene maps of the chromosomes in any species. From an evolutionary standpoint, genetic recombination is the single most important process for the production of genotypic diversity in all life forms. Indeed, it is difficult to conceive of meaningful rates of evolutionary change in the absence of recombination.

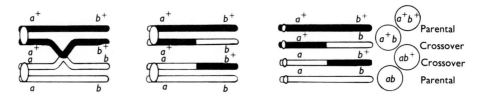

Figure 5.7. Crossing over in a pair of replicated homologous chromosomes heterozygous for two pairs of linked alleles. The synapsed chromosomes (each subdivided into a pair of chromatids) undergo an exchange involving one chromatid of each chromosome. The crossover leads to new allelic combinations in the two exchanged chromatids, but the noncrossover chromatids remain unchanged (a^+b^+ and ab). (From *Understanding Genetics,* 4th ed., by Norman Rothwell. Copyright © 1988 by Oxford University Press. Reprinted by permission)

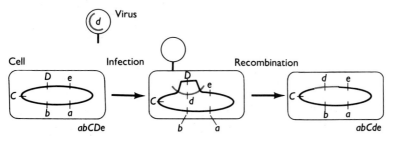

Figure 5.8. Gene transfer from one bacterial cell to another takes place by transduction by means of a virus vector. The virus carrying DNA with gene *d* from one cell (donor) may transfer this gene to another cell (recipient) during infection. Through recombination, allele *d* may be incorporated in place of allele *D* in the recipient's genome, thus changing the cellular genotype from *abCDe* to *abCde*.

Suppose there are three different populations of a bacterial species, each carrying genes that determine the response to antibiotics. Population 1 is resistant to penicillin *(pen-r)*, but sensitive to streptomycin *(str-s)* and tetracycline *(tet-s)*; its genotype is therefore *pen-r str-s tet-s*. Population 2 has the genotype *pen-s str-r tet-s*, and population 3 has the genotype *pen-s str-s tet-r*. Each population is thus resistant to a different one of the three antibiotics, but sensitive to the other two, and will flourish as long as its environment does not include the drugs to which it is sensitive. Any population can become resistant to all three drugs by mutation of the sensitive allele to the resistant alternative at each gene locus and by transmission of these alleles to progeny. Each mutation is an independent event, however, and the occurrence by chance of mutations at all loci in the same cell is calculated as the product of the separate probabilities for each mutation. For an average mutation rate of 10^{-6} for each gene, the chance of a cell becoming resistant to streptomycin and tetracycline and becoming genotypically *pen-r str-r tet-r* is $10^{-6} \times 10^{-6}$, or 10^{-12}. For a cell sensitive to all three drugs, the chance of its gaining resistant alleles by independent mutations at each gene locus is $(10^{-6})^3$, or 10^{-18}. These are very long odds even for bacteria, whose population size can be very large.

New genotypes can arise very quickly, however, by recombination between genes from different cell lineages. In bacteria, which lack sexual mechanisms by which different genomes come together in fertilization or other forms of gamete fusion, genes from one cell lineage can be introduced into another cell by a variety of transfer routes. Genetic studies showed that gene transfer may occur by **conjugation,** as in *E. coli;* by **transduction** by means of virus vectors carrying genes from one cell to another; and by **DNA transformation,** as in pneumococcus and a few other bacterial types. Transduction is the most widespread means for gene transfer between bacteria (Fig. 5.8). Regardless of the transfer route, only a relatively small segment of DNA from *donor* cells is transmitted to *recipient* cells, in which genetic recombination takes place.

Recombination between viral genomes takes place inside infected cells, where viral genomes are replicated and accessible to one another in the interval before the nucleic acids are enclosed in the protein coat of the mature virus particle. In eukaryotic species, the principal interval during which genetic recombination occurs is *meiosis of sexual reproduction.* Gametes produced by meiosis or shortly afterward have recombinant genotypes of enormous diversity. When gamete fusion takes place to initiate the next sexual generation, the number of possible combinations of alleles in the progeny genotypes is virtually astronomical. Each member of each generation can be genotypically different. Genotypic variety in sexual eukaryotes is thus generated regularly by meiosis, when genetic recombination and independent assortment occur, and by combinations of gamete chromosomes in unlimited variety during fertilization in the sexual cycle.

Genetic recombination can also occur through crossing over between chromosomes in somatic cells, although in considerably lower frequency and in more random fashion than in meiotic cells. In asexual eukaryotes, some degree of genotypic diversity can arise from *somatic-cell recombination,* but only if the recombinant genomes are included in cells that propagate the species. In fungi, for example, recombinant genomes in body cells or in asexual spores may give rise to new individuals of the next generation.

Sexual reproduction must be included among the processes accounting for the variety and success of eukaryotic life, principally because of the regularity with which a high level of genetic diversity is achieved in every generation. It is unlikely that sexual reproduction alone is responsible for eukaryote evolutionary success because the truly explosive increase in organism diversity occurred in the multicellular kingdoms. Protists existed for hundreds of millions of years with a variety of life styles and cellular forms, but as unicells, before multicellularity appeared. Most of the protists are sexually reproducing and fully capable of generating considerable genetic diversity. The achievement of multicellularity in sexual eukaryotes apparently provided the push that led to new evolutionary opportunities over the past 600 million years.

Changes in Chromosome Number and Structure

A wealth of information on changes in whole chromosomes or genomes has been collected over many decades of classical cytological, genetic, and cytogenetic analysis, and more recently by the application of molecular methods. For convenience, we can consider three general categories of changes: (1) increase in genome number, leading to **polyploids,** and changes involving the addition or subtraction of individual chromosomes, leading to **aneuploids;** (2) change in chromosome number due to the structural modification of individual chromosomes of a genome; and (3) rearrangement of the genes in a chromosome through deletion, duplication, inversion, and translocation, but with no change in chromosome number. The relative effect on genetic diversity varies from one category to another and from one group of eukaryotes to another.

Polyploidy is relatively common in plants, but is less common in animals, except in some invertebrates. The presumption is that polyploid animals would be at a considerable disadvantage because of disturbances in the distribution and number of sex chromosomes and ensuing hormonal irregularities. Very few plants are known to have sex chromosomes, and, indeed, most plants are hermaphroditic; that is, one individual has both male and female sexual structures. About 50 percent of the flowering plants are estimated to be polyploid, and it is not unusual for different species of a genus to have different numbers of genomes. Various species of wheat, for example, are diploid, tetraploid, and hexaploid. Some other important cultivated plants— such as cotton, tobacco, and potato—are polyploid, but many others are exclusively diploid.

Successful sexual polyploids usually have an even number of genomes, which are distributed in equal numbers to sexual progeny at meiosis and fertilization. Such tetraploids, hexaploids, and other even-numbered polyploids produce viable gametes and can give rise to successive sexual generations that are also fertile and maintain a constant chromosome number. Triploids, pentaploids, and other polyploids with an odd number of genomes are almost always sterile because of irregularities at meiosis and the consequent inviability of gametes (Fig. 5.9). Sterile polyploids have little or no evolutionary importance, but they may be desirable for ornamental purposes. Many prize orchids, for example, are sterile triploids or pentaploids that flower for a long period of time in the absence of otherwise competing processes of seed and fruit formation.

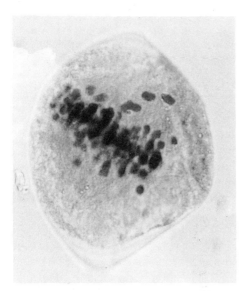

Figure 5.9. Irregularities in the distribution of chromosomes occur regularly in polyploids that have an odd number of genomes, leading to reproductive sterility as a consequence of inviable gametes. (Photograph by N. V. Rothwell)

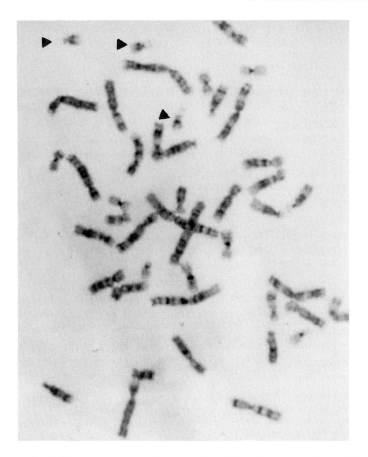

Figure 5.10. The aneuploid chromosome complement of a male with trisomy 21, or Down syndrome. Instead of the usual pair of chromosomes 21, there are three copies. Even if it is not possible to distinguish chromosomes 21 and 22, both of which are small and acrocentric, the total of five such chromosomes instead of four provides the evidence for a chromosomal basis for the clinical symptoms. (Photograph courtesy of T. R. Tegenkamp)

Aneuploids are uncommon in natural populations, presumably because of disadvantages due to genomic imbalance. The loss of a chromosome leaves a possibly recessive allele unmasked on the remaining partner chromosome, and the addition of a copy of one or more chromosomes also appears to be detrimental in most cases. In human beings, the presence of an extra copy of chromosome 21 (trisomy 21) is associated with the development of clinical symptoms of Down syndrome (Fig. 5.10). Other **trisomies** (three copies of a chromosome) are lethal early in a pregnancy or very shortly after birth, except for anomalies involving the X chromosome. Triplo-X females *(47,XXX)* and those with four X chromosomes *(48,XXXX)* suffer some degree of mental retardation and other developmental defects. Males with extra copies of the X chromosome, called Klinefelter males, are anatomically feminized to a certain extent and show increasingly severe mental retardation with increasing number of X chromosomes. Although neither **monosomies** ($2n - 1$) nor tri-somies ($2n + 1$) are well tolerated in general, an extra chromosome appears to be less devastating than a missing chromosome in all the species that have been studied.

The relative success of polyploidy among flowering plants can be gauged in both short-term and long-term evolutionary advantages. In the short term, the greater genotypic diversity in populations with more than two copies of the genome may not be matched by greater phenotypic diversity. The expression of recessive traits in diploids requires only two recessive alleles, one on each chromosome of a pair. But four and six recessive alleles are needed for the expression of recessive traits in tetraploids and hexaploids, respectively. In fact, we expect less phenotypic diversity in polyploids than in diploids, which may be advantageous in fluctuating environmental conditions. The relatively few successful individuals would produce progeny like themselves, which also would enjoy success in the environment. In the long-term evolutionary perspective, however, the higher potential genotypic diversity in polyploids could be expressed as mutations occurred and led to allelic and genic divergence.

The low success of aneuploidy stands in contrast to changes in chromosome number that result from chromosome fusion or fragmentation. In these cases, the chromosome number is changed but little or no gene loss occurs as a consequence of the structural modifications involving two or more individual chromosomes of the genome. The close genetic relationship among members of an **aneuploid series** of species can be demonstrated by genetic and cytological methods. For example, different species of *Drosophila* have three, four, five, and six pairs of chromosomes (Fig. 5.11). Studies of the banding patterns in salivary-gland chromosomes reveal the presence of virtually the

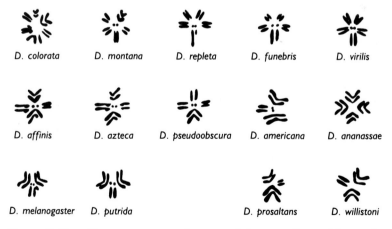

D. colorata   *D. montana*   *D. repleta*   *D. funebris*   *D. virilis*

D. affinis   *D. azteca*   *D. pseudoobscura*   *D. americana*   *D. ananassae*

D. melanogaster   *D. putrida*   *D. prosaltans*   *D. willistoni*

Figure 5.11. Chromosome complements of fourteen *Drosophila* species, showing the aneuploid series of $n = 6$ to $n = 3$. These numerical variations have arisen by means of centric fusion and other events in the evolution of the group. The X and Y chromosomes of these male genomes are at the bottom of each drawing.

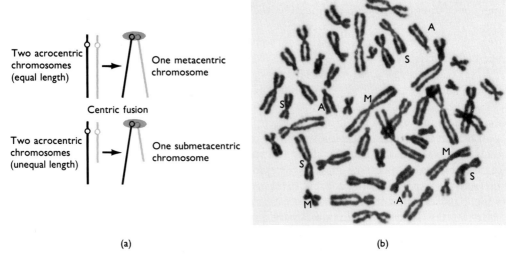

(a)   (b)

Figure 5.12. Aneuploidy by means of centric fusion. (a) Two acrocentric chromosomes may become joined at their centromere regions and produce a metacentric or submetacentric chromosome, with little or no loss of material. Both centromeres usually are retained in the common centromere region of the fusion product. (b) Human chromosomes include metacentrics (M), submetacentrics (S), and acrocentrics (A). At least one large metacentric is presumed to have arisen by the centric fusion of two small acrocentrics since the divergence of the human and the great ape lineages. Human chromosome 2 is a large metacentric not found in ape genomes, but its banding pattern coincides with the patterns of two small acrocentrics in ape genomes that are absent in the human genome.

entire genome in every species, but distributed among different numbers of chromosomes making up the genome. Genetic analysis provides substantiating evidence for the retention of virtually all the genes shared by these species. Similar aneuploid series are known to occur in various plant and animal groups.

The most common process leading to aneuploidy without substantial gain or loss of material is called **centric fusion**. Usually, two smaller acrocentric chromosomes suffer breaks in the short arm and rejoin to produce a larger metacentric or submetacentric chromosome with two centromeres in a common centromere region (Fig. 5.12). Neither the mechanism for centric fusion nor the mechanism for dissociation is known, other than standard processes involving the breakage and repair of DNA by enzymatic activities.

The evolutionary significance of the usual trend toward reduction in chromosome number lies in the reduced release of potential genotypic variety. Genes that assorted independently when in separate chromosomes are linked in the new larger chromosomes and recombine only when crossing over occurs. Species with fewer chromosomes tend to be less variable than related species with the same genes in more chromosomes. A homogeneous environmental condition would favor low genotypic variability. In such a habitat, centric fusions would be retained in populations where genetic constancy would be more selectively advantageous than genetic diversity. The short-term evolutionary advantages could be mitigated in the long term, though, because less diversity would reduce the chances for success in changed environments or conditions.

The third category of chromosome changes involves structural rearrangements of chromosomes, usually without a change in chromosome number, as a result of chromosome breakage and subsequent rejoining of broken ends. Four general types of structural change are recognized, and each has been studied for many decades by cytogenetic methods. In **deletion,** or deficiency, a portion of a chromosome is lost; **duplications** are repeats of chromosome segments; **inversions** involve the excision and reinsertion of a chromosome segment in reverse order, having been rotated 180° from the original order; and **translocations** involve the transfer of segments between nonhomologous chromosomes in the genome. The frequency of occurrence and the evolutionary significance vary from one type of chromosome aberration to another.

Deletions are usually harmful or lethal, depending on the extent of chromosome loss. Losses of genetic material are barely tolerated, even in heterozygotes. This may be due to genetic imbalance, unmasking of harmful recessive alleles, or other factors as yet unknown. Deletions may occur because of splicing of broken ends without the incorporation of the fragment produced by the break, miscopying during DNA replication, or unequal crossing over in which duplications also are produced. A classic example of unequal crossing over leading to deletion and duplication is the bar-eye mutation in *Drosophila melanogaster*. The recombination of flanking marker genes on either side of the bar locus shows that crossing over is involved, rather than mutation or another genetic process (Fig. 5.13). The bar-eye phenotype varies in relation to the number of gene copies on the X chromosome and the total number of copies in the cell. For example, females with two gene copies on each X chromosome have more eye facets (68) than females with three of the four copies on one X chromosome and the fourth copy on the other X chromosome (45 facets). Wild-type females with one normal gene copy on each X chromosome have almost 800 facets per eye. This kind of phenotypic variance also is an example of a *position effect,* whereby gene expression differs in relation to physical location in the genome and not to DNA sequence.

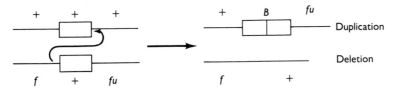

Figure 5.13. Unequal crossing over in *Drosophila melanogaster* produces modified chromosome products, such as the duplicated locus associated with the bar-eye phenotype. Recombination of flanking *f* and *fu* gene markers clearly shows the occurrence of a recombinant crossover event and not a mutation at the bar-eye site on the X chromosome.

Gene duplication is widespread in eukaryotes, particularly in the more highly evolved groups. Its significance in evolution is very great indeed, primarily in providing additional DNA whose sequences ultimately diverge to code for new or modified functions. Exon duplication and shuffling of duplicated sequences have already been discussed as important features leading to speedier protein evolution than occurs by mutation alone. Repetitive DNA, whether coding or noncoding, is another widespread eukaryotic feature that must have arisen as the result of duplication through unequal crossing over, copying errors during DNA replication, and assorted breaks and splices that may have situated the duplicated segments in the same chromosome or in different chromosomes of the genome.

Inversion and translocation are commonly occurring rearrangements that require two breaks and the rejoining of the broken ends in an order or a location other than the original. Inversion involves rearrangement in one chromosome, whereas translocation nearly always involves two or more nonhomologous chromosomes (Fig. 5.14). Each aberration can be identified cytologically by band patterns in stained chromosomes from somatic cells, and even more easily by unusual chromosome configurations during the first meiotic division in heterozygotes with one set of changed and one set of unchanged chromosomes. The reordering of chromosome segments can also be determined in homozygotes by standard transmission genetics. Different orders of linked genes can be mapped in inversion homozygotes, and translocation homozygotes show independent assortment of normally linked genes as well as linkage of normally reassorting genes.

Inversion is a barrier to successful reproduction between individuals or populations carrying different inverted or noninverted chromosomes. Heterozygotes produced by parents with different chromosome inversions, or by one parent with inversions and the other without, encounter serious difficulties at meiosis and usually produce few viable gametes. Synapsis at meiosis between homologous chromosome segments leads to the looping out of the heterozygous region. A crossover within this region produces chromosomes with deletions and duplications, and gametes that contain these chromosomes tend to be inviable (Fig. 5.15). As a reproductive barrier between chromosomally different populations, inversion is an important component of speciation because it prevents gene mixing between populations and thereby reproductively isolates them from each other. It also leads to reduced genotypic diversity in any population whose members propagate more nonrecombinants than recombinants, as is true for groups with some members that have inversions and other members that do not. Reduced genotypic diversity may be adaptive and maintained by the preservation of blocks of genes, or **supergenes.** In spite of a high level of allelic heterozygosity in inverted chromosomes, inversions preserve these adaptive gene complexes through their effect in reducing crossover gametes.

Translocation also reduces or prevents reproduction between individuals or populations that differ in chromosome organization. Translocation hetero-

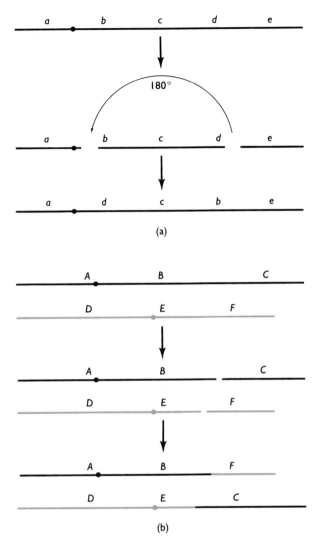

Figure 5.14. Chromosome breakage and rejoining in new arrangements may produce (a) inversions in a single chromosome, or (b) translocations that may involve exchanges between two nonhomologous chromosomes. The letters designate gene loci.

zygotes often produce gametes that are inviable due to gene duplication and deletion that arise from irregularities at meiosis (Fig. 5.16). Nonrecombinants outnumber recombinants to a significant degree in mixed populations, leading to reduced production of diverse genotypes. Translocation may have genetic advantages if certain gene combinations are retained to a greater degree when linked on a chromosome than when assorted independently, and these altered combinations may increase the chances for success of a population. Con-

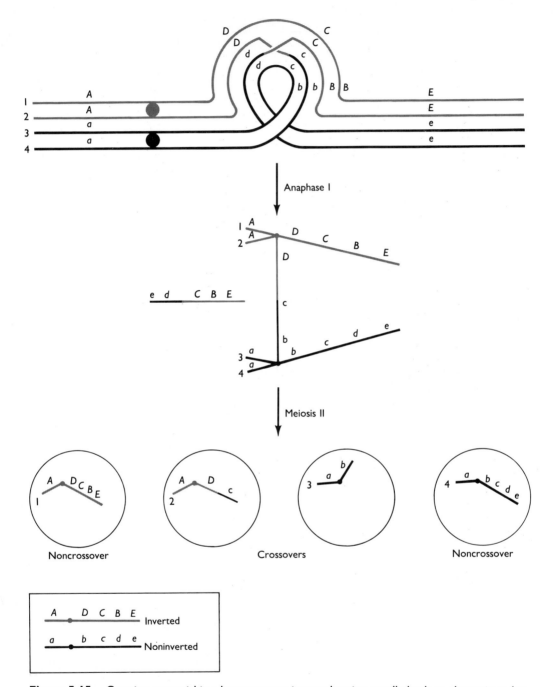

Figure 5.15. Crossing over within a heterozygous inverted region usually leads to aberrations during meiosis, such as an acentric fragment (usually lost) and a chromosome bridge (later broken) between separated chromosomes in anaphase nuclei. Gametes with aberrant crossover chromosome products are generally inviable, but noncrossover gametes have a complete set of genes and are viable.

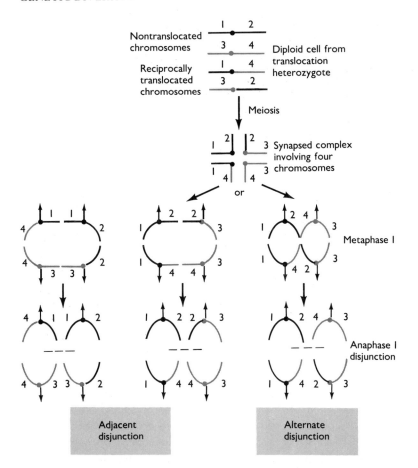

Figure 5.16. Meiosis in a translocation heterozygote, showing the consequences of three different chromosome alignments at metaphase I, leading to adjacent chromosomes or alternating chromosomes moving to the same pole at anaphase I. Chromosome arms are numbered, showing that deletions and duplications arise in adjacent disjunction but not in alternate disjunction.

versely, the greater production of gene combinations by independent assortment than by genetic recombination may be valuable in certain cases and help the population toward greater evolutionary success. The relative advantages of higher or lower genotypic diversity must be judged in both the short-term and the long-term perspectives of evolution.

It is relevant to note here that new genotypes may arise as the result of gene transfer from one species or population to another through mediation by plasmids, viruses, and other vectors. Such *horizontal gene transfers* may occur by breakage of the resident chromosome and subsequent integration of the transferred gene into that chromosome, as well as by events during DNA replication in the recipient.

Phenotypic Diversity by the Regulation of Gene Expression

From genetic and molecular studies, it is clear that differential gene expression is the single most important phenomenon responsible for the development of different phenotypes among genetically identical individuals and populations. Since the discovery of regulatory genetic elements and their influence on gene expression, by François Jacob and Jacques Monod in 1961, geneticists have accumulated data on a wide variety of mechanisms and elements by which gene expression may be regulated. This variety is greater for eukaryotes than for prokaryotes and is of fundamental importance in the evolution of multicellular organisms.

Differential Gene Expression

Gene expression is the transcription and translation of coded genetic information. **Differential gene expression** refers to the same genes being turned on in some cells or at some times and turned off in other cells or at other times. Differential gene expression is responsible for the synthesis of globin in the reticulocyte precursors of red blood cells and for the absence of hemoglobins in other cells in which these genes are not transcribed or translated, even though they are present. In addition to on–off controls, amounts of gene products may be regulated differentially in different cells or at different times.

The most common controls over gene expression are mechanisms that regulate the transcription of DNA into messenger RNA. Transcription may be prevented or induced by proteins or other molecules that interact directly or indirectly with DNA, and either prevent or permit the movement of RNA polymerase along the DNA template to catalyze the transcription of structural genes. In bacteria, as was shown by Jacob and Monod in classical genetic studies of *E. coli*, a cluster of related structural genes exists in an **operon** that also includes a single **operator** and a single **promoter,** which are regulatory elements of the package. For transcription to occur, RNA polymerase must bind to the promoter and proceed downstream (in the 3′ direction) to catalyze mRNA synthesis from structural-gene sequences. If **repressor** protein is bound to the operator, however, transcription is prevented (Fig. 5.17). Molecular analysis has shown that the large repressor protein covers up the RNA polymerase binding site of the promoter, as well as the immediately adjacent operator sequence. Thus the RNA polymerase molecule can bind to recognition sites in the promoter only if repressor protein is not bound to the operator. This form of **transcription control** is usually referred to as *operator–repressor control* of transcription. The repressor protein is the product of a separate repressor gene, and not usually part of the operon with which it interacts.

Bacterial operons coding for enzymes of amino acid biosynthesis may also be subject to *attenuation control*. In this system, transcription begins but may

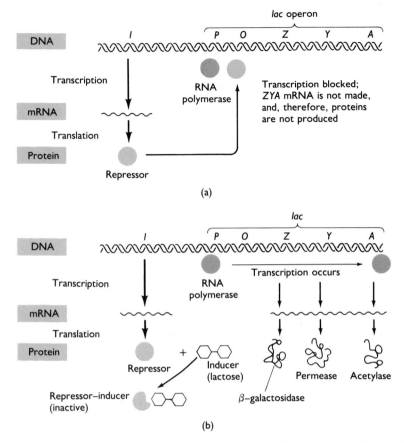

Figure 5.17. The operon model of control over *lac* gene expression for three inducible proteins of lactose metabolism in *E. coli*. (a) In the absence of inducer (lactose), the *lac* repressor protein binds to *lacO* (operator) and blocks the movement of RNA polymerase toward *lacZYA* genes. Proteins are not synthesized because mRNA transcripts are not available for translation. (b) In the presence of inducer, repressor protein binds to lactose and is rendered incapable of binding to *lacO*. RNA polymerase can proceed along the DNA template, from its *lacP* (promoter) binding site, and can catalyze transcription of *lacZYA* sequences. The translation of these transcripts results in the synthesis of the three inducible proteins.

be terminated prematurely—before the RNA polymerase actually reaches the structural-gene sequences. Attenuation control is similar to operator–repressor control in that both mechanisms are *negative* controls over transcription; the RNA polymerase can act only if the block to transcription is removed. In the case of *positive* transcription controls, the binding of polymerase is helped or made more efficient by the binding of activator proteins to specific regions in the upstream portion of the promoter. This stimulates transcription by contacts with RNA polymerase (Fig. 5.18).

Transcription controls in eukaryotes differ in many respects from those in

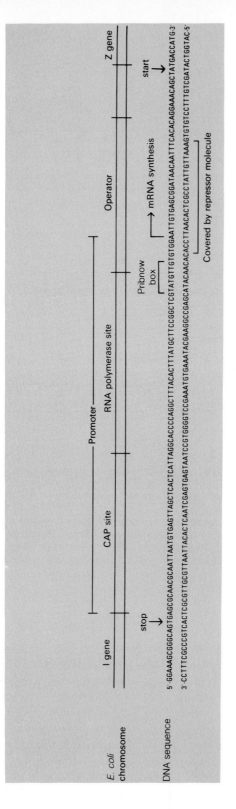

Figure 5.18. DNA base sequence of the regulatory region of the *lac* operon in *E. coli*. The end of the repressor gene (I) and all of the promoter and operator are shown, along with the start of the *β*-galactosidase gene (Z). Binding sites for an activator protein (CAP), RNA polymerase, and the *lac* repressor protein show some overlaps. (From *Biochemistry*, 2nd ed., by Frank B. Armstrong. Copyright © 1983 by Oxford University Press. Reprinted by permission)

bacteria. No evidence exists for the presence of operators or repressors, so eukaryotes have neither operator–repressor control nor attentuation control systems. A variety of transcription controls in eukaryotes is known, however, some of which depend on particular DNA sequences and others on the packing density of chromosome regions or chemical modifications of these regions.

The promoter is organized along similar lines in prokaryotes and eukaryotes. Each promoter contains an upstream region to which proteins bind and stimulate transcription by helping RNA polymerase recognize the sequence in preparation for mRNA synthesis. Once the polymerase reaches the conserved downstream promoter sequence, called the *TATA box* in eukaryotes and the *Pribnow box* in prokaryotes, it binds tightly and correctly initiates transcription at the $+1$ nucleotide, or the startpoint of transcription (Fig. 5.19). Molecular studies done since 1980 have revealed the existence in eukaryotes and their viruses of upstream DNA sequences called **enhancers**. Although the dividing line between enhancers and other upstream promoter elements is somewhat blurred, enhancers can be identified by several experimental criteria. Enhancers function only in the same DNA molecule containing the target gene; that is, they are *cis*-acting elements. They can activate transcription independent of their orientation in the DNA molecule, and at distances of more than 1000 bases upstream or downstream of the startpoint of transcription of the target gene. Enhancer elements have not been found in prokaryotes or their viruses.

The versatility of positive transcription control in relation to enhancers is evident from studies of various genes, such as those that code for *metallothioneins*, small proteins that bind heavy metals and thereby protect cells against their toxic effects. In the well-known metallothionein genes studied in mouse and human cells, different upstream enhancer sites induce gene transcription in response to heavy metals, glucocorticoid hormones, bacterial lipopolysaccharides, and the positively acting protein transcription factor Sp1. These enhancer sequences are not overlapping, and each appears to be responsive specifically to its own inducing factor.

Enhancers seem to be components of networks that control gene expression during development in multicellular organisms. In the mouse, three different enhancer elements are part of a system that regulates transcription of the alpha-fetoprotein gene in specific tissues at specific times and in specific amounts during fetal development. Alpha-fetoprotein is the major plasma protein in the fetus, and its production is repressed after birth, when serum albumin synthesis takes over. In *Drosophila melanogaster*, the yolk protein 1 gene *(yp1)* is transcribed only in the presence of a specific enhancer sequence of 125 bases located nearly 200 bases upstream of the startpoint of transcription of *yp1*. Transcripts are found only in fat bodies of adult female flies, so enhancement is tissue-specific, time-specific, and sex-specific (Fig. 5.20). A different upstream enhancer sequence stimulates transcription of the *yp1* gene in ovarian follicle cells of adult females. It is very likely that different tissue-specific factors (such as hormones) interact with each of these two enhancers,

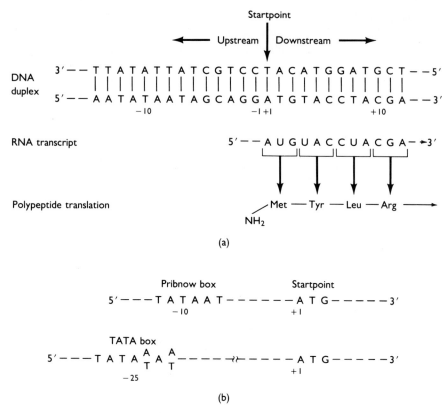

Figure 5.19. The correct initiation of transcription by RNA polymerase depends in large part on its interaction with promoter elements upstream of the startpoint of transcription (+ I nucleotide). (a) Regions upstream of the startpoint of transcription are given consecutive minus numbers, and include (b) the Pribnow box, a hexanucleotide − 10 sequence, in prokaryotes, or the TATA box, a heptanucleotide − 25 sequence, in eukaryotes. The polymerase binds tightly to DNA at these highly conserved sequences and correctly initiates transcription, beginning with the + I nucleotide. DNA sequences are usually shown in 5′→3′ orientation, because they are then identical to 5′→3′ mRNA, except that U is present in RNA where T occurs in DNA.

leading to differential enhancer activation of the same gene in different cells and tissues. The absence of these inducing factors in other cells, in earlier stages of development, and in males would explain the absence of *yp1* transcripts in these instances, even though these enhancers are present in every cell at every stage of development in both sexes. When the appropriate inducing factor is present and binds to the enhancer, a high rate of transcription is stimulated through binding of a high density of RNA polymerase II (the polymerase that transcribes mRNA in eukaryotes) over the gene.

Another realm of transcription controls in eukaryotes concerns modifications in the chemical and organizational features of the chromosome. In transcriptionally active chromatin, the chromosome region engaged in mRNA synthesis is unfolded to some degree and thereby provides access to template DNA for the binding and activity of RNA polymerase II. The most dramatic evidence for organizational differentiation between transcribing and nontranscribing chromatin has been known for more than 40 years from studies of giant salivary-gland chromosomes in larvae of *Drosophila, Chironomus,* and other members of the order of two-winged insects, or Diptera. Large, diffuse puffs appear and disappear along the chromosomes during larval development, each puff being a locally unfolded chromosome segment that is actively engaged in transcription (Fig. 5.21). The puffs regress when transcription stops, returning to the tightly folded inactive chromatin configuration.

Molecular analysis of chromosomes in other cells and other organisms has indicated the unfolding of transcriptionally active chromatin, even though these chromosomes lack the size and detail of those that allow microscopical studies of insect larvae. By the application of DNase I (deoxyribonuclease I), geneticists have found that active chromatin is much more sensitive than inactive chromatin to nuclease digestion. These results indicate accessibility of active chromatin to the enzyme, presumably because of its degree of local unfolding. Another line of evidence for the unfolding of active chromatin comes from digestion studies using *low* concentrations of DNase I and other DNA-cutting enzymes, which generate sets of specific fragments that can be identified by gel electrophoresis. These fragments result only when chroma-

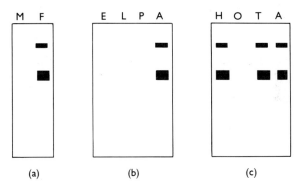

Figure 5.20. Cloned *ypl* genes containing one of the two enhancer sequences in *Drosophila melanogaster* were introduced into experimental flies, and RNA transcripts of the gene were detected by gel electrophoresis. These highly simplified gel diagrams show that RNA transcripts of *ypl* were present (a) only in females (F), not in males (M); (b) only in adults (A), not in the egg (E), larval (L), or pupal (P) stages of development; and (c) in tissues of the head (H), thorax (T), and abdomen (A), but not in ovaries (O), even though all these tissues contain fat cells. Ovarian gene expression is activated by an enhancer that was absent from the cloned *ypl* DNA used in these studies. Activation of *ypl* gene expression is thus sex-specific, stage-specific, and tissue-specific. (Based on M. J. Garabedian et al., 1985, Independent control elements that determine yolk protein expression in alternative Drosophila tissues. *Proc. Nat. Acad. Sci. U.S.* 82:1396)

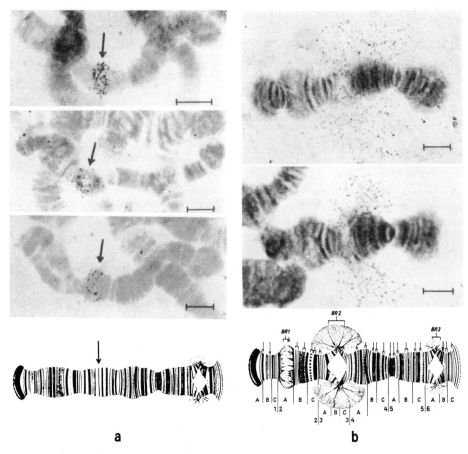

Figure 5.21. Differential gene activation as seen in puffing of chromosome IV in the midge *Chironomus tentans*. (a) Polytene chromosome IV from rectum cell nuclei undergoes little puffing at the BR2 locus (arrow), which is identified by hybridization with radioactively labeled BR2 mRNA. (b) In salivary gland nuclei from the same larvae, however, a large puff is evident at the BR2 locus, and its expanded DNA conformation is evident from silver grains of the hybridized, labeled BR2 mRNA. The transcriptionally active chromatin in (b) is conformationally distinct from inactive chromatin at the BR2 locus in (a). (From B. Lambert, 1975, The chromosomal distribution of Balbiani ring DNA in *Chironomus tentans. Chromosoma* 50:193)

tin is digested, not when purified DNA is digested under the same conditions, indicating that chromosome nucleoprotein organization is involved. The sites of cleavage by enzymes, called **hypersensitive sites,** have been mapped in the promoter/enhancer region extending for several hundred bases upstream of the startpoint of transcription of the relevant gene (Fig. 5.22). Accessible sites for the binding of inducing factors to enhancers and other promoter sequences may thus be made available, leading to the movement of RNA polymerase II to these regions and its subsequent binding to the TATA box for the initiation of transcription. Hypersensitive sites occur in the promoter/ enhancer region of an activated gene, but not when the same gene is inactive in the same or in different cells of the organism. At the present time, however, the mechanism that generates hypersensitive sites remains unclear, but some possibilities are under study.

One of several proposed mechanisms for chemical modifications that characterize activation or inactivation of transcription is **DNA methylation,** specifically the methylation of cytosine to form 5-methlycytosine residues. By appropriate enzymatic digestion assays of gene sequences, DNA fragments can be resolved and identified by gel electrophoresis as being methylated or not (Fig. 5.23). In a wide variety of genes from different organisms, it has been found that some sites in the DNA sequences are not methylated or are only partially methylated in tissues where the genes are transcribed. In these tissues, genes not engaged in transcription are methylated at these same DNA sites. For example, human γ-globin genes are transcribed and translated in erythroid tissue during fetal development but not after birth. Nonmethylated sites in γ-globin DNA are found in fetal erythroid tissue, but many sites are methylated in γ-globin DNA in adult erythroid tissue and virtually all the sites are methylated in nonerythroid tissue where γ-globin genes are not transcribed. Significantly, the distribution of nonmethylated and undermethylated sites coincides with that of DNase I–sensitive sites in transcriptionally active chromatin. Unfortunately, little is known at present about the mechanisms for DNA methylation in chromatin, leading to folding and unfolding of the nucleoprotein fiber of the eukaryotic chromosome.

In eukaryotes, much more than in prokaryotes, gene expression may be regulated by **posttranscriptional controls.** This may be due at least in part to the different logistics involved in transcription and translation in eukaryotes and prokaryotes. In prokaryotes, translation begins while transcription is still in progress in the noncompartmentalized cell. Little opportunity is thus available for any extensive modification or processing, and, furthermore, the mRNAs have no introns to be excised or exons to be spliced. In eukaryotes, on the contrary, information flow from DNA to protein is more complex and

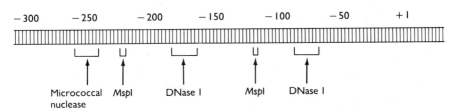

Figure 5.22. The hypersensitive region of the human β-globin gene extends over a stretch of about 250 base pairs upstream of the startpoint of transcription (+1 nucleotide). Sites sensitive to the DNA-cutting enzymes DNase I, micrococcal nuclease, and *MspI* have been mapped and are labeled just below the DNA duplex. Hypersensitive sites can be demonstrated by enzymatic tests to be present in transcriptionally active chromatin, but not in nontranscribing chromatin.

more leisurely, which is reflected in the greater variety of controls. Transcription takes place in the nucleus; mRNA transcripts are processed in the nucleus; and the mature mRNAs are transported from the nucleus to the cytoplasm, where translation occurs. Eukaryote gene expression may be modulated at any of the posttranscriptional stages, from steps for processing pre-mRNA transcripts to events governing the use and turnover of mature mRNAs prior to translation (Fig. 5.24).

Eukaryote **translational controls** are of various types, but an important device is embodied in the relative stability of mRNA. Long-lived mRNAs can engage in numerous rounds of translation, leading to continued protein syn-

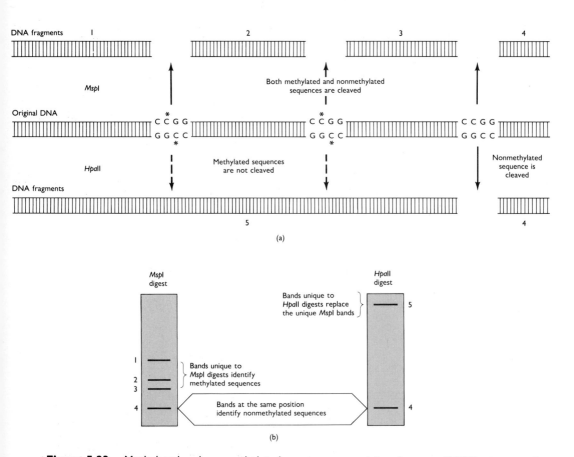

Figure 5.23. Methylated and nonmethylated sequences containing the same CCGG target can be distinguished by restriction-fragment assays following the use of restriction enzymes *Mspl* and *Hpall*. (a) *Mspl* cleaves both methylated and nonmethylated CCGG sequences, whereas *Hpall* cleaves only nonmethylated CCGG sequences. Thus any bands of restriction fragments present in gels for both enzyme digests will identify nonmethylated CCGG. (b) Methylated CCGG sequences in the gel are identified by unique bands in the *Mspl* digest, which are replaced by one or more unique bands of larger fragments in the *Hpall* digest. DNA methylation is thought to inactivate transcription.

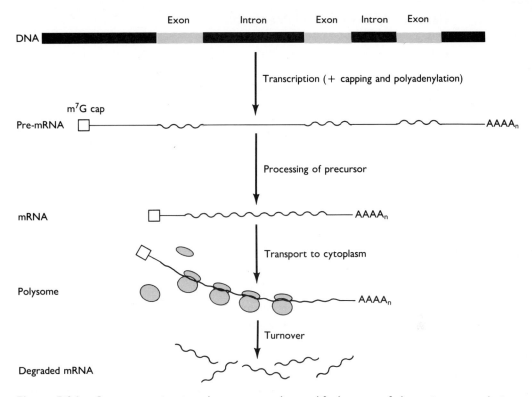

Figure 5.24. Gene expression in eukaryotes may be modified at any of the various steps during information flow from gene to protein. Control may be exerted such that transcription may or may not occur, or it may occur during the posttranscriptional processes of capping, polyadenylation, intron excision, and exon splicing of the pre-mRNA. In addition, gene expression is influenced by the relative efficiency of mRNA transport from nucleus to cytoplasm, the availability of ribosome sub-units for translation, the turnover of mRNA, and other cellular activities.

thesis even in the absence of mRNA production. For example, globin mRNAs continue to guide hemoglobin synthesis in mammalian reticulocytes long after these cells have lost their nuclei and can no longer make new transcripts. In the silk gland of silkworm larvae *(Bombyx mori)*, a single *fibroin* gene can control the synthesis of about 10 billion molecules of this protein in just a few days. About 100 million mRNA molecules may be transcribed from the gene, and each stable mRNA can guide the translation of 100,000 fibroin molecules for silk production.

All these controls and others act by *modulation* of gene expression without concomitant change in the gene sequence, location, or copy number. Some of the mechanisms for modifications of gene expression through *alterations* in the genome will now be discussed. These controls govern important changes in viruses and cells and are of particular relevance to the diversity of phenotypes that may be expressed during development in multicellular organisms.

Genomic Alterations and Phenotypic Variety

Genomic changes associated with expressions of phenotypic diversity fall into different categories, three of which are (1) **gene amplification,** in which some genes replicate at the same time that other genes do not; (2) **gene rearrangements,** by which many different proteins may be synthesized from many combinations of relatively few genes; and (3) genetic alterations due to movements of **transposable genetic elements** from place to place in the genome. The first two types of modification are intimately associated with eukaryote development, but the third category produces phenotypic variety in viruses, prokaryotes, and eukaryotes.

Differential gene replication underlies gene amplification, whereby a large amount of some product can be synthesized from a greatly increased number of copies of a gene sequence at some time in the history of the cell. In contrast to translational control by means of stable mRNAs, which permits a specialized cell to concentrate for its lifetime on making one or a few proteins, gene amplification is characteristic of more generalized cells that provide a special service only at a specified time during development. The best known but not the only example concerns amplification of the major rRNA gene, whose precursor transcript is processed to produce ribosomal 18S, 28S, and 5.8S RNA molecules. The phenomenon is characteristic of protozoan and animal species, but seems not to occur in the relatively few plants that have been studied.

The major rRNA gene exists in hundreds or thousands of repeats arranged in tandem and localized in the *nucleolar-organizing region* of a chromosome; this collection of repeated rRNA genes is referred to as **ribosomal DNA, or rDNA.** In the toad *Xenopus laevis,* the original 900 copies of the rRNA gene in diploid cells is increased to 600 to 1600 *times* 900 only in the oocyte and only during an early stage of meiosis (Fig. 5.25). The amplified rDNA is transcribed into rRNA, which becomes part of vast numbers of new ribosomes. The extra rRNA genes become inactive after oogenesis, but the myriad ribosomes formed earlier serve as the sites for all protein synthesis in the egg and embryo up to the time of gastrulation.

In amphibians and other species whose rDNA is amplified, replication proceeds very rapidly by a *rolling circle mechanism.* Multiple lengths of rDNA are cut into individual units by the action of nucleases. Transcription occurs, and the precursor rRNA is then processed and incorporated into new ribosomes (Fig. 5.26). The signals by which gene amplification is turned on and off at specific gene loci in specific cells at specific intervals during development are unclear. Nevertheless, the process provides the cell with a large amount of gene product for a limited period of time, and permits the cell to continue all its later activities unencumbered with a long-term commitment to modified gene expression.

The most astonishing program for generating phenotypic diversity from a relatively small number of genes by means of gene rearrangements occurs in the B lymphocyte of the immune system. From fewer than 1000 genes, an

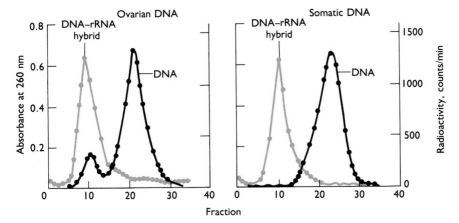

Figure 5.25. Identification of a nuclear DNA component as rDNA, and amplification of rDNA in oocyte nuclei of the toad *Xenopus laevis*. DNA–rRNA hybrid duplexes sediment in the same region of the gradient as a satellite DNA from oocytes in ovarian preparations, indicating that the satellite is rDNA. Although the satellite is not evident in the absorbance curve for somatic nuclear preparations, rDNA is present; molecular hybrids between rRNA and satellite DNA do form and do sediment where expected in the gradient. Gene amplification is clearly evident in oocyte nuclei, from the greater amount of rDNA in these cells compared with somatic cells.

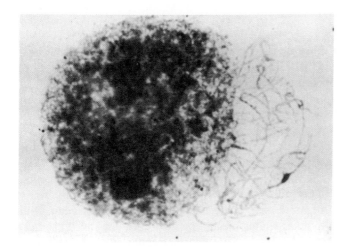

Figure 5.26. Light micrograph of an oocyte nucleus from the beetle *Dytiscus marginalis*. A large cap of amplified rDNA is at the left, and it constitutes about 90 percent of the nuclear DNA at this stage of early meiotic prophase I. ×1000. (Photograph courtesy of J. G. Gall, from J. G. Gall and J.-D. Rochaix, 1974. *Proc. Nat. Acad. Sci. U.S.* 71:1819, Fig. 1)

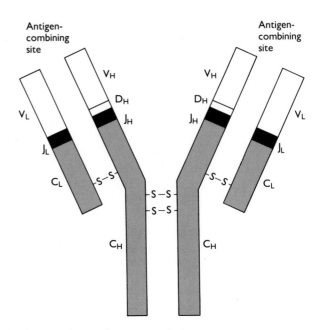

Figure 5.27. The immunoglobulin molecule is made up of one pair of identical heavy chains and one pair of identical light chains, all held together by disulfide bonds. Each heavy chain has a variable (V_H), diversity (D_H), joining (J_H), and constant (C_H) segment, and each light chain has three such segments (V_L, J_L, C_L) but lacks a diversity region. The two identical antigen-combining sites are located in the variable regions, at the N-terminals of the four chains. The greater molecular weight of the heavy chain is due to its constant segment being longer than the equivalent segment in the light chain.

animal may generate more than 1 million **immunoglobulins,** or **antibodies,** to confront the more than 1 million different **antigens** it may encounter in its lifetime. Immunoglobulins (Ig) bind to foreign antigenic substances in immune responses, which constitute a major body defense against infection and disease. Different rearrangement events take place in B lymphocytes during their differentiation and, later, when they are challenged by antigens.

Each Ig molecule consists of two identical heavy (H) polypeptide chains and two identical light (L) chains, all held together by disulfide bonds (Fig. 5.27). Each of the four functionally distinct segments of an H chain is coded by a different gene from a **multigene family** clustered on one chromosome. Each of the three segments of an L chain is coded by a different gene from either one of two multigene families located on two different chromosomes. The particular set of variable, joining, and constant genes of the L chain (V_L, J_L, C_L) and of variable, diversity, joining, and constant genes of the H chain (V_H, D_H, J_H, C_H) differs from one B lymphocyte to another. Any one lymphocyte contains one version of each gene set, but many different versions may be present in a population of these cells in an animal.

The assembly of genes coded for a particular L chain or H chain takes place by gene rearrangements that involve the excision of all the alternatives but one from specific multigene families, and the splicing of the retained V_L and J_L or V_H, D_H, and J_H to the entire C gene region and the spacer preceding it. This rearranged DNA is transcribed into pre-mRNA, which, in turn, is processed by excision and splicing to produce a mature mRNA with one gene copy for each segment of an L chain or H chain of the Ig molecule (Fig. 5.28). The specificity of interaction with an antigen resides in the variable segments at the N-terminals of the H and L chains, which are very diverse. The functions of the five classes of Ig, however, depend on the particular constant segment of the H chain (Table 5.3). In immature B lymphocytes, the C_μ or C_δ gene is transcribed, leading to translation of IgM or IgD molecules. This situation may change, by means of heavy-chain class switching in germ-line DNA, after lymphocytes are exposed to antigens.

Heavy-chain class switching involves deletions and rearrangements of DNA in activated B lymphocytes, or **plasma cells.** These rearrangements may lead to a switch from production of IgM or IgD to that of IgG, IgE, or IgA, which have different functions in the immune system. If most of the C_H gene cluster is excised, leaving C_ϵ and C_α, the plasma cells will secrete IgE from mRNA transcribed along the rearranged gene sequence (Fig. 5.29). No change occurs in the *VDJ* sequence established in the immature lymphocyte, so no change occurs in the antigenic specificity of the particular cell. The only change is in the constant region of the H chain.

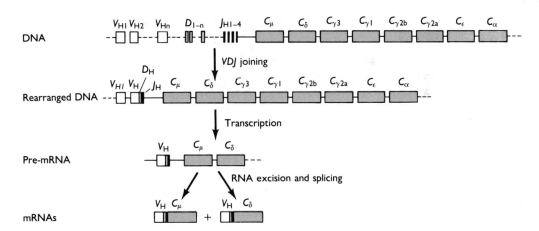

Figure 5.28. Heavy-chain gene assembly. In the differentiating B lymphocyte, germ-line heavy-chain genes are rearranged by deletion and subsequent rejoining of one each of the V, D, and J genes to the set of C genes in the multigene family. Transcription of rearranged DNA into pre-mRNA is followed by differential excision and splicing reactions to produce two classes of heavy-chain mRNA. These messengers guide the synthesis of either IgM or IgD heavy chains, but any one lymphocyte makes only one of these types of heavy chains.

TABLE 5.3
Types of Human Immunoglobulins

Immunoglobulin Class	Heavy Chain Present	Functions
IgA	α (alpha)	Main antibody in saliva and intestinal fluids
IgD	δ (delta)	Cell-surface receptor on immature B lymphocytes
IgE	ϵ (epsilon)	Antiparasitic immune response; releases histamine from mast cells (allergic response)
IgG	γ (gamma)	Main serum antibody; activates complement
IgM	μ (mu)	Cell-surface receptor; serum antibody (early); activates complement

B lymphocytes may circulate in the lymphatic system for years without change. On being stimulated by specific antigens, particular B lymphocytes proliferate to produce clones geared to the synthesis of a large quantity of the Ig molecule that engages with an antigen to produce an immune response in the organism. By virtue of gene rearrangements involving relatively few genes, the animal is genetically prepared to defend itself against almost any antigen it may encounter at any time in its life. Any antigenically specific Ig

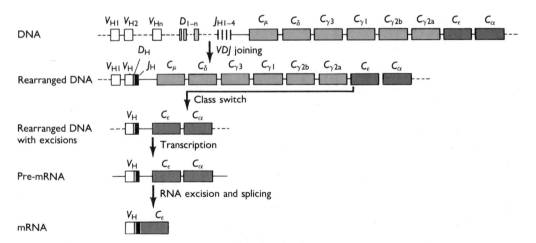

Figure 5.29. Heavy-chain class switching. Antigenic stimulation in developing B lymphocytes leads to the secretion of IgC, IgE, or IgA in cells previously active in making IgM. To switch expression from IgM to another class, deletion of C genes and splicing of remaining C segments to VDJ genes must take place. In the example shown, six C genes are excised, and the two remaining C genes are spliced to VDJ in a rearranged DNA sequence. After transcription of pre-mRNA, the transcript is processed further to produce a functional mRNA that will guide translation of IgE heavy-chain polypeptide.

may undergo a change in biological function by heavy-chain class switching and act in different parts of the body by different kinds of responses to the same antigen. The generation of phenotypic diversity in expression of immune system DNA is surely a major adaptation in animal evolution.

Transposable genetic elements are unique DNA segments that move from place to place in the genome through insertion and excision events governed by particular features of their sequence organization. These mobile elements occur in genomes of viruses, prokaryotes, and eukaryotes, and they may influence gene expression, lead to rearrangements of genomic sequences, and sometimes induce mutations or unstable states near the site of transposition. Only a few highlights of these systems will be presented here, but the literature is extensive.

Transposons (Tn) are the major type of mobile genetic element, and several classes of transposons have been suggested. In general, transposons are about 4500 to 7000 base pairs long, exist in families of dispersed repeat sequences in the genome, and include two genes of their own as well as one or more genes carried as passengers. One indigenous transposon gene codes for a *transposase* enzyme, which initiates the insertion or excision of the transposon into or out of a chromosome, and the other gene codes for a *resolvase* enzyme, which helps complete the insertion or excision (Fig. 5.30). The entire sequence of transposon genes is flanked by various repeat sequences of particular lengths and orientations at each terminus of the mobile element.

Many of the known transposons studied in bacteria carry alleles that confer antibiotic resistance on the cells. In fact, multiple antibiotic resistances can be spread very quickly as strains of pathogenic bacteria pick up one or more of these genes from transposons inserted into bacterial DNA by plasmid vectors.

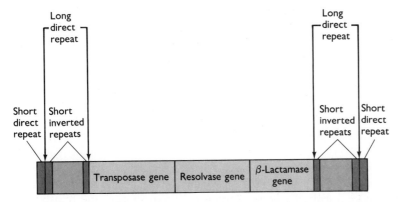

Figure 5.30. Transposable genetic elements and integrated retroviral DNA are organized in very similar fashion. Three kinds of repeat sequences border the gene region at each terminus in both kinds of DNA. Transposons carry a transposase gene and a resolvase gene, which encode enzymes that aid in transposition, plus one or more other genes that are not associated with transposability. In bacterial transposon 3, the β-lactamase gene encodes the enzyme that inactivates the β-lactam antibiotic ampicillin.

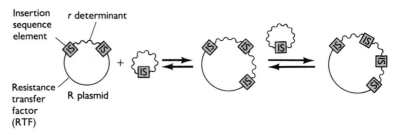

Figure 5.31. R plasmids consist of one or more *r* determinants, which carry antibiotic-resistance genes, and a resistance transfer factor (RTF), which mediates plasmid transfer between cells. The *r* determinants are transposons with insertion sequences (IS) at their terminals. Insertion or removal of *r* determinants may lead to R plasmids carrying different numbers and kinds of drug-resistance genes.

R plasmids exist as DNA circles in the cytoplasm or as integrated parts of bacterial chromosomes, as do other plasmids. R plasmids, however, are composed of a *resistance transfer factor* (RTF), which is a segment that mediates intercellular transfers, and one or more *r determinants,* which are transposons that may carry antibiotic-resistance alleles. The R plasmids may incorporate more than one transposon, or *r* determinant, in the DNA and thereby pick up and transfer to the recipient bacterial cells one or more genes for resistance to particular antibiotics (Fig. 5.31). Bacterial populations may acquire drug resistance very rapidly and produce newly evolved strains very rapidly. From the human viewpoint, however, the sudden and increasing ineffectiveness of antibiotic therapy can present serious problems in the treatment of infections.

Little is known as yet about the functional and evolutionary significance of transposable genetic elements in general, particularly in eukaryotes. They may wander from place to place in a genome, but in many cases no particular proteins are known to be encoded or no cellular function exists for any proteins that have been shown to be translations of transposon sequences. In a few cases, the insertion of a transposon may lead to a sharp increase in the mutation rate of a gene adjacent to the transposition site. It has been suggested that the effects of transposon insertion and excision are related to the regulation of gene expression during development, rather than to short-term contributions based on a few coded polypeptides. Whatever the significance of mobile genetic elements may be, they keep the genome in a state of dynamic flux.

Phenotypic diversity clearly is expanded by rearrangements of genomic DNA, but existing data point to the regulation of transcription and translation as the primary control over gene expression. Transcription and translational controls over the steps involved in information flow from DNA to protein are the workhorse systems producing phenotypic diversity in all organisms. Only a small percentage of variety arises from genomic alterations, but it remains a distinct possibility that these changes may have a substantial impact on development and, perhaps, on evolutionary pathways, particularly in multicellular organisms.

The Storage of Diversity

So far, we have considered a variety of processes by which genic, genotypic, and phenotypic diversity may be produced in cells that make up an individual and in populations of organisms. But how are these resources of diversity managed in biological systems—that is, how can diversity be stored and released in populations?

Diploidy

The major source of new genetic information in all life forms is mutation, by which new alleles and new genes may arise. Mutations are rare, random events and are usually detrimental when expressed in the individual. The problem of retaining new variation, the raw material for evolution, and of protecting the population from the immediate expression of harmful mutant alleles is reduced in **diploid** organisms. The wild-type allele is usually dominant over mutant alleles, and recessive variation is masked from immediate expression in **heterozygotes,** which carry two genomes and two copies of the gene.

Diploidy is a characteristic of sexually reproducing organisms, in which gamete fusion produces a *zygote* with one whole genome from each parent lineage. The interval between the production of the diploid zygote and the restoration of haploidy in the life cycle varies from one group of eukaryotes to another (Fig. 5.32). At one extreme are sexual forms in which the zygote is the only diploid element because meiosis takes place in this cell and initiates the haploid phase for all other elements of the life cycle. Many protists, fungi, and algae fall into this category. At the other extreme exist sexual organisms in which the gametes are the only haploid components because they are the immediate products of meiosis and their fusion restores the diploid phase at once. In some protists and algae and in all animals, the predominant phase in the life cycle is diploidy. Land plants show a clear evolutionary progression in which diploidy becomes increasingly predominant in the life cycle. From these comparisons and from theoretical premises, it is obvious that extension of the diploid phase in the life cycle provides a considerable evolutionary advantage.

As more of the life cycle is included in the diploid phase, the organism produces fewer vulnerable haploid components. For example, mosses and liverworts are persistent haploids that give rise to diploid spore-producing

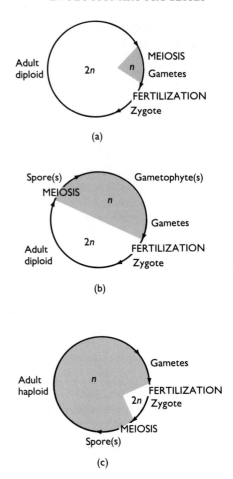

Figure 5.32. Any sexual cycle is punctuated by meiosis and fertilization, but their relative timing varies so the haploid and diploid proportions of the cycle also vary. (a) In animals, meiosis is followed immediately by fertilization, so the diploid phase is predominant. (b) A more extended haploid phase is typical of plants, whose gametophyte stage occurs between meiosis and fertilization. (c) Many protists, fungi, and algae have a brief diploid phase if the zygote undergoes meiosis; thus the zygote may be the only diploid cell in the entire cycle. Regardless of the duration of each phase, the chromosome number of the species remains constant. Reduction in chromosome number in meiosis and doubling of chromosome number in fertilization are compensating events in a sexual life cycle.

structures that are only transiently associated with the predominant haploid plants. This potentially hazardous arrangement is sharply reduced in the vascular plants, all of which have a predominant diploid phase. In ferns, gymnosperms, and flowering plants, the familiar vegetation consists of diploids exclusively. The haploid phase in ferns includes a tiny, free-living form that exists independently of the diploid. This haploid is the gamete-producing component, or *gametophyte*, and it is exposed to all the hazards of existence.

In gymnosperms and even more so in flowering plants, however, the haploid gametophyte has evolved into a microscopic assembly of relatively few cells that is produced and nurtured for its brief existence entirely within the predominant diploid plant (Fig. 5.33). The advantage of diploidy is thus extended for an increasing interval during development.

Even among animals, all of which are diploid, except for the gametes they produce, there is a clear evolutionary progression of greater protection during development of the organism before it is first exposed to the rigors of existence. Among vertebrates, most amphibian hatchlings emerge in an embryonic state, whereas some frog and all reptile and bird eggs hatch out forms that have developed within the egg to an adult stage, albeit in miniature. The mammalian systems include placental species, whose live-born young are nurtured and protected during an interval of gestation within the mother's body. These comparisons indicate that diploidy per se is not an adequate device for the storage of diversity, but that once diploidy has been established it can be reinforced as more of the life cycle is devoted to the diploid state. The fullest capability for the retention of new diversity and its protection from immediate expression occurs when diploidy is maximized, as in the most highly evolved plants and animals.

Many predominantly haploid species have an alternative mechanism to heterozygosity by which genic diversity can be retained and protected from immediate expression. Organisms whose cells are multinucleate, such as most of the fungi, may carry different alleles in different haploid nuclei of a cell. It is possible for the expression of recessive alleles to be masked in the presence of dominant alleles in other nuclei of the cell. Individuals whose nuclei have different alleles of the same genes are called **heterokaryons;** homokaryons have nuclei with identical alleles (Fig. 5.34). The irregularity of the number of nuclei per cell, of the sorting of nuclei into other cells, and of other features in heterokaryons makes this system of storing diversity less advantageous than diploidy.

Controls Governing the Release of Genotypic Diversity

The major mechanism for the production of genotypic diversity in all life forms is genetic recombination by means of crossing over. Sexual eukaryotes have the additional mechanism of independent assortment of members of different pairs of alleles during reproduction. As we discussed earlier, species with all their genes in fewer chromosomes produce fewer genotypes by independent assortment than do species with the same genes distributed among a larger number of chromosomes. One of the evolutionary devices by which genotypic diversity may be released slowly, despite much allelic diversity in the gene pool, is by the reduction of chromosome number after centric fusion or other processes of chromosome restructuring. Similarly, reduced numbers of recombinant genotypes are produced in populations carrying inversions

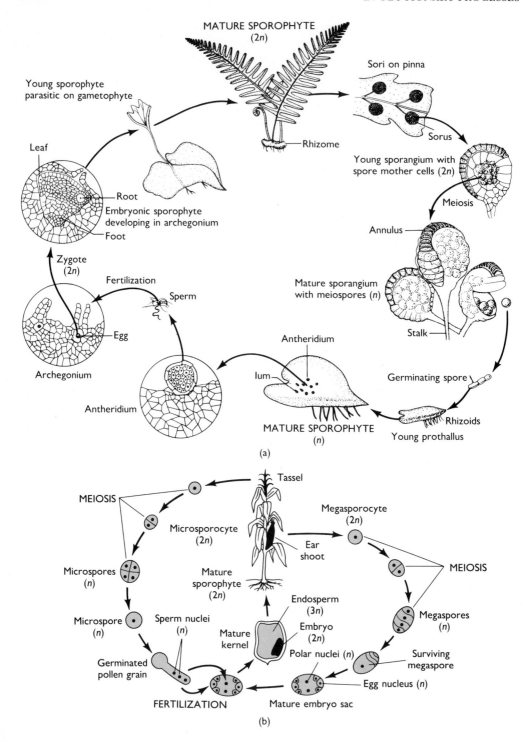

(a)

(b)

Figure 5.33.   Life cycles of plants are characterized by alternating *n* gametophyte and 2*n* sporo-phyte phases. (a) Ferns produce a prominent sporophyte, which produces haploid spores that ger-minate and develop into tiny but independent green gametophytes. Water is required for sperm to swim to the egg-containing organs in the same or in another gametophyte. The fertilized egg develops into a sporophyte plant of the next generation. (b) In flowering plants, illustrated by maize *(Zea mays)*, reproductive cells and structures in the flowers contribute to seed and fruit formation. In maize, male and female flowers are separate but are produced on the same plant (tassel and ear, respectively). The fruit is a kernel, which contains a single seed. Each seed includes a 2*n* embryo, from which the new maize plant will develop, and 3*n* endosperm tissue, which nourishes the growing embryo and young seedling during germination. The sporophyte is conspicuous, whereas the male and female gamete-producing systems are microscopic and totally dependent on the sporophyte for nutrition and protection. This reproductive strategy is less precarious than that of ferns. ([a] from *Biology: A Human Approach*, 3rd ed., by Irwin Sherman and Vilia Sherman. Copyright © 1983 by Oxford University Press. Reprinted by permission. [b] from C. J. Avers, 1986, *Molecular Cell Biology*. Menlo Park, Calif.: Benjamin-Cummings)

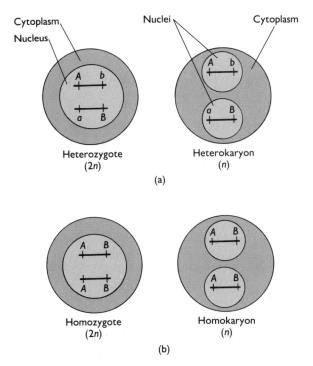

Figure 5.34.   (a) Different alleles of two sets of genes are present in one nucleus in diploid hetero-zygotes *(AaBb)*, but each haploid nucleus in a heterokaryon has either one allele or the other of a gene locus (*Ab* + *aB*). (b) Diploid homozygotes have identical alleles of two sets of genes per nucleus *(AABB)*, but each nucleus in a homokaryon has only one set of genes and carries the same allele for each locus (*AB* + *AB*).

and translocations in the genome. We have discussed many processes by which genetic diversity is generated. Are there, then, advantages in reducing the amount of potential diversity that is actually released and expressed in populations? Or are there disadvantages in damping the release of potential diversity? The answers to these questions are yes; that is, minimizing the release of genetic diversity has both advantages and disadvantages, depending on the cases in question.

On a long-term evolutionary basis, more diversity presumably provides greater opportunities for continued change in the face of fluctuating environments and other challenges of existence. On this premise, therefore, a release of maximum potential diversity may allow some rare but more advantageous genotypes to be expressed under conditions in which previously successful and more abundant genotypes may have become disadvantageous. On a short-term basis, however, an abundance of genotypes released under conditions that are unsuited to their expression would only reduce the chances for the perpetuation of the successful population, in which relatively few potential genotypes would allow the population to prosper and all other genotypes would not.

As an example of the relative advantages and disadvantages of different rates of release of diversity, we can look at populations of the ciliated protozoan *Paramecium,* an asexually and sexually reproducing, predominantly diploid, unicellular eukaryote. One or a few paramecia entering a new habitat for which they are genotypically suited can produce large numbers of identical cells, all with the genotype appropriate for this habitat, by asexual fission. The release of minimum genotypic diversity is ensured by asexual reproduction. If the habitat becomes inhospitable to the paramecia, because of reduced nutrients or some other cause, the limited genotypic diversity is unlikely to allow the continued success of the population. It is at such times of stress that paramecia are triggered to undergo sexual reproduction through meiosis and nuclear exchange to produce new diploid genotypes, one or a few of which may be more successful in the changed conditions. Paramecia are unusual, but not unique, in combining the advantages and disadvantages of high and low levels of released diversity through sexual or asexual reproduction under different living conditions. In most cases, species have a genetic system characterized by a high or a low level of released diversity at all times.

Species that have undergone genomic alterations that limit the release of diversity have opted for the short-term evolutionary benefits of genetic constancy. Species that release a great deal of their potential diversity may produce many offspring with genotypes that are not suited to present conditions, but the populations can face the challenges of change in their environment more effectively over the long term of evolutionary time. These variations in genetic systems will be considered in greater detail in Chapter 6.

Now that we have touched on the production and release of genetic diversity, these phenomena must be viewed in the context of natural selection, which acts on genetic diversity and often leads to evolutionary adaptation.

Indeed, the production and retention of genetic diversity and the variable rates of its release cannot be placed in proper perspective in the absence of natural selection, the single most important unifying principle in all of evolution and, indeed, in all of biology.

Summary

Inherited variation is the raw material of evolution. Genetic diversity may be genic, genotypic, or phenotypic, and may be produced by a number of genetic processes and mechanisms. Genic diversity is represented by different alleles of existing genes, produced by mutations in both structural and regulatory gene sequences, and by new genes carrying exons recruited from other genes through recombination involving intron segments or through divergence following duplication. Genotypic diversity is generated in all organisms by genetic recombination through crossing over between linked genes and, additionally, in sexual species by independent assortment of members of pairs of unlinked genes. Both independent assortment and genetic recombination occur regularly during meiosis in sexual organisms. In asexual life forms, the genes from different lineages may be brought together by conjugation or DNA transformation, but mainly by transduction involving virus or plasmid vectors. Recombination is an irregular event in these cases.

Phenotypic diversity in genetically identical cells or populations involves differential gene expression regulated by one or more controls. Transcription controls are most common, but different specific controls are active in prokaryotes and eukaryotes. Operator–repressor and attenuation controls are typically bacterial, but absent in eukaryotes, which have alternative control mechanisms by which transcription is turned on or off in specific cells at specific times during development. In addition, chemical and organizational modifications in eukaryotic chromosomes are associated with transcriptionally active chromatin. Posttranscriptional and translational controls also regulate gene expression without changing gene sequence, number, or location in the genome, and are like transcription controls in this respect. Another set of eukaryotic controls regulate gene expression by means of alterations in the genome. Among these mechanisms are gene amplification, gene rearrangements, and transposable (mobile) genetic elements, such as transposons.

Diploidy provides a means for the storage of new genetic material, particularly by the masking of usually detrimental mutant alleles from immediate expression in heterozygotes. Sexual species have alternating diploid and haploid phases in their life cycle, and more highly evolved organisms have a greatly prolonged diploid phase and a brief, microscopic haploid component. Although greater genetic diversity is equated with more abundant raw materials for evolutionary change, controls that govern the release of genotypic diversity permit both long-term and short-term advantages for populations, usually in accord with habitat conditions and other features of the life style.

References and Additional Readings

Beermann, W., and U. Clever. 1964. Chromosome puffs. *Sci. Amer.* 210(4):50.

Borst, P., and D. R. Greaves. 1987. Programmed gene rearrangements altering gene expression. *Science* 235:658.

Brown, D. D. 1981. Gene expression in eukaryotes. *Science* 211:667.

Brown, D. D., and I. B. Dawid. 1968. Specific gene amplification in oocytes. *Science* 160:272.

Brown, M. S., and J. L. Goldstein. 1986. A receptor-mediated pathway for cholesterol homeostasis [Nobel lecture]. *Science* 232:34.

Busslinger, M., J. Hurst, and R. A. Flavell. 1983. DNA methylation and the regulation of gene expression. *Cell* 34:197.

Carlson, T. A., and B. K. Chelm. 1986. Apparent eukaryotic origin of glutamine synthetase II from the bacterium *Bradyrhizobium japonicum. Nature* 322:568.

Clowes, R. C. 1973. The molecule of infectious drug resistance. *Sci. Amer.* 229(4):18.

Cohen, S. N., and J. A. Shapiro. 1980. Transposable genetic elements. *Sci. Amer.* 242(2):40.

Crow, J. F. 1979. Genes that violate Mendel's rules. *Sci. Amer.* 240(2):134.

Darnell, J. E., Jr. 1982. Variety in the level of gene control in eukaryotic cells. *Nature* 297:365.

D'Eustachio, P., and F. H. Ruddle. 1983. Somatic cell genetics and gene families. *Science* 220:919.

Dickson, R. C. et al. 1975. Genetic regulation: the lac control region. *Science* 187:27.

Dobzhansky, T. 1970. *Genetics of the Evolutionary Process.* New York: Columbia University Press.

Donelson, J. E., and M. J. Turner. 1985. How the trypanosome changes its coat. *Sci. Amer.* 252(2):44.

Doolittle, W. F. 1987. What introns have to tell us: hierarchy in genome evolution. *Cold Spring Harbor Sympos. Quant. Biol.* 52:907.

Doolittle, W. F., and C. Sapienza. 1980. Selfish genes, the phenotype paradigm, and genome evolution. *Nature* 284:601.

Dunn, L. C. 1965. *A Short History of Genetics.* New York: McGraw-Hill.

Elgin, S. C. R. 1981. DNAse I–hypersensitive sites of chromatin. *Cell* 27:413.

Federoff, N. V. 1984. Transposable genetic elements in maize. *Sci. Amer.* 250(6):84.

Felsenfeld, G., and J. McGhee. 1982. Methylation and gene control. *Nature* 296:602.

Garabedian, M. J., M.-C. Hung, and P. C. Wensink. 1985. Independent control elements that determine yolk protein gene expression in alternative *Drosophila* tissues. *Proc. Natl. Acad. Sci. U.S.* 82:1396.

Gilbert, W. 1978. Why genes in pieces? *Nature* 271:501.

Gilbert, W., and D. Dressler. 1968. DNA replication: the rolling circle model. *Cold Spring Harbor Sympos. Quant. Biol.* 33:473.

Hentze, M. W. et al. 1987. Identification of the iron-responsive element for the translational regulation of human ferritin mRNA. *Science* 238:1570.

Jacob, F. 1966. Genetics of the bacterial cell [Nobel lecture]. *Science* 152:1470.

John, B., and G. Mikos, eds. 1987. *The Eukaryote Genome in Development and Evolution*. Winchester, Mass.: Allen and Unwin.

Korge, G. 1977. Direct correlation between a chromosome puff and the synthesis of a larval salivary protein in *Drosophila melanogaster*. *Chromosoma* 62:155.

Lambert, B. 1975. The chromosomal distribution of Balbiani ring DNA in *Chironomus tentans*. *Chromosoma* 50:193.

LaPorte, D. C. 1984. Antisense DNA: a new mechanism for the control of gene expression. *Trends Biochem. Sci.* 9:463.

Leder, P. 1982. The genetics of antibody diversity. *Sci. Amer.* 246(5):102.

Ley, T. J. et al. 1982. 5-Azacytidine selectively increases γ-globin synthesis in a patient with β^+ thalassemia. *New Engl. Jour. Med.* 307:1469.

Lonberg, N., and W. Gilbert. 1985. Intron/exon structure of the chicken pyruvate kinase gene. *Cell* 40:81.

Maniatis, T., S. Goodbourn, and J. A. Fischer. 1987. Regulation of inducible and tissue-specific gene expression. *Science* 236:1237.

Martin, G. R. 1982. X-chromosome inactivation in mammals. *Cell* 29:721.

McClintock, B. 1984. The significance of responses of the genome to challenge [Nobel lecture]. *Science* 226:792.

McKnight, S. L., and R. Kingsbury. 1982. Transcriptional control signals of a eukaryotic protein-coding gene. *Science* 217:316.

Muller, H. J. 1927. Artificial transmutation of the gene. *Science* 66:84.

Muller, H. J. 1947. The production of mutations [Nobel lecture]. *Jour. Hered.* 38:259.

North, G. 1984. Multiple levels of gene control in eukaryotic cells. *Nature* 312:308.

Novick, R. D. 1980. Plasmids. *Sci. Amer.* 243(6):102.

Ohno, S. 1987. Early genes that were oligomeric repeats generated a number of divergent domains on their own. *Proc. Natl. Acad. Sci. U.S.* 84:6486.

Orgel, L. E., and F. H. C. Crick. 1980. Selfish DNA: the ultimate parasite. *Nature* 284:604.

Pacquin, C., and J. Adams. 1983. Frequency of fixation of adaptive mutations is higher in evolving diploid than haploid yeast populations. *Nature* 302:495.

Ptashne, K., and S. N. Cohen. 1975. Occurrence of insertion sequence (IS) regions on plasmid deoxyribonucleic acid as direct and inverted nucleotide sequence duplications. *Jour. Bacteriol.* 122:776.

Ptashne, M. 1986. *A Genetic Switch in a Bacterial Virus*. New York: Freeman.

Ravin, A. W. 1965. *The Evolution of Genetics*. New York: Academic Press.

Reudelhuber, T. 1984. Upstream and downstream control of eukaryotic genes. *Nature* 312:700.

Schimke, R. T. 1980. Gene amplification and drug resistance. *Sci. Amer.* 243(5):60.

Scholer, H. et al. 1986. In vivo competition between a metallothionein regulatory element and the SV40 enhancer. *Science* 232:76.

Spradling, A. C., and A. P. Mahowald. 1980. Amplification of genes for chorion proteins during oogenesis in *Drosophila melanogaster*. *Proc. Natl. Acad. Sci. U.S.* 77:1096.

Stadler, L. J. 1928. Mutations in barley induced by x-rays and radium. *Science* 68:186.

Stahl, F. J. 1987. Genetic recombination. *Sci. Amer.* 256(2):90.

Starlinger, P. 1984. Transposable elements. *Trends Biochem. Sci.* 9:125.

Stebbins, G. L. 1971. *Chromosomal Evolution in Higher Plants*. Reading, Mass.: Addison-Wesley.

Südhof, T. C. et al. 1985. The LDL receptor gene: a mosaic of exons shared with different proteins. *Science* 228:815.

Suzuki, S. et al. 1986. cDNA and amino acid sequences of the cell adhesion protein receptor recognizing fibronectin reveal a transmembrane domain and homologies with other adhesion protein receptors. *Proc. Natl. Acad. Sci. U.S.* 83:8614.

Tonegawa, S. 1983. Somatic generation of antibody diversity. *Nature* 302:575.

Watson, J. D. et al. 1988. *Molecular Biology of the Gene*, 4th ed. Menlo Park, Calif.: Benjamin/Cummings.

Weintraub, H., and M. Groudine. 1976. Chromosomal subunits in active genes have an altered conformation. *Science* 193:848.

White, M. J. D. 1973. *Animal Cytology and Evolution*, 3rd ed. New York: Cambridge University Press.

Wills, C. 1970. Genetic load. *Sci. Amer.* 222(3):98.

Wolter, F. P. 1988. *rbcS* genes in *Solanum tuberosum:* conservation of transit peptide and exon shuffling during evolution. *Proc. Natl. Acad. Sci. U.S.* 85:846.

Yanofsky, C. 1981. Attentuation in the control of expression of bacterial operons. *Nature* 289:751.

Zakour, R. A., and L. A. Loeb. 1982. Site-specific mutagenesis by error-directed DNA synthesis. *Nature* 295:708.

CHAPTER 6

Natural Selection and Adaptation

In spite of alternating flurries of praise and scorn over the years for Darwin's principle of natural selection, it has weathered the different climates of opinion very well indeed and serves today as a foundation for biology. Selection is generally viewed as the guiding force by which variation unsuited to particular conditions is diminished and more suitable variation is established in populations over the course of time. In this chapter, a number of aspects of the principle of natural selection will be examined, as well as certain features of evolutionary patterns that reveal the versatility of selection. We will also pay attention to some recent challenges to classic selectionist notions and to the relativity of adaptation as the evolutionary outcome of selection acting on biological diversity.

The Concept of Selection

A comparison of living organisms and of life forms preserved over time in the fossil record clearly reveals a variety of evolutionary progressions from simpler to more complex and sophisticated systems. It seems impossible that rare, random, generally detrimental mutations can have been responsible for these evolutionary novelties and sequences. It was Darwin's intellectual contribution to biology of an incisive theory of natural selection that provided the first approach to resolving the paradox, long before much was known about heredity. Since Darwin's time, a huge and varied body of evidence has been gathered in support of natural selection, particularly from genetics.

Basic Principles of Natural Selection

The formal theory that **natural selection** guides inherited variation in the general direction of adaptation was first proposed by Charles Darwin and Alfred Russel Wallace in 1858. The two British naturalists had come to similar conclusions based on independent studies, and their communications with each other led to their being invited to present a joint report to the Linnaean Society in London. Darwin is credited almost entirely as author of the theory, however, for various reasons, primarily because of his monumental book *On the Origin of Species* and his other books and research publications that appeared until his death in 1882. *On the Origin of Species by Means of Natural Selection* was published in 1859, and went through another five editions. It has never been out of print and has been translated into about 30 languages, which bespeak its status as a scientific classic.

Both Darwin and Wallace were influenced by the Malthusian principle that population size remains relatively constant because of competition for finite resources, even though the potential exists for a geometric increase in numbers of individuals in each generation. Darwin proposed a "struggle for existence," in which the more fit individuals would leave more descendants (like themselves) than would the less fit individuals. These differences in inherited **fitness** lead to differences in populations over time; that is, evolution takes place by the action of natural selection on individuals with varied hereditary characteristics (Fig. 6.1). *Descent with modification,* or evolution, occurs gradually by the accumulation of small inherited differences that make some individuals more fit and thus more likely to succeed in the competition for finite resources, transmitting these favored properties to their offspring generation after generation. "Survival of the fittest," as Darwin's proposal was dubbed

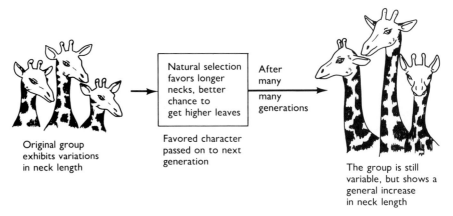

Figure 6.1. Charles Darwin proposed that, by means of natural selection, differences in the inherited fitness of individuals eventually would lead to differences in populations over evolutionary time.

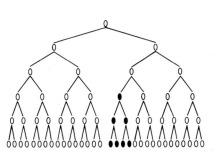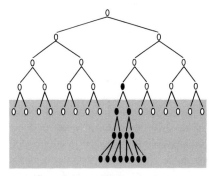

Growth in normal environment   Growth in modified environment

Figure 6.2. Natural selection can be viewed as differential reproduction of genetically diverse individuals, relative to their fitness in particular environments and to other conditions for successful existence. Changes in the proportion of diverse types in population leads to evolutionary modifications in species.

by Herbert Spencer, provides the *mechanism* by which evolution occurs; the *fact* that evolution had occurred was a separate matter that Darwin addressed and documented in great detail in his book. Two different questions were presented and discussed by Darwin: Has evolution occurred, and, if so, by what mechanism have organisms changed and diversified over time?

Darwin's data and arguments for the fact of evolution, rather than special creation, not only were persuasive but also had a receptive educated audience. His proposed mechanism of evolution by natural selection, however, was controversial in his day and remains open to challenges. A large body of genetic and other lines of evidence, much of which was not available in Darwin's time, supports the theory of natural selection. In fact, the rediscovery of Gregor Mendel's work in 1900 and the incorporation of genetics into selection theory have given the greatest boost to the acceptance of Darwin's basic ideas.

Natural selection can be defined as the *differential reproduction of genetically diverse organisms* (Fig. 6.2). This definition accepts Darwin's emphasis on selection at the level of the individual organism. There is no consensus even today, however, on the basic unit of evolutionary change. It ranges from the single individual to entire species, but the majority of biologists accept the organism as the unit of evolutionary change. As will be discussed in Chapter 7, selection theory has been fleshed out by the concept of the gene pool, or total allelic diversity in populations. Individual organisms contribute to and draw from the gene pool through reproduction over successive generations. The reproductive consequences of selection on the organism are thus translated into consequences for populations and species over evolutionary time.

As Sir Ronald Fisher, one of the founders of population genetic analysis of evolution, stated succinctly, natural selection provides the mechanism for generating an exceedingly high degree of improbability. It offers the means by which rare beneficial changes may be incorporated and become predominant in a population at the same time that harmful changes are weeded out or greatly diminished. The improbability of direction toward more complex and sophisticated organisms, given the nature of mutations, is rendered highly probable by the differential reproduction of genetically diverse individuals. Even though Darwin wrote that *nature selects* the more fit over the less fit, he was making a literary comparison with the known changes introduced into domesticated plant and animal varieties by artificial selection practiced by people; that is, *humans select* and thereby direct changes along a desired pathway. It is more appropriate to view natural selection in terms of differential reproduction based on inherent fitness relative to particular living conditions. We have no notion of an active physical force, called nature, that chooses some organisms and not others to be successful. Fitness is an outcome of inheritance, which is expressed as differential reproduction. Those inherently better fit are more likely to predominate because of the nature of their inheritance, not because of deliberate action by the totality we refer to as nature.

An excellent example of evolution in action and the workings of natural selection in guiding genetically diverse populations in particular directions is provided by studies of the peppered moths *(Biston betularia)* of Great Britain. Before the onset and spread of the Industrial Revolution, light-colored moths predominated, and dark, or melanistic-phase, moths were extremely rare, as seen in private collections of the time (Fig. 6.3). As factory chimneys belched filth into the air and increasing amounts of soot settled on tree trunks and other places where these moths alight, a corresponding increase of melanistic-phase moths occurred. In areas that remained unindustrialized, light moths remained predominant. The change in frequencies of the dark and the light alleles of the body-color gene was a direct outcome of the advantages of protective coloration in different environments. Light moths are conspicuous against the darkened backgrounds in industrialized regions and are subject to heavy predation by birds, but their body color remains advantageous in regions in which trees and buildings are not sooty. Melanistic-phase moths have advantages on darkened backgrounds, but suffer heavy predation by birds in the areas that allow light moths to flourish. Each expression of protective coloration is advantageous in one set of conditions but not in the other. Interestingly, the frequency of melanistic-phase moths in England decreased significantly after a program was introduced to reduce the emission of soot and other components of air pollution. Change in allelic frequencies thus accounts for evolutionary change guided by natural selection acting on allelic diversity. Allelic diversity allows dark and light moths to reproduce in both kinds of environments, but the fitness of each genotype is relative to the environment. Examples of industrial melanism exist in other insects and in spiders in different regions of the world.

Figure 6.3. Protective coloration of dark-phase and light-phase peppered moths (*Biston betularia*) in different environments. Each kind of moth is less subject to predation in a particular environment, because of its camouflage against a dark or light background. Each kind thus comes to predominate over the other kind in its own more suitable environment. These genetic color variants provide a good example of evolutionary guidance by natural selection. (Reprinted from Colin Patterson, *Evolution,* Copyright © Trustees of the British Museum (Natural History) 1978. Used by permission of the publisher, Cornell University Press)

Lamarckism and Its Experimental Disproof

Jean-Baptiste Lamarck wrote the first historically important proposal for the occurrence of evolution of more complex organisms from simpler ancestors. In his book *Philosophie zoologique,* published in 1809, Lamarck compiled a systematic and thorough collection of observations that documented his thesis of evolution by descent. The mechanism by which evolutionary diversity and direction occur was proposed to be based on the common theme of changes in the organism taking place in response to environmental stimuli or condi-

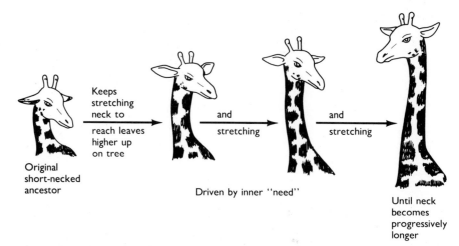

Keeps stretching neck to reach leaves higher up on tree

Original short-necked ancestor

and stretching

Driven by inner "need"

and stretching

Until neck becomes progressively longer

Figure 6.4. Jean-Baptiste Lamarck proposed that evolutionary changes occur in response to the needs of an organism in particular environments, and that such changes acquired during the lifetime of individuals will be inherited by their descendants (compare Fig. 6.1).

tions. Each change benefits the organism, and the acquired traits are thereafter inherited by its offspring. These changes accumulate as the organism evolves toward higher status, which Lamarck believed to be the direction and ultimate achievement of evolution. Lamarck also emphasized changes resulting from the use and disuse of body parts; parts that are used are retained and even enhanced, whereas parts that are not used are diminished and eventually lost (Fig. 6.4). Lamarck's theory of evolution is called the **theory of inheritance of acquired characteristics,** or the **theory of use and disuse,** according to his proposals.

Many, but not all, biologists accepted the major premises of Lamarck's theory, but its validity was eroded by the advent of genetic studies after 1900. Mutations are random and usually harmful, whether they are spontaneous or induced, and clearly are not directed toward a specific modification that would benefit the organism. On a molecular basis, biologists know of no way by which information from the environment can be transmitted and received by a particular gene and lead that gene to change, whether for good or ill. Despite the overwhelming evidence against Lamarckism, it is so persuasive a philosophical doctrine that it is accepted even today by some scientists and by the majority of nonscientists.

One of the last major holdouts for Lamarckism existed in microbiology, as late as the 1940s. Bacteria initially sensitive to drugs and viruses might be altered overnight into populations that are resistant to these agents, and resistance is retained in their progeny even in drug-free and virus-free cultures. The speed and direction of change, which is of benefit to the organism, and the inheritance of this altered trait by the offspring smacked of Lamarckism. Was it possible that bacterial studies revealed the truth of Lamarckism, which was not apparent from genetic analysis of plants and animals? The problem was addressed in landmark experiments conducted by Salvador Luria and Max Delbrück in 1943.

The experiments were designed to determine whether beneficial changes are the result of specifically directed mutations, in accordance with Lamarckism, or of natural selection acting on preexisting spontaneous mutants in a modified environment. Virus-sensitive *E. coli* cells were inoculated into culture media containing viruses and were allowed to grow for 24 hours. Samples of each culture were then spread on solid medium to permit each cell to develop into a colony on plates containing viruses in the nutrient agar media. Control cultures were grown in flasks of virus-free nutrient medium.

Mutually exclusive predictions were made for the two possible sets of results (Fig. 6.5). If mutants were induced in direct response to the presence of the virus, approximately the same number or proportion of mutants should be found in all the experimental populations because all were exposed for the same amount of time to the same environment. If natural selection is the evolutionary process, however, the number of virus-resistant colonies would fluctuate from one population to another because of the random nature of mutation. Some cultures by chance would have no mutants, and other cultures might have a few mutants or many mutants. Populations lacking mutants would be those that happened to have included no spontaneous mutants in the original inoculum, as well as no mutants that had arisen by chance during the initial 24-hour growth period before plating. Relatively few mutants would be found in populations of sensitive cells in which a mutation had occurred during growth, and in the few progeny cells produced by the spontaneously mutated parent cell. Many mutants would be present if the original inoculum had contained a virus-resistant cell that had produced many progeny cells carrying the same transmitted trait. The **fluctuation test** experiment thus allowed Luria and Delbrück to compare the results observed with the results expected on the basis of each of the alternative theories. Their data provided evidence for natural selection of spontaneously occurring mutation, which might be amplified by differential reproduction in a suitable environment.

These experimental results were widely accepted as proof of natural selection and disproof of Lamarckism, which was what most scientists had believed to be true all along. The undercurrent of uncertainty that continued to exist among microbiologists was due to their wariness of conclusions based largely on statistical analysis, particularly when great emphasis was placed on negative results—the absence of mutants in many of the sampled populations. In addition, all the cells in experimental populations had been continuously exposed to viruses. Was it possible that mutations had been specifically induced, but that the detection of Lamarckian change was obscured by the experimental method and the limited knowlege or ignorance of relevant biological processes? Various experimental studies were pursued after 1943, but the definitive experimental method needed to find virus-resistant mutants in virus-free environments was not reported until 1952. These experiments,

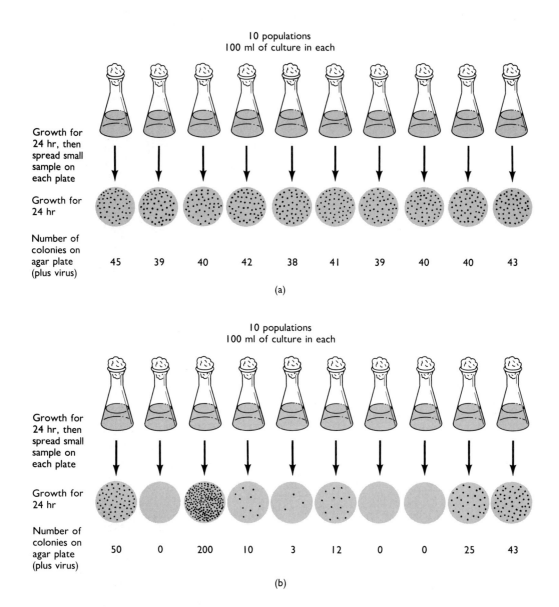

Figure 6.5. Design of the fluctuation test to distinguish between induction of new mutants versus natural selection of preexisting mutants in populations subjected to a hostile environment. A population of bacteria is divided, and each subpopulation is analyzed after growth in the presence of viruses (or other agents) to which the bacterial cells are sensitive. (a) The number of colonies of resistant cells sampled from each subpopulation will be more or less the same if the environmental agent induced mutations to resistance, because the probability of mutation would be the same in each subpopulation of cells. (b) The number of colonies of resistant cells will fluctuate between none and hundreds per sampled subpopulation, depending on the number of preexisting mutants that happened to be present, if changes reflect the guidance of selection during the experiment.

conducted by Joshua Lederberg and Esther Lederberg, convinced micro-biologists and others that Lamarckism was invalid as an evolutionary theory.

The Lederbergs designed a **replica plating test,** in which virus-sensitive *E. coli* populations were spread onto virus-free medium and allowed to grow throughout the experiment. Samples were removed from this medium for replica plating on virus-containing medium in order to identify resistant mutants from the original culture. The transfers were made by pressing a piece of velvet to the plated cells, and then pressing the velvet onto the virus-containing medium. Cells adhering to the nap of the velvet were pressed onto the agar, where they could grow, and the orientations of original virus-free plates and replica virus-containing plates were marked by pins that left impressions in both sets of plates.

By reference to the pin markings, cells were picked from the virus-free plates that corresponded in location to resistant cells that had grown on the replica plates. These cells were inoculated into fresh virus-free medium and allowed to grow for further rounds of replica plating (Fig. 6.6). Eventually, the cells were sufficiently diluted to obtain well-isolated colonies on the virus-free plates, each colony consisting of the progeny of a single cell. Now, for the first time, cells from virus-free medium were replica-plated to virus-containing medium to verify their virus-resistant trait and the resistance inherited by all the progeny of a cell in an individual colony.

The results of the replica plating test clearly showed that spontaneous mutations arise even in the absence of specific inducing agents. If the mutations are of little or no immediate advantage, the mutants remain relatively infrequent until the environment changes. In the Lederbergs' experiments, the presence of viruses in the environment conferred great advantage on spontaneously resistant mutants. They proliferated at the same time that virus-sensitive cells failed to reproduce, and the populations changed over-night as a result of the differential reproduction of resistant and sensitive bacteria. The rapidity of change is due largely to the high rate of reproduction in bacteria, many of which have a generation time of less than 1 hour. One cell-doubling every 30 minutes by an original cell and its offspring could produce a population of 2^{48} identical cells in 24 hours. This astronomical number is never reached, fortunately, because of limited space and resources for proliferation, but a flask containing even one mutant could produce billions of descendants overnight if the environment was suitable.

The fluctuation test and replica plating test experiments were conducted with bacteria and focused on their resistance to viruses. The same principles of mutation and selection extend to all life forms. Evolutionary diversity is the result of natural selection acting on genetically diverse organisms, usually in the direction of adaptation. As will be discussed in the next sections, different patterns of selection lead to different patterns of evolutionary diversity. Selection acts on diversity in ways that are relative to conditions for existence, not in some absolute parameter of evolutionary change.

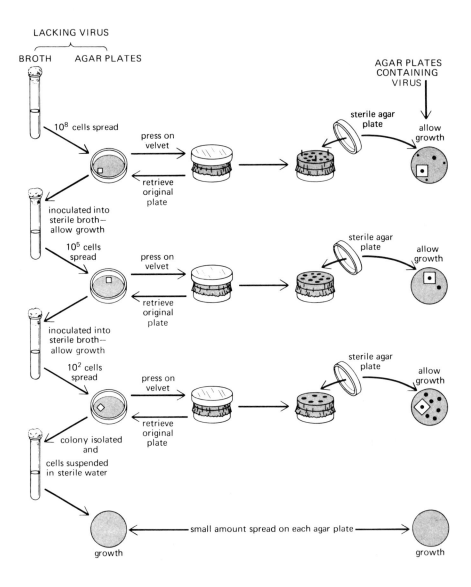

Figure 6.6. Design of the replica plating test to distinguish between induction of new mutations in response to environmental factors versus natural selection of preexisting spontaneous mutants in populations. The presence of rare preexisting virus-resistant mutants in a virus-free environment is detected by successive rounds of replica plating. The virus-resistant mutants that are eventually isolated are never in contact with viruses, but their identical progeny cells growing on virus-containing replica plates provide the information needed to pinpoint the rare mutants in a preponderantly virus-sensitive cell population growing on virus-free medium.

Selection Strategies

Darwin referred to the "struggle for existence" among individuals, by which he meant the competition for the finite resources on which successful existence depends. By natural selection, the more fit compete successfully for these resources, and the less fit are left farther and farther behind in the metaphorical contest among genetically different individuals. Conditions for life vary from one place to another and from one time to another, so the effects of selection must be considered in relation to the space–time mosaic of the environment. Different modes, or strategies, of selection acting on diversity have different effects on the quantity and quality of adaptation that may be achieved. In this discussion, we will use the simplest models, in which the different phenotypes are expressions of alleles at a single gene locus.

Strategies that Reduce Diversity

The great majority of mutations are detrimental, and their accumulation in the gene pool is impeded by **stabilizing,** or *normalizing,* **selection**—the most common of all the selection strategies (Fig. 6.7). In the case of a dominant mutant allele that lowers fitness, the mutant phenotype is expressed in heterozygotes and is thus immediately exposed to selection. The predominant homozygous recessive genotype is maintained as selection acts to reduce diversity at this gene locus, by weeding out the new mutant phenotype or by reducing its frequency through differential reproduction. In instances that have been analyzed quantitatively, it is clear that the frequency of carriers of a dominant mutant allele is lower than the frequency expected according to the mutation rate. This reduction in frequency is the outcome of selection against the mutant phenotype, which leads to the maintenance of a reduced level of diversity at this gene locus. The population is stabilized, or normalized, as a consequence of this selection strategy.

In the case of a recessive mutation, the mutant allele is sheltered in heterozygotes whose wild-type, "normal" allele masks expression of the recessive alternative. Recessive mutations can thus accumulate in populations, until the frequency of heterozygotes increases and matings between heterozygotes produce the less fit or unfit recessives. In stabilizing selection, some equilibrium eventually is reached such that the number of mutant alleles eliminated equals the number of mutations arising in each generation. The delay in exposure of harmful recessive alleles to selection is responsible for the *genetic load* carried in populations of human beings and other diploid sexual species, as mentioned in Chapter 5.

Any change in the fitness of the mutant phenotype will lead to a change in the frequency of the harmful recessive allele in populations. Children born with the recessive phenotype of *phenylketonuria* suffer the effects of a meta-

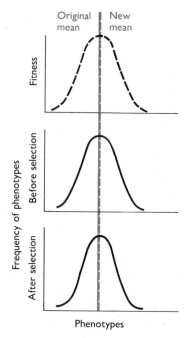

Figure 6.7. Stabilizing (normalizing) selection produces little change in phenotype frequencies, because an equilibrium is reached eventually as the number of detrimental alleles eliminated comes to equal the number of new mutations at the locus in each generation. The original mean (solid vertical line) and the eventual new mean (dashed vertical line) of phenotype frequencies are about the same value, centering on the abundant heterozygotes.

bolic disorder that causes brain damage leading to gross mental retardation. The disorder is due to the child's inability to enzymatically convert phenyl-alanine into tyrosine during the processing of dietary proteins (Fig. 6.8). If the child is restricted to a suitable low-protein diet during the first 6 years, little phenylalanine is available for metabolism, and the devastating effects of brain damage can be avoided. Youngsters older than 6 are in less danger of developing the disorder. The recessive genotype remains unchanged, but the fitness of the expressed genotype is increased by the dietary program. The frequency of the recessive allele is also increased because recessive homozy-gotes and heterozygotes may transmit the allele to their offspring.

The relative fitness of a genotype may vary from one environment to another and from one time to another. When conditions are relatively unchanging, stabilizing selection reduces genetic diversity in populations to a maintenance level. When environments change, one homozygous genotype may become more fit as the alternative genotype becomes less fit. In these cases, differential reproduction leads to the pattern of **directional selection**, in which one homozygous genotype is favored and the other is not (Fig. 6.9). The overall consequence is a reduction in diversity, as the less fit genotype is sharply reduced or even eliminated. Fewer of these homozygotes will be produced, and fewer heterozygotes as well, as the highly fit homozygotes become the predominant or exclusive element in the population.

Examples of directional selection are among the most spectacular because we have been witness to the events in our own lifetimes. Changes in the frequencies of light and dark peppered moths and other examples of industrial melanism, as mentioned earlier, are well-documented cases of shifts in the frequency of one homozygous genotype relative to the alternative genotype in changing surroundings. Equally dramatic have been the increases in frequency of antibiotic-, pesticide-, and herbicide resistance in species that significantly affect human welfare. In many cases, the change takes place in the frequency of alternative alleles at a single gene locus, in response to changed selection pressures in new environments. In other cases, such as pesticide resistance, many different mutations at the same or different gene loci can produce resistance.

The scenario is very much the same in all these examples of resistant strains replacing the previously predominant sensitive strains of organisms. The initial challenge of a new chemical leads to a reasonably satisfactory level of control with a low dose or infrequent treatment. As time goes by, a higher dose or more frequent treatment is required for adequate control; ultimately, the amount needed for effective results becomes uneconomical or unsafe for human beings. By the relentless action of directional selection, the alleles for sensitivity are virtually wiped out, and resistant genotypes become widespread and firmly established. These genotypes will persist even if the chemical is removed, unless the resistant forms have lowered fitness in environments that have reverted to a chemical-free condition. Even so, the effects of directional selection are such that allelic diversity is reduced for a long time, and the battle against many pests and pathogens may continue to be frustrating and counterproductive.

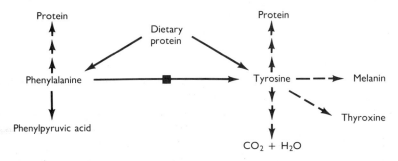

Figure 6.8. The normal pathway of processing phenylalanine from dietary protein is altered in phenylketonuric children. Thus the production of tyrosine is blocked, and phenylpyruvic acid is produced (and excreted) in excess. The child can develop normally if restricted to a low-protein diet, but the recessive alleles remain unchanged even though phenotypic fitness is increased in the individual.

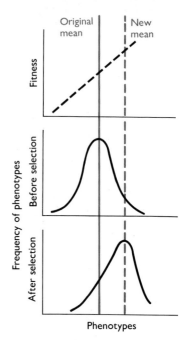

Figure 6.9. Directional selection favors one homozygous genotype over the other in a changed environment, leading to reduced diversity in a population as the frequency of the less fit genotype is diminished.

Strategies that Increase Diversity

Three particular strategies—diversifying selection, frequency-dependent selection, and balancing selection—engender increased diversity by favoring different genotypes in a heterogeneous environment in which the populations live. These strategies are not necessarily mutually exclusive, and all are associated with the phenomenon of **polymorphism.** Polymorphic populations are characterized by the presence of two or more variants that occur in frequencies too high to be due to recurrent mutation alone. Populations are considered to be monomorphic if at least two genotypic or phenotypic classes cannot be distinguished; that is, only one allele appears to exist at a particular gene locus in all members of the population.

In **diversifying selection,** or *disruptive selection,* two or more genotypes have relatively high fitness in different habitats of a heterogeneous environment, which is, after all, the usual nature of a species range. The favored genotypes are usually homozygotes, so that heterozygotes become infrequent in all or most of the mosaic of surroundings, and one or another of the homozygous classes predominates in an environmental niche. The evolutionary pattern called *Batesian mimicry,* named for the English naturalist H. W. Bates, is believed to develop by the agency of diversifying selection. Batesian mimicry is exemplified by the evolved phenotypic similarity between a distasteful species and an edible species selected visually by a predator (Fig. 6.10). The predator learns to avoid the distasteful species by its pattern, and it will also avoid mimic species with a very similar pattern. Mimicry succeeds if the distasteful species far outnumbers the edible mimic, which ensures that the predator learns its lesson well and no longer pursues the pattern it associates with an undesirable food.

One or a few alleles are believed to become predominant by means of diversifying selection, and subsequent selection for *modifier genes* contributes to

Figure 6.10. Batesian mimicry in African butterflies. Three distasteful species (left) are mimicked by strains of the species *Papilio dardanus* (center) and *Hypolimnas* (right). (From W. Wickler, 1968. *Mimicry in Plants and Animals.* Used with permission of McGraw-Hill Book Company, New York)

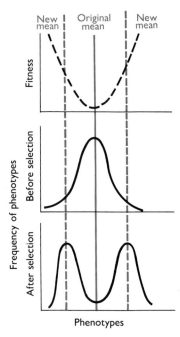

Figure 6.11. Diversifying (disruptive) selection acts against less fit heterozygotes, leading to increased diversity of the more fit homozygous phenotypes in populations.

the production of several mimic phenotypes in a species. Experimental crosses between differently patterned mimics produce phenotypically intermediate offspring, which leads us to expect the elimination of less fit heterozygous intermediates and the preservation of homozygous mimic genotypes in a range of habitats (Fig. 6.11). In this situation, diversifying selection acts against heterozygotes but leads to increased diversity in the polymorphic homozygous classes. In other situations, it is quite possible for diversity to be distributed across a very broad spectrum of heterozygotes, which carry different alleles of a gene with more than two alternative forms. Each diploid individual has two copies of the gene, but different combinations may be produced in populations with more than two alleles for a particular gene locus.

Batesian mimicry can equally well arise through **frequency-dependent selection,** which favors a particular phenotype when it is uncommon, but acts against the same phenotype when it is more common. A mimic phenotype is most successful when it is not so common that the predator learns it is a tasty meal, even if it also makes the mistake of occasionally eating a distasteful look-alike. Another example of the general theme that predation helps maintain polymorphism in prey is provided by frequency-dependent selection acting on three color forms of the bug *Sigara distincta,* which is preyed on by fish. Each cryptically colored form of the bug is subject to the significantly greatest predation when it is the most common form. The particular degree of cryptic coloration is unimportant, because the best camouflaged bugs are eaten just as voraciously as the other forms if they are the most common type in the habitat (Fig. 6.12).

Balancing selection, or *heterozygote advantage,* is a strategy by which stable polymorphism is maintained in populations. In the best known example involving a human gene, the high frequency of individuals with the recessive blood disorder sickle-cell anemia in some parts of the world is directly correlated with the incidence of falciparum malaria in these regions (Fig. 6.13). Heterozygotes for the sickle-cell allele are more fit than are homozygous dominants in a malarial environment because of their resistance to infection by the protozoan pathogen *Plasmodium falciparum.* The homozygous dominants are more fit than the recessives, who, although resistant to malaria, suffer the more debilitating effects of sickle-cell anemia. The high frequency of the recessive genotype in these populations is the direct outcome of matings between heterozygotes. In populations of American blacks, whose ancestors came mainly from West Africa during the centuries of the inhuman trade in slaves, the frequency of sickle-cell anemia has fallen steadily from an initially high estimated frequency of several percent. Today, about 1 in 400 American blacks suffers from sickle-cell anemia, which is a homozygous recessive fre-

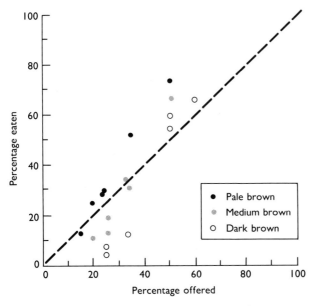

Figure 6.12. Frequency-dependent selection favors a particular phenotype when it is uncommon, but acts against the phenotype when it is more common. Results of experiments with the bug *Sigara distincta* show that the degree of cryptic coloration (three shades of brown) is not a significant factor in determining the pattern of predation by fishes. Each phenotypic group of insects is subject to the highest level of predation when it is the most common form. Polymorphism is thus maintained in the prey population of insects.

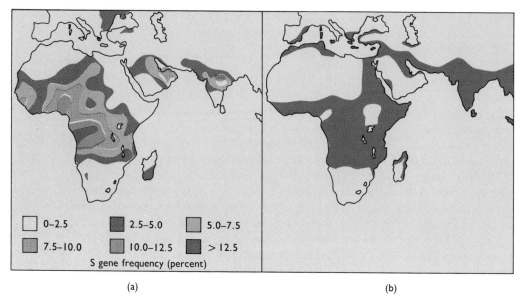

Figure 6.13. Distribution of the sickle-cell allele and of falciparum malaria in Africa and parts of Europe and Asia. (a) The sickle-cell allele is most common in central Africa, but occurs elsewhere as well, and (b) the areas where falciparum malaria is prevalent coincide with the areas in which the sickle-cell allele occurs. A stable polymorphism is thus maintained in malarial regions by the strategy of balancing selection. (From *Biology: A Human Approach,* 3rd ed., by Irwin Sherman and Vilia Sherman. Copyright © 1983 by Oxford University Press. Reprinted by permission)

quency of 0.25 percent. Heterozygotes enjoy no advantage over homozygous dominants in our nonmalarial environment, and the reduction in frequency of heterozygotes has led directly to the reduction in frequency of the recessive genotype. About 1 in 95 to 100 American blacks is heterozygous for the sickle-cell allele today.

Similar examples of balancing selection based on heterozygote advantage in polymorphic populations appear to characterize some other inherited human blood disorders (Fig. 6.14). They may be due to mutant alleles of the β-globin gene, like sickle-cell anemia, the α-globin gene, or various nonglobin genes. The resistance to infection of heterozygotes and the correlation with the incidence of malaria in the regions lead to the conclusion that heterozygote advantage is responsible for the polymorphism and for the disproportionately high frequency of heterozygotes and relatively unfit homozygous recessives.

Many other examples of polymorphisms are known, some of which are clearly based on selection pressures and others for which selection does not appear to be the explanation. Some flowering plants have self-incompatibility alleles at one or more gene loci, which help to ensure cross-fertilization. Pollen fails to germinate or to grow into the stylar tissue of flowers on the same plant, which are allelically identical, but can successfully fertilize flowers of plants with different incompatibility alleles. The greater diversity of cross-fertilized than of self-fertilized populations must surely underlie the selective advantages of this allelic polymorphism. In land snails, ladybird beetles, and a number of other invertebrates, a variety of protective-coloration patterns provide camouflage in a variety of habitats. The sum total of success in a varied environment is much more likely for a polymorphic population than for one with a single pattern of cryptic coloration.

In *Drosophila pseudoobscura* populations of the southwestern United States, different inversions have been described for chromosome 3, and different inversion types may coexist in the same populations (Fig. 6.15). From studies of natural populations, it is clear that the frequencies of inversion types and of the different alleles carried at some loci in the inverted regions of the chromosomes fluctuate with the seasons and probably with localities where temperatures vary. The *Standard* type is more common than the *Chiricahua* inversion in early spring and summer, but *Chiricahua* is as frequent or more frequent in May and June. Laboratory studies by Theodosius Dobzhansky revealed equal fitness for homozygous *Standard* and *Chiricahua* types at 16°C, but at 25°C the fitness of *Chiricahua* is less than half the value calculated for

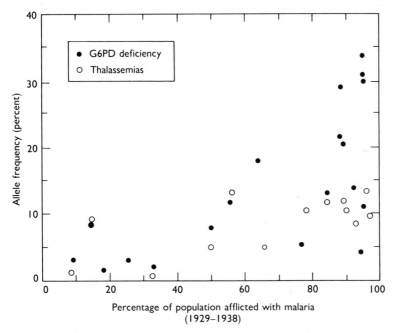

Figure 6.14. Correlation between the incidence of malaria and the frequencies of genes for thalassemias (α- or β-globin defects) and for glucose 6-phosphate dehydrogenase (G6PD) deficiency (an X-linked nonglobin defect) in Sardinian villages. Heterozygotes enjoy an advantage over homozygous dominants, because they are more resistant to malaria. Heterozygote advantage seems to be responsible for the maintenance of polymorphisms by means of a balancing selection strategy in these populations. (Based on studies by M. Siniscalco et al., 1961, Favism and thalassemia in Sardinia and their relationship to malaria. *Nature* 190:1179. Copyright © 1961 by Macmillan Journals Limited)

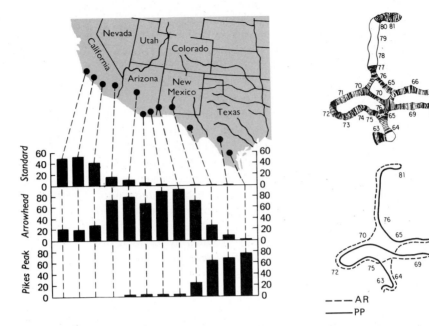

Figure 6.15. Relative frequencies of three different gene arrangements (inversions) in chromosome 3 of *Drosophila pseudoobscura* in the southwestern United States. The inversions in the *Standard, Arrowhead* (AR), and *Pikes Peak* (PP) chromosome types can be identified by salivary-chromosome band patterns, and heterozygotes can be distinguished from homozygotes by their inversion loops and by the bands on each chromosome 3 in the diploid cells. Paired chromosomes 3 in a nucleus heterozygous for AR and PP inversions are shown at the upper right, and below it is an interpretive diagram of these chromosomes. The numbers refer to bands on chromosome 3. Balancing selection helps to maintain inversion polymorphisms in these and similar populations, relative to advantages in relation to temperature, altitude, and other variations in the environments. (Based on studies by Th. Dobzhansky)

Standard. Balancing selection appears to account for the stable chromosome polymorphism, with each type favored over another at one time or one place.

Much more puzzling polymorphisms have been identified at the molecular level by **gel electrophoresis** of enzymes and by the sequencing of their amino acids or of DNA. About one-third of the proteins that have been studied by gel electrophoresis are polymorphic, being the products of two or more allelic forms of the gene. For the majority of these enzymes, only two alleles are known at a gene locus, one of which is responsible for an allozyme that migrates relatively rapidly in the gel and the other for an allozyme that moves more slowly in the gel under identical conditions. These **allozymes,** or variant forms of an enzyme, are referred to as the "fast" and "slow" forms; the alleles may also be called "fast" and "slow." A few enzymes, such as various esterases, are rampantly polymorphic and exist in dozens of allozymic forms encoded by dozens of different alleles of the gene (Fig. 6.16). Even in such highly polymorphic enzymes, however, two of the alleles usually occur in high frequency and the others in very low frequency.

By definition, these variants are polymorphic if they exist in higher frequencies than can be accounted for by recurrent mutation. The implication is that some selection strategy is responsible for maintaining the observed frequencies and diversity of allozymes and the alleles that encode them. Unfortunately, few data are available to support the presumption of different degrees of fitness for these variants and, therefore, little evidence for the maintenance of all these polymorphisms by selection. The riddle of allozymic variation is a major basis for proposals that chance rather than selection is at work; that is, most of the variation is selectively neutral and largely unimportant to population fitness. The neutralist–selectionist controversy is an important segment of current debate over the relative significance of Darwinian and non-Darwinian mechanisms in fixing evolutionary variation. Neutralists do not argue against natural selection as a potent evolutionary process, but they do argue that much of the variation at the *molecular* level is due only to chance.

We will discuss neutral evolution versus natural selection later in the chapter, but we can certainly conclude from the evidence presented here that natural selection is a major guiding force that acts to fix adaptive diversity in populations. In some situations, the effects of selection reduce diversity; in other situations, diversity is enhanced. In both cases, fitness is relative to conditions in which the organisms live. The same trait may be fit in one time or place and unfit in another, so adaptation itself is relative to particular circumstances. Whether accrued advantages serve the population immediately or over a long term, the patterns and directions of phenotypic evolutionary change can be traced primarily to natural selection acting on inherently diverse organisms.

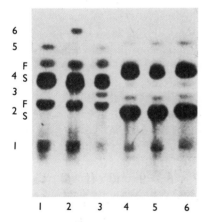

Figure 6.16. Gel electrophoretograms of allozymic forms of six esterases in six individuals from an allelically polymorphic population of *Drosophila virilis*. Slow-migrating allozymes (S) and fast-moving allozymes (F) of esterases 2 and 4 are clearly separated in the gel.

Group Selection Versus Individual Advantage

The effects of selection on the organism are such that an individual with an enhanced genetic advantage is more likely than one without the traits to survive and transmit this advantage to its progeny. Each offspring, in turn, passes down the alleles responsible for the advantageous traits, and whole populations may change as the result of selection acting on its individual members. Biologists tend to agree with Darwin that the fitness of the individuals accounts for evolutionary changes in the whole population and, ultimately, in the species as a collection of populations sharing a *gene pool*. In some cases, however, the success of a group is increased by the activities of individuals that may appear to derive little or no benefit for themselves. The greater reproductive efficiency of a group at the expense of individuals whose own reproductive success is reduced or canceled is called **group selection.** When close relatives, rather than nonrelatives, are benefited, the selection program is referred to as **kin selection.** In either case, the strategy is based on natural selection.

Group selection theory is very controversial, but it has been invoked most often in explaining the evolution of *altruistic behavior,* by which individuals are benefited at the apparent expense of the altruist itself. A bird that gives an alarm call may be at great risk of being noticed by a predator, but its behavior helps the rest of the flock by increasing their chances for escape. A parent may expose itself to danger by defending its young against a predator, or large males may protect other members of the herd by placing themselves at risk of bodily harm or death. The clearest examples of inherent altruistic behavior come from the social insects, which include ants, termites, and some wasps and bees. There is a genetic input to the behavior of social insects, as is evident from their stereotyped, nonspontaneous traits and activities.

In a colony of social insects, sterile worker or soldier castes labor selflessly for the benefit of the whole colony, while the queen and fertile males perpetuate the colony but do little else once it is established. The sterile and sexual castes of a colony are closely related and may even be the offspring of the same two parents. The sterile groups increase the probability for transmission of their own genes to the next generation by enhancing the chances for success of the sexual caste with which they share these genes. Each sibling worker or soldier shares 50 percent of its alleles with its full siblings and with each parent; each would share 50 percent of its alleles with its own offspring, if this were possible. By raising the overall fitness of the colony, the individual altruists increase the chances for their own inherent fitness to be represented in the gene pool of the new generation. The more siblings that are thus benefited, the more copies of the altruist's genes will be present in the next generation. Altruistic behavior thereby comes under the heading of differential reproductive success by means of natural selection. The important difference is that reproductive success embraces the entire colony and not the individual directly, by the mechanism of kin selection.

Most of the literature on group selection concerns theoretical aspects, rather than experimental and field studies to test the model. It is often difficult to determine that a behavior is indeed altruistic, even if it seems to be because of the observer's experience or bias. In addition, it is not clear if the members of a group are related and, if they are, just how close the genetic relatedness might be. Many reported studies therefore make an assumption of altruistic behavior, which may not be valid, and base their conclusions on it. Furthermore, even if the behavior is altrusitic, one may not be able to distinguish between related and unrelated beneficiaries and, therefore, between kin selection and group selection.

A carefully designed field study in which assumptions and hypotheses were tested adequately was reported in 1977 by Paul Sherman, on Belding's ground squirrels *(Spermophilus beldingi)* in California. In a 3-year study of these diurnal rodents, Sherman and his co-workers logged over 3000 hours of observations of the behavior of nearly 2000 animals whose genetic relatedness was determined in this and earlier work. The questions were whether or not the alarm call announcing the presence of terrestrial predators was indeed an altruistic behavior, and, if so, whether the beneficiaries were kin to the altruists. The behavior would benefit the alarm-caller directly—and, perhaps, other group members indirectly—if its vocalization diverted the predator's attention to other prey, discouraged the predator from pursuing it, or reduced later attacks by the predator. If the alarm had none of these beneficial effects on the caller directly, it still had to be determined if some of or all the other members of the group derived benefits, and then to see whether the beneficiaries were related to the caller.

The detailed observations were subjected to statistical analysis, and the data were interpreted as showing that the caller was behaving altruistically rather than selfishly (Fig. 6.17). The beneficiaries proved to be related to the caller, which most often was an adult female signaling to her offspring. The data therefore supported the predictions made for the selective advantages of altruism in this case and for its probable institution by kin selection rather than by selection of the individual directly.

A different aspect of behavior evolution involves *cooperation* among members of a group, a *reciprocal* relationship that is rarely, if ever, involved in clear cases of altruism. Cooperation requires interactions among individuals, all of which benefit as a result of one another's behaviors. Thus it is a system that enhances the relative success of a group, but only by directly benefiting its individual members. The evolution of various social behaviors by individual advantage rather than group selection has been the focus of theoretical study by many investigators. A central concept of evolutionarily stable strategies developed by John Maynard Smith in the 1970s has been invoked in a wide number of proposals. A strategy is "a specification of what an individual will do in any situation in which it may find itself." An **evolutionarily stable strategy,** or ESS, is defined by Maynard Smith as "a strategy such that, if all the members of a population adopt it, then no mutant strategy could invade the population under the influence of natural selection."

In 1981, Robert Axelrod and William Hamilton proposed the **tit for tat** theory of cooperative behavior, which conforms to the requirements for ESS; that is, once established, the behavior is stabilized against changes influenced by natural selection. Their theory, like many of the others, is based on mathematical game theory, and the model was tested by computer-tournament analysis. The theme of tit for tat is illustrated by the Prisoner's Dilemma game, in which each player can either cooperate or defect (Fig. 6.18). The payoff is greatest (higher fitness) for each player to selfishly defect; but if both defect, they do worse than if they had cooperated. The dilemma exists because it pays for each player to defect no matter what the other player does, but the punishment for mutual defection is greater than the reward for cooperation.

When individuals have one-time-only encounters, the stable strategy is mutual defection. Repeated contacts between the same individuals, however, favor the stable strategy of cooperation. Mutual benefit through cooperation can replace individual benefit through defection, if the strategy is one of coop-

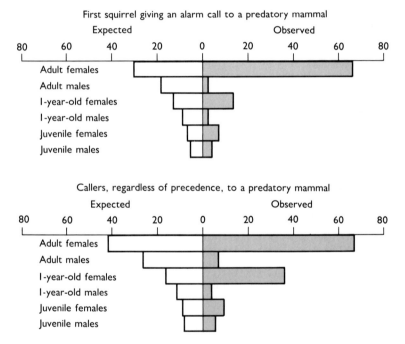

Figure 6.17. Expected and observed frequencies of alarm calling by various sex and age classes of Belding's ground squirrels. The overall signficance of both sets of comparisons is due largely to females calling more often and males less often than expected if calls are made randomly in proportion to the number of times the animals are present when a predatory mammal appears. These data, from 102 interactions between ground squirrels and predators over 3 years, indicate altruistic behavior. Kin selection is implicated because the beneficiaries proved to be related to the caller, which most often was an adult female signaling to her offspring. (From P. W. Sherman, 1977, Nepotism and the evolution of alarm calls. *Science* 197:1246, Fig. 3. Copyright © 1977 by the American Association for the Advancement of Science)

		Player 2 Cooperation	Player 2 Defection
Player 1	Cooperation	R = 3 Reward for mutual cooperation	S = 0 Sucker's payoff
Player 1	Defection	T = 5 Temptation to defect	P = 1 Punishment for mutual defection

Figure 6.18. The Prisoner's Dilemma game illustrates the tit for tat theory for a strategy that can lead to establishing a stable cooperative behavior pattern among individuals who repeatedly interact. The greatest payoff (highest fitness) and most stable strategy for one-time-only encounters are for each individual to selfishly defect. When interactions occur repeatedly, however, each individual benefits most from mutual cooperation if the strategy is one of cooperating on the first move and then doing whatever the other player did on the previous move. These conclusions were based on scores for players who participated in computer-tournament games.

erating on the first move and then doing whatever the other player did on the preceding move. Tit for tat is thus a strategy of cooperation based on reciprocity between individuals that meet repeatedly. This model can explain the existence of stable asocial or social groups and the evolution of sociality from asociality. Axelrod and Hamilton provided mathematical analysis for various evolutionary features of the tit for tat model, including high fitness for the strategy and its resistance to invasion by a mutant strategy. Their analysis led them to conclude that tit for tat indeed fits the requirements for an evolutionarily stable strategy, as defined by Maynard Smith. Neither group selection nor genetic relatedness need be invoked; individual advantage in the classic Darwinian framework of natural selection is the key factor in this evolutionary theory for cooperation and sociality. More analysis and test variety are required, however, before such a simple explanation for a complex phenomenon can be completely accepted.

Adaptation

Adaptation refers to the evident fit between organisms and their environments, and it is regarded as the evolutionary outcome of natural selection acting on genetically diverse individuals. Differential reproductive success is the predicted result of differential fitness, and those individuals that are best adapted to their environments will leave more offspring than will less fit types in a given set of circumstances. Biologists tend to look for the adaptive quality of each characteristic in the organism, fully expecting each feature to exist because it conferred benefit in the past and has been amplified in successive generations by means of natural selection. This viewpoint, of optimizing the organism in relation to its living conditions, has been referred to by Stephen Jay Gould and Richard Lewontin as the **adaptationist program.** As will be discussed in the following sections, the argument of adaptationists that all features are the result of natural selection appears to be seriously flawed. The goal of a nonadaptationist perspective is not to discredit natural selection, but to place it into a broader framework of evolutionary theory.

Appraisal of the Adaptationist Program

There is no shortage of clear-cut adaptations due to natural selection, as was described earlier in this chapter and in others. Industrial melanism is readily explained by selection acting on different genetic variants, as are instances of increasingly greater resistance to agents that damage or kill pests and pathogens. These examples, in particular, are readily understood because we know the history of evolutionary change and its genetic basis. The difficulties arise for characteristics that are easily described but are otherwise little known or understood. The adaptationist presumes that a feature has adaptive value and focuses on determining the basis for the adaptation, rather than seeking information to learn whether the trait is indeed adaptive. The insistence on adaptation is based on the predication that each feature is the outcome of selection, which not only biases an investigation but also directs a study away from any consideration of alternative evolutionary mechanisms that may underlie a trait. The adaptationist view is that the world presents problems and that adaptations are the solutions to these problems faced by organisms in their part of the world.

Human infants and ape infants lack a prominent chin, but human adults develop a prominent chin, whereas adult apes do not. We tend to equate a more prominent chin, particularly in males, with greater and more noble human virtues and fully expect heroes to have a well-developed chin. The chin distinguishes humans not only from apes but also from one another, and it was long believed that the human chin was an adaptive trait. No basis was ever found for this belief, and eventually it was discovered that differential growth, or **allometry,** not selection per se, is responsible for chin development.

Human and ape skulls are very similar at the fetal and neonatal stages of development, but differ considerably in adults (Fig. 6.19). The human skull retains much of its earlier proportions, except for the chin, but the ape skull undergoes much morphological change during growth to the adult form, and a chin does not develop. The retention of fetal patterns of morphological development into adulthood is called **neoteny,** and it characterizes the two

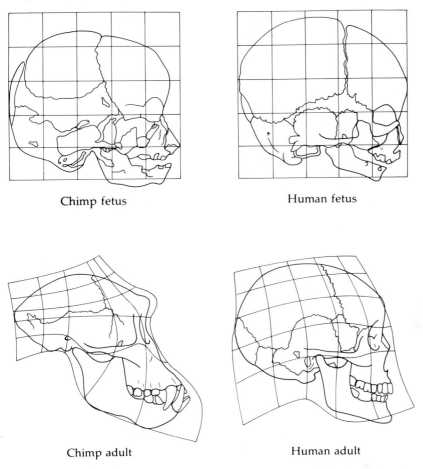

Chimp fetus

Human fetus

Chimp adult

Human adult

Figure 6.19. When plotted on transformed coordinates, which show the relative displacement of each part, it is evident that growth of the chimpanzee skull and human skull are more similar in the fetal stage than in the adult stage. The adult human skull resembles its fetal form far more than the adult ape skull resembles its fetal form, except for the chin, which becomes relatively larger in human beings. The retention of fetal patterns of morphological development into adulthood, or neoteny, characterizes differential (allometric) growth of the alveolar and dentary fields in the human lower jaw. (From D. Starck and B. Kummer, 1962, Zur Ontogeneses des Schimpansenschädels. *Anthrop. Anz.* 25:204)

growth fields in the human lower jaw: the alveolar field, in which the teeth are set, and the dentary field of the bony structure of the jaw. Both growth fields have become smaller in human evolution, but the dentary field has shrunk at a somewhat slower rate than the alveolar field, with the resulting production of the chin during development. As a consequence of differential growth, the chin no longer requires an adaptive explanation. The process responsible for the chin, as well as some other traits, is allometry, not selection for a prominent chin. Natural selection for neoteny, however, is reasonably well understood and comes under the heading of adaptation (see Chapter 9).

The particular explanation offered for a patently adaptive feature may not accurately reflect the past history of selection leading to the feature, especially when it serves more than one function. A gaudy color or an ornate body part in an animal may serve as a recognition signal for appropriate sexual encounters, a warning to others of the same sex, or a means for diverting the attention of a predator. Which of these functions do we choose in reconstructing the selection program that established the adaptation? It is impossible in many cases to decide which function was primary and which others were secondary and not the target of selection or the basis for greater reproductive fitness and success. Not every adaptation is the result of natural selection. Some advantageous features are *opportunistic;* that is, they are chance results of selection acting on a different function of the same trait.

In a similar vein, changes in a gene may have several different effects on an organism; such unrelated effects are carried along if one of them is advantageous and enhances the reproductive success of the organism. This common feature of gene expression is **pleiotropy,** in which a cascade of effects can often be traced to the physiological or morphological or behavioral consequences of a molecular modification in a gene product (Fig. 6.20). The amino acid tyrosine may be processed in different ways in the body, each pathway being under the control of enzymes coded by genes. A change in one or more genes in one of these pathways may facilitate the processing of tyrosine and lead to the production of the black pigment melanin at the same time. The pigment may or may not provide some benefit, but it is carried along if the primary effect on tyrosine is selected. In this regard, a trait with no current advantage may become advantageous later in the history of a lineage. Such opportunistic adaptive features are referred to as **preadaptations.** For example, ancestral forms of the potato are often resistant to a fungal blight, even though the pathogen is absent from the environment. The trait proved to be adaptive in descendant forms of the potato, which live in environments that contain the fungal pathogen. Little is known about the factors involved in many complex traits, so biologists may often be misled into concentrating on a more evident but secondary effect of gene action rather than on the primary focus of the selection process.

If adaptations are assumed to continually improve the chances for success of the organism in its environment, descendant species would be expected to exist for longer times than their less fit ancestors. This expectation is not borne out by comparison of extinction rates, however, which shows that they are more or less constant for each group (Fig. 6.21). Leigh Van Valen suggested an explanation for this paradox in his **Red Queen hypothesis,** named for the character in *Through the Looking-Glass* who has to keep running just to stay in the same place. This hypothesis states that the environment of existing organisms is constantly changing, so natural selection acts to maintain adaptive states rather than improve them. Organisms become extinct if they have insufficient genetic variation to keep up with changes in their environ-

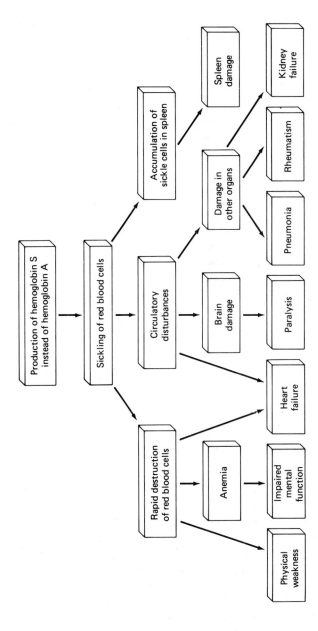

Figure 6.20. Pleiotropic effects of the sickle-cell hemoglobin allele in homozygous recessive individuals. The synthesis of sickle-cell hemoglobin is under direct genetic control, but the cascade of clinical effects is due to the abnormal behavior of red blood cells containing hemoglobin S molecules.

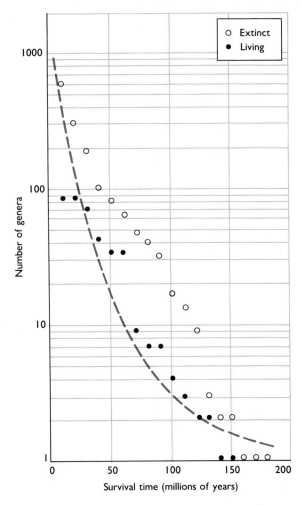

Figure 6.21. Comparison of extinction rates for living and extinct genera of Echinoidea, a class of marine invertebrates to which sea urchins belong. The points fall along relatively straight lines, indicating constant rates of extinction in both groups. If there were a lower probability of extinction for long-lived genera, as expected if natural selection led to continual improved adaptations and longer duration of more fit descendant species than their less fit ancestors, the points would fall along a concave curve (dashed line). (Based on studies by L. Van Valen)

ment. Because genetic resources are finite, environmental conditions eventually change so rapidly that any species is ultimately doomed to extinction.

Although the Red Queen hypothesis provides one explanation for observed rates of extinction and ties them to limitations on adaptation, it fails to explain the wealth of diversity that does exist in an environment. The oceans teem with life forms that range from simple bacteria to warm-blooded whales. The relative success of many of these organisms is clearly due to adaptation, as seen in the convergent body plans of marine vertebrates. The similarities in adaptation reflect the outcome of similar selection pressures; the streamlined body shape and limbs modified into flippers or paddles are hardly due to coincidence (Fig. 6.22). These adaptations are limited to some degree by **developmental constraints.** Aquatic descendants of terrestrial ancestors, such as whales and penguins, have modified limbs rather than fins. Sea snakes are as limbless as their terrestrial relatives and move easily through the water by virtue of a body that is flattened in cross section. These adaptations represent modifications of later stages in development, which are tolerated far more than are changes early in embryology. The dislocation of earlier stages in development would have far-ranging effects on the whole organism, whereas changes later in development would not interfere with laying down the basic structures and functions of the individual. Biologists may not be able to predict the kinds of adaptations that will arise in the future, but can predict that all or most of them will be accommodated within the existing framework of form and function that has been established step by step during evolution.

We most admire adaptations that lead to a most exquisite and detailed fit

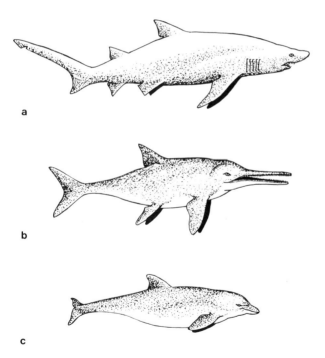

Figure 6.22. Convergent body plans of three marine vertebrates. The (a) shark, (b) ichthyosaur, an extinct reptile, and (c) dolphin, a mammal, display a streamlined form well suited to aquatic life. Similarities in these adaptations among distantly related species reflect the influence of similar selection pressures in similar environments during evolution. (From *Understanding Evolution,* by Earl D. Hanson. Copyright © 1981 by Oxford University Press. Reprinted by permission)

between the organism and its living conditions. The flower that lures carrion flies by its scent of decaying meat is thereby assured of pollination. A few of these flowers may actually be a trap for these insects, which pollinate the flowers as they move about inside them. When the flower collapses 2 or 3 days after pollination, the flies can escape and move on to pollinate other flowers of the species. The reproductive success of these plants testifies to their high level of fitness, or adaptedness, to specific conditions. Should these conditions change in any way that prohibits pollination, the plants may become extinct if their genetic resources are inadequate to cope by alternative means. By contrast, flowers pollinated by various insects may not be as exquisitely adapted to their environment, but they are more adaptable in the long run if they continue to reproduce successfully despite the loss of one or more pollinating agents.

Adaptedness, or genetic fitness, provides a major focus for many adaptationist programs. **Adaptability,** or genetic flexibility, is often ignored or may even be viewed as a failure of natural selection to channel the organism into pathways of increasingly greater fit to its living conditions. When viewed in the short run of evolutionary time, highly adapted organisms may exploit their environments very successfully indeed. In the long run, however, adaptedness may foreclose some of the very options that could prove beneficial if conditions change or new opportunities arise. A horse is well adapted to running, but the modifications in its legs that suit it to this activity also preclude the horse from climbing a tree to escape a predator. Humans cannot run as fast as a horse or climb a tree as swiftly as a squirrel, but can manage both actions, even if somewhat clumsily. The retention of all five digits and of the skeletal construction of the human arm during our evolution from terrestrial ancestors has contributed greatly to our success as a species. The manipulative skills and dexterity of the primate hand have been put to new uses made possible by greater intelligence during human evolution. The new opportunities could be better exploited by a more generalized anatomy than by one that had become "specialized" for some particular activity or function. If all things were equal, adaptability would yield a bigger payoff in evolutionary terms, but only on the basis of probability in the long run. Adaptedness may yield immediate and great success, but with less chance for evolutionary benefit in the long run.

The Neutralist–Selectionist Controversy

In contrast to the adaptationist program, which presumes a selective basis for all or most of the organism's phenotype, the theory of **neutral evolution** stipulates the virtual absence of selective forces at the molecular level. In the view of Motoo Kimura, who proposed the neutral theory in 1968, and of Thomas Jukes and others who have supported and strengthened the theory, molecular evolution proceeds primarily by mutation and random fixation or loss of alleles rather than by selection. The neutral theory distinguishes change at the molecular level from change at the phenotypic level, which neutralists and selectionists agree is influenced more by natural selection than by mutation or chance. Selectionists believe that molecular change as well as phenotypic change is guided primarily by natural selection. The neutralist–selectionist controversy is debated as vigorously today as it was more than 20 years ago.

Kimura proposed the neutral theory to explain certain features of molecular variation observed in comparisons among different evolutionary lineages. In a given protein, the rate at which amino acids are substituted for one another is about the same in many diverse lineages. In the α-globin chain of hemoglobin, for example, the number of amino acid differences is progressively greater for vertebrates that have evolved over progressively greater intervals, but if any one of these is compared with any other vertebrate, the number of amino acid substitutions is constant (Fig. 6.23). The shark α globin differs by about 80 amino acids from the α globin of mammals; the carp α globin differs by about 70 amino acids from the α globin of mammals; and so forth. The approximate uniformity of protein evolution in very different organisms is more in line with the neutral theory than with selection theory. According to the neutral theory, the fixation of alleles depends on the mutation rate and random extinction or fixation, which leads to a constant rate of molecular evolutionary change. Different proteins evolve at different rates; the *molecular clock* itself is not absolute. The prediction based on selection, however, is that a different number of amino acids would be substituted in any one protein shared by different lineages, due to different levels of benefit arising from changes in different organisms subject to different selection pressures in changing environments over different intervals of time.

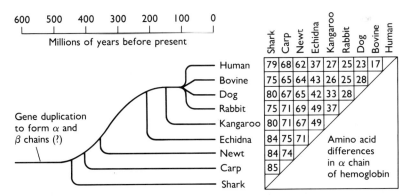

Figure 6.23. Comparison of amino acid differences in the vertebrate α-globin chain. Any one of these species differs from the others to approximately the same extent; for example, shark α globin differs by about 80 amino acids from the α globin of carp (a fish), newt (an amphibian), or any of the six mammals. Such constancy of protein difference suggests the more uniform consequences predicted by the neutral theory than by the natural selection theory of evolution. (Adapted from M. Kimura, 1982, in *Molecular Evolution, Protein Polymorphism, and the Neutral Theory*, ed. M. Kimura, pp. 3–56. Berlin: Springer-Verlag)

Kimura's analysis revealed that the substitutions are random rather than patterned and that the overall rate of change in the DNA sequence itself is very high. The rate of substitution of at least 1 nucleotide per genome every 2 years is the calculated value for a mammalian lineage. The high rate of mutation could thus account for considerable diversity, but the constant rate of molecular change and the absence of pattern indicate that a great deal of the diversity is fixed or lost at random in the population.

An appropriate test of the opposing neutral and selection theories can be made by analyzing enzyme polymorphisms. Whether allozyme frequencies in populations reflect neutral evolution or selection based on differential fitness can be analyzed experimentally or in natural populations that inhabit measurably different environments. One of the problems of such studies, however, is a significant lack of knowledge about important properties of the enzymes themselves. Some of the difficulties involved in studies of polymorphic enzymes can be illustrated by alcohol dehydrogenase in *Drosophila melanogaster,* which is probably the most intensively studied enzymatic system.

Alcohol dehydrogenase (ADH), which catalyzes both the detoxification of ethanol and its utilization as a carbon source, exists mainly as fast and slow allozymic forms. The fast and slow alleles, Adh^F and Adh^S, have combined frequencies of about 90 percent in any population; the rag-tag assortment of other alleles at this gene locus can be effectively disregarded. The enzyme is probably important in the larval stage particularly, because of high alcohol levels in the rotting fruit eaten by the larvae. Physicochemical studies of the fast and slow allozymes have shown variations in their kinetics on different substrates, in their thermostabilities, and in their catalytic properties. Together with the very well-known genetics, the relative ease of experimental manipulation, and the wide geographical distribution of *Drosophila,* the ADH system would seem to deserve all the attention it has received.

The higher thermostability of the slow ADH would lead us to predict a positive correlation between high Adh^S frequency in natural populations and high maximum monthly temperature in or low geographical latitude of their habitats, if this allele conferred greater fitness than its alternative at higher temperatures. Adh^S frequency increases progressively in populations in North America, Asia, and Australia the progressively closer their habitats are to the equator. There is no correlation between the frequency of Adh^S in Asian and Australian populations and the maximum monthly temperature of their habitats, but there is with North American populations. These conflicting sets of data, like many others of a similar nature, do not permit a choice between the alternative theories. Natural selection cannot be invoked in one case and not in the other when dealing with the same property of the same protein in both cases.

Less ambiguous results have been obtained in some experimental studies of ADH allozyme frequencies when larvae have been reared in a medium containing ethanol. Such an environment should elicit selection pressure on the locus itself and lead to a change in allelic frequency if one allele confers greater fitness than the other. In many such experiments, it was found that one of the allozymes does increase the survival of larvae in this medium. Even if these data are accepted as evidence for differential fitness, we do not know whether selection is exerted on the *Adh* locus directly or on other genes whose activities may influence *Adh* expression. The effect may be due to regulation of *Adh* expression, leading to different concentrations of the enzyme. Homozygotes for the fast allele are known to have a higher cellular concentration of ADH than do homozygotes for the slow allele, and this difference in concentration is due to differential synthesis of the protein through regulatory gene control. The apparent difference in fitness may thus be due to other genes closely linked to the *Adh* locus, and not to the *Adh* gene itself.

The problem just described can be resolved if biologists know the linked genes involved, their individual frequencies, and the frequency of associations of these alleles. If alleles at two or more loci exist in higher or lower frequencies when they are associated than would be predicted from their individual frequencies, the alleles are in **linkage disequilibrium.** Lewontin described this evolutionary genetic phenomenon and indicated its importance in any determination of selective fitness conferred by a particular gene locus. When tracking the frequency differences for some particular target gene, it is necessary to determine whether its frequencies vary as a direct result of its own properties or because its ups and downs simply coincide with the changes in frequency of different genes and gene products. Linkage disequilibrium is often difficult to determine, even in organisms whose genes have been mapped in detail. It is even more difficult in organisms whose genes are barely known. In the case of ADH in *Drosophila melanogaster,* it remains uncertain whether the observed effects of selection are due to functional differences between its allozymic variants or to differences in regulatory variants that are closely linked to the *Adh* locus.

Protein polymorphisms could be studied before DNA sequencing became possible in the mid-1970s, and they continue to be the focus of many studies. Once it became available, DNA sequencing provided another method for studying variation in even greater detail. *Silent mutations* were found in abundance, as changes to synonymous codons and to codons specifying an amino acid that does not change the electrical charge or activity of the protein, as were diverse sequences in noncoding introns and intergenic spacers (Table 6.1). Many of these mutations clearly are neutral and must be maintained by the rate of mutation and random drift toward fixation or exclusion.

Noncoding DNA is much more variable than coding DNA, and differences in the amount of variation characterize different parts of a gene specifying different parts of a protein. Regions of a protein that are essential for function are less diverse than those that are less involved or not involved in function. When the protein is folded into its three-dimensional active form, it can be seen that amino acids vary more at the surface of the molecule than in its interior, where the active site or essential binding sites are usually located (Table 6.2). As a consequence of these **molecular constraints,** a portion of

TABLE 6.1
Rates of Nonsynonymous and Synonymous Nucleotide Substitutions in 20 Genes, Based on Comparisons Among Different Mammalian Orders that Are Assumed to Have Diverged 80 Million Years Ago

Genes	Number of Codons Compared	Substitution Rate ($\pm$ S.E.)*	
		Nonsynonymous	Synonymous
γ interferon	136	2.80 $\pm$ 0.31	8.59 $\pm$ 2.56
α1 interferon	166	1.41 $\pm$ 0.13	3.53 $\pm$ 0.61
Prolactin	197	1.29 $\pm$ 0.12	5.59 $\pm$ 0.93
Alpha-fetoprotein	586	1.21 $\pm$ 0.08	4.90 $\pm$ 0.45
β2 microglobulin	99	1.21 $\pm$ 0.20	11.77 $\pm$ 9.91
Immunoglobulin V$_H$	100	1.07 $\pm$ 0.19	5.67 $\pm$ 1.36
Growth hormone	189	0.95 $\pm$ 0.11	4.37 $\pm$ 0.56
Albumin	590	0.92 $\pm$ 0.07	6.72 $\pm$ 0.62
β globin	144	0.87 $\pm$ 0.11	2.96 $\pm$ 0.46
Metallothionein I	60	0.61 $\pm$ 0.12	3.38 $\pm$ 0.76
α globin	141	0.56 $\pm$ 0.09	3.94 $\pm$ 0.60
γ fibrinogen	411	0.55 $\pm$ 0.06	5.82 $\pm$ 0.67
Metallothionein II	60	0.43 $\pm$ 0.14	2.83 $\pm$ 0.92
Insulin	51	0.16 $\pm$ 0.09	5.41 $\pm$ 1.98
Histone H2A	126	0.079 $\pm$ 0.02	2.06 $\pm$ 0.44
Histone H2B	120	0.075 $\pm$ 0.04	3.59 $\pm$ 0.69
Histone H4	101	0.027 $\pm$ 0.03	6.13 $\pm$ 1.32
α actin	376	0.014 $\pm$ 0.01	3.67 $\pm$ 0.43
Somatostatin-28	28	0.000 $\pm$ 0.00	4.23 $\pm$ 2.84
Histone H3	135	0.000 $\pm$ 0.00	6.38 $\pm$ 1.19
Average†		0.88 $\pm$ 0.75	4.65 $\pm$ 2.06

*All rates are in 10^{-9} substitutions per site per year.
†Based on the above 20 genes plus 22 other genes.
Source: W.-H. Li, C.-I. Wu, and C.-C. Luo, 1985. A new method for estimating synonymous and nonsynonymous rates of nucleotide substitution considering the relative likelihood of nucleotide and codon changes. *Mol. Biol. Evol.* 2:150.

TABLE 6.2
Evolutionary Rates of Amino Acid Replacements (in units of 10^{-9} replacements per amino acid site per year) at the Surface and in the Heme Pocket of the α- and β-globin Chains of Hemoglobin

Region	α Globin	β Globin
Surface	1.35	2.73
Heme pocket	0.165	0.236

Source: M. Kimura and T. Ohta, 1973. Mutation and evolution at the molecular level. *Genetics* 73(Suppl.):19.

the gene sequence is influenced more by natural selection based on fitness than by neutral evolution. But vast stretches of the average sequence are irrelevant to fitness, and these parts of the gene surely are subject to neutral changes during evolution.

Molecular constraints on protein evolution are apparent when different molecules, as well as parts of one molecule, are compared. Richard Dickerson and others have compared rates of evolutionary change in molecules with different degrees of sequence constraint and shown clearly that the rate of molecular change corresponds to the relative freedom of molecules to diversify. Fibrinopeptides show the fastest rate of evolution and histones, the slowest (Fig. 6.24). Fibrinopeptides have little or no function after their cleavage from fibrinogen to yield the blood-clotting protein fibrin. Histones bind tightly to DNA all along their length and to one another in forming and maintaining the nucleosomal organization of the chromosome. Changes almost anywhere in the amino acid sequence of a histone molecule could interfere with protein binding and chromosome function, and little change has actually

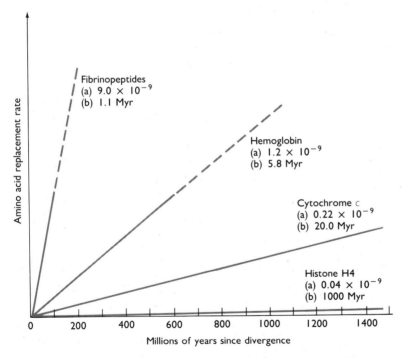

Figure 6.24. Comparison of evolutionary rates of four proteins. Two commonly used comparative measures are (a) number of replacements per amino acid site per year, and (b) average time required for 1 replacement per 100 amino acid sites (1% change). Progressively greater molecular restraints are correlated with progressively slower evolutionary rates of amino acid substitution. (Adapted from R. E. Dickerson, 1971, The structure of cytochrome c and the rates of molecular evolution. *Jour. Mol. Evol.* 1:26)

S. *purpuratus* mRNA	GAU	AAC	AUC	CAA	GGC	AUC	ACC	AAG	CCU	GCA
L. *pictus* mRNA	GAU	AAC	AUC	CAA	GGA	AUA	ACU	AAA	CCG	GCA

Histone H4 amino acid sequence in both species	Asp	Asn	Ile	Gln	Gly	Ile	Thr	Lys	Pro	Ala
	24	25	26	27	28	29	30	31	32	33

Figure 6.25. Base sequences for amino acids 24 to 33 in histone H4 from the sea urchin species *Strongylocentrotus purpuratus* and *Lytechinus pictus*. Five synonymous codons (at arrows) are present, differing only in the 3′ base, and they specify the same amino acids in the proteins of both species. (From M. Kimura, 1983, The neutral theory of molecular evolution, in *Evolution of Genes and Proteins,* ed. M. Nei and R. K. Koehn, pp. 208–233. Sunderland, Mass.: Sinauer)

occurred in histones over more than 1 billion years of eukaryote evolution. Fibrinopeptides, on the contrary, are free to diversify without much effect on the organism. Interestingly, a substantial number of synonymous codon changes have been found in histone gene sequences (Fig. 6.25). The rate of change in the third nucleotide is rather rapid, whereas amino acid substitutions occur at a very slow pace. Neutral evolution may well be involved in substitutions of nucleotides at the third position of many histone codons, and in other genes as well.

Studies of protein polymorphisms and DNA polymorphisms clearly reveal an immense amount of genetic diversity in organisms. Some of this variation can be accounted for by natural selection, but some must surely be due to neutral evolution. Even those polymorphic variants that are equally fit under one set of environmental conditions may not be so under a different set of conditions, according to a number of experimental studies. Thus enough diversity exists in populations to account for observed rates of evolution, and many mutations may influence fitness in some environments and not in others. Neutral alleles are a genetic reserve that may be important to reproductive success in new environments in future times, regardless of their current lack of influence. We can safely predict that life will meet future challenges and continue to diversify in the future as it has in the past and is in the present, through the combined processes of natural selection and neutral evolution acting on the ample stores of genetic diversity.

Summary

Natural selection is the mechanism proposed by Charles Darwin to explain observed patterns of evolution, and it serves as a unifying principle for all of biology. The more inherently fit individuals leave more descendants (like themselves) than do the less fit, leading to changes toward greater adaptation in populations and species. Fitness is relative, varying with conditions of existence at different times and in different places. Experimental proof for Darwinian selection, involving differential reproduction of genetically diverse individuals, and against the Lamarckian theory of inheritance of acquired characteristics, was provided in the 1940s and 1950s by the fluctuation test and the replica plating test.

Selection strategies may reduce allelic and genotypic diversity by impeding the accumulation of detrimental alleles and stabilizing the level of diversity (stabilizing, or normalizing, selection) or by favoring one homozygous genotype over its alternative (directional selection). Strategies that increase diversity may favor various homozygous genotypes over heterozygotes (diversifying, or disruptive, selection), favor any number of uncommon phenotypes over more common ones (frequency-dependent selection), or stabilize the proportions of homozygous and heterozygous types, often by balancing selection (heterozygote advantage). Strategies that increase diversity act to establish and maintain chromosomal, allelic, and molecular polymorphisms in populations, although chance may be more influential than selection in some situations.

Kin selection, one type of group selection strategy, has been invoked to explain the evolution of altruistic behavior. By its unselfish behavior, the altruist benefits its close relatives and may thereby increase the chances for the transmission of shared genes to the next generation. Reciprocal interactions that benefit the interacting individuals directly may enhance the chances for success of the whole group as a consequence of cooperation among its members, regardless of their genetic relatedness. The tit for tat theory is an evolutionarily stable strategy that has been proposed to explain the establishment and maintenance of sociality, based on advantages to cooperating individuals.

Adherents of the adaptationist viewpoint infer the existence of adaptive qualities for each characteristic, expecting past benefit to be the explanation for the institution and spread of a trait in populations. Many traits can be explained as the consequence of selection, but others cannot. Mechanisms responsible for particular traits may include allometry, neoteny, pleiotropy, developmental constraints, and molecular constraints, all of which may or may not have a selective basis. Adaptationists focus on the quality of adaptedness, or high fitness, whereas the quality of adaptability, or genetic flexibility and moderate fitness, may prove more advantageous in the long run of evolution. The neutral theory of molecular evolution assumes the occurrence of constant rates of change due to random processes of mutation and fixation, rather than to selection, but only at the molecular level and only in parts of molecules. The tests for neutral or selected evolutionary changes include analyses of allozymic polymorphisms. These and other tests have not provided unambiguous data on which all molecular biologists can agree, so the neutralist–selectionist controversy continues to be debated.

References and Additional Readings

Axelrod, R. 1986. *The Evolution of Cooperation*. New York: Basic Books.

Axelrod, R., and W. D. Hamilton. 1981. The evolution of cooperation. *Science* 211:1390.

Ayala, F. J. 1978. The mechanisms of evolution. *Sci. Amer.* 239(3):56.

Barrett, S. C. H. 1987. Mimicry in plants. *Sci. Amer.* 257(3):76.

Boag, P. T., and P. R. Grant. 1981. Intense natural selection in a population of Darwin's finches (Geospizinae) in the Galápagos. *Science* 214:82.

Brandon, R. N., and R. M. Burian, eds. 1984. *Genes, Organisms, Populations*. Cambridge, Mass.: MIT Press.

Brent, L. 1981. Supposed Lamarckian inheritance of immunological tolerance. *Nature* 290:508.

Clarke, B. 1975. The causes of biological diversity. *Sci. Amer.* 233(2):50.

Darwin, C. 1859. *On the Origin of Species by Means of Natural Selection*. London: John Murray.

Dickerson, R. E. 1971. The structure of cytochrome *c* and the rates of molecular evolution. *Jour. Mol. Evol.* 1:26.

Dobzhansky, T. 1968. Adaptedness and fitness. In *Population Biology and Evolution*, ed. R. C. Lewontin, p. 109. Syracuse, N.Y.: Syracuse University Press.

Dobzhansky, T. 1970. *Genetics of the Evolutionary Process*. New York: Columbia University Press.

Dobzhansky, T., and F. J. Ayala. 1973. Temporal frequency changes of enzyme and chromosomal polymorphisms in natural populations of *Drosophila*. *Proc. Natl. Acad. Sci. U.S.* 70:680.

Gilbert, L. E. 1982. Coevolution of a butterfly and a vine. *Sci. Amer.* 247(2):110.

Gould, S. J. 1977. *Ontogeny and Phylogeny*. Cambridge, Mass.: Harvard University Press.

Gould, S. J., and R. C. Lewontin. 1978. The spandrels of San Marco and the Panglossian paradigm: a critique of the adaptationist programme. *Proc. Royal Soc. London B* 205:581.

Gustafson, J. P., G. L. Stebbins, and F. J. Ayala, eds. 1986. *Genetics, Development, and Evolution*. New York: Plenum Press.

Huxley, J. 1953. *Evolution in Action*. New York: Harper & Row.

Jacob, F. 1977. Evolution and tinkering. *Science* 196:1161.

Kettlewell, H. B. D. 1955. Selection experiments on industrial melanism in the lepidoptera. *Heredity* 9:323.

Kimura, M. 1979. The neutral theory of evolution. *Sci. Amer.* 241(5):98.

Kimura, M., ed. 1982. *Molecular Evolution, Protein Polymorphisms, and the Neutral Theory*. Berlin: Springer-Verlag.

Kimura, M. 1983. *The Neutral Theory of Evolution*. New York: Cambridge University Press.

Kimura, M. 1983. The neutral theory of molecular evolution. In *Evolution of Genes and Proteins*, ed. M. Nei and R. K. Koehn, p. 208. Sunderland, Mass.: Sinauer.

Kimura, M. 1985. Natural selection and neutral evolution. In *What Darwin Began*, ed. L. R. Godfrey, p. 73. Boston: Allyn and Bacon.

Kimura, M., and T. Ohta. 1973. Mutation and evolution at the molecular level. *Genetics* 73(Suppl.):19.

King, J. L., and T. H. Jukes. 1969. Non-Darwinian evolution. *Science* 164:788.

Koehn, R. K., and T. J. Hilbish. 1987. The adaptive importance of genetic variation. *Amer. Sci.* 75:134.

Koehn, R. K., A. J. Zera, and J. G. Hall. 1983. Enzyme polymorphisms and natural selection. In *Evolution of Genes and Proteins*, ed. M. Nei and R. K. Koehn, p. 115. Sunderland, Mass.: Sinauer.

Lederberg, J., and E. M. Lederberg. 1952. Replica plating and indirect selection of bacterial mutants. *Jour. Bacteriol.* 63:399.

Levinton, J. S. 1986. Developmental constraints and evolutionary saltations: a discussion and critique. In *Genetics, Development and Evolution*, ed. J. P. Gustafson, G. L. Stebbins, and F. J. Ayala, p. 253. New York: Plenum Press.

Lewontin, R. C. 1974. *The Genetic Basis of Evolutionary Change*. New York: Columbia University Press.

Lewontin, R. C. 1978. Adaptation. *Sci. Amer.* 239(3):56.

Lewontin, R. C. 1979. Sociobiology as an adaptationist program. *Behav. Sci.* 24:5.

Lewontin, R. C. 1985. Population genetics. *Ann. Rev. Genet.* 19:81.

Li, W.-H., C.-I. Wu, and C.-C. Luo. 1985. A new method for estimating synonymous and nonsynonymous rates of nucleotide substitution considering the relative likelihood of nucleotide and codon changes. *Mol. Biol. Evol.* 2:150.

Luria, S. E., and M. Delbrück. 1943. Mutations of bacteria from virus sensitivity to virus resistance. *Genetics* 28:491.

Maynard Smith, J. 1976. Group selection. *Quart. Rev. Biol.* 51:277.

Maynard Smith, J. et al. 1985. Developmental constraints and evolution. *Quart. Rev. Biol.* 60:265.

Monod, J. 1971. *Chance and Necessity*. New York: Knopf.

Provine, W. B. 1986. *Sewall Wright and Evolutionary Theory*. Chicago: University of Chicago Press.

Saunders, S. R. 1985. The inheritance of acquired characteristics: a concept that will not die. In *What Darwin Began*, ed. L. R. Godfrey, p. 148. Boston: Allyn and Bacon.

Sherman, P. W. 1977. Nepotism and the evolution of alarm calls. *Science* 197:1246.

Siniscalco, M. et al. 1961. Favism and thalassemia in Sardinia and their relationship to malaria. *Nature* 190:1179.

Sober, E. 1984. *The Nature of Selection*. Cambridge, Mass.: MIT Press.

Sober E., ed. 1984. *Conceptual Issues in Evolutionary Biology*. Cambridge, Mass.: MIT Press.

Starck, D., and B. Kummer. 1962. Zur Ontogeneses des Schimpansenschädels. *Anthrop. Anz.* 25:204.

Stebbins, G. L., and F. J. Ayala. 1985. The evolution of Darwinism. *Sci. Amer.* 253(1):72.

Templeton, A. R. 1982. Adaptation and the integration of evolutionary forces. In *Perspectives on Evolution*, ed. R. Milkman, p. 15. Sunderland, Mass.: Sinauer.

Thomson, K. S. 1985. Essay review: the relationship between development and evolution. In *Oxford Surveys in Evolutionary Biology*, ed. R. Dawkins and M. Ridley, vol. 2, p. 220. New York: Oxford University Press.

Van Valen, L. 1973. A new evolutionary law. *Evol. Theory* 1:1.

Wickler, W. 1968. *Mimicry in Plants and Animals*. New York: McGraw-Hill.

Wright, S. 1931. Evolution in Mendelian populations. *Genetics* 16:97.

CHAPTER 7

Gene Frequencies in Populations

The application of quantitative methods to analyze the gene and its transmission patterns in controlled laboratory populations laid the foundation for understanding the basic features of genetic systems. From these studies were determined dominance and recessiveness of alleles at a gene locus, mutation rates and the consequences of mutation for an organism, locations of genes in mapped chromosomes, and many other genetic features. Biologists cannot control events in natural populations, however, so alternative methods must be used to analyze the patterns of genetic change in successive generations. In this chapter, we will survey a few of the methods from **population genetics** by which evolutionary changes in natural populations can be analyzed. In particular, the emphasis will be on the major genetic processes that influence changes in **gene frequencies.**

The Gene Pool

One of the most fruitful conventions of population genetics is the model of a **gene pool,** the collection of alleles shared by members of a population and transmitted in the gametes from parents to offspring. Each offspring is regarded as a random sample of one egg and one sperm from this pool, and each parent is assumed to contribute equally to a theoretically infinite number of gametes carrying alleles from the gene pool. Each allele is chosen randomly from the pool during gamete formation, and each genotype in the offspring is a random combination of two parental sets of alleles in the fertilized egg. The gene pool model permits formulas to be derived for the relation between allele (gene) and genotype frequencies, for selection, and for other genetic parameters of evolution.

The Hardy-Weinberg Principle

Classic Mendelian ratios are obtained repeatedly in laboratory populations, but are rarely encountered in natural populations. The failure to find predicted Mendelian ratios in natural populations led to uncertainty about the relationship between genes and evolutonary change, and even to the possibility that all populations were moving toward these ratios and would stop changing once they were established. If this were true, how could genes be involved in the expressions of diversity that characterize evolution? This quandary was resolved in 1908 by a simple principle stated independently by the British mathematician G. H. Hardy and the German biologist Wilhelm Weinberg.

The simple premise of the **Hardy-Weinberg principle** is based on the application of the binomial theorem: the proportions of genotypes and phenotypes observed in a population depend on the proportions of the alleles responsible for those genotypes and phenotypes in that population. If the proportion of allele A is represented by p, and that of allele a by q, the three possible genotypes will be present in proportions, or frequencies, derived from the expansion of the binomial $(p + q)^2$, which is $p^2 + 2pq + q^2$. The expected frequency of the AA homozygous class is p^2; the Aa heterozygotes, $2pq$; and the aa homozygous class, q^2 (Fig. 7.1). In the simplest case of only two alleles for a gene locus, $p + q = 1$, and $p^2 + 2pq + q^2 = 1$. The same principle applies for multiple alleles at a locus; that is, the sum of allele frequencies is 1, and the sum of all the possible genotype frequencies also is 1. The Hardy-Weinberg principle thus serves for any gene locus.

In a laboratory population composed entirely of F_1 heterozygotes with the Aa genotype, each of the two alleles has an equal probability of being included in any gamete; thus half the gametes carry allele A and half the gametes carry a, on the average. Each F_2 offspring is the product of random fertilization, so the predicted frequencies of the three possible genotypes in the F_2 progeny are $(0.5 + 0.5)^2 = (0.5)^2 \, AA + 2(0.5 \times 0.5) \, Aa + (0.5)^2 \, aa$, or $0.25 \, AA : 0.50 \, Aa : 0.25 \, aa$. The expected F_2 genotype ratio of 1 AA : 2 Aa : 1 aa is the

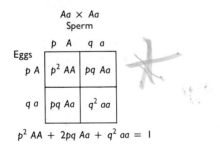

$$p^2 \, AA + 2pq \, Aa + q^2 \, aa = 1$$

Figure 7.1. The proportions of genotypes observed in a population depend on the frequencies (p, q) of the responsible alleles of a gene (A, a) in that population. The three possible genotypes occur in the frequencies of $p^2 + 2pq + q^2$, derived from expansion of the binomial $(p + q)^2 = 1$.

outcome of random fusions between gametes from a pool containing 0.5 A and 0.5 a frequencies. If random matings take place between members of the F_2 generation to produce the F_3 generation, the allele frequencies and the consequent genotype frequencies remain unchanged (Table 7.1). The randomly mating populations have reached **equilibrium** in one generation, and no change will take place in subsequent generations so long as the frequency remains 0.5 for each of the two alternative forms of the gene.

We can substitute any values other than 0.5 for p and q and show that any particular pair of allele frequencies will be established in equilibrium in a randomly mating population in the same time frame. As Hardy and Weinberg showed, the proportions of possible genotypes in a population reflect only the proportions of the alleles at the locus, and dominance or recessiveness of these alleles is irrelevant to the frequencies actually observed (Table 7.2). The actual allele frequencies and their existence in equilibrium in populations depend on the maintenance of specific conditions.

1. *Mating is random;* that is, mates are chosen without regard to the genotype at the gene locus under consideration.
2. The alleles and genotypes show *no differences in fitness;* that is, selection is not in operation at this locus, and the different alleles of the gene are replaced at the same rate.

TABLE 7.1
Frequencies of Random Matings and Offspring in an F_2 Population, where $p(A) = q(a) = 0.5$ and the Genotype Frequencies Are 0.25 AA:0.50 Aa:0.25 aa

Type of F_2 Mating	Frequency of F_2 Mating	Frequency of F_3 Genotypic Classes		
		AA	Aa	aa
AA × AA	¼ × ¼ = ⅟₁₆	⅟₁₆	0	0
AA × Aa	¼ × ½ = ⅛	⅟₁₆	⅟₁₆	0
Aa × AA	½ × ¼ = ⅛	⅟₁₆	⅟₁₆	0
AA × aa	¼ × ¼ = ⅟₁₆	0	⅟₁₆	0
aa × AA	¼ × ¼ = ⅟₁₆	0	⅟₁₆	0
Aa × Aa	½ × ½ = ¼	⅟₁₆	²⁄₁₆	⅟₁₆
Aa × aa	½ × ¼ = ⅛	0	⅟₁₆	⅟₁₆
aa × Aa	¼ × ½ = ⅛	0	⅟₁₆	⅟₁₆
aa × aa	¼ × ¼ = ⅟₁₆	0	0	⅟₁₆
		⁴⁄₁₆	⁸⁄₁₆	⁴⁄₁₆

$$F_3 \quad p^2 + 2pq + q^2 = 1$$
$$0.25\ AA + 0.50\ Aa + 0.25\ aa = 1$$
$$p(A) = 0.5 \quad p + q = 1$$
$$q(a) = 0.5$$

TABLE 7.2
Frequencies of Random Matings and Progeny Types in a Population at Equilibrium for Alleles A, a, where $(p(A) + q(a))^2 = (0.9 + 0.1)^2 = 0.81$ AA $+ 0.18$ Aa $+ 0.01$ aa $= 1$

Type of Mating	Frequency of Mating	Frequency of Progeny Types		
		AA	Aa	aa
0.81 AA $\times$ 0.81 AA	0.6561	0.6561	0	0
0.81 AA $\times$ 0.18 Aa	0.1458	0.0729	0.0729	0
0.18 Aa $\times$ 0.81 AA	0.1458	0.0729	0.0729	0
0.81 AA $\times$ 0.01 aa	0.0081	0	0.0081	0
0.01 aa $\times$ 0.81 AA	0.0081	0	0.0081	0
0.18 Aa $\times$ 0.18 Aa	0.0324	0.0081	0.0162	0.0081
0.18 Aa $\times$ 0.01 aa	0.0018	0	0.0009	0.0009
0.01 aa $\times$ 0.18 Aa	0.0018	0	0.0009	0.0009
0.01 aa $\times$ 0.01 aa	0.0001	0	0	0.0001
		0.81	0.18	0.01

$$\text{and } p(A) = 0.9$$
$$q(a) = 0.1$$
$$p(A) + q(a) = 0.9 + 0.1 = 1$$

3. Copies of the alleles are not increased or decreased by *gene flow*; that is, there is no migration of genes from one population into another.
4. *Mutation* is absent or so negligible that allele copy numbers remain constant.
5. *Population size* is effectively infinite, so sampling errors are of no consequence and changes will not occur by chance events; that is, *random genetic drift* is not in operation.

A departure from any of these conditions may lead to changes in gene frequencies rather than the status quo of equilibrium. Before we can discuss the influence of these factors on gene frequencies, it is necessary to establish whether observations in real populations fit the theoretical premises set by the Hardy-Weinberg principle.

Analysis of Gene Frequencies in Natural Populations

A good example for testing the Hardy-Weinberg theorem is the gene locus for the human MN blood groups. The two alleles, L^M and L^N, are codominant, so each phenotype is equated with a particular one of the three possible genotypes. Phenotype identification is unambiguous and is readily made by serological tests of cross-reactions with anti-M and anti-N sera against a small sample of blood. There are no known selective differences for the alleles or genotypes, and mates are chosen without regard to their MN blood type. The populations picked for analysis are appropriately large and reasonably isolated from gene flow between populations. As in any case involving a large number of individuals, the whole population is analyzed by extrapolation from data obtained in an appropriate sample of the population.

In a sample of 613 white Americans, 179 individuals were of blood group M, 304 were MN, and 130 were N (Table 7.3). The frequencies of alleles L^M and L^N in this population sample can be determined by using the Hardy-Weinberg theorem, from the proportion of L^M alleles in type M ($L^M L^M$) plus one-half MN ($L^M L^N$) in the total sample

$$\text{number M} + \tfrac{1}{2} \text{ number MN/total} = 179 + \tfrac{1}{2}(304)/613 = 0.54 \ L^M$$

or from the combined percentages of M and ½ MN individuals

$$0.292 \text{ M} + \tfrac{1}{2}(0.496) \text{ MN} = 0.54 \ L^M$$

and

$$1 - 0.54 = 0.46 \ L^N$$

Based on the same premises of relationship between genotype and allele frequencies, p can be derived from the square root of p^2. Once p is known, q is obtained from $1 - p$ (or from $\sqrt{q^2}$)

$$p^2(L^M L^M) = 0.292, \text{ and } p = \sqrt{0.292} = 0.54; \text{ and } q = 1 - 0.54 = 0.46$$

or

$$q^2(L^N L^N) = 0.212, \text{ and } q = \sqrt{0.212} = 0.46$$

Once p and q are known, the expected Hardy-Weinberg frequencies can

TABLE 7.3
Observed and Expected Numbers and Frequencies of MN Blood Types in a Sample Population of 613 White Americans

		Numbers and Percentages of Blood Types			Allele Frequencies	
		M	MN	N	$p(L^M)$	$q(L^N)$
Observed	No.	179	304	130	0.54	0.46
	%	29.2	49.6	21.2		
Expected	No.	178	306	129		
	%	29	50	21		

TABLE 7.4
Frequencies of MN Blood Types in Various Human Populations

Population	Sample	Number Observed			Allele Frequencies	
		M	MN	N	$p(L^M)$	$q(L^N)$
New Guinea highlands	303	2	32	269	0.059	0.941
Australian Aborigines	730	22	216	492	0.178	0.822
New York City whites	954	287	481	186	0.553	0.447
Guatemalan Indians	203	112	74	17	0.734	0.266
Navaho Indians (U.S.)	361	305	52	4	0.917	0.083

be calculated for the three genotypes in a sample of 613 persons, from $p^2 + 2pq + q^2 = 1$

$$p^2 = (0.54)^2 = 0.29; \text{ and } 613 \times 0.29 = 178 \ L^M L^M \text{ individuals}$$
$$2pq = 2(0.54 \times 0.46) = 0.50; \text{ and } 613 \times 0.50 = 306 \ L^M L^N \text{ individuals}$$
$$q^2 = (0.46)^2 = 0.21; \text{ and } 613 \times 0.21 = 129 \ L^N L^N \text{ individuals}$$

These predicted genotype frequencies are essentially the same as those observed. No statistical test is required, but a chi-square test or another statistical measure can be used to compare observed and expected frequencies where needed.

When human populations in various regions are sampled for MN blood type frequencies, substantial differences are found (Table 7.4). In each population, however, the observed frequencies are essentially the same as those expected for Hardy-Weinberg ratios (Fig. 7.2). Whether or not these allele frequencies are in equilibrium cannot be determined by these samples alone. Data are needed from two or more generations of these populations in order to determine whether the frequencies are unchanging. These data do show, however, that allele and genotype frequencies have nothing to do with dominance, codominance, or recessiveness, as Hardy and Weinberg pointed out in 1908. The L^M and L^N alleles are the same codominant alternatives in all the populations, regardless of their absolute frequencies in each.

The strikingly different frequencies of the MN blood group alleles in human populations may have been established by chance, because blood types are believed to be selectively neutral and not subject to mating preferences. If these populations were started by relatively few *founder* individuals, the allele frequencies of the founders might be maintained. In smaller populations, chance alone might be responsible for a change in frequency from some initial values to new values, which would then be perpetuated in equilibrium as Hardy-Weinberg proportions.

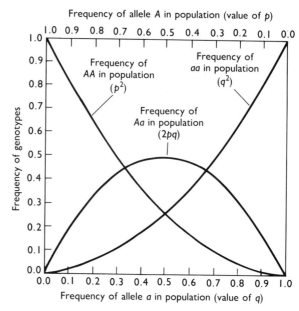

Figure 7.2. The particular frequencies of the genotypes AA, Aa, and aa in a population can be predicted from the particular frequencies (p, q) of the alleles A and a in that population, according to the Hardy-Weinberg principle.

    The MN gene is located on an autosome and exists in only two allelic forms. The Hardy-Weinberg principle can readily be extended to genes on the sex chromosomes; genes with more than two alleles; alleles that behave as dominants, codominants, or recessives; continuously variable and discontinuous traits; and molecular as well as phenotypic variation. The basic principle also permits us to analyze nonequilibrium gene frequencies, and thus to evaluate the kinds and directions of evolutionary change under the influence of selection, mutation, gene flow, mating systems, and random genetic drift.

Major Influences on Gene Frequencies

    Any locus with allele frequencies in equilibrium is not evolving, because there is no genetic change over time. Changes in gene frequencies indicate evolution in progress, but changes may take place at some gene loci and not at others in any population at any time. We will now see how the Hardy-Weinberg theorem can be used to obtain quantitative information on evolutionary processes, and how different factors influence the nature and pace of genetic modification in evolving populations.

The Effects of Selection on Gene Frequencies

Selection plays a major role in adaptive evolution, as has been stated in earlier chapters. The quantitative methods of population genetics allow measurements of the effect of selection on gene frequencies, the pace of elimination of harmful alleles, and the reason for the persistence of some alleles in a population despite their patently harmful effects on the organism.

Suppose we start with a population of zygotes (fertilized eggs) of known genotypes, which are present in the frequencies of 0.25 AA, 0.50 Aa, and 0.25 aa. The allele frequencies are $p_0 = q_0 = 0.5$ in this initial generation. We observe the survival and reproduction of each genotype class and find that for every 100 AA or Aa zygotes that survive and reproduce, only 80 aa zygotes survive and reproduce. These genotypes can be described in terms of **fitness, w,** which is the ability to survive and reproduce, and of **selective disadvantage,** which is reduction in fitness, represented by the **selection coefficient, s.** If the fitness of AA and Aa is set at 1 (100%), then the relative fitness of the aa genotype is only 0.8 (80%) of this level. The formula for determining fitness is $w = 1 - s$, and for the aa genotype $0.8 = 1 - s$, so $s = 0.2$.

The allele frequencies in the gene pool of the next generation will be different from that of the initial generation, because all the AA and Aa genotypes but only a fraction of the aa genotypes contribute to the gene pool (Table 7.5). The selection coefficient s is incorporated into the calculation to determine the quantitative effect of selection against the aa genotype, expressed by new allele frequencies in the gene pool. The new allele frequencies, $p_1(A)$ and $q_1(a)$, will differ from the initial frequencies, $p_0(A)$ and $q_0(a)$, by an increment of change for both alleles, but the sum of their frequencies remains 1. Once the new value of p is known, q can be obtained from $1 - p$.

The change in p, or Δp, is obtained with the formula

$$\Delta p = \frac{s p_0 q_0^2}{1 - s q_0^2} \qquad [7.1]$$

TABLE 7.5
Change in Genotype Frequencies in One Generation, where $s > 0$ for Recessive (aa) Genotype

Genotype	Initial Frequency	Fitness (w)	Contribution to the Gene Pool
AA	p^2	1	$1 \times p^2$
Aa	$2pq$	1	$1 \times 2pq$
aa	q^2	$1 - s$	$(1 - s)q^2 = q^2 - sq^2$
	$p^2 + 2pq + q^2$		$p^2 + 2pq + q^2 - sq^2$
Sum:	$= 1$		$= 1 - sq^2$

Substituting the specific values for p_0, q_0, and s

$$\Delta p = \frac{(0.2)(0.5)(0.5)^2}{1 - (0.2)(0.5)^2} = \frac{0.025}{1 - 0.05}$$
$$\Delta p = 0.03$$

Thus

$$p_1 = p_0 + \Delta p = 0.50 + 0.03$$
$$= 0.53$$

and

$$q_1 = 0.47$$

The new genotype frequencies are calculated from the expansion of the new binomial, or

$$(p_1 + q_1)^2 = p_1^2 + 2p_1q_1 + q_1^2 = 1$$

and

$$(0.53 + 0.47)^2 = (0.53)^2 + 2(0.53 \times 0.47) + (0.47)^2$$
$$= 0.28\ AA + 0.50\ Aa + 0.22\ aa$$

The new genotype frequencies show a significant increase in AA at the expense of aa, but virtually no change in heterozygotes. This is an example of *directional selection* against a somewhat deleterious allele. As selection continues over successive generations, we expect a substantial reduction in aa, a substantial increase in AA, and a slower rate of decrease in Aa genotypes. In this case of relatively weak selection, the effects are strung out over thousands of generations. The same pattern of change can be shown more simply and clearly if we look at a case involving *complete selection* against a harmful allele.

In the case of an aa genotype with 100 percent disadvantage ($w = 0$, $s = 1$)—that is, a lethal recessive genotype that does not survive and reproduce—we can estimate the rate of evolutionary change under strong selection pressure. For these purposes, the fitness of AA and Aa genotypes will remain 1, the starting frequencies of the two alleles will be the same $p_0 = q_0 = 0.5$, and the three genotypes will be in Hardy-Weinberg proportions of $0.25\ AA + 0.50\ Aa + 0.25\ aa$.

The recessive aa individuals make no contribution to the gene pool for the next generation, and the only source of a alleles are the heterozygotes in $AA \times Aa$ and $Aa \times Aa$ matings. In order to determine the expected q_1^2 (and q_1), the effective heterozygote frequency must be known because the aa genotype in the next generation arises from only $Aa \times Aa$ matings. The gene pool is produced by only $p_0^2 + 2p_0q_0$ (q_0^2 is eliminated), and the effective heterozygote frequency is

$$\frac{2p_0q_0}{p_0^2 + 2p_0q_0} = \frac{2q_0}{p_0 + 2q_0} = \frac{2q_0}{p_0 + q_0 + q_0} = \frac{2q_0}{1 + q_0}$$

The probability of mating between heterozygotes is the product of the separate probabilites, or heterozygote frequencies, but only one-quarter of their offspring are expected to be *aa*, according to the standard 1:2:1 Mendelian genotype ratio. The expected frequency of the *aa* genotype, or $q^2{}_1$, is therefore

$$q_1^2 = \frac{2q_0}{1 + q_0} \times \frac{2q_0}{1 + q_0} \times \frac{1}{4} = \frac{q_0^2}{(1 + q_0)^2}$$

and

$$q_1 = \sqrt{\frac{q_0^2}{(1 + q_0)^2}} = \frac{q_0}{1 + q_0}$$

Substituting the values in this example

$$q_1^2 = \frac{(0.5)^2}{(1.5)^2} = \frac{0.25}{2.25} = 0.11 \qquad \text{and} \qquad q_1 = 0.33$$

To determine the pattern of change in gene frequencies over time, t, as measured in generations, we can use the formula

$$q_t = \frac{q_0}{1 + tq_0} \qquad\qquad [7.2]$$

For example, in generation 10, it can be predicted that the initial value of $p_0 = 0.5$ will have changed to $p_{10} = 0.083$

$$q_{10} = \frac{0.5}{1 + 10(0.5)} = 0.083$$

As shown in Table 7.6, complete selection leads to a rapid initial decrease in the frequency of the unfit allele and genotype, but to a slower and slower

TABLE 7.6
Changes in the Frequency of $q(a)$ in 100 Generations, where $s = 1$ for aa Genotype and $s = 0$ for AA and Aa Genotypes. The Population Begins with $p_0 = q_0 = 0.5$ and with 0.25 AA + 0.50 Aa + 0.25 aa as Genotype Frequencies

Generation (t)	$q^2(aa)$	$q(a)$
0	0.25	0.50
1	0.11	0.33
2	0.06	0.25
3	0.04	0.20
4	0.03	0.17
5	0.02	0.14
10	0.007	0.08
20	0.002	0.05
50	0.0004	0.02
100	0.0001	0.01

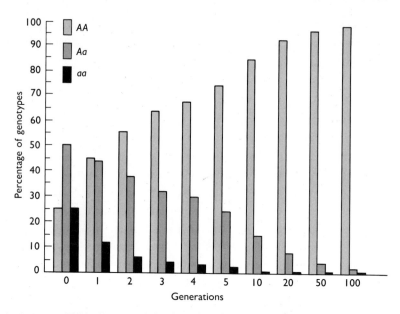

Figure 7.3. The values shown in Table 7.6 are plotted in the form of a histogram, showing the course of complete selection against *aa* genotypes in a population initially composed of 0.25 *AA*, 0.5 *Aa*, and 0.25 *aa* genotypes. The frequency of allele *a* decreases ever slower with time, as more and more of the *a* alleles are protected in heterozygous genotypes and as fewer *aa* genotypes are exposed to selection.

rate of decrease over longer periods of time. If the expected frequencies of all three genotypes over time are plotted in the form of a histogram, it is clear that the decline in heterozygote frequency is not as great or as fast as the decline in recessives (Fig. 7.3). The effects of directional selection on the unfit allele are dampened by protection of the *a* allele in heterozygotes, which here express the dominant phenotype.

By rearranging the formula for q_t, we can solve for t directly and predict the time required to reduce the initial frequency to some specific value. For example, how long would it take a program of complete selection against the recessives to reduce $q(a)$ from 0.5 to 0.01, or from 50 percent to 1 percent of the gene pool?

$$q_t = \frac{q_0}{1 + tq_0}$$

is rearranged to

$$t = \frac{q_0 - q_t}{q_0 q_t} \quad \text{and} \quad t = \frac{0.5 - 0.01}{(0.5)(0.01)} = 98$$

It would take almost 100 generations to reduce the allele frequency to 1 per 100 gametes carrying the gene, and about 1000 generations before it was 1

TABLE 7.7
Frequencies of Homozygotes and of Heterozygous Carriers for Various Recessive Human Disorders

Inherited Disorder	Frequency of Homozygotes (aa)	Frequency of Carriers (Aa)	Ratio of Carriers to Homozygotes
Sickle-cell anemia	I in 400	I in 10	40:1
Cystic fibrosis	I in 1600	I in 20	80:1
Phenylketonuria (PKU)	I in 40,000	I in 100	400:1
Tay-Sachs disease	I in 100,000	I in 160	625:1
Alkaptonuria	I in 1,000,000	I in 500	2000:1

per 1000 (0.001). After nearly 100,000 generations, the lethal allele would be reduced to a frequency of 10^{-5}, which is also a fairly typical frequency of mutation from a wild-type allele, such as A, to a recessive alternative. The balance between selection and mutation pressures will be discussed shortly.

In most populations, the frequencies of lethal alleles are very low, but they persist because selection is relatively ineffective when p, q, or $s = 0$ or very low values. As q gets smaller and smaller over time, and the lethal allele is protected in heterozygotes, any further significant reduction is unlikely. The proportion of heterozygous carriers of harmful alleles is astonishingly high, even for rare recessive genotypes (Table 7.7). The genetic load of harmful alleles carried in diploid sexually reproducing populations is largely the outcome of ineffective selection pressure when q is very low, even if $s = 1$.

Selection against harmful dominant or codominant alleles is very effective, however, because the alleles are expressed in the phenotype of the organism. Similarly, advantageous dominant or codominant alleles increase more rapidly than recessive alleles that confer greater fitness (Fig. 7.4). In general, it is clear from equation 7.2 that the rate of change in allele frequencies due to selection depends on the frequencies of the alleles, the degree of dominance, and the extent of the differences in fitness among genotypes. Different modes of selection, as discussed in Chapter 6, will lead to different patterns of change in allele frequencies and different proportions of genotypes. The particular patterns and rates of change can be predicted by the application of mathematical models based on the Hardy-Weinberg principle.

Even in populations that are very large, mate randomly, and experience no gene flow, alleles with different degrees of fitness will reach equilibrium frequenies that reflect a balance between the opposing forces of selection and mutation. For a recessive allele with some disadvantage ($s > 0$), for example, mutation from A to a replenishes allele a at the same time that selection reduces its frequency in the gene pool. The balance between selection and forward mutation from A to a is shown by $spq^2 = up$, where u is the mutation rate of $A \rightarrow a$. We can ignore the effect of reverse mutation, $a \rightarrow A$, which

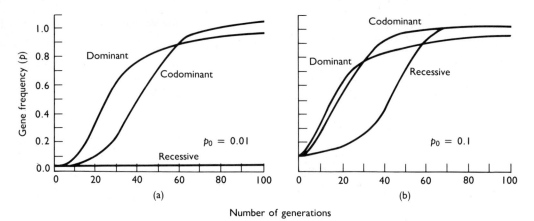

Figure 7.4. Increase in frequency over 100 generations of an advantageous dominant, codominant, or recessive allele from an initial value of (a) $p_0 = 0.01$, or (b) $p_0 = 0.1$. The advantageous codominant allele reaches fixation before either an advantageous dominant or an advantageous recessive, because the advantageous dominant reaches fixation only as the disadvantageous recessive is very slowly eliminated. At a low initial frequency, an advantageous recessive increases very little, because few homozygous recessives are exposed to selection. At a higher frequency, either initially or long after an advantageous recessive mutation has arisen in a population, the advantageous recessive allele reaches fixation very rapidly. This is due to the very rapid elimination of the deleterious dominant or codominant alternative allele at the gene locus. (Adapted from D. J. Futuyma, 1986, *Evolutionary Biology*, chap. 6, Fig. 4, p. 157. Sunderland, Mass.: Sinauer)

usually occurs at a much lower rate than forward mutation, and derive the equilibrium frequency of the *aa* genotype, or $\hat{q}^2$, and of the *a* allele, or $\hat{q}$

$$spq^2 = up$$
$$sq^2 = u$$
$$\hat{q}^2 = u/s \quad \text{and} \quad \hat{q} = \sqrt{u/s}$$

If 10 percent of *aa* genotypes fail to survive and reproduce ($s = 0.1$), the equilibrium frequency for alleles at a locus with a typical forward mutation rate of 10^{-5} will be

$$\hat{q}^2 = 10^{-5}/10^{-1} = 10^{-4} \quad \text{and} \quad \hat{q} = \sqrt{10^{-4}} = 10^{-2}, \text{ or } 0.01$$

For a lethal genotype ($s = 1$)

$$\hat{q}^2 = 10^{-5}/1 = 10^{-5} \quad \text{and} \quad \hat{q} = \sqrt{10^{-5}} = 0.003$$

The equilibrium frequency of A, or $\hat{p}$, of course, will be $1 - \hat{q}$.

These calculations show that a 10-fold change in s produces a 10-fold change in the *aa* genotype, and a 3-fold change in the frequency of *a* in the gene pool. These differences in allele frequencies at equilibrium are also reflected in the heterozygote frequency, as seen by

$$2\hat{p}\hat{q} = 2(0.99 \times 0.01) = 0.02$$

versus

$$2(0.997 \times 0.003) = 0.006$$

The proportion of heterozygous carriers of the harmful allele is more than three times higher when $s = 0.1$ than when $s = 1$ for *aa* genotype. The higher proportion of carriers leads to a higher proportion of *aa* individuals produced by matings between heterozygotes, as seen by

$(2pq \times 2pq) \times ¼$ (expected production of *aa* offspring)
$(0.02 \times 0.02)(¼) = 0.0001$, versus $(0.006 \times 0.006)(¼) = 0.000009$

The probability, or risk, of an *aa* offspring produced by heterozygote matings rises from 9 per million when $s = 1$, to 100 per million when $s = 0.1$, at equilibrium.

These comparisons have some bearing on changes in human populations due to medical improvements that reduce the mortality rate among individuals with inherited disorders. A longer life expectancy, which may allow the individual to survive and reproduce, will lead to an increase of the deleterious allele and genotype in the population (Fig. 7.5). Although we welcome therapies that increase chances for survival and a better quality of life, the economic burden of continued medical care for afflicted persons and their families must be taken into consideration. For disorders that require

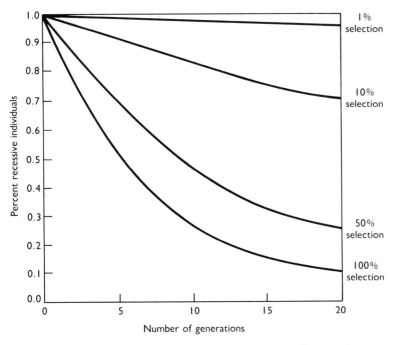

Figure 7.5. Individuals with the recessive genotype are eliminated more and more effectively as the intensity of selection increases. Beginning with a population containing 1 percent recessives, the recessives are reduced in frequency to about 0.1 percent in 20 generations under a regime of complete selection. The change in recessive frequency is much slower, however, when selection intensity is lower. (From *Biology: A Human Approach*, 3rd ed., by Irwin Sherman and Vilia Sherman. Copyright © 1983 by Oxford University Press. Reprinted by permission)

institutionalization, the economic burden is increased for a broad cross section of the society through taxation to pay for the costs of public support. Problems of this sort pose ethical and moral questions, which may or may not be given due consideration in every society.

As seen in the previous examples, equilibrium gene frequencies will be established even though the same selection pressure continues to be exerted for or against some genotype. Should conditions change in any way to alter selection pressure, however, gene frequencies will again change until a new equilibrium frequency is attained. Although some geneticists believe that mutation rates may be modified under certain conditions, current information points to relatively constant rates of mutation over time. Evolutionary changes in gene frequencies would thus appear to be influenced far more by selection than by mutation. In the absence of selection and other factors, however, mutation may be the major influence on gene frequencies. This is a major thrust of the neutral theory of molecular evolution.

Mutation and Gene Frequencies

Recurrent forward mutations, $A \rightarrow a$, reduce the frequency of A and increase that of a in the gene pool, and reverse mutations, $a \rightarrow A$, have the opposite effect. The rate of forward mutation, u, and the rate of reverse mutation, v, are expressed as proportions of alleles mutating per generation. The actual value of total alleles of a gene changing from A to a in one generation is up; from a to A, it is vq.

The extent of the mutation effect depends on the initial frequencies of the alleles as well as the mutation rates, which is implicit in the terms up and vq. Suppose we start with allele A at a frequency of 0.8 and a mutation rate of 10^{-5}, and allele a at a frequency of 0.2 and a mutation rate of 10^{-8} (reverse mutation rates generally are much lower than forward rates). The net change in the frequency of A, or Δp, from $p_0 = 0.8$ to the new value in the next generation is

$$\Delta p = vq - up \qquad\qquad [7.3]$$

Substituting these values in the equation

$$\begin{aligned}
\Delta p &= (1 \times 10^{-8})(0.2) - (1 \times 10^{-5})(0.8) \\
&= (2 \times 10^{-9}) - (8 \times 10^{-6}) \\
&= -8 \times 10^{-6} \text{ (rounded off from } -7.998 \times 10^{-6})
\end{aligned}$$

The new frequency of allele A, or p_1, is the sum of the initial frequency plus the change in frequency of A

$$p_1 = p_0 + \Delta p$$

then

$$\begin{aligned}
p_1 &= 0.8 + (-8 \times 10^{-6}) \\
&= 0.799992
\end{aligned}$$

and

$$q_1 = 0.200008$$

The net effect of mutation in one generation in this case is only eight more *a* alleles (and eight fewer *A*) per million gametes making up the gene pool for the new generation. The allele frequencies ultimately reach an equilibrium when no further change occurs as the result of mutation alone. This is expressed as

$$\Delta p = up - vq = 0$$

or

$$u\hat{p} = v\hat{q} \qquad [7.4]$$

where $\hat{p}$ and $\hat{q}$ are the equilibrium frequencies of *A* and *a*, respectively. The equilibrium frequencies of the two alleles can be determined by

$$u\hat{p} = v\hat{q}$$
$$u\hat{p} = v(1 - \hat{p})$$
$$u\hat{p} = v - v\hat{p}$$
$$\hat{p}(u + v) = v$$
$$\hat{p} = v/(u + v)$$

similarly

$$\hat{q} = u/(u + v)$$

The equilibrium values $\hat{p}$ and $\hat{q}$ thus depend on only the mutation rates and are independent of the initial frequencies p_0 and q_0, including zero and 1. The time at which the alleles reach equilibrium will vary, of course, according to their particular starting frequencies. In this example, the stable equilibrium frequencies will be $\hat{p} = 0.001$ and $\hat{q} = 0.999$. It will take many tens of thousands of generations, however, before these frequencies become established under the influence of mutation alone.

It can thus be seen that a selectively neutral allele can become fixed in the gene pool as the consequence of mutation, but the process is very slow in large, randomly mating populations not undergoing gene flow. Mutation is regarded as a relatively weak force in evolution because of the slow pace of change effected. But mutation is the primary means by which most *new genetic information* is produced, and many alleles may exist for a gene locus because each nucleotide substitution may give rise to a new gene sequence. The rate of forward mutation is much higher than the rate of reverse mutation back to the original sequence, so any allele can be rare but many different alleles may exist in the gene pool. In view of the apparent absence of distinguishable phenotypic effects among different alleles for many molecular polymorphisms, it would seem that mutation may be the principal influence on the relative allele frequencies at these loci. As mentioned in Chapter 6, however, alleles with equal effects under one set of conditions may confer differences in fitness under another set of conditions. The reservoir of allelic diversity provides greater evolutionary flexibility in populations over the long run of time and in changing environments.

Inbreeding and Outbreeding Systems of Mating

Departures from random mating are the rule in virtually every species that has been studied; that is, mate selection is rarely **panmictic.** In a panmictic population, any two individuals of opposite sex may mate anywhere within the range of that population. In most species, all potential partners in separate subpopulations do not have equal access to one another. Even within a subpopulation, it is more likely that individuals will mate with those closer to themselves than with others farther away. Any of these populations may be outbreeding; that is, the mating partners are not genetically related. Inbreeding occurs when mates are more closely related than they would be if they were chosen at random. These different mating systems should have different consequences on the amount of genetic variation in populations, and they do.

In the most extreme case of self-fertilization, calculations can be made for the effect of inbreeding on the genotype frequencies in successive generations. Suppose the initial population consists of all three possible genotypes for the pair of alleles A and a at a gene locus. A fraction D are homozygous dominants (AA); a fraction H are heterozygous (Aa); and R are homozygous recessives (aa). The homozygous genotypes produce only their own kind after self-fertilization, but the progeny of the heterozygotes are $0.25AA$, $0.5Aa$, and $0.25aa$. We must therefore add one-quarter of the heterozygotes' progeny to each of the homozygous classes, and do so for each generation. The proportion of heterozygotes is reduced to half of its previous value in each generation, and the remaining half is divided equally between the two homozygous classes (Fig. 7.6). The reduction by one-half of heterozygotes in each generation is independent of the initial proportion and occurs whether or not the genotypes existed in Hardy-Weinberg ratios.

Instead of tediously calculating the genotype frequencies by keeping track of the numbers in each generation, the effects of inbreeding can be measured by the standard term F, the **inbreeding coefficient.** The most useful definition of F is that it is the proportion of heterozygosity reduced by inbreeding, relative to the degree of heterozygosity in a randomly mating population with the same allele frequencies (Table 7.8). Hardy-Weinberg proportions exist when inbreeding is absent ($F = 0$); only homozygous classes exist in extreme programs of inbreeding by self-fertilization ($F = 1$); and when inbreeding is between these two extremes ($F = 0 < 1$), heterozygosity is decreased in proportion to the value of F.

Inbreeding by self-fertilization is not mandatory in hermaphroditic species, in which each individual can produce both male and female gametes. Self-fertilization is impossible in species whose members are of one sex or the

Generation	Proportion of genotypes		
	AA	Aa	aa
0	D	H	R
1	$D + \dfrac{H}{4}$	$\dfrac{H}{2}$	$R + \dfrac{H}{4}$
2	$D + \dfrac{H}{4} + \dfrac{H}{8}$	$\dfrac{H}{4}$	$R + \dfrac{H}{4} + \dfrac{H}{8}$
t	$D + \dfrac{H}{2}[1 - (\tfrac{1}{2})^t]$	$H(\tfrac{1}{2})^t$	$R + \dfrac{H}{2}[1 - (\tfrac{1}{2})^t]$
∞	$D + \dfrac{H}{2}$	0	$R + \dfrac{H}{2}$

Figure 7.6. The effects of inbreeding on genotype frequencies in successive generations of self-fertilizing populations. The proportion of heterozygotes is reduced to one-half its previous value in each generation, and the remaining half is divided equally between the two homozygous classes. The proportions at any time, t, and the eventual composition of the population (at infinity, ∞) are shown. The reduction by one-half of heterozygous genotypes in each generation is independent of the initial proportion, H. (Adapted from J. F. Crow, 1986, *Basic Concepts in Population, Quantitative, and Evolutionary Genetics*, Fig. 2-1, p. 31. New York: Freeman)

other. Heterozygosity decreases when relatives mate, the rate of decrease varying according to the degree of relatedness between the mates (Fig. 7.7). In human societies, the general taboo includes matings between parent and offspring and between brother and sister (sibs), but usually does not extend to matings between cousins. In some cultures, brother–sister matings were the rule, as among royalty in ancient Egypt and in the Aztec and Hawaiian societies more recently. In modern Western societies, about 1 marriage in 200 is between cousins. Matings between closer relatives occur at an insignificant level, according to official records.

Self-fertilization and sib mating are deliberate programs that produce almost completely homozygous strains for purposes of research, medicine, and agriculture. Highly homozygous strains in normally outbred species are generally characterized by poor survival, reduced vigor, and low fertility. They may be homozygous for the dominant allele at some loci and for the recessive allele at other loci in the genome. The deleterious effects of inbreeding are believed to be due primarily to the unmasking of harmful recessive alleles previously unexpressed in heterozygotes. Although harmful dominant alleles are effectively eliminated by natural selection, recessive mutant alleles are sheltered from selection in heterozygotes.

TABLE 7.8
Effects of Different Degrees of Inbreeding on Genotype Frequencies in Populations

Inbreeding		Genotype Frequencies		
Proportion	Coefficient (F)	AA	Aa	aa
0	0	p^2	$2pq$	q^2
$0 < 1$	$1 - (\frac{1}{2})^{t*}$	$p^2 + Fpq$	$2pq - 2Fpq$	$q^2 + Fpq$
1	1	$p^2 + pq$	0	$q^2 + pq$

*t is the number of generations of inbreeding; for example, $1 - (\frac{1}{2})^2 = \frac{3}{4}$.

What is the effect of first-cousin matings on the production of deleterious recessives in their offspring compared with the offspring of unrelated parents? We can calculate the probability of an offspring with the *aa* genotype by the use of the term $q^2 + Fpq$, as shown in Table 7.8, when there is some degree of inbreeding. For offspring of first cousins, the coefficient of inbreeding is $\frac{1}{16}$, as determined by tracing the probability that the two alleles at a locus are *identical by descent* (Fig. 7.8). The probability that two unrelated parents will have an offspring of the *aa* genotype is q^2. For the comparison, we can use the example of cystic fibrosis, a recessive disorder that occurs in about 1 per 10,000 human births: $q^2 = \frac{1}{10,000}$, $q = \frac{1}{100}$, and $p = \frac{99}{100}$.

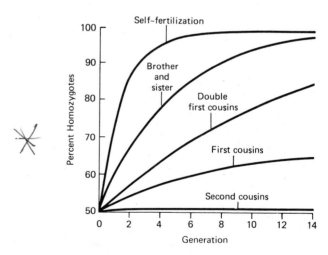

Figure 7.7. The percentage of homozygous individuals in successive generations under different degrees of inbreeding, according to the degree of relatedness between parents. (From S. Wright, 1921. *Genetics* 6:172)

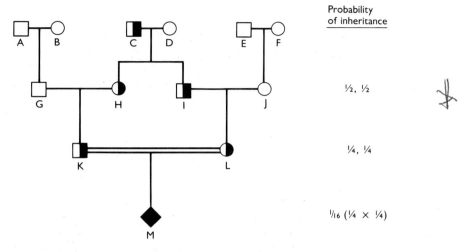

Figure 7.8. The probability of inheritance, or coefficient of inbreeding, is determined by tracing the probability that the two alleles at a locus are identical by descent. The probability is ¹⁄₁₆ for expression of a recessive genotype in individual M, the offspring of a consanguineous mating between first cousins K and L. This value is based on a probability of ¼ for each of these first cousins having inherited the identical recessive allele from their common grandfather, C, through parent H or I, who are brother and sister offspring of C. The product of the separate probabilities ¼ × ¼ is ¹⁄₁₆.

The probability, *P*, that first-cousin parents will have an offspring with the recessive disorder of cystic fibrosis is

$$P_{aa} = q^2 + Fpq$$
$$= (1/10,000) + (1/16)(99/100)(1/100) \quad [7.5]$$
$$= 23/32,000 = 7.2/10,000$$

The chance is about seven times greater that first cousins will have a child with cystic fibrosis than in the population at large. The increase in homozygous recessives by inbreeding is much greater for rare alleles than for more common harmful alleles. If the frequency were 1 per million instead of 1 per 10,000 ($q^2 = 1/10^6$; $q = 1/10^3$; $p = 999/10^3$), first-cousin matings would have more than 63 times the risk of producing an offspring with the disorder than matings between unrelated individuals in the general population.

The frequency of homozygous recessives thus increases only slightly for the more common recessive alleles, but some very rare human disorders have been found only in children of *consanguineous* matings, usually between first cousins (Fig. 7.9). In fact, if a mutation had occurred only once in the history of a population, it could be homozygous only by inbreeding. The overall effect of consanguineous matings is estimated to be about 50 percent, based on studies in various human populations. For example, if the incidence of severe disease were 0.01, a 50 percent increase due to first-cousin marriages would change the incidence to 0.015, or from 10 per 1000 to 15 per 1000. The

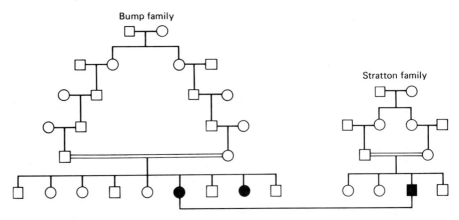

Figure 7.9. Lavinia Bump and Charles Stratton ("General Tom Thumb") were midgets in the famous P. T. Barnum circus in the late 1800s. They were from unrelated families, but each of them was the child of parents who were cousins (note the double horizontal line connecting the parents in each case). Such rare autosomal recessive inheritance may not be phenotypically evident for generations, but recessive offspring are more likely to be produced in consanguineous matings.

occasional first-cousin marriages thus add relatively little to the incidence of harmful recessive disorders in a population.

Occasional inbreeding in otherwise randomly mating outbreeding populations has little effect on increasing the frequency of harmful recessive traits, but a consistent pattern of inbreeding would be expected to weaken the population and increase the chance of extinction, because of the known deleterious effects of many recessive alleles. How, then, can we explain the evolution of self-fertilizing species from outbreeding ancestors, which is considered to be the pattern in a number of plants that have been studied? This problem was solved by Sir Ronald Fisher, one of the founders of population genetics. The population can evolve from random mating to self-fertilization only if the self-fertilizers produce more progeny than the outbreeders, and if this excess in progeny is greater than the reduction in fitness of the inbred progeny.

To illustrate the evolutionary pattern, consider a cross-fertilizing species that is prevented by some genetic mechanism from self-fertilization. If the size of the population is constant, each plant transmits one sperm nucleus in a pollen grain and one egg to the gametic pool, on the average. If one plant can fertilize its own eggs by virtue of a gene change, but still produce pollen that fertilizes other plants, the number of offspring produced by cross-fertilization will not be reduced significantly because a great amount of pollen is produced. The mutant plant, however, will contribute three gametes, on the average, instead of only two gametes to the next generation: the egg and cross-fertilizing pollen, plus the self-fertilizing pollen.

The alleles for self-fertilization should thus increase rapidly in the population, but self-fertilization also causes lethality. The genes for self-fertilization will increase only if the 50 percent excess in progeny is greater than the lowered fitness of the self-fertilized offspring. Suppose that an average plant has a recessive lethal allele at each of two independently assorting gene loci (*AaBb*). If the plant is self-fertilized, $\frac{1}{16}$ of the progeny will be homozygous for both lethals (*aabb*), $\frac{6}{16}$ will be homozygous for one or the other lethal (*aaB-* and *A-bb*), and $\frac{9}{16}$ will not be homozygous for either lethal (*A-B-*), according to conventional Mendelian ratios for dihybrid inheritance. The number of self-fertilizing gametes contributed by the surviving *A-B-* progeny will be $2(\frac{9}{16}) = \frac{9}{8}$, which is greater than 1. The genes for self-fertilization will therefore tend to increase.

If the recessive alleles causing reduced fitness are few enough that self-fertilization causes less than a 50 percent reduction in reproductive capacity, genes for self-fertilization tend to increase in frequency, and the proportion of self-fertilizers will reach 100 percent. If not, the self-fertilizers cannot become established, and their proportion will decrease to zero. Most plant species are therefore expected to be either predominantly self-fertilizing or predominantly cross-fertilizing, and very few to exhibit both types of reproduction. This is indeed the pattern found in natural populations of plant species.

In this situation, as in others previously discussed, the direction of evolutionary change is the outcome of the greater advantages of immediate reproductive success, even if the long-term prospects are reduced. Self-fertilization can replace cross-fertilization, but inbreeding leads to the continued production of detrimental recessive traits at gene loci and the increasingly lower vitality and fertility of later generations. The particular benefits of self-fertilization are expected to be great enough to counteract the deleterious effects, at least under prevailing conditions. Should these conditions change, however, the species is at greater risk of extinction or at least of severe reduction in population size, range of habitats, or other elements of its existence.

Gene Flow Between Subpopulations

Natural populations of virtually all species are usually subdivided into local *subpopulations*, each of which tends to be randomly mating but to exchange genes occasionally as migrants move among local units. If this **gene flow** is extensive, the whole population may be close to existing as a panmictic unit. If little migration occurs, the subpopulations may be genetically quite different from one another because they tend to diverge under the influences of random mutation and varied selection pressures in heterogeneous environments. The *amount* of migration is therefore a determining factor in the level of genetic differentiation among subpopulations of a species.

The new allele frequencies in a population that receives migrant genes through interbreeding to produce the first generation are calculated with the formula

$$p_1 = p_0(1 - m) + p_m m \qquad [7.6]$$

where p_0 is the allele frequency in the recipient population, p_m is the frequency of the same allele for the migrant gene, and m is the fraction of migrant genes in the population, or the **migration coefficient.**

Suppose a population with the allele frequencies $p(A) = 0.9$ and $q(a) = 0.1$ receives 10 percent migrant genes whose allele frequencies are $p(A) = 0.1$ and $q(a) = 0.9$. The new allele frequencies in the next generation will be

$$p_1 = (0.9)(1 - 0.1) + (0.1)(0.1)$$
$$p_1 = 0.82$$

and

$$q_1 = 0.18$$

The frequency of A has decreased by nearly 10 percent, and the frequency of a has increased by almost 100 percent, from 0.1 to 0.18, in the recipient population in one generation. If the influx of migrants continues, the allele frequencies in the recipient population will be converted to the allele frequencies in the population contributing the migrants. The speed and extent of change in allele frequencies clearly vary with the original frequencies in the two populations, the proportion of migrants, the number of generations involved, and the degree of interbreeding between residents and migrants.

In a study reported in 1969, the frequency of the Duffy blood group allele Fy^a served as an index of overall gene flow in populations of American whites and blacks in various parts of the United States. The recessive fy^a allele governs the absence of Duffy blood antigen a, and homozygous recessives are resistant to the malarial parasite *Plasmodium vivax*, which is endemic in West Africa. The frequency of the dominant Fy^a allele is virtually zero in West African populations, from which the ancestors of most American blacks came about 10 generations ago. At the time of the study, the Fy^a frequency was 0.422 in Georgia whites and 0.045 in Georgia blacks. The proportion of Fy^a introduced into the Georgia black population can be calculated and can serve as an index of migration of white genes into this black population. The formula is

$$p_t - p_m = (1 - m)^t(p_0 - p_m) \qquad [7.7]$$

In this example, the time in number of generations of gene flow, t, is 10; the initial frequency of the Fy^a allele in the black population is $p_0 = 0$; the allele frequency after 10 generations is $p_{10} = 0.045$; and the migrant gene frequency in the white population is $p_m = 0.422$. We can solve for m by rearranging the formula to get

$$(1 - m)^{10} = (p_{10} - p_m)(p_0 - p_m)$$
$$(1 - m)^{10} = (0.045 - 0.422)(0 - 0.422)$$
$$m = 0.011$$

The average rate of migration of white genes into this black population has been about 1 percent per generation.

From the Duffy allele frequency, the proportion of white genes in this black Georgia population is $0.045/0.422 = 0.11$. Somewhat higher values, of about 20 to 25 percent, were found in similar studies of Oakland, Detroit, and New York black populations. The introduction of African genes into the white population produced a smaller effect, because of the larger populations of whites than blacks and because persons of mixed background tend to be classified and to classify themselves as blacks.

Each population would eventually have allele frequencies that were weighted averages of the frequencies of the starting populations, if gene flow continued to occur between these populations. The value of m is the determining factor for the rate of change leading to the eventual gene frequencies in these populations, and any others experiencing gene flow. Other processes become more influential over a considerable period of time, because p_t gets closer and closer to p_m as t gets larger.

Random Genetic Drift in Relation to Population Size

The concept of changes in gene frequencies due to chance alone, or **random genetic drift,** was proposed by Sewall Wright in the early 1930s. The significance of random drift in evolution is still a controversial matter, but many evolutionists believe that it is relatively unimportant, except for very small populations. The difficulty in their accepting random drift as a significant component in evolution is that most evolutionists are dedicated to Darwinian selection and cannot acknowledge a concept that places chance over selection in any meaningful evolutionary context. This problem is the heart of the controversy over neutral evolution, as discussed in Chapter 6.

The loss or fixation of an allele by chance alone is an accepted premise in relation to populations small enough to be subject to random sampling errors in the formation of the gene pool for the next generation. Suppose we take an extreme example of a single self-fertilizing heterozygous plant (Aa), which produces only one surviving offspring. The population size, N, is 1, and remains the same in each generation. The single surviving offspring of the Aa self-fertilizing plant could be any one of the three possible genotypes. The chance is 0.5 that the offspring is Aa, in which case the frequencies will remain $p(A) = q(a) = 0.5$. If the surviving offspring by chance is one of the two possible homozygous genotypes, the allele frequencies would change to $p(A) = 1$ and $q(a) = 0$ for an AA offspring, or $p(A) = 0$ and $q(a) = 1$ for an aa offspring. The fixation or loss of an allele thus depends only on chance and not on the relative fitness of the genotypes in this case.

Suppose there is a number N of diploid individuals in a population, but only a proportion of them are breeders and contribute to the gene pool. The **effective population number,** N_e, is the number of breeders rather than the total census of individuals. In an idealized population, each parent is assumed to have the same expected number of offspring, because of random variability, even though each may not produce exactly the same number of progeny in any particular population. The number of genes in a diploid population is $2N$. The probability of an offspring receiving two identical alleles from the gene pool is $1/2N$, and the probability of its receiving two different alleles is $1 - 1/2N$. The reduction in heterozygosity depends on the amount of inbreeding, represented by the inbreeding coefficient F, and the effective size of the population, N_e. In any generation, t, this will be

$$F_t = \frac{1}{2N} + \left(1 - \frac{1}{2N}\right) F_{t-1} \qquad [7.8]$$

We can rewrite this equation as

$$1 - F_t = \left(1 - \frac{1}{2N}\right)(1 - F_{t-1}) \qquad [7.9]$$

Since $1 - F$ is proportional to heterozygosity, H, we substitute and have

$$H_t = \left(1 - \frac{1}{2N_e}\right) H_{t-1} \qquad [7.10]$$

This equation shows that in a population of effective size N_e, heterozygosity is reduced in each generation by a factor $1/2N_e$. Genetic drift continuing over t generations will lead to a reduction in heterozygosity, which will be

$$H_t = \left(1 - \frac{1}{2N_e}\right)^t H_0 \qquad [7.11]$$

Genetic drift is more effective in smaller than in larger populations, which is evident from the presence of population size in the denominator of the fraction. In a population of $N_e = 20$, an average of 1 gene locus in 20 would become homozygous in 1 generation; for $N_e = 2000$, an average of 1 in 2000 heterozygous loci would become homozygous in 1 generation (Fig. 7.10).

From these considerations, it is clear that in populations of fluctuating size, the times of small N_e are the most influential in changing the gene frequencies. Such **population bottlenecks** may have a long-lasting effect and cause a significant reduction in genetic variability many generations after N_e has been increased to a substantial number (Fig. 7.11). An example of a severe bottleneck effect is provided by the northern elephant seal *(Mirounga angustirostris)*. As a result of unrestrained hunting, only about 20 animals remained by the turn of the century. After the passage of protective legislation in 1922 and later, the seal population off the coast of Mexico and California grew and is now over 100,000. From a study made in 1974, it is clear that the present northern seal population has a low level of genetic variability. The electrophoretic analysis revealed that each of 24 gene loci is homozygous for a single allozyme. These genes are polymorphic in other animal species, including the southern elephant seal, which did not experience a bottleneck. The lingering effects of the population bottleneck are still evident among the tens of thousands of northern seals living today.

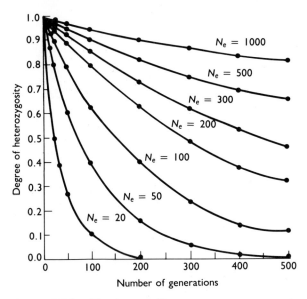

Figure 7.10. The degree of heterozygosity remaining, or homozygosity attained, in populations of different effective size (N_e) after a given number of generations of random mating, starting with 1 as the initial level of heterozygosity. Genetic drift is more effective in smaller than in larger populations. The formula for these calculations is given in equation 7.11. (Adapted with permission of Macmillan Publishing Company from M. W. Strickberger, 1968, *Genetics*, Fig. 33–2, p. 786. Copyright © 1968 by M. W. Strickberger)

Another example of startlingly low genetic variability is the cheetah *(Acinonyx jubatus)*, a member of the cat family. The species once was worldwide and numerous, but only about 20,000 cheetahs are believed to remain today, and they are restricted to a few pockets in Africa. The steady decline in the number of cheetahs led to attempts to breed the animals in captivity, but with very little success. Stephen O'Brien found that captive cheetahs from southern Africa have very low sperm counts, and nearly 80 percent of their sperm are malformed. He analyzed 52 proteins by gel electrophoresis and found no variation whatsoever—all the captive cheetahs were identical and homozygous. Studies of hundreds of other species over many years have shown that, in contrast to such a high degree of monomorphic loci and of homozygosity in the cheetah, 10 to 60 percent of surveyed genes are polymorphic, and 1 to 36 percent of the loci in an average individual are heterozygous.

Other kinds of studies confirmed the highly homozygous and almost invariant genotype of these cheetahs, including their ability to accept skin grafts as readily from one another as from their own bodies. These captive cheetahs were genetically like identical siblings. This most unusual situation in a presumably outbred mammalian species almost certainly reflects the past occurrence of one or more population bottlenecks, whose number and time of

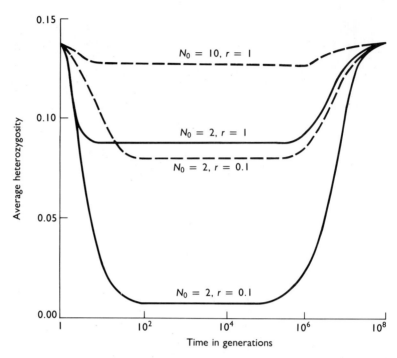

Figure 7.11. Changes in average heterozygosity when populations pass through a bottleneck, and the size of the original founder population (N_0) is 2 versus 10 individuals. A lower growth rate ($r = 0.1$) leads to a more substantial decline in heterozygosity, because of the higher probability of inbreeding following the bottleneck. The original and final heterozygosity is 0.138 (13.8%). (From M. Nei et al., 1975. *Evolution* 29:1)

occurrence are not known. Nor is it known how many survivors founded the present cheetah populations in Africa.

The peril of bottlenecks is based on the relatively high genetic load in diploid populations. Deleterious recessive alleles are fixed by chance alone in very small populations, and matings take place between close relatives, especially if the population expands slowly. Inbreeding leads to reduced heterozygosity, causing lower vigor and fertility as harmful recessive traits are expressed and no longer masked in heterozygotes. The great concern about endangered species is due as much to this problem as to numerical depletion of breeding individuals to zero and consequent extinction of a species. Once a locus has become homozygous in a population, the only means to restore a lost allele and increase variation is by mutation or by gene flow from other populations. Interbreeding of captive animals from different subpopulations of an endangered species can increase variation rapidly, if different loci are homozygous as well as if more of the genes are heterozygous in the subpopulations.

Apart from the severe reduction in heterozygosity due to serious population bottlenecks and the slow increase in size of the population afterward, should perpetually small populations be expected to lose heterozygosity steadily by genetic drift and eventually become homozygous at all or most gene loci? Drift toward homozygosity does not continue indefinitely. Heterozygosity reaches an equilibrium value as the force of genetic drift comes into balance with the effects of recurrent mutation and of gene flow.

Mutation and Gene Flow Balance Genetic Drift

The influence of mutation in curbing the relentless progress of genetic drift can be calculated by modification of equation 7.8, which shows that heterozygosity is reduced as the inbreeding coefficient increases. The relationship indicates the probability that two alleles in an offspring are identical by descent. This will not be true if either allele has just mutated and if it is assumed that the new allele does not already exist in the population. If u is the rate of mutation of an allele to any other possible allele, the probability that a gene has not mutated is $1 - u$, so the probability that neither of the two loci has mutated is $(1 - u)^2$. Mutation is incorporated into equation 7.8, as follows

$$F_t = \left[\frac{1}{2N_e} + \left(1 - \frac{1}{2N_e} \right) F_{t-1} \right] (1 - u)^2 \qquad [7.12]$$

An equilibrium is eventually reached when the loss of alleles by genetic drift is balanced by the gain in new alleles by mutation. The equilibrium value of F at this time is $\hat{F} = F_t = F_{t-1}$, and the equation is rearranged to get

$$\hat{F} = \frac{(1 - u)^2}{2N_e - (2N_e - 1)(1 - u)^2} \qquad [7.13]$$

We can ignore u^2 because u is often 10^{-5} or less, so

$$\hat{F} = \frac{1 - 2u}{4N_e u - 2u + 1} \qquad [7.14]$$

and the relation is approximately

$$\hat{F} \approx \frac{1}{4N_e u + 1} \qquad [7.15]$$

and because heterozygosity $H = 1 - F$, the heterozygosity at equilibrium is

$$\hat{H} \approx \frac{4N_e u}{4N_e u + 1} \qquad [7.16]$$

The crucial quantity in determining $\hat{H}$ is $4N_e u$. If it is near zero, the population is predominantly monomorphic and individuals are homozygous at most loci because genetic drift reduces heterozygosity to a very low level. If $4N_e u$ is as large as 10 or more, the population is almost entirely heterozygous.

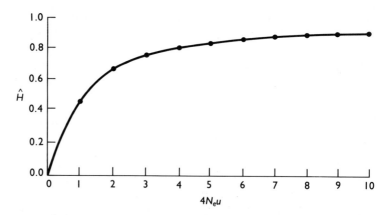

Figure 7.12. Mean heterozygosity, $\hat{H}$, is a function of $4N_e u$, where N_e is the effective population number, and u is the mutation rate at the gene locus. (From J. F. Crow, 1986, *Basic Concepts in Population, Quantitative, and Evolutionary Genetics*, Fig. 2-7, p. 48. New York: Freeman)

This relationship of $\hat{H}$ to $4N_e u$ is shown in Fig. 7.12, from which it can also be seen that the average heterozygosity of 0.01 to 0.36 calculated for polymorphic loci means that $4N_e u$ is generally 1 or less. Biologists know of no population that approaches a heterozygosity anywhere near 1 (Table 7.9). Thus the model is not strictly correct in including the assumptions that the alleles are selectively neutral and that each mutant is a new allele in an unlimited spectrum of possible alleles (the "infinite-allele model").

TABLE 7.9
Average Heterozygosity for Electrophoretically Detectable Proteins in Various Organisms

Organism	Number of Species	Average Total Heterozygosity
Flowering plants	40	0.052
Mammals	184	0.041
Birds	46	0.051
Reptiles	75	0.055
Amphibians	61	0.067
Mollusks	46	0.148
Crustaceans	122	0.082
Drosophila	34	0.123
Other insects	122	0.089

Source: E. Nevo et al., 1984. *Lecture Notes in Biomathematics* **53**:13.

To calculate the effect of gene flow on genetic drift toward homozygosity, equation 7.8 is modified in the same way as for mutation. The subpopulation has an effective number of N_e, and it exchanges a fraction m of its genes for migrant genes drawn at random from the rest of the population, which we assume is very large. In the equation, the term G can be used instead of F, but the two are analogous in representing homozygosity. In this case, G is the probability that two alleles chosen at random in a subpopulation in generation t are identical (F is the same probability that two alleles are identical by descent). If one of the alleles is a migrant, the two alleles are not identical. The probability that neither of the alleles is a migrant is $(1 - m)^2$. We include migration in equation 7.8 and get

$$G_t = \left[\frac{1}{2N_e} + \left(1 - \frac{1}{2N_e} \right) G_{t-1} \right] (1 - m)^2 \qquad [7.17]$$

At equilibrium $\hat{G} = G_t = G_{t-1}$, and the equation is rearranged to get

$$\hat{G} = \frac{(1 - m)^2}{2N_e - (2N_e - 1)(1 - m)^2} \qquad [7.18]$$

and the relation at equilibrium for homozygosity and for heterozygosity is approximately

$$\hat{G} \approx \frac{1}{4N_e m + 1} \text{ and } \hat{H} \approx \frac{4N_e m}{4N_e m + 1} \qquad [7.19]$$

If $4N_e m = 1$, then $\hat{H} = 0.5$. If the number is greater than 1, the subpopulation tends to be heterozygous; if less than 1, the subpopulation tends to be homozygous for the gene locus in question. The crucial value, therefore, is 1, by which is meant one migrant allele per generation.

The influences of mutation and gene flow thus balance genetic drift, so diversity within a subpopulation rarely reaches zero or nearly zero. All three of these processes contribute to genetic diversity in the species as a whole, which is composed of some number of subpopulations. Different random mutations, different migrant alleles and different proportions of these alleles, and different alleles fixed at the same loci in different subpopulations by genetic drift lead to genetic variation from one subpopulation to another within a species (Table 7.10). Mathematical formulas are available to analyze the combined effects of these influential factors and the levels of diversity that can be achieved theoretically.

Whether it is maintained by selection or by neutral evolutionary forces, genetic diversity is the raw material for continued evolution in changing conditions. Genetic differentiation of subpopulations of a species provides a greater likelihood that some one or more of these subpopulations may carry the whole species over times of stress. Whether or not the species will flourish afterward will depend on many factors, including the possibility of a population bottleneck, the time and rate of expansion from a low N_e to larger

TABLE 7.10
Analysis of Genetic Diversity Among Different Populations of Various Species

Species	Number of Populations	Number of Gene Loci	Heterozygosity Total (H_T)	Heterozygosity Average ($\overline{H}_S$)	Coefficient of Gene Differentiation (G_{ST})*
Human—major races	3	62	0.148	0.135	0.088†
Yanomamo Indians	37	15	0.039	0.036	0.069
Mus musculus (house mouse)	4	40	0.097	0.086	0.119
Dipodomys ordii (kangaroo rat)	9	18	0.037	0.012	0.674
Brown trout	38	35	0.040	0.025	0.292
Drosophila equinoxialis	5	27	0.201	0.179	0.109
Horseshoe crab	4	25	0.066	0.061	0.072
Lycopodium lucidulum (club moss)	4	13	0.071	0.051	0.284
Escherichia coli (colon bacillus)	3	12	0.518	0.499	0.036

*The coefficient of gene differentiation between populations is obtained by

$$G_{ST} = \frac{H_T - \overline{H}_S}{H_T}$$

†The value of 8.8% represents the diversity between populations, so 91.2% of the total variability occurs within populations. Each value in this column is interpreted in the same way.
Source: M. Nei, 1987. *Molecular Evolutionary Genetics*. New York: Columbia University Press.

effective size, and the particular alleles fixed at various loci. In addition, genetically diverse subpopulations may include the springboards from which new species evolve, the topic to which we now turn.

Summary

The quantitative methods of population genetics are devoted to the analysis of evolutionary processes that influence changes in gene frequencies in populations. By reference to the model of the gene pool (the collection of alleles shared by members of a population and transmitted in gametes from parents to offspring), mathematical formulas have been devised to analyze changes and factors that influence the nature and pace of genetic modifications. The Hardy-Weinberg principle states that the proportions of genotypes and phenotypes observed are dependent on the proportions of the alleles at the responsible gene locus, or $(p + q)^2 = p^2 + 2pq + q^2 = 1$. Equilibrium gene frequencies can be achieved and maintained in the gene pool if the major influences of selection, mutation, nonrandom mating, gene flow, and random genetic drift (chance loss or fixation of alleles) are not in operation. The effect of one or more of these influences can be measured by quantitative mathematical methods, in terms of changes in gene frequencies.

The effect of selection is determined by fitness and selective disadvantage ($w = 1 - s$), and proves to be a greater influence than mutation on gene frequencies. Mutation is a generally weak force in adaptive evolution, because of the slow pace of change, but it is the source of new genetic information and a large allelic reservoir of diversity in the gene pool. The amount of outbreeding and inbreeding of a mating system influences the level of genetic variation in populations. Decrease in heterozygosity is proportionate to a higher inbreeding coefficient (F), which is characteristic of consanguineous matings in human populations and is represented in its most extreme form in self-fertilizing plants.

The amount of gene flow, or migration, among subpopulations can be calculated by the use of a migration coefficient (m) to determine the rate and extent of change in gene frequencies, and often is an important influence on the gene pool. Random genetic drift is more effective in smaller than in larger populations, as shown by the reduction in heterozygosity in each generation by a factor of $1/2N_e$, where N_e represents the number of breeders in a population of any size. In populations that fluctuate in size, the times of small N_e are the most influential in changing gene frequencies. Such population bottlenecks may lead to a significant reduction in genetic variability many generations after a sizable increase in N_e has been attained. The peril of bottlenecks is based on the relatively high genetic load of detrimental recessive alleles carried in sexual diploid populations. The problem of endangered species is due to the higher chances for loss of beneficial alleles and fixation of detrimental alleles, as well as the small number of breeders to perpetuate the population. The force of random genetic drift comes into balance with the effects of recurrent mutation and gene flow, however, so heterozygosity may reach some equilibrium value rather than being lost entirely. The proportion of heterozygosity in the gene pool at equilibrium is determined by the fraction $4N_e u$, where u is the rate of recurrent mutation.

References and Additional Readings

Bonnell, M. L., and R. K. Selander. 1974. Elephant seals: genetic variation and near extinction. *Science* 184:908.

Box, J. 1978. *R. A. Fisher: The Life of a Scientist*. New York: Wiley.

Cavalli-Sforza, L. L. 1969. "Genetic drift" in an Italian population. *Sci. Amer.* 221(2):30.

Clark, R. W. 1969. *JBS: The Life and Work of J. B. S. Haldane*. New York: Coward-McCann.

Crow, J. F. 1986. *Basic Concepts in Population, Quantitative, and Evolutionary Genetics*. New York: Freeman.

Crow, J. F. 1987. Population genetics history: a personal view. *Ann. Rev. Genet.* 21:1.

Fisher, R. A. 1930. *The Genetical Theory of Natural Selection.* New York: Oxford University Press.

Haldane, J. B. S. 1932. *The Causes of Evolution.* New York: Harper.

Hardy, G. H. 1908. Mendelian proportions in a mixed population. *Science* 28:49.

Hartl, D. L. 1988. *A Primer of Population Genetics,* 2nd ed. Sunderland, Mass.: Sinauer.

Hartl, D. L., and D. E. Dykhuizen. 1984. The population genetics of *Escherichia coli. Ann. Rev. Genet.* 18:31.

Hedrick, P. W. 1983. *The Genetics of Populations.* Boston: Jones and Bartlett.

Hill, W. G., ed. 1984. *Quantitative Genetics.* Benchmark Papers in Genetics. New York: Van Nostrand Reinhold.

Istock, C. A. 1983. The extent and consequences of heritable variation for fitness characters. In *Population Biology: Retrospect and Prospect,* ed. C. R. King and P. W. Dawson, p. 61. New York: Columbia University Press.

Koehn, R. K., R. I. E. Newell, and F. Immermann. 1980. Maintenance of an aminopeptidase allele frequency cline by natural selection. *Proc. Natl. Acad. Sci. U.S.* 77:5385.

Levin, D. A. 1981. Dispersal versus gene flow in plants. *Annals Missouri Bot. Gard.* 68:233.

Muller, H. J. 1950. Our load of mutations. *Amer. Jour. Human Genet.* 2:111.

Neel, J. V., and E. D. Rothman. 1978. Indirect estimates of mutation rates in tribal Amerindians. *Proc. Natl. Acad. Sci. U.S.* 75:5585.

Nei, M. 1987. *Molecular Evolutionary Genetics.* New York: Columbia University Press.

Nei, M., and M. W. Feldman. 1972. Identity of genes by descent within and between populations under mutation and migration pressures. *Theoret. Popul. Biol.* 3:460.

Nei, M., and Y. Imaizumi. 1966. Genetic structure of human populations. II. Differentiation of blood group frequencies among isolated populations. *Heredity* 21:183.

Nei, M., T. Maruyama, and R. Chakraborty. 1975. The bottleneck effect and genetic variability in populations. *Evolution* 29:1.

O'Brien, S. J. et al. 1985. Genetic basis for species vulnerability in the cheetah. *Science* 227:1428.

O'Brien, S. J., D. E. Wildt, and M. Bush. 1985. The cheetah in genetic peril. *Sci. Amer.* 254(5):84.

Ohta, T. 1982. Linkage disequilibrium due to random genetic drift in finite subdivided populations. *Proc. Natl. Acad. Sci. U.S.* 79:1940.

Ohta, T., and K.-I. Aoki, eds. 1986. *Population Genetics and Molecular Evolution.* Berlin: Springer-Verlag.

Provine, W. B. 1986. *Sewall Wright and Evolutionary Theory.* Chicago: University of Chicago Press.

Reed, T. E. 1969. Caucasian genes in American Negroes. *Science* 165:762.

Schoen, D. J. 1982. The breeding system of *Gilia achilleifolia:* variation in floral characteristics and outcrossing rate. *Evolution* 36:352.

Slatkin, M. 1985. Gene flow in natural populations. *Ann. Rev. Ecol. Syst.* 16:393.

Weinberg, W. 1908. Über den Nachweis der Vererbung beim Menschen. (English translation in *Papers on Human Genetics,* ed. S. H. Boyer, p. 4. Englewood Cliffs: N.J.: Prentice-Hall, 1963.

Wright, S. 1921. Systems of mating. *Genetics* 6:111.

Wright, S. 1922. Coefficients of inbreeding and relationship. *Amer. Nat.* 56:330.

Wright, S. 1931. Evolution in Mendelian populations. *Genetics* 16:97.

Wright, S. 1937. The distribution of gene frequencies in populations. *Proc. Natl. Acad. Sci. U.S.* 23:307.

Wright, S. 1968. *Evolution and the Genetics of Populations*, vol. 1: *Genetic and Biometric Foundations*. Chicago: University of Chicago Press.

Wright, S. 1969. *Evolution and the Genetics of Populations*, vol. 2: *The Theory of Gene Frequencies*. Chicago: University of Chicago Press.

Wright, S. 1977. *Evolution and the Genetics of Populations*, vol. 3: *Experimental Results and Evolutionary Deductions*. Chicago: University of Chicago Press.

Wright, S. 1978. *Evolution and the Genetics of Populations*, vol. 4: *Variability Within and Among Natural Populations*. Chicago: University of Chicago Press.

Wright, S. 1980. Genic and organismic selection. *Evolution* 34:825.

CHAPTER 8

Tempo and Mode in Speciation

At least 2 to 3 million species of organisms exist in the world today, and many more have been estimated to be present but as yet are unidentified. The wealth of forms is believed to represent less than 1 percent of all the life that has ever existed, all the others having become extinct over evolutionary time. The formation of new species, or **speciation,** is an important theme in evolutionary biology, but one that is controversial in many aspects. Of course, the evolution of new species from preexisting species is the accepted thesis today, not the pre-Darwinian notion of the fixity of species in unchanging form throughout time.

In this chapter, we will sort through theory and evidence bearing on some of the major questions regarding speciation. What processes lead to speciation? How do species maintain their integrity and distinctiveness from one another? What is the tempo of speciation, and how are the tempo and mode of speciation related? Are the same processes involved in both the formation of species and the evolution of taxonomic categories above the species level, such as genera and families? There is no dearth of theories pertaining to these questions. There are great difficulties, however, in relating data to theory, and even in obtaining basic data in some cases.

Modes of Speciation

According to the Darwinian view, evolution is the conversion of variation among individuals into variation among populations and species in both time and space. The mode of speciation—that is, the processes and influences that lead one population of organisms to diverge into two or more populations of different organisms—is not universally agreed on. A large part of the problem is the absence of a definition of a species on which everyone can agree. In addition, the same data may be given different interpretations, or very different emphases may be given to the importance of particular elements of information.

The Species Concept

The most widely accepted modern definition of a **species** is that it is a group of individuals that can successfully breed with one another but not with members of other groups. The emphasis is on **reproductive isolation,** based on genetic incompatibility among species. The thrust of this concept, first articulated in the 1930s and 1940s by Theodosius Dobzhansky, Ernst Mayr, and others, is that new species are descended from ancestral populations by genetic processes that hinder or prevent interbreeding. The idea stimulated a broad search in both natural and experimental populations for the barriers to gene exchange, also called *isolating mechanisms,* and for the conditions and influences that underlie speciation.

As valuable as this concept proved to be for evolutionary study, the test of reproductive isolation cannot be applied to all organisms. Asexual life forms do not interbreed, and each *clone* represents the descendants of a single individual that is reproductively self-sufficient. Genetic divergence clearly occurs in asexual life, and new species have emerged, but not as a result of the erection of barriers to gene exchange between previously compatible populations. Fossils can be classifed as belonging to one species or another, but little or nothing is known about reproductive isolation among these groups. Even in living sexual species, there often are practical problems of various kinds that rule out hybridization as a direct test for reproductive isolation of populations. In these situations, phenotypic criteria provide the basis for the assignment of individuals to a particular species.

Although some species definitions may include the criterion of substantial and unambiguous differences in morphology, biochemistry, and other phenotypic traits, it is difficult to apply in many cases. *Drosophila*, rodents, flowering plants, and other groups include **sibling species,** which are populations of genetically related organisms that are morphologically almost identical but cannot interbreed. At the other extreme are morphologically distinctive forms that can freely interbreed to produce healthy, fertile progeny. The California king snake *(Lampropeltis getulus)* occurs in two very distinctive color forms that were originally assigned to separate species. The two forms interbreed freely, however, and the color patterns appear to be governed by a single gene. The sycamore trees of the eastern United States *(Platanus occidentalis)* and the eastern Mediterranean *(P. orientalis)* are geographically separated and do not interbreed for this reason. When these trees are planted close together, however, they can produce fertile progeny that were even given their own species name *(P. acerifolia)* before their origin was known.

The difficulty in applying a universal criterion of reproductive isolation or phenotypic variation as the index of genetic differentiation among species is also encountered in dealing with categories below the rank of species (*infraspecific units*). A number of such categories could be recognized simply because no two populations are exactly alike. It is widely believed, however, that the most useful and biologically significant taxon of this nature is the subspecies. **Subspecies** of a species are populations that share most of their characteristics but differ in a few traits, inhabit different geographical or ecological subdivisions of the entire range of the species, and can freely interbreed with one another. Subspecies are not so genetically differentiated as to preclude gene exchange, but they usually intergrade only at their common range boundaries, where partners are more accessible to each other (Fig. 8.1).

The identification of geographically or ecologically distinct subspecies has genetic validity. All the populations are genetically closely related and genetically compatible, but interbreeding is minimized by habitat separation. The fact that subspecies are phenotypically or ecologically distinguishable to some extent indicates that some genetic differentiation has occurred. Whether or not these populations are incipient species and will differentiate further to species status depends on future events. If spatial separation breaks down and gene flow increases, the subspecies may merge into a larger panmictic population and thereby lose their identity. If the subspecies continue to diverge genetically, they may eventually be classified as different species, particularly if they become reproductively isolated (Fig. 8.2). The subspecies is thus considered to be a valid biological entity, because of its potential to evolve into a new species.

A species is a genetically differentiated reproductive community, but is it necessarily reproductively isolated from all other reproductive communities? Different species of many kinds of plants, particularly trees and other perennials, and such animals as amphibians rarely interbreed but may be genetically compatible nevertheless. Various oak species *(Quercus)*, the sycamores mentioned earlier, and other long-lived flowering plants can interbreed but rarely do. The occasional hybrid that may be produced is considered to be less important than the complex of differences in morphology, habitat, breeding time or pollinating agent, and other inherited traits that readily differentiate one species from another.

Many amphibian species have the ability to exchange genes, according to controlled studies, but rarely, if ever, interbreed in natural populations. The American toad *(Bufo americanus)* and Fowler's toad *(B. fowleri)* are morphologically dissimilar, and the males of each species have a distinctive mating call (a trilling call for the former and a retching sound for the latter). These toads breed in the same ponds in the spring, but the American toad completes its mating at about the time that Fowler's toad begins its reproductive activities.

A single species may consist of a graded spectrum of phenotypes, or a **cline,** that represents different but connected populations along a geographical

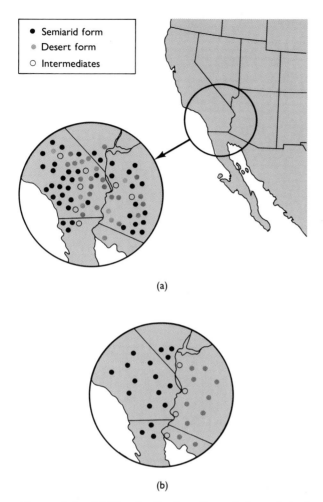

(a)

(b)

Figure 8.1. (a) The distribution of populations of the long-nosed snake *(Rhinocheilus lecontei)* in ecologically distinct semiarid and desert regions of the southwestern United States and northwestern Mexico. Hybridization is apparent from the presence of intermediates between these two forms. (b) If the two forms inhabited different geographical regions and produced fertile hybrids at the border of their ranges, both forms would be classified as subspecies of *R. lecontei.* ([a] from J. M. Savage, 1977, *Evolution,* 3rd ed., Fig. 9-1, p. 121. New York: Holt, Rinehart and Winston)

range. Populations at the extremes of the geographical range usually cannot interbreed in the laboratory, nor do they have the opportunity for gene exchange in nature. Neighboring populations, however, usually interbreed in the wild as well as in the laboratory (Fig. 8.3). Considered as a whole, such a collection of populations makes up a single species whose members are capable of interbreeding. Taken separately, particular populations can be regarded as subspecies or even separate species. It would be confusing to identify sep-

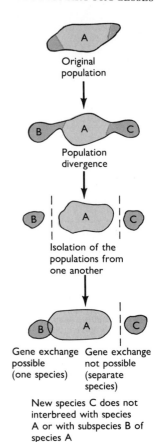

Figure 8.2. Hypothetical population A undergoing divergence and speciation. Given a sufficient length of time in isolation, subpopulations A and C, may become unable to exchange genes, which indicates the attainment of separate species status. If gene exchange remains possible, as between A and B, the two subpopulations constitute members of one species.

arate entities in such a collection, but more important, it would be biologically meaningless. The complex is evolving, and the certain signs of genetic differentiation include phenotypic variations and the gradation of possibilities for gene flow among the connected populations (Box 8.1).

Is reproductive isolation the inevitable outcome of speciation? The many exceptions, including the examples just discussed, would appear to indicate that it is not—at least, not in every case. This brings into question the nature of the genetic changes that are responsible for speciation. The implication has been that the changes underlying phenotypic differentiation also bring about the incompatibility underlying reproductive isolation. In fact, very little evidence exists to support this implied correspondence. There is only a vague idea of the genetic basis for reproductive isolation, but it is unlikely to be due

entirely to gene substitutions or allele frequency differences that govern structural loci. Genetic differentiation would be expected to involve regulatory
genes that govern the expression of structural genes, chromosome rearrangements that may lead to little phenotypic variation but that characterize genomic changes, and other sources of genetic variation (see Chapter 5). One of
the least understood but most important aspects of speciation is the relationship between reproductive isolation and phenotypic variation. Because of this
impasse, some biologists do not consider reproductive isolation to be the corollary of speciation, although it may be the case in various organisms. They
are therefore reluctant to accept the criterion of reproductive isolation as the
universal and distinctive feature of species differences.

Categories above the species level, such as genus, family, order, class, and
phylum, can at best be described only in subjective terms. A genus consists of
one or more species that have one or more basic traits that distinguish them
from species of any other genus (Table 8.1). The characters that define one
genus may be the same as those that define a whole family (one or more
related genera), so the decisions are often somewhat arbitrary. All the species
of oaks are classified as members of the genus *Quercus* on the basis of their
producing acorns as fruits. In the rose family (Rosaceae), however, the genera
of apples, cherries, pears, roses, and others produce a fruit type called a pome.

Figure 8.3. A single species may consist of a graded spectrum of phenotypes, or a cline of different
populations connected by gene flow between neighboring units along a geographical range. Interbreeding does not occur between populations far apart from each other, because of the lack of
opportunity. In some cases, members of distant populations cannot interbreed even when provided
with the opportunity in laboratory settings.

Box 8.1   Genetic Distances Between Related Taxa

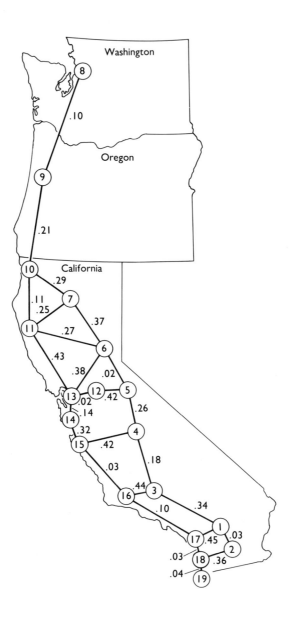

The levels of evolutionary divergence that distinguish taxa of different ranks may be estimated by quantitative measures of **genetic distance, D,** and **genetic identity,** based on electrophoretic analysis of allozyme polymorphisms at various gene loci. As will be discussed in more detail in Chapter 9, the value D is interpreted as the average number of electrophoretically detectable codon substitutions per gene locus that have accumulated in the populations since they diverged from their last common ancestor. Genetic identity values are interpreted as the average number of gene products that are not distinguishable by their electrophoretic behavior.

No single scale of D values serves for all organisms. But a scale of values for the *Drosophila willistoni* group shows the general range of genetic differences and similarities for taxa at four levels of evolutionary divergence.

These values can be compared with a much broader range found in 19 populations of several subspecies of the plethodontid salamander *Ensatina eschscholtzii* in California. This study of 26 proteins showed a range in D values (next to the lines connecting numbered populations in the figure) from 0.021 to 0.765. The level of morphological differentiation among the salamander populations corresponded only very roughly with the degree of genetic divergence.

Level of Divergence	Genetic Distance	Genetic Identity
Local populations	0.031 ± 0.007	0.970 ± 0.006
Subspecies	0.230 ± 0.016	0.795 ± 0.013
Sibling species	0.581 ± 0.039	0.563 ± 0.023
Nonsibling species (congeneric)	1.056 ± 0.068	0.352 ± 0.023

Source: D. B. Wake and K. P. Yanev, 1986. Geographic variation in allozymes in a "ring species," the plethodontid salamander *Ensatina eschscholtzii* of western North America. *Evolution* 40:702.

TABLE 8.1
Classification of Four Species

Category	Haircap Moss	White Oak	Housefly	Human
	Taxon			
Kingdom	Plantae	Plantae	Animalia	Animalia
Phylum or division	Bryophyta	Anthophyta	Arthropoda	Chordata
Class	Musci	Dicotyledones	Insecta	Mammalia
Order	Bryales	Fagales	Diptera	Primates
Family	Polytrichaceae	Fagaceae	Muscidae	Hominidae
Genus	*Polytrichum*	*Quercus*	*Musca*	*Homo*
Species*	*P. commune*	*Q. alba*	*M. domestica*	*H. sapiens*

*The species is always referred to by its binomial designation of genus plus the species names. If the genus has not been cited, the full binomial is given *(Homo sapiens)*; if the genus has been cited, the binomial may be abbreviated *(H. sapiens)*.

Variations in this kind of fruit, along with other characters, provide the basis for recognizing different genera.

Regardless of phenotypic differentiation, individuals belonging to different genera or higher taxonomic categories are reproductively isolated. Members of these groups probably have evolved along divergent pathways for a long enough time to have established discrete gene pools no longer connected by evolutionary intermediate populations. The genetic dissimilarities and the absence of connections that would permit gene flow are adequate to isolate these taxa reproductively. If intergeneric hybrids can be produced, the classification usually is revised to include the taxa in one genus. Even so, the boundaries among genera are not very distinct in some of the more complex families, such as the grass family (Graminineae) and the orchid family (Orchidaceae).

From an evolutionary standpoint, populations of organisms are expected to diverge genetically over time because of **stochastic** (random) **processes** of mutation and genetic drift. The particular direction of their evolutionary change also depends on **deterministic processes,** particularly natural selection and gene flow. Subspecies have taken steps toward differentiation into new species, but gene flow connects them within the framework of a single species with a shared gene pool. Genera and higher categories are not only genetically differentiated but also reproductively isolated groups unconnected to one another by transitional forms. These taxa could continue to evolve even greater levels of genetic distinctness and perhaps produce new families, classes, or phyla of organisms. This neo-Darwinian view of evolution is less clear and more widely argued with regard to species than to subspecies or genera. It is generally agreed that the species is a natural evolutionary unit, but whether or not its genetic differentiation is coupled to reproductive incompatibility remains controversial.

Spatial Isolation as a Prelude to Speciation

Mutation and genetic drift underlie divergence, but all the breeders have access to a common gene pool in a randomly mating system. Even in groups divided into subpopulations, gene flow reduces the chances that significant genetic differentiation will be established among groups living in the same geographical space **(sympatry)**. If subpopulations are geographically isolated from one another **(allopatry)**, gene flow is absent and different alleles or gene substitutions may become fixed at various loci. These considerations provide the basis for Mayr's theory of **allopatric speciation,** according to which geographical isolation is an essential prerequisite for virtually all episodes of speciation.

Most of the evidence for allopatric speciation is biogeographical, and includes two major observations. First, there is a correspondence between the extent of species diversification and the amount and kinds of topographically discontinuous space. Second, the same kinds of inherited differences characterize populations that are reproductively isolated as well as populations that are genetically compatible but ordinarily do not interbreed.

Particularly persuasive evidence for a correspondence between speciation and geographical isolation is found on islands. Species of *Drosophila* occur worldwide, but nearly one-third of the known species are found exclusively in the Hawaiian Islands. Similarly, Darwin's finches (Geospizinae) are a subfamily of birds found only in the Galápagos Islands, which straddle the equator 600 miles west of Ecuador, and on Cocos Island, north of the Galápagos (Fig. 8.4). These Pacific Ocean islands were never connected to the South or Central American mainland, and they have been forming for about 4 or 5 million years. The ancestral finch species, which probably lives in Panama, flew or was blown out into the Pacific Ocean. Once on the islands, the ancestral stock evolved into 13 species in the Galápagos Archipelago, but only 1 species of Darwin's finches evolved on Cocos Island, hundreds of miles away (Fig. 8.5). In addition to these and other birds, a variety of unique species of plants and animals evolved on these islands, from ancestral forms known to be native to the Americas and other regions. This pattern is characteristic of other island groups around the world, such as Hawaii and the Seychelles. The variety as well as uniqueness of island species corresponds to the variety and amount of discontinuous spaces.

The second major observation in support of allopatric speciation is that noninterbreeding populations, which either cannot or do not exchange genes, are generally characterized by the same kinds of genetic traits. These similarities are interpreted as an indication that the kinds of genetic changes leading to new species are largely the same as those responsible for eventual reproductive isolation, and they are therefore referred to as **isolating mechanisms.**

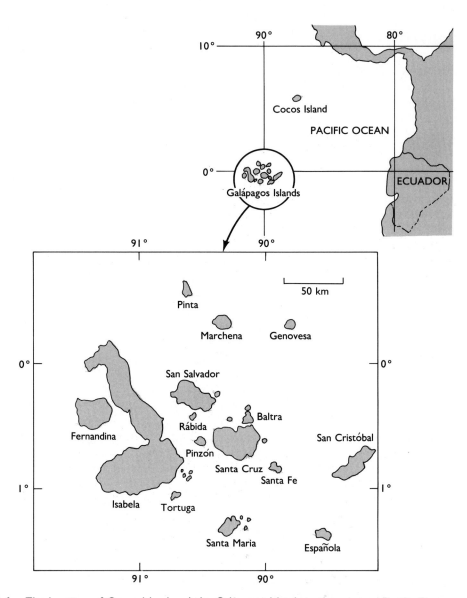

Figure 8.4. The location of Cocos Island and the Galápagos Islands in the western Pacific Ocean, showing their positions relative to each other and to the Central and South American mainland. The Galápagos Islands are about 600 miles from the coastline of Ecuador.

Alternatives to Allopatric Speciation

Speciation due to the establishment of barriers to interbreeding within a panmictic population, in the absence of spatial isolation, is called **sympatric speciation.** The only noncontroversial model for this mode of speciation involves polyploidy, which immediately produces a reproductive barrier on account of the different number of chromosome sets in the original and the polyploid forms. A new tetraploid and the parental diploid produce triploid hybrid offspring, for example, which are sterile because of aberrant pairing and distribution of the three sets of chromosomes during meiosis. Gene exchange is prevented because neither hybridization between triploids nor their backcrossing to either parent will take place.

Speciation by polyploidy is common in plants but not in animals, unless an animal has some means for asexual reproduction. Polyploids can become established and multiply in the same habitats as their parents if they can reproduce by vegetative means, which is a common characteristic of perennial plants in particular. If the polyploids can multiply and compete successfully with the parents for living space initially, they may maintain themselves later by breeding with each other as well as by vegetative propagation. Polyploids of self-fertilizing species, of course, can multiply sexually from the beginning of their existence, whether or not they also possess asexual means.

A number of sympatric speciation models are based on disruptive natural selection, particularly involving polymorphic loci that govern or relate to mate preference and host preference. In these and similar situations, **assortative mating** occurs; that is, matings take place between individuals with similar phenotypes rather than randomly. Because the phenotype is determined by the genotype, assortative mating has some of the same consequences as inbreeding (Fig. 8.7). Mathematical models of assortative mating for multifactorial genes indicate, however, that populations tend to become genetically more variable rather than more homozygous. Divergence toward different adaptive niches leads to greater genetic differentiation and may ultimately result in reproductive isolation between the derivative populations.

It has been argued from a genetic standpoint that sympatric speciation is unlikely to occur if several or many loci govern mate or host preference, but it is theoretically possible mainly if such a feature is controlled by a single gene. If one polymorphic locus governs mate preference, different monomorphic populations may become established only if other genes conferring fitness exist in linkage disequilibrium with the major monomorphisms. If several loci control mate preference, recombination would greatly hinder selection to establish linkage disequilibrium among all the relevant loci and would therefore reduce the likelihood of reproductive isolation between populations. The Hawaiian species *Drosophila silvestris* and *D. heteroneura* may be just such a case of sympatric speciation based on different mate preferences due to a single gene governing head shape and a number of interacting autosomal genes in linkage disequilibrium with the major gene.

There are at least two difficulties in deciding whether sympatric speciation

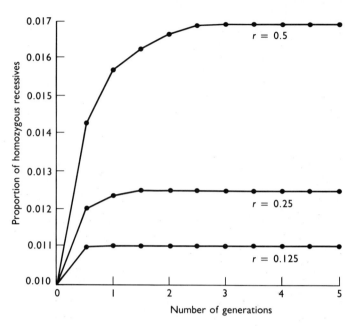

Figure 8.7. Change in the proportion of homozygous recessives according to different degrees of assortative mating (r) for the same allele frequency ($q = 0.1$). The effect of assortative mating is fairly small, unless r is high. Equilibrium is attained in a few generations, and the population never becomes completely homozygous (unless $r = 1$), unlike most forms of inbreeding. (From J. F. Crow, 1986, *Basic Concepts in Population, Quantitative, and Evolutionary Genetics*, Fig. 2-9, p. 53. New York: Freeman)

underlies divergence in experimental as well as natural populations. In experimental studies, the programs of disruptive selection within a population can lead to assortative mating between different emergent phenotypic groups. These data are not given much weight, however, because it is considered unlikely that selection in sympatric natural populations would be as severe as in experimental populations. A second difficulty is that the observed differences among most species either are too genetically complex to fit the theoretical models for sympatric speciation or represent only a fraction of the total variation distinguishing species that live in the same area. In general, sympatric speciation may occur, but it is not considered to be a signficant general evolutionary device for the differentiation of reproductively isolated populations across the broad spectrum of sexual organisms.

The establishment of reproductive isolation between adjacent populations has been proposed in the model of **parapatric speciation,** according to which selection favoring different alleles in parapatric populations leads to a cline in allele frequencies. Given strong selection pressures on genes that enhance reproductive isolation, adjacent populations could diverge to become different species. Much of the empirical evidence is somewhat ambiguous, however, and the model is not favored by many evolutionists. The most persuasive data in support of parapatric speciation has been collected for grasses and other plant species that grow in areas high in toxic metal wastes near mines or factories and for closely related species in adjacent nontoxic fields. Some of these parapatric species differ in such features as self-incompatibility or flowering time, as well as tolerance to heavy metals. These populations are therefore reproductively isolated from one another, at least to some degree.

Whether or not speciation may occur occasionally in sympatric or parapatric populations, the most prevalent mode is believed to be allopatric speciation. The general requirement of initial geographical isolation fits with the accepted idea that genetic divergence from one gene pool to two or more reproductively isolated gene pools is most likely to occur in different spaces. Stochastic processes of mutation and genetic drift, coupled with differences in selection and other deterministic processes, ultimately lead to sufficient genetic differentiation so that the separated populations can no longer interbreed successfully. Reproductive isolation is thus viewed as the eventual outcome of ongoing genetic divergence, which is unlikely to become established in panmictic populations.

Patterns of Speciation

In the preceding discussion, speciation was portrayed as an evolutionary phenomenon based on the gradual accumulation of selectively favorable genetic variation in spatially isolated populations, leading to genetically incompatible groups descended from an ancestral gene pool. This view of *gradual divergence* is different in tempo and genetic mode from alternatives that emphasize a rapid pace of speciation based more on random processes of mutation and genetic drift than on selection. These alternatives go by various names, including *genetic revolution* and *punctuated equilibrium.* They are similar in stressing a rapid rate of speciation and in emphasizing the origin of new species by the splitting of one or more subpopulations into reproductively isolated groups that ultimately may replace the ancestral population. In addition, the proponents of punctuated equilibrium imply that some of the genetic processes leading to speciation may be different from the conventional processes producing variation within a species. These models of speciation continue to generate highly charged debates and undoubtedly will continue to do so for some time to come.

Speciation by Gradual Divergence in Ancestral Populations

A very effective image with which to discuss and compare the various patterns of speciation in relation to time and space is that of an **adaptive landscape,** such as Sewall Wright described in the 1930s. According to this representa-

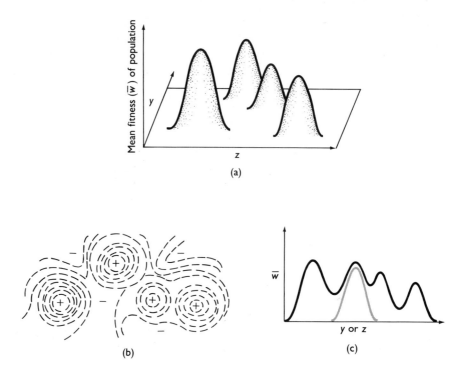

Figure 8.8. The adaptive landscape of heterogeneous environments occupied by genotypically variable populations. (a) Three-dimensional display of adaptive peaks of higher fitness and valleys of lower fitness for genotypic combinations of the characters y and z. (b) Two-dimensional representation of an adaptive landscape of adaptive peaks of higher fitness (+) and valleys of lower fitness (−), as first suggested by Sewall Wright. (c) An alternative notation for peaks and valleys of an adaptive landscape, showing a population (gray line) occupying one of a number of available fitness peaks.

tion, we can have a three-dimensional display of species fitness, designated by $\overline{w}$ (mean fitness) on the vertical axis, based on the fitness of any combinations of characters, y and z on the other two axes, in different environmental conditions (Fig. 8.8). The relative overall fitness of any local population will depend on the combined fitness of the particular characters, which may vary from one location to another within the heterogeneous environment in which the population and its subpopulations reside. The adaptive peaks in the species-fitness landscape represent levels of higher fitness, and the valleys represent levels of lower fitness at one time or another in the history of the species or in particular subpopulations. Natural selection will tend to move the population to the tip of the adaptive peak—that is, to a level of optimum fitness at a given time and place.

In changing environments, different character combinations are favored, changing the value of $\overline{w}$ for points y and z. The position and form of the peak may vary, therefore, but selection pressures in the prevailing environment will largely determine the nature of the adaptive landscape. An evolving population undergoes **peak shifts** as selection leads it from the fitness summit of one peak to the summit of another. The force of selection is balanced by mutation and other processes of genetic change and by genetic drift. These random factors may by chance cause an increase in the mean fitness of the population and move it up one peak or another or reduce mean fitness and cause a descent from a peak.

The genetic model of speciation by gradual divergence depicts the transition of the ancestral population to a new form as a series of moves from an earlier adaptive peak to another adaptive peak under different conditions at a later time (Fig. 8.9). As the gene pool changes with time, the population may evolve into a different phenotypic entity and earn a new species name. If the population becomes mired in a valley of lower fitness and fails to reach a new fitness peak, perhaps because of inadequate or inappropriate genetic reserves of variation, it may become extinct.

This speciation model easily includes the concept of geographical isolation as a prerequisite to speciation. Isolated populations would become genetically differentiated as the result of different genetic changes acted on by different selection pressures, reaching different peaks of optimum fitness if they are successful. Even if the subpopulations, or subspecies, could hybridize at their

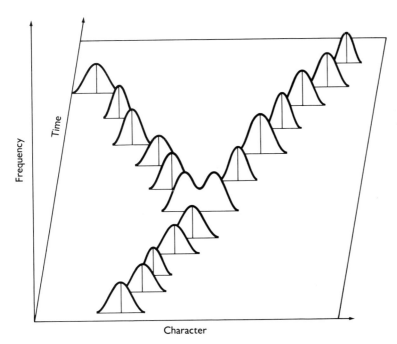

Figure 8.9. Speciation by gradual divergence, as a population moves from one adaptive peak to another under different conditions at later times. Divergence of an ancestral population may lead to new species that occupy different fitness peaks in the same adaptive landscape, as the gene pool changes over time.

common border the hybrids would probably be less fit than either parent population and fail to become established or to continue to interbreed with the more fit parent populations. These subpopulations would merge into a single panmictic population if they did come to live in the same space, moving toward a particular adaptive peak for the population as a whole rather than continuing to occupy different fitness peaks in the same environment. If the subpopulations become reproductively isolated, however, the different species could occupy the same living space, but almost certainly would be adapted to different features of their common environment. For example, they might use different food sources, ecologically different space, or some other differential feature of their habitat.

The gradualist view of speciation predicts a relatively slow pace of transition from one ancestral population to a descendant population. The population would presumably move rather quickly up the slope of a new adaptive peak, pushed by the strong forces of selection toward maximum fitness. The time spent between peaks, however, was long believed to represent a relatively slow process of transition. It seemed reasonable to expect a laborious move down the slope of a peak, because the population would be struggling against the forces of selection as its fitness decreased during the descent from the summit of optimum adaptive status. In very small populations, however, random genetic drift might prove to be the major influence on the gene pool and perhaps cause more rapid peak shifts.

The arguments against gradual divergence as the major speciation pattern are based on at least three sets of observations that do not conform to the model's predictions. First, there should be a series of **transitional forms** that connect the ancestral and descendant species, but they are often missing in both living and fossil groups. Indeed, the existence of gaps in the fossil record, or "missing links," is characteristic of the entire recorded history of life. Second, the fossil record seems to display a common pattern of a very long period of **evolutionary stasis** between the time a species appears and the time it vanishes. These intervals of little significant phenotypic change may extend over millions of years, a long enough time for transitions to occur and not be missed in the fossil record regardless of its fragmentary nature. Third, the model predicts a relatively leisurely pace of speciation, yet bursts of new species seem to have appeared relatively suddenly in the fossil record and in groups of organisms still in existence.

Two of the alternatives to gradual divergence as the major pattern of speciation seek to account for one or more of the apparent contradictions to gradual transition from one species into another.

Speciation by Genetic Revolutions

The central argument of theories to explain speciation patterns characterized by substantial and rapid genetic differentiation, or **genetic revolutions,** is that genic or chromosomal variations are fixed in small populations by the combined effects of genetic drift and natural selection and that the low fitness of heterozygotes prevents significant gene flow between such populations. The problem to be solved is how an ancestral population occupying one adaptive peak can diverge into two or more populations occupying different adaptive peaks in the *same* adaptive landscape. The solution to this problem begins with the assumption that the ancestral population is potentially capable of occupying more than one peak in a particular environment. The question can then center on the most likely pathway by which this population becomes subdivided into genetically different but optimally fit descendant populations that undergo little or no gene exchange.

Suppose there is a population with high equilibrium frequencies for certain alleles A_1 at one or more loci conferring fitness, but this population would be equally fit if alleles A_2 at these loci happened to be present in high frequencies. If a subpopulation experienced a slight increase in A_2 frequencies, it would quickly return to the high equilibrium frequencies of the predominant A_1 alleles, by the influence of selection (Fig. 8.10). If genetic drift caused a more substantial shift from A_1 to A_2 in the subpopulation, however, it would move to the A_2 adaptive peak under the influence of natural selection. The shift from the A_1 peak to the A_2 peak is most effective if the heterozygotes have low fitness. If this were true, however, selection would strongly oppose

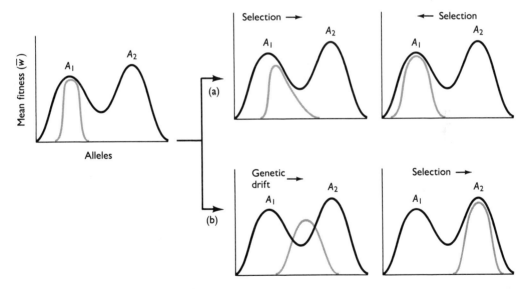

Figure 8.10. Movement of a population or subpopulation (gray line) from one adaptive peak to another (black line) is thought to depend on the relative increase in frequencies of alleles (A_1, A_2) conferring equal fitness in the adaptive landscape. (a) A slight increase in A_2 frequency results in a return to the high equilibrium frequency of the predominant A_1 alleles under the influence of selection. (b) A more substantial shift from A_1 to A_2 by means of genetic drift may be followed by a move to the A_2 peak under the subsequent influence of natural selection.

genetic drift, except in very small populations where chance influences are great.

If the mean fitness at the A_1 peak is designated as $\overline{w}_1$ and the mean fitness at the low point of the valley between the A_1 and A_2 peaks as $\overline{w}_{1-2}$, then the probability of a shift from peak A_1 to peak A_2 is proportional to $(\overline{w}_{1-2}/\overline{w}_1)^{2N_e}$, where N_e is the effective population size. The lower the fitness value between peaks, the lower the probability that a subpopulation will cross the valley to the new adaptive peak. But the chance is much greater if the population size is very small. Arbitrary values can be used to illustrate this point.

Suppose the average fitnesses are $\overline{w}_1 = 1$ and $\overline{w}_{1-2} = 0.8$ in a population of five breeders. The probability of a shift to a new adaptive peak is $(0.8/1)^{10} = 0.11$. If $\overline{w}_{1-2}$ has a lower fitness value of 0.4, the probability of a peak shift is reduced to $(0.4/1)^{10} = 0.0001$. If the population has only two breeders, however, the chance of a peak shift at the low fitness of $\overline{w}_{1-2} = 0.4$ is $(0.4/1)^4 = 0.025$, compared with 0.0001 for an effective population size of five breeders.

In 1954, Ernst Mayr proposed a mode of **peripatric speciation,** by which small isolated populations at the periphery of the main population may undergo peak shifts largely under the initial influence of genetic drift. This is likely to occur if the new populations are established by a few *founder* individuals and if there is no gene flow between the peripheral isolates and the main population. Each peripheral population would occupy a small area at first and so be subject to selection pressures different from those acting on the other isolates because the local environment is likely to be relatively uniform but vary from one place to another. Different alleles at the same loci would be fixed by genetic drift in peripheral isolates initially, but selection would subsequently lead to substitutions and changes in allele frequencies at many different loci. The result would be extensive and diverse modificaton of the gene pools, or a "genetic revolution." The model is often referred to as the **founder effect** because all events are set in motion by a few individuals isolated from the main population. In essence, the model predicts that in small peripheral isolates, genetic drift destabilizes an adaptive gene complex to the point where selection can lead to a cascade of genetic changes. These changes result in a new adaptive complex that is incompatible with the ancestral one.

A considerable body of biogeographical evidence appears to support peripatric speciation based on the founder effect. Mayr particularly noted various examples of animals in the main range having little obvious variation in major characters, whereas closely related populations in isolated sections at the periphery of the main range are quite variable for many of these traits. For many of these examples, however, there are no data on reproductive isolation between peripheral and main populations. Mayr's theory concerns the origin of species, which should include their being reproductively isolated if their different adaptive gene complexes cause genetic incompatibility.

Mayr's proposed founder effect in peripatric speciation has been the basis for somewhat similar speciation models proposed by Alan Templeton and by Hampton Carson, based on experimental data as well as biogeography. Their models differ from Mayr's particularly in regard to the extent of genetic change that would occur in the peripheral isolates. Whereas Mayr proposed that initial changes by genetic drift would lead to a cascade of alterations at many other loci, both Templeton and Carson proposed that initial genetic drift as well as subsequent selection would affect only one or a very few important phenotypic traits.

According to Templeton's model, one or a few loci with a major phenotypic effect could undergo changes in allele frequencies by genetic drift at the time a few founders initiate a population. Once these frequencies were established, selection would act on polygenic systems that modify the phenotypic character to a new adaptive peak. Loci that undergo significant changes in allele frequencies would provide the basis for divergence, but the population could climb to a new peak only if these major traits were molded and stabilized by **modifier alleles** at other interactive loci. Templeton's data showing that the Hawaiian species *Drosophila silvestris* and *D. heteroneura* differ in the alleles of the single gene governing head shape, but that the phenotypic trait is modified and stabilized by at least 10 autosomal loci, support his speciation model.

Carson's model differs from the others in positing a requirement for *cycles* of severe reduction in population size and of exponential increase in size—that is, **population bottlenecks** and **population flushes.** According to this model, some blocks of genes with strong interactions are relatively resistant to selection, even though many other loci change readily by selection. Recombinants for alleles at the resistant blocks of loci would have low fitness, so the ongoing population would tend to retain its particular fit combinations of alleles and resist change. When a few founders of a new population reproduced exponentially in a population flush, the selection-resistant blocks of loci would become destabilized. Under these conditions of relaxed selection, recombinants that had low fitness in the main population could now increase substantially. If the population size was severely constricted following a flush, genetic drift and selection would act together to pick out the new recombinants that would persist. The population isolate would thereby be brought nearer to a different adaptive peak from the one occupied by the main population. Cycles of population flushes and bottlenecks would thus lead to the differentiation of new populations from the main population in the adaptive landscape.

All three models based on the founder effect propose that speciation may be rapid and extensive and that new species would be reproductively isolated as the result of genetic incompatibility engendered by different sets of changes initiated by genetic drift and subsequently stabilized by natural selection. Both biogeographical and experimental evidence support all these models, but there is not enough compelling evidence to choose among them. Recent studies by Edwin Bryant on the effects of population bottlenecks and flushes in creating phenotypically variable isolates from a main stock of houseflies have provided important data that seem to fit better with the underlying population genetics of Templeton's and Carson's models than with Mayr's model.

Bryant showed that housefly populations founded by 1, 4, and 16 pairs of breeders exhibited greater genetic variance than existed in the main stock, as measured for 8 major phenotypic traits among about 3000 descendants. The boost in genetic variance was especially marked in the bottleneck populations started with 4 and 16 pairs of breeders (Fig. 8.11). These results are more in accord with Templeton's and Carson's models of significant modification in a few major traits, than with Mayr's model of changes at numerous loci during the phase of population increase following the bottleneck beginning. Even more, Mayr assumed significant decrease in variance in the initial isolates compared with the main population, because any isolate would have only a fraction of the total variation carried in the main gene pool. Mayr's model thus predicts that any isolate would be at great initial risk because of low variance at the time of the bottleneck, but once past this phase, the variance would increase greatly as selection acted on many genes.

Mayr's proposal of initial catastrophic loss of variance was disputed by several population geneticists on theoretical grounds, especially by Russell Lande in 1980. Lande showed mathematically that the loss of variance is $1/2N_e$ in any generation. Thus even if one male and one female ($N_e = 2$) founded a population, only ¼ of the variance would be lost and ¾ would be retained. With as few as four pairs of founders ($N_e = 8$), similar to one of Bryant's experimental populations, only $^{1}/_{16}$ of the original variance would be lost and $^{15}/_{16}$ would be retained. Bryant's data revealed that variance is enhanced in the post-bottleneck populations, however, so suitable genetic models have yet to be found for the founder effect. In addition, we must get definitive information about the kinds of alleles present, the ways in which the alleles interact, and the means by which the alleles and their interactions become modified through a bottleneck. The broader question that remains to be answered by any speciation theory concerns the major hallmarks of the origin of new species: What is the relationship between genetic differentiation among species and the incompatibility that produces reproductive isolation, presumably as corollaries of each other?

Genetic revolutions have been proposed to explain rapid bursts of speciation, a pattern seen in both living and fossil groups of organisms. The punctuated equilibrium theory is, however, derived more from studies of the fossil record than from studies of experimental and natural populations of living species.

Punctuated Equilibrium and Speciation Phenomena

In 1972, Niles Eldredge and Steven Jay Gould published the first of a series of papers presenting a view of speciation based largely on their interpretations of the fossil record. As paleontologists, Eldredge and Gould regard the fossil record as a reasonably reliable history of speciation events, not as a distorted picture of the past. In their view, the absence of transitional, or intermediate,

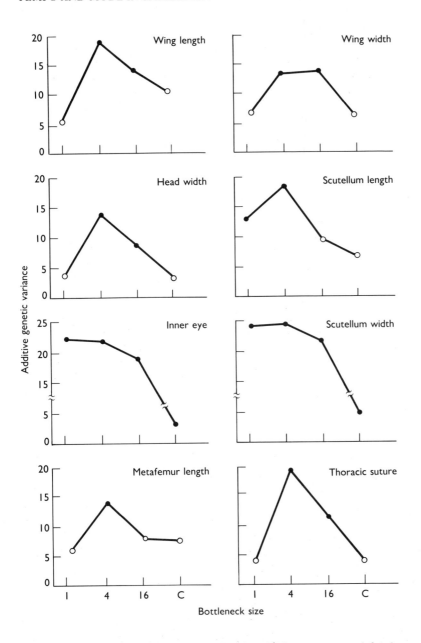

Figure 8.11. Genetic variance (multiplied by 10^4, for convenience) for 8 measured morphological traits in the housefly averaged over the 4 replicate lines within each bottleneck size (1, 4, and 16 pairs of breeders) and the control. Statistically significant differences in variance ($P < 0.05$) are shown as filled circles, and insignificant differences ($P > 0.05$) as open circles in each graph. Increased variance is evident in every case for bottleneck populations started with 4 breeder pairs, and in 5 of the 8 comparisons for bottlenecks involving 16 breeder pairs, when compared with controls. (Adapted from E. H. Bryant et al., 1986, The effect of an experimental bottleneck upon quantitative genetic variation in the housefly. *Genetics* 114:1191, Fig. 2)

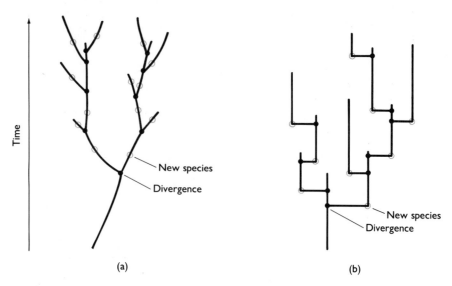

(a)   (b)

Figure 8.12. Speciation by gradual divergence (a), as compared with the pattern of evolutionary stasis and bursts of speciation postulated by the punctuation equilibrium model (b).

forms in most of the record is due to their not having occurred, rather than their having been erased from an otherwise accurate record because of geologic accidents and other reasons. Eldredge and Gould pointed out in particular an apparent pattern of relatively brief times of speciation episodes, which were separated by very long periods of species stability during which little significant change occurred. This pattern of evolutionary stasis punctuated by rapid bursts of speciation is the starting point for their theory of **punctuated equilibrium,** a term that describes the pattern but not the processes involved (Fig. 8.12).

The traditional neo-Darwinian explanation for the rarity of a series of transitional forms between ancestral and descendant species in the fossil record is rooted in natural selection and population genetics. Transitional forms bridging the difference between an optimally fit ancestral population and its optimally fit descendant population are assumed to have lower fitness than either of them and to exist in relatively small populations for brief periods of time. Only a tiny fraction of former life becomes fossilized, and transitional forms are not expected to be found routinely, because of the very low probability that individuals from small populations with a fleeting existence would be preserved in the rare instances of fossilization. Once a species had attained a peak of optimal fitness and proceeded to flourish, however, it would undergo little significant change unless selection pressures changed and a new fitness optimum emerged. An established species would therefore be more likely to exist in large, stable populations and be represented among the rare samples of former life that make up the fossil record. The fossil record is thus incomplete and full of gaps; it is a disproportionate sampling of former life. Even the best preserved segments of the record include gaps representing tens of thousands of years. Such an interval is only the blink of an eye in geologic terms, but it represents thousands of generations to geneticists and other biologists.

Eldredge and Gould emphasize the relationship between tempo and mode in speciation, as enunciated by George Gaylord Simpson in his classic book *Tempo and Mode in Evolution*, published in 1944. The pattern of speciation bursts in a short time was attributed to speciation by means of **splitting** of an ancestral population into two or more descendant populations that are genetically altered and reproductively isolated. Eldredge and Gould accept the mode of splitting rather than of gradual divergence and also incorporate many elements of Mayr's model of peripatric speciation into their theory. Just as the fossil record indicates the replacement of an ancestral species by descendant species, Eldredge and Gould proposed that small local populations that split from a major population can replace the ancestral species as they move into the main range, expand, and compete successfully for similar resources (Fig. 8.13).

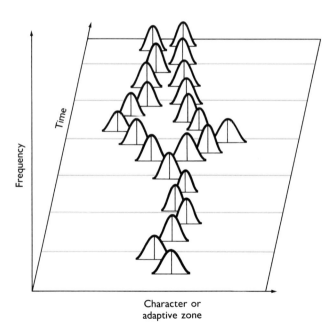

Character or
adaptive zone

Figure 8.13. Speciation by "splitting" of an ancestral population into two (or more) genetically distinct descendant populations. According to this model, small localized populations that separate from a major population could replace the ancestral species in the same adaptive landscape after successful competition for similar resources.

Neo-Darwinian processes of mutation, selection, and genetic drift are not irrelevant in evolution by punctuated equilibrium, as some would believe. They are considered to be responsible for variation within a species, but not necessarily for the origin of species. Eldredge and Gould have offered a variety of suggestions and some modifications of the original concept since 1972, but still have not come to grips with the vital issues of speciation processes. They do not subscribe to the notion of quantum jumps to new species, or *saltation,* nor do they necessarily embrace the concept of *macromutations,* which cause major phenotypic shifts through modifications in one or a few genes governing major morphological characters. But they do not rule out the possibility of macromutations occurring along with conventional and familiar *micromutations,* which produce small changes in many genes. They believe it unlikely, however, that micromutations alone lead to speciation because such genetic changes would cause gradual divergence to new species, which, they say, is not evident in much of the fossil record.

In 1981, Peter Williamson published a detailed report on his studies and analysis of 13 lineages of fresh-water mollusks from a fairly complete sequence found in the Turkana Basin of northern Kenya. The deposits are about 400 meters thick, cover a period of several million years of recent times, and include species with existing modern relatives with which comparisons could be made. Williamson concluded that new species appeared relatively quickly, but intervals between speciation events were long in duration and characterized by little morphological change (Fig. 8.14). The pattern conforms to the one predicted by punctuated equilibrium theory, and was presented as detailed evidence in support of the theory and against gradual divergence to new species.

The publication evoked immediate responses from major figures in evolutionary biology, both pro and con, including questions about the duration of intermediate forms and the time required for reproductive isolation to be established. For example, intermediate forms existed in these invertebrate lineages for between 5000 and 50,000 years, which corresponds to an average of 20,000 generations in these mollusks. Experimental studies with *Drosophila,* mice, corn, and other organisms have shown that significant morphological change and incipient reproductive isolation can occur in much shorter times that those in the molluscan record. Once again, what is gradual change to the geneticist appears to be sudden change to the paleontologist. No consensus was reached on the interpretations of the mollusk study at the time, and they remain to be resolved.

Mathematical models from population genetics rule out the need to invoke new and unnamed genetic processes as explanations. In 1985, both Russell Lande and Charles Newman and his colleagues showed mathematically that downhill movement from an adaptive peak can be very rapid, a point not previously appreciated. Uphill progress toward a peak is expected to be rapid because of forceful natural selection for optimum fitness, a point well understood for a long time. By showing that downhill movement also occurs rap-

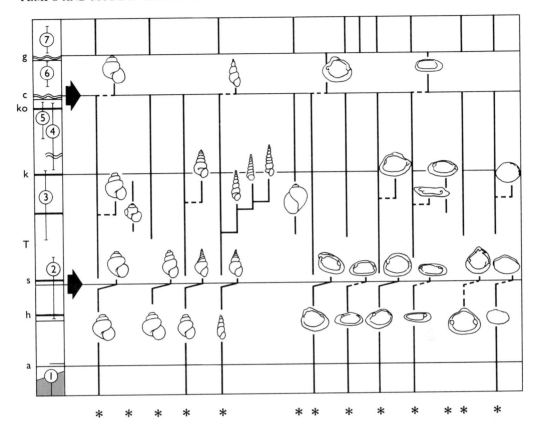

Figure 8.14. Generalized summary of patterns of evolutionary change in the Turkana Basin sequence of 13 lineages (asterisks) of fresh-water mollusks. The temporal sequence of strata, at the left, represents several million years of recent times. The depicted events represent interpretations of the data, and show simultaneous speciation events in all lineages at two (time) levels of basin strata (heavy arrows). The pattern conforms to the punctuation equilibrium model of evolutionary stasis interrupted by relatively brief episodes of speciation. (Modified from P. G. Williamson, 1981, Paleontological documentation of speciation in Cenozoic molluscs from Turkana Basin. *Nature* 293:437, Fig. 4. Copyright by Macmillian Journals Limited)

idly, Lande and Newman explained that speciation will usually take place rapidly by conventional processes.

Movement down an adaptive peak had been considered a slow process, because of the struggle against the forces of natural selection, and downward drift is easily reversed by selection. If the journey down the slope is rapid, however, the population becomes less fit for a particular environment for only a brief time and, therefore, has a better chance of circumvencing the effects of selection. Lande and Newman showed mathematically that *transitions* between adaptive peaks can be rapid (Fig. 8.15). Of equal importance was

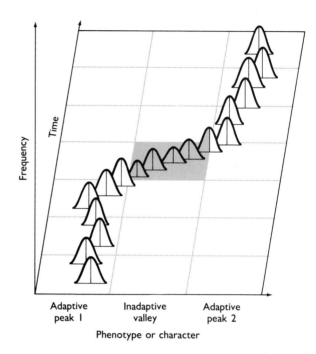

Figure 8.15. Movement in an adaptive landscape from one adaptive peak to another may take place more quickly than previously thought, according to mathematical models used by Russell Lande and Charles Newman. The journey down from one adaptive peak may involve a brief time of transition through the less adaptive valley, as well as the quick trip up to another adaptive peak as previously accepted. Rapid speciation can thus occur by means of conventional evolutionary processes of genetic drift and selection.

their demonstration that rapid downhill drift is as likely to take place in moderately sized populations as in very small ones. Very large populations, however, are unlikely to experience peak shifts because of the counteracting influence of selection and the negligible influence of random genetic drift. Such large populations are thus subject to little evolutionary change.

These mathematical models, based on Wright's "shifting balance" model of population change, predict stasis and punctuation for small to moderately sized populations, but only stasis for large populations. In other words, punctuated equilibrium is a speciation pattern expected on traditional grounds and not a special problem requiring new evolutionary theory. Although these mathematical models can account for stasis and punctuations in a relatively constant environment, they do not account for the pattern in a changing environment. They do not explain to everyone's satisfaction the evolutionary stability of a species over millions of years, during which time environments and living conditions undoubtedly change. It remains unclear whether or not evolutionary biologists can invoke stabilizing selection and preservation of a successful norm for long-term stasis.

Varied Tempos of Speciation

From the fossil record and from living species whose lineages extend into the known past and can therefore be reconstructed, it is quite clear that species originate at greatly different rates. Some species seem to appear within tens of thousands of years, whereas others seem not to have changed or to have diversified for tens or hundreds of millions of years. The tempo of speciation may vary within as well as among groups of organisms, and may change during the evolutionary history of a group. Here we will be concerned only with times of divergence of closely related species. Divergence above the species level and long-term trends of evolutionary diversification will be discussed in Chapter 10.

Extremes of Speciation Tempo

The most obvious evidence for different rates of species divergence comes from comparisons of organisms showing the slowest rates of change and those showing high rates of speciation. Species often referred to as **living fossils** are characterized by little or no significant morphological change over tens or hundreds of millions of years and are the only remaining representatives of a group that was highly diversified in the past. At the other extreme is the pattern of **adaptive radiation,** which represents episodes of very rapid and abundant speciation, usually from one or very few ancestral forms.

Living fossils exist in various groups of organisms, but among the best examples are certain vertebrate animals and ancient groups of land plants. The lobe-finned fishes are a primitive group that flourished in great diversity during the mid-Paleozoic Era, about 360 to 400 million years ago. Most of these species were fresh-water forms, but the coelacanth group took up a marine life. Living coelacanths are restricted to only one genus of deep-water oceanic forms, and were believed to be extinct until a series of accidental discoveries in the Indian Ocean were first reported in 1938. The relatively few living specimens seen since 1938 are quite similar to fossil lobe-finned species that lived almost 400 million years ago (Fig. 8.16).

Lungfishes also were numerous and diverse in the middle of the Paleozoic Era, but are now reduced to only three genera in widely separated parts of the world. The living Australian genus *Neoceratodus* very closely resembles the fossil *Ceratodus,* which lived in the Early Mesozoic Era, about 200 million years ago. The familiar marsupial opossum, only one of which remains in North America, is very similar to the abundant forms of the Late Mesozoic Era, about 100 million years ago. The tuatara *(Sphenodon punctatum)* is the only living representative of the rhyncocephalians, a primitive group of reptiles that flourished during the Early Mesozoic Era but is now restricted to a few islands off the coast of New Zealand (Fig. 8.17).

Some living fossil species are so similar to fossils known from the geologic

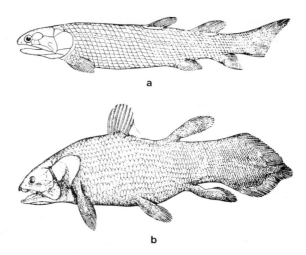

a

b

Figure 8.16. (a) The Middle Paleozoic fossil lobe-finned fish *Osteolepis* closely resembles (b) the living marine coelacanth *(Latimeria chalumnae)*, thought to be extinct until discovered in 1938 and seen since by deep-sea fishermen. (From A. S. Romer, 1970, *The Vertebrate Body*, 4th ed. Philadelphia: Saunders)

record that all of them can easily be classified as members of the same species. Two such examples are the bowfin fish *(Amia calva)*, which has persisted for 60 million years, or most of the Cenozoic Era, and the gymnospern (naked-seeded) plant *Gingko biloba*, whose leaves are identical to fossil specimens 60 million years old. The gingko is the only living species of a widespread and diverse group that made up a sizable portion of ancient forests until the end of the Paleozoic Era, about 250 million years ago. Living gingkos were discovered in China many years ago, and have since been planted as ornamental trees in many parts of the world. The redwoods *(Sequoia)* are another gymnosperm group greatly restricted today but widespread in ancient times.

If these are species that have endured for exceptionally long periods of time, what is the average duration for a species? The estimates vary from one group of organisms to another, but average duration times can be determined from the fossil record and other data. Diatoms are single-celled protists with silica walls whose average duration is 25 million years, whereas species of mosses and their relatives exist for about 15 million years, and seed-bearing plant species for about 6 million years. Among animals, fresh-water fishes endure for about 6 million years, and an estimate of 2 million years has been made for species of beetles as well as species of mammals.

Spectacular examples of adaptive radiations leading to scores or even hundreds of new species from one or a few ancestral forms are found on island groups and in isolated deep-water lakes of fairly recent origin, as dated by geologic evidence. We earlier mentioned the evolution of 14 species of Darwin's finches in less than 5 million years in the Galápagos Archipelago and on Cocos Island. The ancestral finch species still lives in Panama and has not evolved to any great extent compared with its descendant species on the

Galápagos. Even more astonishing is the honeycreeper group of finches (Drepanidae) endemic to the Hawaiian Islands, 2000 miles from any continental landmass. Nearly 40 species of honeycreepers were present when Captain James Cook arrived on the islands in 1778, and many more existed before then. One ancestral finch species that accidentally reached the islands is believed to have diversified into all the known honeycreepers. These birds are highly diverse, and include species that are more like warblers, thrushes, parrots, and other birds than the standard stout-beaked finch (Fig. 8.18). The Laysan finch is believed to most closely resemble the founder of the group.

Island groups of relatively recent age are the locale for adaptive radiations in certain groups of plants and insects as well as birds, but not amphibians, reptiles, or mammals, although a few unique endemic species may occur. The 400 species of *Drosophila* on the Hawaiian Islands are believed to have evolved from 1 or 2 ancestral *Drosphila* species that reached the archipelago by accident. Silverswords are plants in the sunflower family, and nearly 30

Figure 8.17. The tuatara of New Zealand is the only living representative of the rhyncocephalians, a primitive reptile group that flourished 200 million years ago in the Early Mesozoic Era. (Photograph courtesy of St. Louis Zoological Park)

0   5 cm

Figure 8.18. Honeycreepers are a group of finches endemic to Hawaii, a chain of volcanic islands in the Pacific Ocean. The conventional finch beak size and shape has been modified considerably in descendant species, only some of which are illustrated. (From D. Lack, 1947, *Darwin's Finches.* Cambridge: Cambridge University Press)

species belonging to only 3 genera have evolved on these islands from a single colonizing species. These species vary from small mat-forming shrubs to vines and trees, and they live in a variety of habitats from the wettest to the most arid spots on the islands.

As in the Galápagos and Hawaiian islands, among others, dramatic examples of adaptive radiations have been found in deep-water lakes that may be less than 1 million years old. Some of the deepest fresh-water lakes in the world are in the Great Rift Valley of East Africa and are known to be relatively young (Fig. 8.19). Although Lake Victoria may be only 500,000 to 750,000 years old, it has more than 150 species of cichlid fishes, a perchlike type, most of which belong to the genus *Haplochromis*. Lake Malawi has scores of cichlid species that have evolved remarkably diverse feeding habits that correspond to the great variety of teeth and other parts of the feeding apparatus in this animals. If it were not for the broad similarities in other morphological features, biologists would hardly suspect that these species are very closely related.

On an even tighter timetable, it is possible that species of pupfish *(Cyprinodon)* have evolved within the past 20,000 to 30,000 years in isolated springs reduced from ancient lakes in Death Valley in California and Nevada. In the East African Lake Nabugabo, there exist several endemic species of cichlids, which must have come into existence within the past few thousand years since this lake became separated from Lake Victoria by a narrow strip of lowland.

How can we account for this spread of speciation rates, extending from tens or hundreds of thousands of years up to tens or hundreds of millions of years? Living fossil species are at the tail end of a long evolutionary history. In earlier phases of these species' history, the groups that became established and diversified were those that replaced less advanced forms because of competitive advantages based on new adaptive gene complexes. As still more advanced groups evolved in turn, they had the competitive edge for finite resources and replaced the older groups in the same environments or were able to exploit new environments.

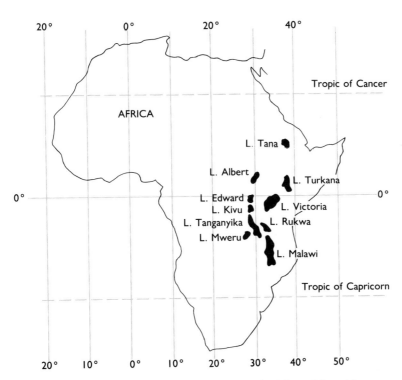

Figure 8.19. Fresh-water lakes in the Great Rift Valley of East Africa are among the deepest lakes in the world, and are often of relatively recent age. Lake Victoria, for example, is much less than 1 million years old, but is home to more than 150 species of cichlid fishes. These species must have evolved in an adaptive radiation in the relatively brief time since the lake was formed.

The appearance of placental mammals in the Cenozoic Era led to the restriction or extinction of previously widespread marsupial mammal species. The main exception was in Australia, which became an isolated landmass at the height of marsupial success and before placental mammals evolved in the region. Marsupials remained predominant there, but declined elsewhere under the competitive onslaught of newly evolved, better adapted placental mammal species. The occasional living fossil species that escaped extinction, such as the North American opossum, remained in evolutionary stasis and would be unlikely to produce new descendant species that could compete with placental species in a new world.

A similar pattern of species depletion is evident among the ancient plant groups, which failed to compete successfully with the new forms for the same resources. The older gymnosperms were replaced by newly evolved and better adapted conifers, such as pines and spruces. The few remaining ancient gymnosperms are generally found in restricted habitats, in evolutionary stasis, where they seem to resist the incursions of their more advanced relatives.

Geography and Speciation Tempo

From the examples of adaptive radiation given above, an outstanding influence on speciation would appear to be the geographical isolation of an entirely new region. Such a region would typically include a great variety of potential habitats, which would become differentiated over time because of geologic, climatic, and biological modifications. A substantial portion of zonal differentiation would be initiated by colonizing plant species. Once established, plants cause changes in soil and microclimatic conditions, and serve as foods for animals. In many of these regions, there are either no predators for many of the life forms or very few such species. The diversity of life styles by which each niche can be exploited successfully and the paucity or absence of predators allow populations to flourish in different ways in different places; that is, they may be able to occupy a large number and variety of adaptive peaks.

The islands and deep-water lakes that support an abundance and diversity of some kinds of organisms are usually characterized by the absence or rarity of many other kinds of organisms. For example, on the Galápagos Archipelago, there are no indigenous amphibians, only one native mammal species, and no native members of the palm family, lily family, and other tropical plant families that would be expected in such an equatorial location. Relatively unique reptiles are found on these islands, including the marine iguana and giant tortoises, but not in any variety of species. Darwin's finches are the major passerine birds there, with only a half-dozen other passerines sharing the islands. The peculiarities of geographical distribution and the richness of species in some groups, but not in others, provide clues to the biological make-up of isolated regions.

Accidental dispersion to islands far away from other land areas would be more successful and probable for spores, seeds, and animals that are relatively small and easily transported over great distances with little mortality of the passengers or strays. Insects and strong fliers in groups of birds are more common than other animal life. By the conventional modes of allopatric and peripatric speciation, the few founders arriving on new island territories can diversify under the influences of natural selection and genetic drift. With few or no predators, ample and diverse foods and habitats, and other amenities of life, new species and an occasional new genus may evolve in a richness that corresponds to the variety and stringency of isolating mechanisms. The profusion of drosophilids, birds, cichlid fishes, and other organisms seems to be correlated with variety of courtship behaviors as well as foods and habitats. All these factors would be expected to separate populations and limit gene flow among them, leading ultimately to reproductive isolation, even if not concurrently with species formation.

It might seem that a deep-water lake is a single environment and unlikely to foster adaptive radiation, unless by the improbable mode of sympatric speciation. Examination of these lakes, however, reveals a variety of distinctly localized living zones. Different temperatures and amounts of light and oxygen at various depths, different food sources at the margins and in shallower and deeper parts of the lake, and other variations contribute to localized but isolated environmental zones, or allopatry and peripatry rather than sympatry. These variations are enhanced by diverse and specific courtship displays, as well as other isolating mechanisms, and a minimum of predators for the species-rich groups.

These riches of biological, ecological, and environmental diversity are not typically encountered outside islands and deep-water lakes, which probably explains the slower and less dramatic speciation rates in most environments. Shallow ponds and lakes contain fewer species with longer duration times than do the myriad living zones of lakes that may be more than 1000 m deep in some places. The apparent proportionality of speciations, times of divergence and endurance, and varieties of living zones and life styles point to the action of a common set of evolutionary processes. There is no need to invoke new evolutionary mechanisms; mutation, natural selection, and genetic drift acting on gene pools in isolation are undoubtedly responsible for species diversification at average rates as well as at slower or faster than average rates.

What is the evolutionary significance of speciation? In general, new species are believed to represent an early step in the development of long-term trends and patterns of the evolutionary history of life. Each new family, order, class, or phylum of organisms evolves from one or more species at the base of a lineage of organisms. In the next two chapters, we will discuss such trends and patterns as interpreted from the fossil record and from comparative studies of living organisms. Once this broader perspective is provided, it should be easier to tie these trends and patterns together with divergence among populations and species in the continuing drama of biological evolution.

Summary

A species is generally considered to be a reproductive community whose members can freely interbreed with one another but not with individuals from other species. The criterion of reproductive isolation cannot be applied to asexual organisms, and it often fails to characterize sexual species in perennial plants and amphibians. It is not clear that the same genes responsible for phenotypic differences among species are also responsible for the genetic incompatibility that underlies reproductive isolation. Allopatric speciation is believed to be the principal mode of species formation, and it requires a prior interval of geographical isolation when gene flow between subpopulations is absent and divergence leads separate gene pools, which may culminate in reproductive isolation. Many examples and models are available for a correspondence between geographical isolation and subsequent speciation, particularly in such island groups as Hawaii and the Galápagos. Genetic barriers to gene exchange, or isolating mechanisms, include differences in habitat, behavior, courtship and breeding, and food sources and preferences. These genetic barriers become more substantial as divergence continues to occur, and may be evident in the gradation of incompatibilities observed as hybrid sterility, hybrid inviability, and lack of hybrid production. Sympatric speciation may characterize production of descendant polyploid species in the same locations as their diploid progenitors, but is otherwise considered to be unlikely or unimportant as a mode of speciation in sexual organisms.

Gradual divergence, genetic revolution, and punctuated equilibrium are the major alternative patterns of speciation that have been theorized, and they are based on perceived differences in the tempo of speciation and the primary genetic processes involved. Gradual divergence postulates a slow pace of divergence and the transition from ancestral to descendant species in accordance with a major influence of natural selection. The paucity of transitional forms, the existence of a species in stasis for much of its duration, and the appearance of new species in bursts, however, have prompted the formulation of the other two theoretical patterns of speciation. Genetic revolutions are most probable in small populations where drift as well as selection are active. These instances of rapid speciation may be initiated by a few founder individuals in peripheral isolated populations, leading either to a cascade of genetic changes or to changes in one or a few major genes in linkage disequilibrium with other genes that modify or influence their expression. The founder effect in episodes of peripatric speciation may involve cycles of population flushes and population bottlenecks, when genetic drift destabilizes an adaptive complex and opens the way to later selective changes. Punctuated equilibrium takes the fossil record at face value, deemphasizing its fragmentary nature and distortions, and suggests that new and unnamed genetic processes may be responsible for brief episodes of bursts of speciation that punctuate long intervals of species stasis. All these patterns of speciation can be depicted graphically by the device of an adaptive landscape, whose shifts in adaptive peaks represent shifts of subpopulations undergoing evolutionary change.

The tempo of speciation varies from the extreme of living fossils, which change very slowly, to the other extreme of adaptive radiations, in which bursts of new species emerge from one or a few ancestral founder populations. All these ranges of tempo can be explained by the conventional mechanisms of mutation, selection, and genetic drift; other processes are not necessarily required to explain differences in time or in amounts of species diversity during evolution.

References and Additional Readings

Atchley, M. R., and D. S. Woodruff, eds. 1981. *Evolution and Speciation: Essays in Honor of M. J. D. White*. New York: Cambridge University Press.

Ayala, F. J. 1978. The mechanisms of evolution. *Sci. Amer.* 239(3):56.

Ayala, F. J. 1982. The genetic structure of species. In *Perspectives on Evolution*, ed. R. Milkman, p. 60. Sunderland, Mass.: Sinauer.

Ayala, F. J. 1985. Microevolution and macroevolution. In *From Molecules to Men*, ed. D. S. Bendall, p. 387. New York: Cambridge University Press.

Barigozzi, C., ed. 1982. *Mechanisms of Speciation*. New York: Liss.

Barton, N. H., and B. Charlesworth. 1984. Genetic revolutions, founder effects, and speciation. *Ann. Rev. Ecol. Syst.* 15:133.

Bryant, E. H., S. A. McCommas, and L. M. Combs. 1986. The effect of an experimental bottleneck upon quantitative genetic variation in the housefly. *Genetics* 114:1191.

Bush, G. L. 1982. What do we really know about speciation? In *Perspectives on Evolution*, ed. R. Milkman, p. 119. Sunderland, Mass.: Sinauer.

Carson, H. L. 1983. Chromosomal sequences and interisland colonizations in Hawaiian *Drosophila*. *Genetics* 103:465.

Carson, H. L. 1987. The genetic system, the deme, and the origin of species. *Ann. Rev. Genet.* 21:405.

Case, T. J., and M. L. Cody. 1987. Testing island biogeographic theories. *Amer. Sci.* 75:402.

Charlesworth, B., R. Lande, and M. Slatkin. 1982. A neo-Darwinian commentary on macroevolution. *Evolution* 36:474.

Crow, J. F. 1986. *Basic Concepts in Population, Quantitative, and Evolutionary Genetics*. New York: Freeman.

Eldredge, N. 1971. The allopatric model and phylogeny in Paleozoic invertebrates. *Evolution* 25:156.

Eldredge, N. 1985. Evolutionary tempos and modes: a paleontological perspective. In *What Darwin Began*, ed. L. R. Godfrey, p. 113. Boston: Allyn and Bacon.

Eldredge, N. 1985. *Unfinished Synthesis: Biological Hierarchies and Modern Evolutonary Thought*. New York: Oxford University Press.

Eldredge, N., and S. J. Gould. 1972. Punctuated equilibria: an alternative to phyletic gradualism. In *Models in Paleobiology*, ed. T. J. M. Schopf, p. 82. San Francisco: Freeman Cooper.

Eldredge, N., and S. Salthe. 1984. Hierarchies and evolution. In *Oxford Surveys in*

Evolutionary Biology, ed. R. Dawkins and M. Ridley, vol. 1, p. 184. New York: Oxford University Press.

Fryer, G., and T. D. Iles. 1972. *The Cichlid Fishes of the Great Lakes of Africa.* Neptune City, N.J.: T. F. H. Public.

Futuyma, D. J., and G. C. Mayer. 1980. Non-allopatric speciation in animals. *Syst. Zool.* 29:254.

Gingerich, P. D. 1983. Rates of evolution: effects of time and temporal scaling. *Science* 222:159.

Gottlieb, L. D. 1984. Genetics and morphological evolution in plants. *Amer. Nat.* 123:681.

Gould, S. J. 1982. Darwinism and the expansion of evolutionary theory. *Science* 216:380.

Gould, S. J., and N. Eldredge. 1977. Punctuated equilibria: the tempo and mode of evolution reconsidered. *Paleobiology* 3:115.

Gould, S. J., N. L. Gilinsky, and R. Z. German. 1987. Asymmetry of lineages and the direction of evolutionary time. *Science* 236:1437.

Grant, P. R. 1981. Speciation and adaptive radiation of Darwin's finches. *Amer. Sci.* 69:653.

Hallam, A., ed. 1977. *Patterns of Evolution as Illustrated by the Fossil Record.* Amsterdam: Elsevier.

Hedrick, P. W. 1983. *Genetics of Populations.* Boston: Jones and Bartlett.

Karlin, S., and E. Nevo, eds. 1986. *Evolutionary Processes and Theory.* New York: Academic Press.

Kimura, M. 1985. Natural selection and neutral evolution. In *What Darwin Began,* ed. L. R. Godfrey, p. 73. Boston: Allyn and Bacon.

Lande, R. 1985. Expected time for random genetic drift of a population between stable phenotypic states. *Proc. Natl. Acad. Sci. U.S.* 82:7641.

Lewontin, R. C., ed. 1968. *Population Biology and Evolution.* Syracuse, N.Y.: Syracuse University Press.

Lewontin, R. C. 1974. *The Genetic Basis of Evolutionary Change.* New York: Columbia University Press.

Lewontin, R. C. 1985. Population genetics. *Ann. Rev. Genet.* 19:81.

Maynard Smith, J. 1983. The genetics of stasis and punctuation. *Ann. Rev. Genet.* 17:11.

Mayr, E. 1942. *Systematics and the Origin of Species.* New York: Columbia University Press.

Mayr, E. 1963. *Animal Species and Evolution.* Cambridge, Mass.: Harvard University Press.

Mayr, E. 1981. Biological classification: toward a synthesis of opposing methodologies. *Science* 214:510.

McNeilly, T. 1968. Evolution in closely adjacent plant populations. III. *Agrostis tenuis* on a small copper mine. *Heredity* 23:99.

Milkman, R., ed. 1982. *Perspectives on Evolution.* Sunderland, Mass.: Sinauer.

Moulton, M. P., and S. L. Pimm. 1983. The introduced Hawaiian avifauna: biogeographic evidence for competition. *Amer. Nat.* 121:669.

Nagylaki, T. 1975. Conditions for the existence of clines. *Genetics* 80:595.

Nei, M., T. Maruyama, and C.-I. Wu. 1983. Models of evolution of reproductive isolation. *Genetics* 103:557.

Newman, C. M., J. E. Cohen, and C. Kipnis. 1985. Neo-darwinian evolution implies punctuated equilibria. *Nature* 315:400.

O'Brien, S. J. et al. 1985. Genetic basis for species vulnerability in the cheetah. *Science* 227:1428.

Rhodes, F. H. T. 1981. Gradualism, punctuated equilibrium, and the *Origin of Species*. *Nature* 305:269.

Rose, M. R., and W. F. Doolittle. 1983. Molecular biological mechanisms of speciation. *Science* 220:157.

Sheldon, P. R. 1987. Parallel gradualistic evolution of Ordovician trilobites. *Nature* 330:561.

Simpson, G. G. 1944. *Tempo and Mode in Evolution*. New York: Columbia University Press.

Stanley, S. M. 1979. A theory of evolution above the species level. *Proc. Natl. Acad. Sci. U.S.* 72:646.

Stanley, S. M. 1986. *Earth and Life Through Time*. New York: Freeman.

Stebbins, G. L. 1950. *Variation and Evolution in Plants*. New York: Columbia University Press.

Stebbins, G. L., and F. J. Ayala. 1981. Is a new evolutionary synthesis necessary? *Science* 213:967.

Stebbins, G. L., and F. J. Ayala. 1985. The evolution of Darwinism. *Sci. Amer.* 253(1):72.

Templeton, A. R. 1980. The theory of speciation via the founder principle. *Genetics* 94:1011.

Templeton, A. R. 1982. Adaptation and the integration of evolutionary forces. In *Perspectives on Evolution*, ed. R. Milkman, p. 15. Sunderland, Mass.: Sinauer.

Tompkins, R. 1978. Genic control of axolotl metamorphosis. *Amer. Zool.* 18:313.

Vrba, E. S. 1980. Evolution, species, and fossils. *So. Afr. Jour. Sci.* 76:61.

Wake, D. B. 1966. Comparative osteology and evolution of the lungless plethodontid salamanders, family Plethodontidae. *Mem. So. Calif. Acad. Sci.* 4:1.

Wake, D. B., and A. Larson. 1987. Multidimensional analysis of an evolving lineage. *Science* 238:42.

Wake, D. B., G. Roth, and M. H. Wake. 1983. On the problem of stasis in organismal evolution. *Jour. Theoret. Biol.* 101:211.

Wake, D. B., and K. P. Yanev. 1986. Geographic variation in allozymes in a "ring species," the plethodontid salamander *Ensatina eschscholtzii* of western North America. *Evolution* 40:702.

Ward, P. D., and P. W. Signor III. 1983. Evolutionary tempo in Jurassic and Cretaceous ammonites. *Paleobiology* 9:183.

White, M. J. D. 1978. *Modes of Speciation*. San Francisco: Freeman.

Williamson, P. G. 1981. Paleontological documentation of speciation in Cenozoic molluscs from Turkana Basin. *Nature* 293:437.

Wright, S. 1931. Evolution in Mendelian populations. *Genetics* 16:97.

Wright, S. 1982. Character change, speciation, and the higher taxa. *Evolution* 36:427.

Wright, S. 1982. The shifting balance theory and macroevolution. *Ann. Rev. Genet.* 16:1.

Zaret, T. M., and R. T. Paine. 1973. Species introduction in a tropical lake. *Science* 182:449.

PART III

Patterns and Trends in Evolution

CHAPTER 9

Phylogenetic Analysis

In the preceding discussions, we established the reality of evolutionary history and concentrated on the mechanisms that cause evolution. We now turn to a consideration of the history of evolutionary changes. The particular focus will be on the genealogies of multicellular plants and animals and the methods used to determine phylogeny. **Phylogeny** means the family history of modern organisms and their ancestors, and **phylogenetic studies** are thus directed at determining these histories and their patterns of change. A large assortment of data must be assembled into evolutionarily accurate genealogies. A variety of methods are available for this task, particularly comparative morphology and molecular analysis of living organisms, because the fossil record alone is too fragmentary. The incorporation of geologic and paleontological data, however, does provide a general timetable and an indication of patterns and trends of plant and animal evolution.

Determining Relationships

The task of phylogenetic analysis is to determine relationships among organisms, which are members of particular species, and to describe patterns of evolutionary change. In order to achieve this goal, it is necessary to establish criteria by which degrees of relationship can be assessed. On the basis of these assessments, species can be combined into genera, families, and other ranks of higher taxa, in an evolutionarily meaningful hierarchy that mirrors descent with modification. There are different schools of systematics, however, that favor the use of quite different criteria for phylogenetic analysis. These contending views have generated a great deal of controversy, which has been quite strident at times.

Different Approaches to Determining Phylogenies

From an evolutionary perspective, any scheme that groups organisms into individual *taxa* and that orders these taxa into a hierarchy of *taxonomic categories* should fulfill two requirements. First, the taxa should be grouped according to their genetic relationships, which reflect the evolutionary history of each taxon. Second, it should be clear how these relationships have been inferred from the data collection and their analysis. Three major schools of systematics advocate different approaches to the problem and often arrive at different interpretations. These approaches are called phylogenetics, phenetics, and cladistics.

The school of **phylogenetics** advocates a traditional method of determining phylogenies, which involves the reconstruction of lineages of organisms traced back to a common ancestor. The resulting evolutionary tree consists of individual branches and twigs representing taxa with a shared evolutionary history. A number of characters is included in an analysis, and the descendants are characterized by both their similarities to and differences from one another and their ancestral forms. The major emphasis is on the history of descent, not on resemblance or dissimilarity per se. Such an evolutionary tree, or phylogenetic hypothesis, depicts directional change within each lineage, or **anagenesis,** and branching episodes of speciation, or **cladogenesis** (Fig. 9.1). Anagenetic changes reflect the transformation of ancestral to descendant forms along a shared pathway of evolutionary change. Cladogenetic changes represent episodes of divergence from an ancestral pathway, leading to new species with altered traits.

Every living species is a mixture of *ancestral* characteristics, which have been conserved with little or no significant change from distant ancestors, and *derived* characteristics, which have changed from the ancestral state at later times in history. The human leg is just such a mosaic because the skeletal structure and the five toes are not significantly different from those of the limb of our ancient tetrapod ancestors, which lived more than 350 million years ago, but the foot has been reorganized more recently into a support structure for bipedal locomotion (Fig. 9.2). Ancestral features tend to be considered as "primitive" and derived features as "advanced." In fact, both sets of features have continued to evolve in the ancestral and descendant lineages, but at very different rates.

The classification scheme that emerges from the phylogenetic approach to systematics may include groups that are monophyletic or paraphyletic. A **monophyletic** group consists of all the descendants derived from a single ancestral stem. The stem group is usually of equal rank to the descendant group; for example, the class Mammalia evolved from the class Reptilia (Fig. 9.3). If some of the descendants of an ancestor are not included, the group is **paraphyletic.** The reptiles are just such a paraphyletic class because their avian and mammalian descendants have been separated into their own classes, on the basis of perceived major differences in their evolved features. Both the monophyletic and the paraphyletic patterns reflect evolutionarily related phy-

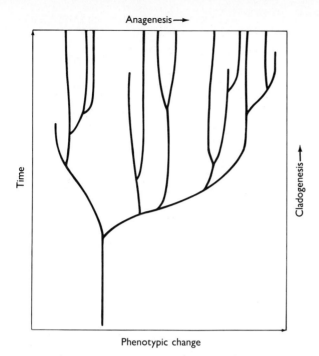

Anagenesis ⟶

Cladogenesis ⟶

Time

Phenotypic change

Figure 9.1. An evolutionary tree indicates directional phenotypic change within each lineage, or anagenesis, and branching episodes of speciation, or cladogenesis. Anagenetic changes reflect descent from shared ancestral forms, and cladogenetic changes represent divergence from ancestral forms to new descendant species.

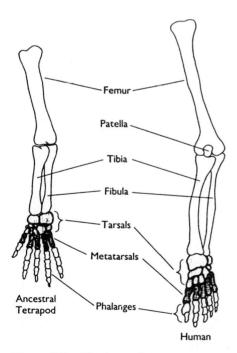

Femur

Patella

Tibia

Fibula

Tarsals

Metatarsals

Ancestral
Tetrapod

Phalanges

Human

Figure 9.2. The human leg is a mosaic composed of retained ancestral skeletal features and derived features, which have been more recently modified from the ancestral state.

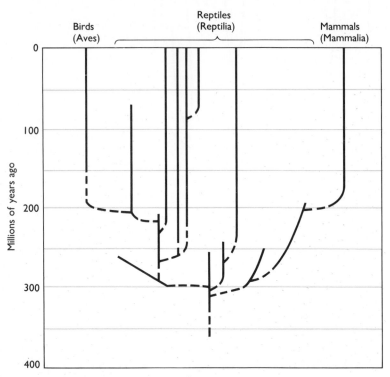

Figure 9.3. A phylogenetic interpretation of vertebrate classification includes the monophyletic classes of birds and mammals, each descended from a single reptilian ancestral stem, and the paraphyletic group of reptiles, whose avian and mammalian descendants are not included in the class Reptilia.

logenies within a group, but taxonomic judgments about the relatively greater importance of similarities or differences between groups leads to the reflection of one pattern or the other in the proposed scheme of taxonomic categories.

Phylogeneticists attempt to reconstruct genealogies that reflect both the degree of common ancestry and the amount of divergence that has occurred—that is, incorporate the similarities of ancestral character states as well as the differences of derived character states. The call for subjective judgments in assessing the relative influence or importance of similarities or dissimilarities in reconstructing evolutionary histories, or the classifications that reflect these histories, has led systematists to develop methods that aim for greater objectivity in phylogenetic analysis and interpretation. In the 1950s, the school of phenetics was established, and in the 1960s, the school of cladistics came into being. Their approaches are quite different, but each school of systematics aims for maximum objectivity in determining the degree of relationship among species and among higher taxa.

The school of **phenetics** excludes differences among taxa and emphasizes their similarities, even if the resulting classification does not strictly reflect common ancestry. The phenetic approach to classification generally involves measurements or numerical scores for morphological resemblances among taxa. The quantitative information about a great many characters can then be compared among different taxa by calculating an index of similarity. By the methods of numerical taxonomy, clusters of taxa with similar morphologies can be grouped into higher taxa, leading theoretically to the inclusion of all organisms in an objective scheme of classification.

The gain in objectivity by phenetic methods may be obtained at the expense of reduced evolutionary significance. Closely related species may be put into different groups if they have diverged morphologically, and unrelated species that happen to be morphologically similar may be included in the same group. The similarities portrayed in the form of a **phenogram** may or may not reflect the effects of evolutionary divergence and convergence or different rates of evolutionary change in the same features in various lineages (Fig. 9.4).

If convergent evolution has led to morphological similarities, the phenetic phylogeny based on such characters will be **polyphyletic** rather than indicating common ancestry. Polyphyletic taxa are considered by evolutionists to be "unnatural" groups, whose common features evolved by different genetic programs along different developmental pathways, perhaps in response to similar selection pressures. If discovered to be polyphyletic—that is, include

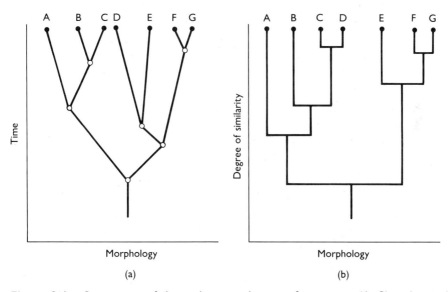

Figure 9.4. Comparison of the evolutionary history of seven taxa (A–G) as depicted by two different approaches. If the ancestry of the morphologically similar taxa C and D is not known, (a) the phylogenetic tree would incorporate differences as well as similarities and perhaps generate a different evolutionary history and, hence, a different classification from interpretations of (b) a phenogram, which is based on the phenetic emphasis on similarities among taxa.

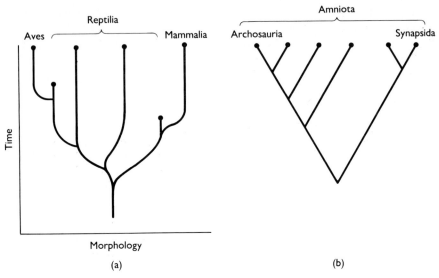

Figure 9.5. Comparison of vertebrate phylogeny and classification according to two schools of systematics. (a) Phylogenetic criteria judge birds, reptiles, and mammals as members of different vertebrate classes, based on a selection of ancestral and derived character states. (b) A cladogram generated from cladistic criteria, however, emphasizes uniquely derived character states that identify each branch as a distinct monophyletic group. The Amniota thus include Archosauria (birds and their reptilian ancestral lineage), Synapsida (mammals and their reptilian ancestral lineage), and reptiles exclusive of the lines that produced the birds and mammals. The members of the Amniota share the character of an amniote egg from which the animal develops (see Fig. 10.23).

more than one ancestral stem—the group is usually reclassified to place the unrelated organisms into separate taxa that more accurately reflect their genetic relationships and independent evolutionary histories.

The school of **cladistics,** which was founded by Willi Hennig in the 1960s, deemphasizes ancestral character states, and thus effectively minimizes overall similarities in determining relationships among organisms. Cladists maintain that the branching of lineages is the critical evolutionary event that offers objective information for classification. Because an anagenetic sequence provides a continuous series of ancestor–descendant populations, the separation of these populations into taxa becomes judgmental rather than objective. Cladists therefore portray a phylogeny in the form of a **cladogram,** in which every branch represents a monophyletic group whose members share uniquely derived character states (Fig. 9.5).

Although objectivity in classification and phylogeny construction can be achieved by cladistic methods, there are some drawbacks in this approach. Species may be grouped together even if they have not reached equivalent stages in evolutionary development, particularly because the common ancestor must be placed in the same taxa as its descendants in a monophyletic branch. Such a stem ancestral line may not have evolved the features that characterize the descendants. Similarly, some groups are split off because of changes that may be less evolutionarily significant than the set of congruous characters that is retained by the lineages. Apart from these and other controversial inferences from cladistics, cladists have been influential in formulating criteria for separating ancestral and descendant lineages and for recognizing sister taxa that branch from a common lineage.

It is clear from the differing premises of the phylogenetic, phenetic, and cladistic schools that different phylogenies may be constructed by them for the same sets of organisms. The more inclusive approach of the phylogeneticists is preferred by many evolutionists, because evolutionary processes and phenomena are taken into account in reconstructing an evolutionary history. The genealogies produced by phenetic and cladistic approaches often minimize evolutionary judgments in favor of greater objectivity, so either similarities (phenetics) or dissimilarities (cladistics) are the central focus, regardless of the evolutionary forces producing them. Whether or not the emphasis on objectivity will lead to a better understanding of patterns of descent and to classification that truly reflects evolutionary history remains to be determined.

Divergence and Convergence

In order to sort out relationships and lineages, it is essential to distinguish resemblances based on a shared genetic heritage from similarities that have arisen independently in different and unrelated evolutionary histories. It is thus necessary to differentiate between **homologous** (genetically related) and **analogous** (similar but unrelated) structures and habits. Homologous features arise during anagenetic and cladogenetic evolutionary episodes of **divergence** within a lineage, often leading to apparent differences that may obscure underlying genetic relatedness. Analogous features, which are the result of evolutionary **convergence**, appear to be related because of similarity, but actually have arisen by independent pathways of development governed by different genetic programs (Fig. 9.6).

A character or character state is homologous in two species when it has been inherited by both from their last common ancestor. Analogous characters are present in two species but were not present in all the ancestors intervening between them and their last common ancestor. It is difficult to recognize homologies when divergence has produced very different degrees of derived character states. The forelimbs of a horse, whale, bat, and monkey are very different structures suited to different locomotor functions and life styles. All these morphologically distinct forelimbs developed from a common skeletal plan, which can be traced back through all the intervening ancestors between these modern mammals and their common tetrapod ancestor, which emerged from the seas to live on land (Fig. 9.7). Similarly, the avian feather is a unique derived character that evolved from the ancestral reptilian scale. The two kinds of skin covering appear to be very different from each other, but developmental analysis clearly shows that the feather is a modified scale.

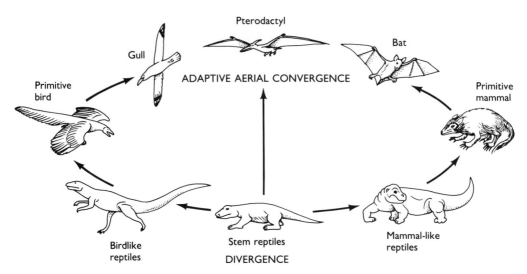

Figure 9.6. Divergence within a lineage leads to differences among descendant lines, but their common heritage is often evident in underlying homologous aspects of their genomes or phenotypes. Convergent evolution produces similarities that appear to be homologies, but can be identified as analogies if independent genetic programs produced superficially similar evolutionary products. Homologies are inherited by descendants from their last common ancestor and are present in all the ancestors intervening between them and their last common ancestor. Analogies are absent in the ancestors intervening between different species and their last common ancestor.

This and other features analyzed in comparative studies and evident in the fossil record indicate that birds are descended from a reptilian lineage, and both groups share many homologies that are grounded in a common fundamental genetic program.

During divergent evolution, novel structures and habits may evolve, or a lineage may acquire an ancestral pattern. Among the reptiles, birds, and mammals, all of which evolved on the land and remain predominantly terrestrial, various species have returned to an aquatic life. Extinct ichthyosaurs and other aquatic reptiles, even more than modern crocodilians, were aquatic species that evolved from terrestrial reptilian lineages. Penguins are flightless birds that are adapted in body form and feather characters to a principally aquatic existence. Whales and porpoises are descended from terrestrial mammalian ancestors, but evolved adaptations to a wholly aquatic life relatively recently. All these aquatic species share a suite of adaptations that might be construed as evidence for their being closely related and descended from a common aquatic ancestor. These features are, of course, analogous, having appeared independently at different times from different (terrestrial) ancestors by convergent evolutionary pathways (Fig. 9.8). There is no unbroken sequence of aquatic ancestors from fishes to aquatic reptiles, birds, and mammals. The analogous development of fins and paddles or flippers, of a streamlined body shape, and of other aquatic adaptations almost certainly resulted from similar selection pressures acting on different suitable genetic programs.

Flight is a novelty among reptiles and mammals, but it is the standard means of locomotion of birds and many groups of insects. Many of the wing structures in these groups are analogous products of convergent evolution (Fig. 9.9). It is clear from both fossil and living lineages that flying vertebrates are not descended from flying insects, nor are they descended from one another or a recent common vertebrate ancestor. Birds evolved from reptiles, but from

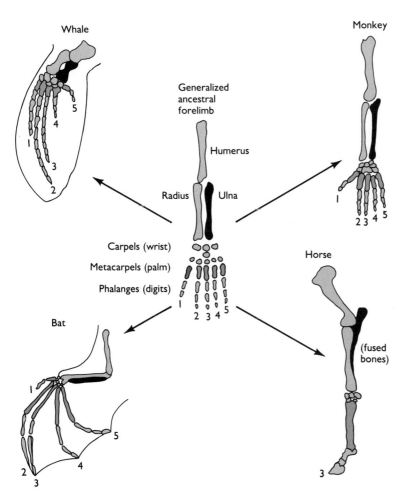

Figure 9.7. Forelimb homologies in four kinds of mammals are evident from their underlying skeletal construction, even though the limbs are adapted to very different modes of activity (swimming, grasping, running, flying). The interpretation of homology is supported by tracing the evolutionary histories of these mammals back through all the intervening ancestors to their last common tetrapod ancestor.

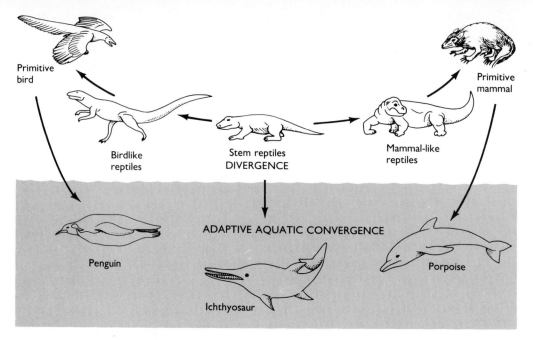

Figure 9.8. Through convergent evolution, some of the descendants of the terrestrial birds, reptiles, and mammals evolved aquatic adaptations that are reminiscent of their ancient fish ancestry. The aquatic life style was not inherited in an unbroken sequence from fish ancestors, but is the result of adaptations acquired independently in three lineages of terrestrial vertebrates.

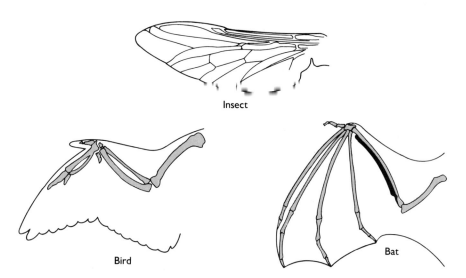

Figure 9.9. Analogous wings of insects, birds, and bats evolved independently as adaptations to flying. The skeletal structures of the developmentally different wings of bats and birds are homologous, however, having been modified from the forelimb skeleton of a common tetrapod ancestor and all intervening ancestors. Insects are not the ancestors of birds, nor are birds the ancestors of bats, even though all three kinds of animals have wings.

a different lineage from the one that produced pterosaurs. Mammals also evolved from reptiles, but bats are descendants of nonflying terrestrial mammals that themselves appeared about 100 million years later than the earliest known mammals.

Analogies may arise in divergent lineages, but only after the different ancestral lines have become established and begun their own independent evolutionary pathways. For a long time, it was inferred that the amazing similarities between various marsupial species in Australia and placental species elsewhere in the world were due to parallel evolution in the two mammalian groups. **Parallel evolution** refers to independent evolutionary modifications of the same kind in closely related groups that have similar developmental programs inherited from a relatively recent common ancestor. More detailed anatomical studies have shown, however, that the two sets of similar organisms are the products of convergent evolution. The koala is not a marsupial version of a bear; the wombat and groundhog occupy similar habitats but are genetically dissimilar; and the Tasmanian wolf resembled members of the dog family, but only superficially (Fig. 9.10). Patterns of convergence are often mistaken for parallel evolution of groups that are known to be closely related. Mistakes of this sort lead to erroneous phylogenetic reconstructions of evolutionary history, particularly when similarities are emphasized in lineage analysis. These errors are much less likely to be made in cladistic phylogenies, which emphasize differences rather than similarities among taxa.

A significant correlation appears to exist between the extent of convergence in past times and the greater or lesser degree of geographical isolation of evolving organisms. Only two major landmasses existed during a substantial segment of the Mesozoic Era, when the reptiles were evolving their greatest levels of diversity. During the succeeding Cenozoic Era, which began about 65 million years ago, many isolated landmasses were created as the Mesozoic supercontinent Pangaea continued to split up and separate by continental drift. Little or no convergence occurred during reptilian evolution in the Mesozoic, but many episodes of convergence typify the splurge of mammalian speciation in the Cenozoic. For example, four orders (groups of families) of ant-eating mammals evolved in the Southern Hemisphere, each adapted to the same life style and having similar but genetically distinct morphological features (Fig. 9.11). The spiny anteater, or echidna (Monotremata), is found in Australia; the ant bear (Edentata) inhabits the South American tropics; the aardvark (Tubulidentata) lives in parts of the African tropics; and the pangolin (Pholiodota) is found in northern and western Africa and in Southeast Asia.

In geographical isolation, convergence toward similar adaptive life styles may occur if similar selection pressures direct suitable genetic programs of somewhat different kinds. In the absence of significant isolation, a single group is more likely to occupy the whole of an area to which it is adapted, and organisms that compete for the same resources are unlikely to become established in the area. Migrations over a continuous landmass (or in a large body of water) may lead to the dispersal of the successful, dominant group

Marsupials

Pouched mouse
(*Sminthopsis*)

Marsupial mole
(*Notoryctes*)

Flying opossum
(*Petaurus*)

Wombat (*Vombatus*)

Eastern native cat
(*Dasyurus*)

Tasmanian wolf
(*Thylacinus*)

Placental mammals

Harvest mouse
(*Mus*)

Common mole
(*Talpa*)

Flying squirrel
(*Petaurista*)

Marmot (*Marmota*)

Serval (*Felis*)

Wolf (*Canis*)

Figure 9.10. Similarities between members of a pair of Australian marsupial and placental mammals were mistakenly assumed to have arisen by parallel evolution. More detailed, recent studies have revealed that these animals are only superficially similar in their life styles, and actually are analogous products of convergent evolution in different ancestral lineages of mammals. (From *The Life of Vertebrates*, 3rd ed., by John Zachary Young. Copyright © 1981 by Oxford University Press. Reprinted by permission)

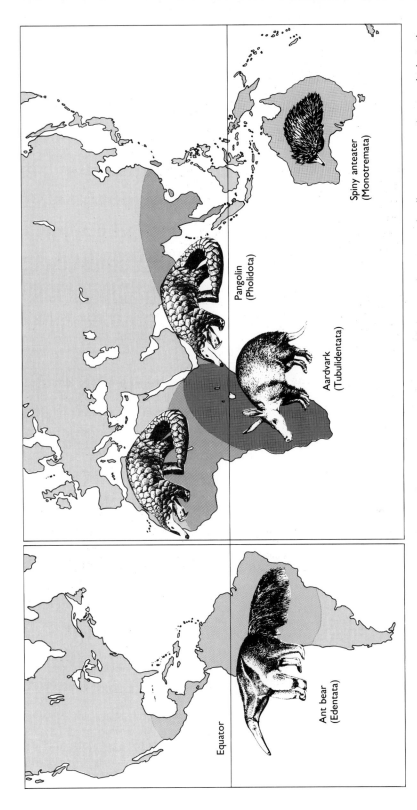

Figure 9.11. The convergent evolution of four orders of ant-eating mammals presumably occurred in geographically separate regions in which similar selection factors influenced similar adaptations in genetically different ancestors. Each group could evolve and flourish in the absence of competition from others using similar natural resources. (From B. Kurtén, "Continental Drift and Evolution." Copyright © 1969 by Scientific American, Inc. All rights reserved)

over that area. Migrations are minimized between separated regions, and in the absence of competition for a suitable habitat, convergent groups may come into existence in separate regions, each exploiting the available resources without interference from similar species.

Distinguishing Between Ancestral and Derived Character States

To determine the direction of evolutionary change in a phylogeny, it is necessary to be able to distinguish between ancestral and derived states. For example, the wingless condition in insects evolved to the derived winged condition in many groups. In some groups, however, winglessness may have reemerged as a derived state from a winged insect ancestor. How can biologists determine whether a particular character state is unchanged from the ancestral form, or has been derived in later episodes of evolution?

The most unambiguous way to ascertain which character state within a group of species is ancestral, or primitive, is by comparing the character in the **ingroup** of species under study with the nature of this character in an **outgroup** of related species. All the living species of *Equus* (horses, zebras, asses) are one-toed, whereas their only living relatives in the order Perissodactyla (rhinoceroses and tapirs) are multitoed. The ingroup of horses thus possesses a single toe as a derived character state from an ancestral multitoed state, which is typical of the related outgroups. In the case of the horse, the fossil record of the past 60 million years corroborates this inference. The trends in horse evolution include the reduction in digit number, as well as other features, which confirms that the multitoed state is ancestral and the single-toed state is derived.

Ernst Haeckel's law that "ontogeny recapitulates phylogeny" was formulated in 1866, and provided a strong basis for inferring the sequences of evolutionary change (phylogeny) from the sequences of embryological development **(ontogeny)**. Haeckel's law of **recapitulation** was based on the idea that early-developing embryonic features represent ancestral character states, and later-developing embryonic features represent derived character states. The basic premise that each stage in embryonic development represents an *adult* stage of one of the ancestral forms served to infer specific ancestry in evolution. Many exceptions were either disregarded or molded to fit the basic premises of recapitulation. It was not until the establishment of Mendelian genetics in the 1900s that the final blows were given to Haeckel's ideas.

The recapitulation thesis required two basic corollaries: (1) evolutionary change occurs by the successive *terminal addition* of stages to an unaltered ancestral ontogeny; and (2) the duration of an ancestral ontogeny must be *continually shortened* (condensed) during the subsequent evolution of its lineage (Fig. 9.12). The second corollary accounted for the relative brevity of ontogenetic development, despite the continued addition of steps, and for the early occurrence and small size of the embryonic stages representing the adult ancestral stages during ontogeny. Genetic studies disposed of the first corollary by showing that all the genes for all the developmental stages are present in the fertilized egg, and any inherited changes are caused by mutational modifications. Any change can occur at any time in development, according to mutation analysis, so there is no a priori reason to postulate that each new feature is added terminally to ontogeny.

The disposition of the second corollary came from genetic studies showing that genes produce enzymes, which control the rates of processes. There is no basis for assuming that acceleration of ontogeny is more fundamental or common than retardation of developmental processes. The many exceptions to orthodox ideas of recapitulation can be properly explained by gene action during development, not by unknown forces that were presumed to speed up development and produce tiny embryonic versions of ancestral adult forms.

Evolutionary changes due to alterations in the timing of development, or **heterochrony,** do underlie changes in ontogeny and in phylogenies that include these alterations in one or more lineages. Developmental modifications often result from **paedomorphosis,** the retention of juvenile (ancestral) features by adults of the species. The acceleration of rates of development, or

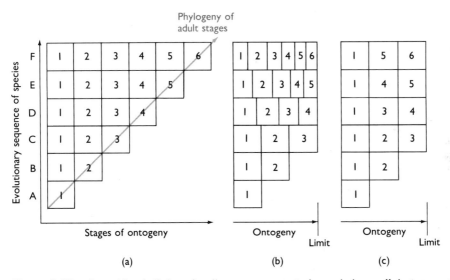

Figure 9.12. Ernst Haeckel's law that "ontogeny recapitulates phylogeny" during species evolution (A → F) postulated that (a) new stages (1 → 6) in ontogenetic development of the individual are added successively at the end of an ancestral sequence of stages. The phylogeny of adult stages thus parallels the ontogeny of the most recent descendant species (F). Haeckel also proposed the occurrence of a condensation or shortening of the duration of ontogeny in successively evolved species, either by (b) acceleration of developmental events in one or more ancestral stages or by (c) deletion of some ancestral stages. In either case, the condensation of ontogeny would explain the brevity and early occurrence of ancestral stages in embryonic development, as well as the relatively small size of the young embryo in recently evolved species of a phylogenetic sequence. (Adapted from S. J. Gould, 1977, *Ontogeny and Phylogeny,* Fig. 7, p. 75. Cambridge, Mass.: Harvard University Press)

progenesis leading to paedomorphosis, results in the rapid attainment of the adult state and an early halt to the developmental program. The retardation of developmental rates, or **neoteny** leading to paedomorphosis, is especially well documented in a number of organisms, and has been suggested to be a significant feature underlying human evolution.

The larger brain of human beings, compared with those of our living ape relatives or of the australopithecine members of the human family, is the outcome of high rates of development in the fetal and infancy stages. The brain (and skull) stops increasing in size at a relatively early stage of infancy in apes, but continues growing for a number of years in human development (see Fig. 6.19). The neotenic brain is largely responsible for significant evolutionary changes based on increased intelligence in *Homo*, as compared with australopithecines or apes. Many of the plethodontid salamanders display neotenic juvenile features that include permanent external gills (Fig. 9.13). These aquatic salamanders undergo retarded somatic development but normal sexual development, so the aquatic larval form becomes sexually mature without reaching the adult stage. In some of these salamander species, however, the aquatic larva has been eliminated entirely from the life cycle, but other juvenile features are retarded in development and persist in the adult stage. Pae-

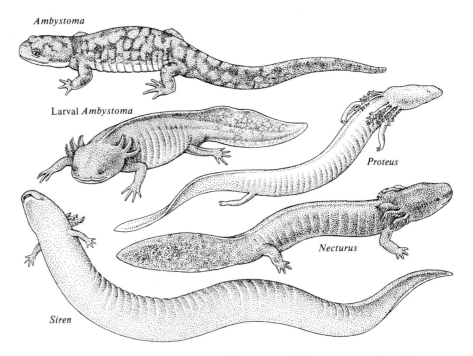

Figure 9.13. The plethodontid salamanders *Proteus*, *Necturus*, and *Siren*, among others, have permanent external gills. That this is a neotenic feature is evident from a comparison with the larval and adult stages of the salamander *Ambystoma*. (From *The Life of Vertebrates*, 3rd ed., by John Zachary Young. Copyright © 1981 by Oxford University Press. Reprinted by permission)

domorphosis has apparently played a dominant role in the evolution of the plethodontid salamanders.

Although gene action determines a developmental program, it is clear that each step in development depends on prior conditions and events; that is, it is **epigenetic.** Interactions among cells and tissues and between organisms and their environment influence the phenotypic outcome of ontogeny. In some cases, the genetic and epigenetic foundations for ontogenetic development are known, which provides a sounder footing in interpreting evolutionary changes in phylogenies. The metamorphosis of larval to adult form in amphibians occurs in response to thyroxine. Thyroxine is produced in the thyroid gland in response to pituitary thyrotropin, which itself is released in response to the production of thyrotropin-releasing hormone in the hypothalamus of the brain. In more recently derived neotenic salamander species, such as the Mexican axolotl *(Ambystoma mexicanum),* an injection of thyroxine can induce metamorphosis to the adult form, which normally fails to develop. This shows that the thyroid and pituitary glands can function normally in the axolotl, but that its neoteny is due to a change in its hypothalamus. Salamanders that have a longer evolutionary history of neoteny, however, do not metamorphose when given thyroxine. Apparently, their tissues no longer can respond to this hormone; the reasons remain unknown at present.

It is clear from these and other examples that a growing body of information from developmental biology and developmental genetics will throw considerable light on many problems of character analysis in the construction of phylogenies. With the information available at present, reasonable phylogenies can be constructed, but they remain subject to modification as new information is obtained.

Reconstructing Phylogenies

An evolutionary tree constructed by phylogenetic, phenetic, or cladistic approaches represents a hypothesis for an evolutionary history. Although some features can be tested or evaluated, others may be difficult or impossible to verify by any current means. More kinds of information are available today by which phylogenies can be assembled and evolutionary histories traced through changes in genes, DNA genomes, and individual proteins, as well as through the continuing traditional data from the morphology of both living and fossil organisms.

Constructing Evolutionary Trees

The first evolutionary tree, published in 1866 by Haeckel, grouped all the known organisms into three kingdoms (Fig. 9.14). Like other phylogenetic trees, Haeckel's was based on inferred patterns of evolutionary descent lead-

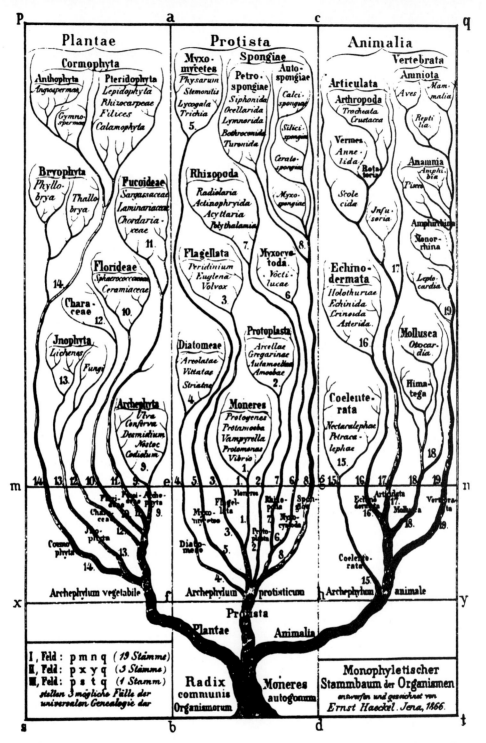

Figure 9.14. The first evolutionary tree of life, published in 1866 by Ernst Haeckel. Some of the arrangements have been retained in modern phylogenies, but many other have been discarded. (From E. Haeckel, 1866, *Evolution*)

ing to similarities and dissimilarities among organisms. Some of these inferences are accepted in modern arrangements for the tree of life. Others have
fallen by the wayside, such as those based on Haeckel's ideas about
recapitulation.

Two levels of descent are recognized among groups of species in a
monophyletic phylogeny. All species that have branched from a recent common ancestor form a **clade** if they share major functional abilities or level of
developmental organization. Portions of the phylogeny that have evolved new
abilities or organization are referred to as **grades** (Fig. 9.15). For example,
the reptiles, birds, and mammals form a monophyletic group. The birds and
mammals have evolved into new grades that are distinct from the reptilian
grade, by virtue of their acquisition of endothermy (warm-bloodedness) and
other significant new features. The evolutionary trees for this group of organisms constructed by the three schools of systematics might be very different

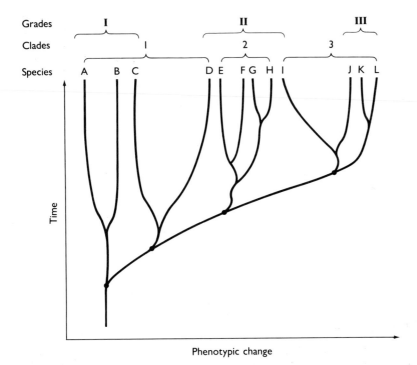

Figure 9.15. Grades (I–III) and clades (1–3) represent two levels of descent among groups of species
(A–L) in a monophyletic phylogeny. Grades are those portions of the phylogeny that have evolved
new abilities or organization. Clades are groups of species that have major similarities and that
branched from a recent common ancestor (circles). Members of a clade may be placed in different
grades if their phenotypic similarities are of grade status, even if they evolved convergently from
different recent ancestors (for example, species C and D of clade I might be assigned to grades I and
II). The classification produced by these criteria could be different from one that takes ancestries into
greater account.

from one another, depending on the characters selected and the inferences made.

To use an extreme example of endothermy as a single defining character, phylogeneticists would separate the three groups into different taxa (classes) because each group has reached grade status and endothermy is inferred to have evolved convergently in birds and mammals. A phenetic tree might put birds and mammals into the same taxon, if their endothermic similarities are emphasized and their pathways of change are disregarded. Such a phylogeny would be polyphyletic. A cladistic tree, which is based strictly on branching episodes and includes the ancestor in the same taxon as its descendants, would group together the birds and those reptiles of a monophyletic lineage, in which all members share a set of uniquely derived character states. The mammals and their reptilian ancestor would be put into another taxon (see Fig. 9.5). The evolutionary significance of new grades may thus be obscured in phenetic or cladistic phylogeny construction, but it is taken into account by the phylogeneticists and, indeed, is a fundamental premise of their tree constructions.

Tree construction according to any of the three approaches is based on a consideration of a number of morphological character states. Given the large number of possible combinations of these features, more than one evolutionary tree for the taxa being studied may be constructed from a matrix of character states. If it were known which character states were primitive and which were derived, the construction of an accurate tree would be much simpler. This information is not always available, however, so it is necessary to judge which is the "best" evolutionary tree of all those that are possible. Many different methods are available for handling this problem, but no one method is always more accurate than another. In addition, some selection techniques work better for trees based on morphological data, and others are more accurate for phylogenies constructed from molecular data.

Maximum parsimony methods are very popular for selecting the most likely tree from among all the possible trees constructed from a matrix of morphological or molecular data. The most parsimonious tree is the one produced by minimizing the number of evolutionary changes required to reconstruct the phylogeny of the taxa under study. Several different algorithms are available for finding the most parsimonious tree, which may be rooted or unrooted (Fig. 9.16). An **unrooted tree,** or network, makes no statement about the direction of evolutionary change, whereas a **rooted tree** indicates the direction of change if the groups of species can be connected to an ancestral stem. An unrooted tree may be rooted if one of the taxa is more distantly related to the others than they are to one another and can serve as an outgroup criterion to establish levels of relationships.

From the equations available, it is clear that the number of possible rooted and unrooted trees becomes unmanageably large if many species are included in an analysis. If only 4 species are involved, 15 rooted and 3 unrooted trees can be constructed from a matrix of character data (Fig. 9.17). If 10 species are included, nearly 35 million rooted trees and more than 2 million unrooted

Species	Character				
	1	2	3	4	5
A	a′	b	c′	d′	e′
B	a′	b	c′	d′	e
C	a′	b	c	d	e
D	a′	b	c′	d	e

(a)

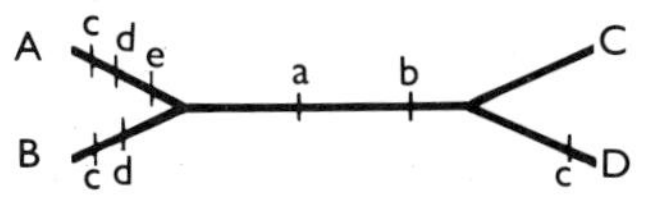

(b)

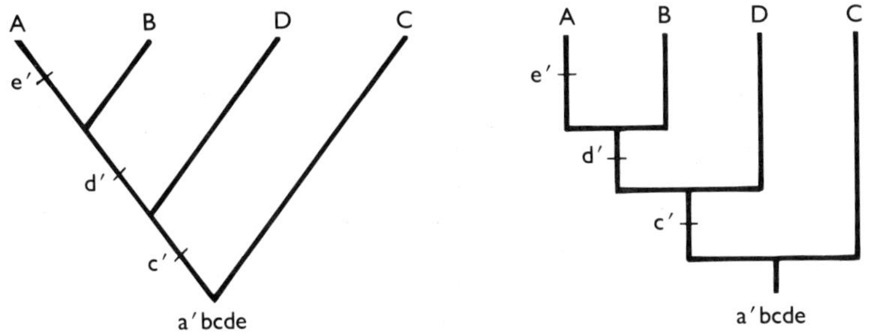

(c)

(d)

Figure 9.16. Trees constructed for four species from (a) a matrix of five characters existing in different character states (c and c′) may be (b) unrooted. Such a tree indicates the branches on which each character changed its state, but not the direction of its change (c → c′ or c′ → c). The tree may be rooted by connection of the species to an ancestral stem, which might show that a′bcde is the ancestral state and that the other three combinations are derived from it. (c) Cladistic analysis places each species on its own branch of a monophyletic lineage, and arranges each branch in order of increasing difference from the ancestral state, as a result of the acquisition of unique derived alterations. (d) Phenetic analysis emphasizes the degrees of similarity among species, and postulates a phylogeny based on the degree of shared similarities. This approach can lead to more specific statements about immediate ancestry and relationships than is provided in the cladogram. Each tree is the most parsimonious, however, because each reconstructs the phylogeny from the minimum possible number of evolutionary changes.

trees become possible. Although most of them can be excluded because of unlikely genetic relationships or other biological information, it is exceedingly difficult to find the true or best evolutionary tree when a large number of species is considered together.

The **compatibility method** is one alternative to maximum parsimony for picking the most likely one of a number of possible evolutionary trees. The compatibility method assumes that the most accurate tree is the one most compatible with the largest number of individual characters, or clique of characters. A tree that is compatible with four characters of a set is thus preferred to an alternative that is compatible with fewer than four characters of this set (Fig. 9.18). The compatibility method may be used for both morphological and molecular phylogenies; for example, a molecular phylogeny based on a set of gene frequencies may be evaluated in the same way as a phylogeny based on a set of morphological character states.

Molecular phylogenetic trees may be constructed by the use of any one of the available **distance matrix methods.** The genetic or evolutionary distance is computed for all pairs of species or populations, and the tree is constructed by considering all the relationships among these distance values.

Regardless of the particular data base, the method for constructing the evolutionary tree, or the method for selecting the best tree, there is little consensus about preferred techniques or data for phylogenetic analysis. Very different phylogenies for the same species may be constructed by the various

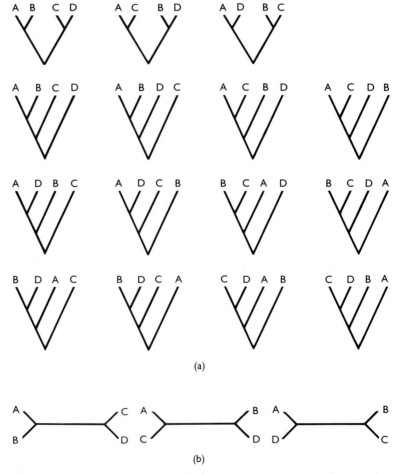

(a)

(b)

Figure 9.17. From a matrix of character data, it is possible to construct a number of phylogenies. For 4 species (A–D), (a) 15 rooted trees and (b) 3 unrooted trees can be assembled. The selection of the most likely one of these possible trees can be made by means of parsimony analysis or an alternative method.

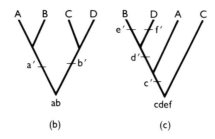

Species	Character					
	1	2	3	4	5	6
A	a'	b	c'	d	e	f
B	a'	b	c'	d'	e'	f
C	a	b'	c	d	e	f
D	a	b'	c'	d'	e	f'

(a) (b) (c)

Figure 9.18. For (a) a set of 6 characters differing in 1 or more character states among 4 species, 2 of the 15 possible rooted trees seem reasonably good phylogenetic interpretations. The tree of choice might be (b) based on two characters (1 and 2), or (c) based on four other characters (3–6). To choose between the two trees by the compatibility method, the most likely one would be compatible with the larger number of individual characters. A tree compatible with all six characters, of course, would be preferred to either of the two trees illustrated.

methods, and there are no objective means for deciding which, if any, is correct. Statistical tests have been applied in some cases, but even they are subject to different interpretations according to the perspectives of the investigators. Trees that do not take evolutionary processes into account are subject to various errors, as are trees that give different weights of importance to one evolutionary process over another. Until there are better theoretical foundations for many tree-making methods, controversies will continue to be generated because of the differences in approach, perspective, and judgments among the various groups of investigators concerned with phylogenetic analysis.

Measures of Genetic Relationships Among Organisms

Morphological characters provide invaluable information for phylogenetic analysis and are virtually the only data for fossil organisms and many groups of living species. These data are genetically relevant because morphology is the outcome of genetic programs that have become modified during evolution, and genetic differentiation is inferred from measures of morphological differentiation. More direct measures of genetic differentiation during evolution can be obtained by examining the genes themselves, nuclear and mitochondrial DNA, and proteins and RNAs that are known to be the direct products of particular genes and their alleles. These molecular characters can be used to trace an evolutionary history by quantitative methods, and they lend themselves very well to statistical measures of their validity. In addition, they provide information that may not be evident from the outward appearance of the organism.

In the 1960s and early 1970s, before it was possible to analyze DNA sequences, various studies were conducted to determine genetic differentiation from data on proteins known to be direct products of specific gene loci. Allozymic variations were analyzed by gel electrophoresis; nonenzyme proteins, by immunological cross-reactions; and individual protein structure, by amino acid sequencing. Genetic differences between species could be ascertained from protein data by the application of equations to determine degrees of protein similarity and dissimilarity.

Electrophoretic data on allozyme polymorphisms can be referred to changes in codons per gene locus by determining average values of genetic identity, I, and of genetic distance, D. **Genetic identity** gives the proportion of genes that are not distinguishably different between pairs of populations, and **genetic distance** is a measure of detectable codon substitutions per gene that have accumulated in the populations since they diverged from a common ancestor. The most widely used formulas are those of Masatoshi Nei, and the mean number of net codon substitutions that have occurred independently is given by

$$D = - \log_e I \qquad\qquad [9.1]$$

where

$$I = \frac{J_{AB}}{\sqrt{J_A J_B}} \qquad\qquad [9.2]$$

and J_{AB}, J_A, and J_B are the arithmetic means over all the gene loci sampled in populations A and B. These estimates of codon similarities and differences are particularly suitable for electrophoretic data of allozyme polymorphisms, but blood group data and some others may also be used under certain conditions.

The mean value of D can range from zero (no genetic differences) to infinity, and the mean value of I can range from zero (complete genetic differentiation) to 1 (complete genetic identity). The phylogeny of a group of species can be reconstructed from a matrix of genetic distances and identities calculated from electrophoretic studies of allozymic variations. For example, John Avise and Francisco Ayala analyzed electrophoretic differences at 24 gene loci in 9 genera of California minnows (Cyprinidae), and calculated values for D and I between the members of all possible pairs of these fishes (Table 9.1). The amount of genetic change varied from a low of $D = 0.055$ between *Hesperoleucus* and *Lavinia*, to a high of $D = 1.118$ between *Pogonichthys* and *Notemigonus*.

These data were used to construct a **dendrogram** of phylogenetic relationships, which included interpretations of lineage and ancestry as well as the amount of genetic change that had occurred in each branch of the tree. For example, the lineage leading to *Notemigonus* accumulated an average of 0.610 electrophoretically detectable nucleotide substitutions per locus since it separated from the last ancestor shared with the other eight genera (Fig. 9.19). The lineage leading to these other eight genera had an average of 0.099 nucleotide substitutions per locus before splitting into the lineage leading to *Pogonichthys* and *Richardsonius* and the lineage leading to the other six genera. The D values of 0.379 for *Pogonichthys* and of 0.140 for *Richardsonius* are the sum of the genetic distance between these two genera ($D = 0.519$) shown in

TABLE 9.1
Genetic Distances (above the diagonal) and Genetic Similarities (below the diagonal)
Among 9 Species of California Minnows, Based on Electrophoretic Studies of Allozymes
Specified by 24 Gene Loci

	1	2	3	4	5	6	7	8	9
1. *Hesperoleucus*	—	0.055	0.095	0.194	0.518	0.705	0.432	0.251	0.901
2. *Lavinia*	0.946	—	0.147	0.216	0.616	0.746	0.519	0.354	0.919
3. *Mylopharodon*	0.909	0.863	—	0.131	0.546	0.600	0.453	0.174	0.790
4. *Ptychocheilus*	0.824	0.806	0.877	—	0.541	0.600	0.526	0.333	0.989
5. *Orthodon*	0.596	0.540	0.579	0.582	—	1.079	0.776	0.518	1.094
6. *Pogonichthys*	0.494	0.474	0.549	0.549	0.340	—	0.519	0.679	1.118
7. *Richardsonius*	0.649	0.595	0.636	0.591	0.460	0.595	—	0.443	0.976
8. *Gila*	0.778	0.701	0.840	0.717	0.596	0.507	0.642	—	0.884
9. *Notemigonus*	0.406	0.399	0.454	0.372	0.335	0.327	0.377	0.413	—

Source: J. C. Avise and F. J. Ayala, 1975. Genetic differentiation in speciose versus depauperate phylads: evidence from the California minnows. *Evolution* 29:411.

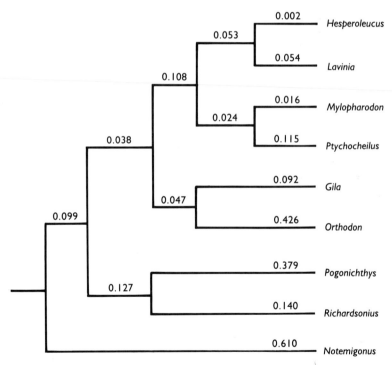

Figure 9.19. Dendrogram of phylogenetic relationships of nine genera of California minnows (Cyprinidae), based on electrophoretic studies of allozymic variations. The allozyme data were used to calculate genetic distance (*D*) measures of detectable codon substitutions per gene since the populations diverged from last common ancestors (see Table 9.1). *D* values are shown on each branch of the dendrogram. (From J. C. Avise and F. J. Ayala, 1975, Genetic differentiation in speciose versus depauperate phylads: evidence from the California minnows. *Evolution* 29:411)

Table 9.1. Similarly, the sum of $D = 0.976$ between *Notemigonus* and *Richardsonius* shown in the table has been apportioned according to the allozyme data to 0.610, 0.099, 0.127, and 0.140 values for the branches of the lineage relating these two species. Branching and common ancestry are indicated, but the specific ancestors cannot be identified from these data.

Similarities between proteins can be estimated by immunological methods, in which a purified protein from a species is injected into a rabbit or another suitable mammal, and the antibodies produced against this protein are tested for cross-reactions with the same protein from species related to the animal under study. The degrees of dissimilarity between the protein used to immunize the rabbit and the protein being tested are measured by the relative strength of the cross-reaction and are expressed as **immunological distance** units (Box 9.1).

Very sensitive immunological methods are efficient when small amounts of purified protein are used in assays. One such technique is *microcomplement fixation*, in which the complement component of vertebrate serum is mixed with the protein of one species, as antigen, and antibodies produced in response to the same protein from a related species. Any complement that does not bind to the antigen–antibody preparation is free to lyse red blood cells later added to the mixture. The larger the number of lysed cells, which is measured spectrophotometrically, the larger the amount of free complement, which indicates a weaker cross-reaction between the antigens and the antibodies raised against related proteins. Immunological distances calculated by this method have been shown to be approximately proportional to the number of amino acid differences between related proteins. These values also provide a basis for estimating the time since the species diverged from their most recent common ancestor.

In a landmark study published in 1967 by Vincent Sarich and Allan Wilson, microcomplement fixation was used for the first time to evaluate differences in serum albumin among humans, apes, and monkeys. The dendrogram constructed from these data showed a close relationship between humans and the African great apes (gorilla and chimpanzee), and their greater, but equidistant, difference from the Asian orangutan (Fig. 9.20). This study provided some of the first molecular data suggesting a significantly earlier separation of the Asian apes from an ancestral form that diverged millions of years later to produce the African apes and hominids. In addition, the evolutionary time frame was compressed into a much shorter interval than had been accepted on the basis of morphological and paleontological studies.

An even more sensitive immunological method is *radioimmune assay* (RIA), which is widely used in medicine to measure small amounts of hormones, enzymes, and other proteins. Since 1980, Jerold Lowenstein has pioneered the use in phylogenetic studies of radioimmune assay for material from fossils as well as from extant species (Fig. 9.21). Tiny amounts of collagen from bones, teeth, or skin and of albumins obtained from muscle can be extracted from fossils that are less than 2 million years old, but occasionally older, and be assayed along with these proteins purified from living species. The iodine-

Box 9.1 Origin and Phylogenetic Relationships of Hawaiian Drosophilids Inferred from Immunological Assays

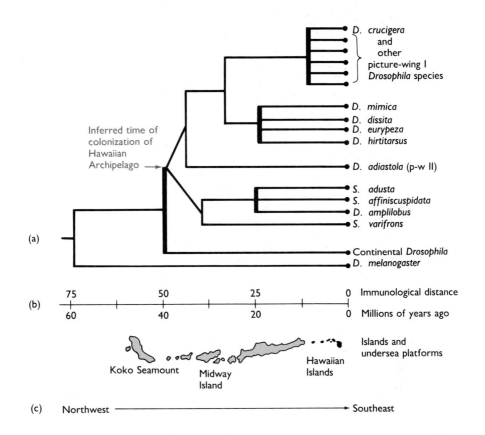

(a)

(b)

(c)

Hawaiian drosophilids (*Drosophila* and *Scaptomyza*) are (a) related to one another and to species on continental landmasses, according to immunological cross-reaction intensities determined by microcomplement fixation. The thick vertical bars denote divergence leading to multiple lineages, among which exact relationships are ambiguous. (b) Immu- nological distances, d_i, were calculated with the formula $d_i = 100 \times \log_{10} I.D.$, where I.D. is the index of dissimilarity between the members of a pair of spe- cies. The relation between d_i and time, t, since the divergence of two species is given by $t = cd_i$, where c is the propor- tionality constant. The proportionality constant varies with the protein studied, and $c = 8 \times 10^5$ for the larval hemo-lymph protein on which this study was based. This value is similar to the constant of 5.5 or 6×10^5 for serum albumin from mammals, which was used by Vincent Sarich and Allan Wilson and by others. The time of colonization of Pacific islands was inferred to be about 40 million years ago, which contradicts earlier inferences of 6 million years based on the age of Kauai in the Hawaiian Islands. (c) The rationalization of these different dates inferred for

colonization comes from the recognition that the present-day Hawaiian Islands (black) are the newest and easternmost in a chain of islands and undersea platforms stretching from southeast to northwest across the Pacific Ocean. The age of the whole system conforms to the earlier colonization time inferred from the immunological studies, and has a direct bearing on times of colonization of other endemic residents of present-day Hawaii (see Chapter 8).

Adapted from S. M. Beverley and A. C. Wilson, 1985. Ancient origin for Hawaiian Drosophilinae inferred from protein comparisons. *Proc. Nat. Acad. Sci. U.S.* 82:4753, Fig. 1.

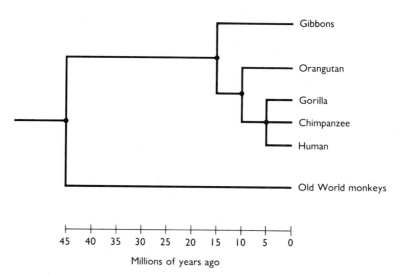

Figure 9.20. Dendrogram constructed from measured differences in serum albumin among higher primates, determined from microcomplement fixation assays by Vincent Sarich and Allan Wilson in 1967. The time of 5 million years since divergence of African apes and the human lineage was much more recent than had been estimated by paleontologists from fossils.

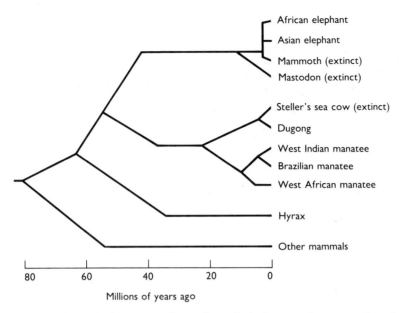

Figure 9.21. Phylogenetic relationships of elephants and sea cows based on immunological data obtained by radioimmune assays of tiny amounts of collagen from both fossil and living species. Immunological distances were translated into times since divergence, according to an equivalency formula (see Box 9.1). (From J. M. Lowenstein, 1983, Molecular approaches to the identification of species. *Amer. Sci.* 73:541, Fig. 3)

125 radioactive label that is bound to antigen–antibody complexes is measured to determine the strength of the immunological binding reaction. The closer the similarity between antiserums and test protein, the stronger the binding of iodine-125, as shown by higher radioactivity counts in the preparation.

In one of his studies, Lowenstein analyzed the phylogenetic relationship of the extinct Tasmanian wolf to South American and Australian marsupials. In contrast to paleontologists' conclusions, a cladistic analysis of dental resemblances suggested that the Tasmanian wolf of Australia was more closely related to marsupial species of a South American lineage than to other Australian marsupial carnivores. Lowenstein attempted to resolve the controversy by radioimmune assay analysis. Using albumin from skin and muscle of museum specimens of the Tasmanian wolf, the last of which had died in a zoo in 1933, Lowenstein found the wolf to be very closely related to two extant Australian marsupial carnivore species (Fig. 9.22). The small immunological distance among these three species implied that their lineages had diverged only 6 to 10 million years ago. These results suggested that the similarity in tooth structure between the Tasmanian wolf and extinct South American species was due to convergent evolution, not to genetic relatedness.

Estimates of phylogenetic relationships based on protein differences can be made more directly and in greater detail by the analysis of the amino acid

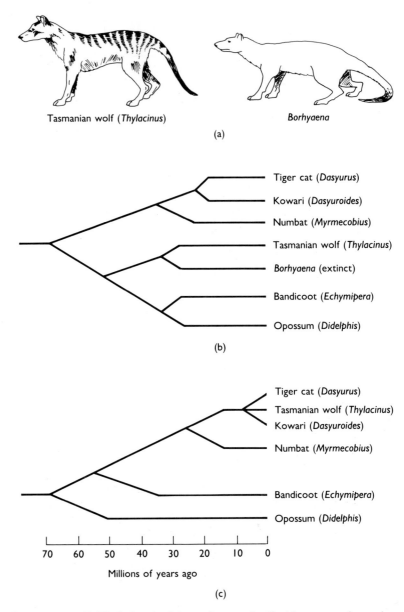

Figure 9.22. (a) The Tasmanian wolf (*Thylacinas*) of Australia was classified by most paleontologists as a closer relative of other Australian marsupial carnivores than of the extinct South American marsupial carnivore *Borhyaena*. (b) A cladistic analysis of dental resemblances, however, placed the Tasmanian wolf closer to *Borhyaena* than to Australian species. (c) Radioimmune assays (RIA) clearly showed that the Tasmanian wolf was more closely related to other Australian species. The analysis would have been more complete with data from *Borhyaena*, which was not available. (From J. M. Lowenstein, 1983, Molecular approaches to the identification of species. *Amer. Sci.* 73:541, Fig. 5)

sequences than by electrophoretic or immunological methods. Changes in the amino acid sequence can be related to changes in codons of the gene sequence, mostly by substitutions of nucleotides but also by additions and deletions of bases or entire codons. Differences in amino acids are translated into units of nucleotide substitutions, usually called the **mutation distances,** according to minimum estimates based on the standard genetic code. Silent changes leading to synonymous codons are therefore excluded, as are reverse mutations from a derived codon back to its original form, because they are undetectable in polypeptides.

Although comparisons are made of homologous proteins (or genes), they may be either orthologous or paralogous. **Orthologous genes** are descendants of an ancestral gene that was present in the last common ancestor of two or more species, whereas **paralogous genes** are descendants of a duplicated ancestral gene. Both kinds of homologous genes (or proteins) can be used to estimate phylogenetic relationships among organisms and to construct phylogenies of the genes themselves.

Comparisons of orthologous proteins in different species provide data for the construction of divergence dendrograms showing both the gene phylogeny and the estimated phylogenetic relationships among the species producing these proteins. The minimum mutation distances that are calculated from alignments of the same protein type in the different species are the basis for a possible dendrogram. This dendrogram is analyzed by the maximum parsimony or an equivalent method to reconstruct the ancestral sequence that gives the fewest mutations over the tree. Composite trees can be constructed from the separate trees for individual orthologous proteins, as shown in Figure 9.23 for globins from soybean, invertebrates, and jawless lamprey (a vertebrate); myoglobin from assorted mammals; β globin from frog, chicken, and various mammals; and α globin from fish, chicken, and representative mammals. This dendrogram also indicates the evolutionary history of divergence among paralogous globin genes, although molecular biologists are more sure about the place of myoglobin, α globin, and β globin as duplicate descendants of an ancient globin (see Chapter 3).

Paralogous proteins are compared in any *one* species to assess the differences that have accumulated since the gene duplicated (see Fig. 3.24). Differences among paralogous proteins in *different* species provide data to estimate the evolutionary history of organisms, as in comparisons of the β-globin gene cluster in primates and selected mammalian outgroups (Fig. 9.24). These data may also be used to construct the phylogenetic history of divergence of the genes themselves, by including time references from the fossil record and from comparative studies of the morphology and other features of living species.

Hemoglobins and other relatively new proteins provide information about closely related species because the genes have been evolving for a relatively short time; vertebrate hemoglobins have existed for only about 500 million years. In addition, amino acid substitutions have been taking place at a reasonably rapid rate, so species can be compared that may have diverged a few million years ago. Orthologous proteins that change slowly are less useful in

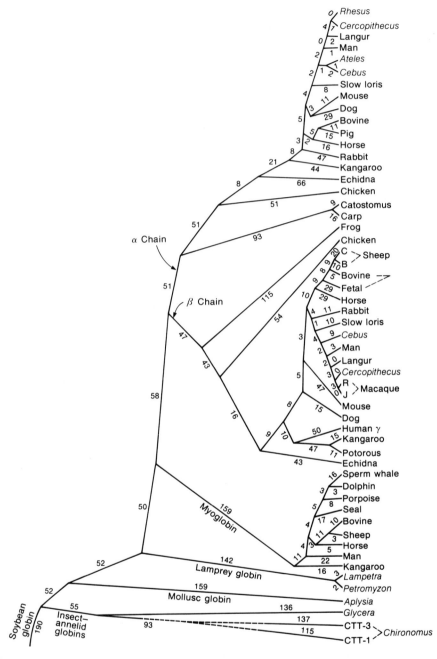

Figure 9.23. Phylogeny of the globin genes and estimated phylogenetic relationships among 55 species producing these polypeptides. The numbers on the branches are the estimated number of nucleotide substitutions during evolution, based on the amino acid sequences. (Adapted from M. Goodman, 1976, Protein sequences in phylogeny, in *Molecular Evolution,* ed. F. J. Ayala, pp. 141–159, Fig. 6. Sunderland, Mass.: Sinauer)

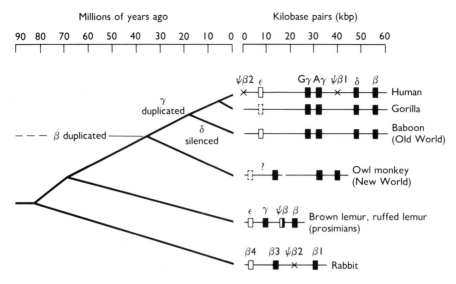

Millions of years ago   Kilobase pairs (kbp)

90 80 70 60 50 40 30 20 10 0 0 10 20 30 40 50 60

Figure 9.24. Changes in the cluster of β-like globin genes in primates and the rabbit outgroup during the past 90 million years. Duplicate β-globin genes were present at least 85 million years ago in the more ancient rabbit lineage. The prosimians, the earliest of the primates, had already undergone evolutionary differentiation of duplicate β-globin genes, as is evident by the presence of the embryonic ε-globin and fetal γ-globin genes and a β-globin gene in a stretch of DNA about 25 kilobase pairs (kbp) long. The New World monkeys have added another β-globin gene duplicate in DNA about 45 kbp long. The closely related Old World primates (monkeys, apes, humans) have virtually identical sets of β-globin and β-like genes spread over a DNA segment about 60 kbp long. Pseudogenes (prefixed by ψ), or inactive duplicates of functional genes, are marked by x in each species. Despite divergence in morphology and base sequences, the organization of the β-like gene cluster has remained unchanged during 35 to 40 million years of evolution of the Old World monkey, ape, and human lineages. (Adapted from A. J. Jeffreys, 1982, Evolution of globin genes, in *Genome Evolution*, ed. G. A. Dover and R. B. Flavell, pp. 157–176, Fig. 1. New York: Academic Press)

comparing recently evolved species because they may have changed little or not at all over tens of millions of years. They are particularly useful, however, in estimating relationships among more ancient lineages in which homologies remain evident after 1 billion years or more of divergent evolution. Studies of cytochrome *c*, the mitochondrial respiratory enzyme, exemplify these points.

From a matrix of the minimum number of amino acid–altering nucleotide substitutions inferred from the amino acid sequences of cytochrome *c* in 20 different organisms (Table 9.2), Walter Fitch and Emanuel Margoliash constructed the most likely divergence dendrogram reflecting phylogenetic relationships among these organisms. This landmark study, conducted in 1967, provided information for species as remotely related as fungi and human beings. The phylogeny agreed fairly well with relationships inferred from the fossil record and other sources, which was remarkable in view of the wide range of organisms studied and the use of only one protein for analysis (Fig. 9.25). Only one amino acid difference characterizes humans and rhesus mon-

TABLE 9.2
Minimum Number of Nucleotide Substitutions Required to Interrelate Amino Acid Sequences in Pairs of Cytochrome c*

Organism	Protein	1	2	3	4	5	6	7	8	9	10	11	12	13	14	15	16	17	18	19	20
Human	1	—																			
Monkey	2	1	—																		
Dog	3	13	12	—																	
Horse	4	17	16	10	—																
Donkey	5	16	15	8	1	—															
Pig	6	13	12	4	5	4	—														
Rabbit	7	12	11	6	11	10	6	—													
Kangaroo	8	12	13	7	11	12	7	7	—												
Duck	9	17	16	12	16	15	13	10	14	—											
Pigeon	10	16	15	12	16	15	13	8	14	3	—										
Chicken	11	18	17	14	16	15	13	11	15	3	4	—									
Penguin	12	18	17	14	17	16	14	11	13	3	4	2	—								
Turtle	13	19	18	13	16	15	13	11	14	7	8	8	8	—							
Snake	14	20	21	30	32	31	30	25	30	24	24	28	28	30	—						
Tuna	15	31	32	29	27	26	25	26	27	26	27	26	27	27	38	—					
Screwworm fly	16	33	32	24	24	25	26	23	26	25	26	26	28	30	40	34	—				
Moth	17	36	35	28	33	32	31	29	31	29	30	31	30	33	41	41	16	—			
Neurospora	18	63	62	64	64	64	64	62	66	61	59	61	62	65	61	72	58	59	—		
Saccharo myces	19	56	57	61	60	59	59	59	58	62	62	62	61	64	61	66	63	60	57	—	
Candida	20	66	65	66	68	67	67	67	68	66	66	66	65	67	69	69	65	61	61	41	—

*These mutation distance values were inferred from the amino acid sequences of the protein.
Source: Adapted from W. M. Fitch and E. Margoliash, 1967. Construction of phylogenetic trees. Science 155:279, Table 3. Copyright © 1967 by the American Association for the Advancement of Science.

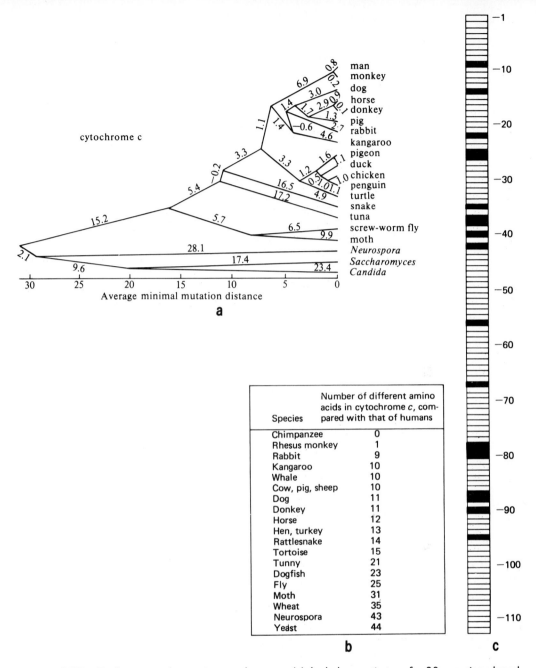

Figure 9.25. Evolutionary changes in cytochrome c. (a) A phylogenetic tree for 20 organisms based on amino acid differences in the protein. The estimated number of nucleotide substitutions that have resulted in amino acid differences during evolution are indicated on the branches of the tree. (From W. M. Fitch and E. Margoliash, 1967, Construction of phylogenetic trees. *Science* 155:279, Fig. 2. Copyright © 1967 by the American Association for the Advancement of Science) (b) Amino acid differences distinguish cytochrome c in humans from that in many, but not all, other species. The more distant the relationship, the greater the number of amino acid differences between species. (c) Diagrammatic representation of the 112 amino acids of the human cytochrome c sequence, showing in black the amino acid sites that are identical in every eukaryote examined thus far.

keys, which diverged at least 30 million years ago from the last common ancestor of the two lineages. There were no differences at all in the amino acid sequence of cytochrome c in humans and chimpanzees, whose lineages may have diverged only 5 to 7 million years ago.

All these and other methods for analyzing proteins rather than DNA were essential before techniques became available for studying the genetic material directly. These approaches remain very useful, and the data they provide can be compared with information obtained independently from DNA homologies determined directly from nucleotide sequences or indirectly by DNA–DNA hybridization methods. Before proceeding to studies of DNA in phylogenetic analysis, we will discuss one of the important features of any molecular analysis—the determination of divergence times among lineages, or molecular clocks of evolution.

Molecular Clocks of Evolution

The concept of a **molecular evolutionary clock** derived from protein phylogenies was first presented in 1962 by Emile Zuckerkandl and Linus Pauling, but it did not gain much momentum until Motoo Kimura proposed his neutral theory of molecular evolution. If allelic substitutions occur at a constant rate, as is claimed for neutral codon changes specifying neutral amino acid substitutions, the rate of evolutionary changes in genes and gene products might approach a constant value and provide a useful base for estimating divergence times. The controversies over these premises have continued to the present day, particularly by evolutionists who emphasize selective forces rather than stochastic mutation and genetic drift as the processes causing change. Many evolutionists infer, however, that the averaging out of selection coefficients and of different rates of substitution at different loci may produce at least apparently constant rates of molecular change when spread over millions of years or generations. A large body of theoretical and computer-tested mathematical formulations by Masatoshi Nei and others has pointed out some of the inherent errors in calculating molecular clocks, and some of the means by which the constancy of a clock may be evaluated. In essence, the expected rate of change is relatively constant, although variations do occur, and molecular clocks prove to be fairly accurate.

Each protein or gene generates its own clock; that is, the rate of mutation to neutral alleles varies from one gene to another (see Table 6.1). By pooling data for several genes (or proteins), the independent estimates of phylogenetic events and their timing would be combined into a more precise and accurate evolutionary clock. Before such a molecular clock can be applied to phylogenetic analysis, its constancy must be determined. Two general procedures are available, one of which requires calibration by reference to geologic or paleontological data (Table 9.3), and the other, the relative rate test, is independent of fossil data.

TABLE 9.3
Times of Divergence for Lineages of Various Organisms and Times of Geologic Events that Have Been Used by Various Investigators to Calibrate Molecular Clocks

	Time (Myr ago)
Lineage divergence	
plant/animal	1000–1200
unicellular/multicellular animals	700
invertebrate/vertebrate animals	500
Echinidae/Strongylocentrotidae sea urchins	65
mouse/rat	10–25
goat/sheep	5–7
goat/cow	12–25
human/orangutan	13–17
New World/Old World monkeys	35–45
monocotyledonous/dicotyledonous plants	100–200
corn/barley	50
Continent separation	
Africa/South America	60–80
Australia/New Zealand	60–80
Island formation	
Kauai	6
Hawaii	0.75
Galápagos	0.5 to 4–5

The **calibration-dependent test** for the constancy of a molecular clock was suggested in 1974 by C. H. Langley and Walter Fitch. They analyzed the amino acid sequences of 4 proteins from 18 vertebrate species and plotted substitutions of nucleotides (inferred from amino acid sequences) against times of divergence estimated from the fossil record. They found in this study and in later studies that the molecular clock is erratic, but these variations are eliminated if the data are averaged for a large number of events over a long period of geologic time.

The **relative rate test** for a constant rate of divergence does depend not on paleontological data and is therefore particularly important for organisms that are poorly represented in the fossil record. The test, first suggested by Sarich and Wilson in their 1967 study, compares any three species of which two are known to be more closely related (ingroups) than either one is to the third (outgroup). If the genetic distance, as determined by any of the molecular methods, is the same between the outgroup and each of the ingroup species, then the two ingroup species must have changed at the same average rate (Fig. 9.26). Although some sequences have evolved at a slower rate than others, the combinations of fast- and slow-evolving proteins or genes produce an average rate that is linear with time for most species.

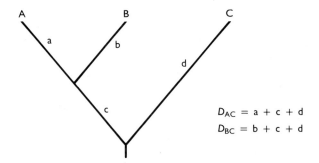

$$D_{AC} = a + c + d$$
$$D_{BC} = b + c + d$$

Figure 9.26. The relative rate test for constant rate of divergence (uniformity of the molecular clock). Rate constancy is indicated if the genetic distance (D) is the same between each ingroup species (A and B) and their common outgroup species (C).

Mutations are believed to occur at a relatively constant rate *per generation,* for two reasons. First, when mutation rates are adjusted for different generation times in organisms, the overall rates are equivalent. Second, if mutation rates were based on chronological time, organisms with a shorter generation time would evolve faster than those with a longer generation time, and there is no evidence that this is true. It is therefore difficult to understand why the molecular clock runs on chronological time rather than generation time. In view of this apparent discrepancy with genetic expectations, some doubts have been raised about the reality of the uniform-rate clock of molecular change.

The resolution of this problem is important for understanding the basic premises of molecular evolution and for evaluating Kimura's neutral theory of molecular evolution, which receives some of its strongest support from the constancy of the molecular clock in chronological time. Some of these matters will be discussed in the next sections, which are concerned with phylogenetic studies based on the analysis of DNA rather than proteins.

Phylogenies Based on DNA Analysis

The estimates of genetic relatedness based on the similarity or dissimilarity of phenotypic characters and on the changes in individual protein products of gene action are indirect assessments of changes that have taken place in nucleotide sequences of whole genomes or of individual gene representatives of a genome. The invention of the DNA–DNA hybridization method by Bernard Hall and Sol Spiegelman in 1961 opened the way to analyzing DNA homologies more directly. Current methods for mapping genes and genomes by restriction enzyme cut sites and for comparing DNA sequences, both of which became available in the 1970s, led to even more independent and specific analytical approaches to determining genetic changes directly at the DNA level. Modern studies of molecular phylogenies rely greatly on all these methods, in which DNA itself is the focus of analysis.

Comparing Genomes by DNA–DNA Hybridization

The basic principle of **DNA–DNA hybridization** is that single strands or single-stranded regions of DNA form stable duplexes only by hydrogen bonding between the complementary bases: guanine with cytosine and thymine with adenine. The better the match between successive complementary bases in the two strands, the higher the temperature required to melt the duplex into its single-stranded components, because more energy is needed to break many hydrogen bonds than to break few bonds. The genetic relatedness of DNAs from different species can thus be determined by allowing the formation of **heteroduplexes** (strands from different species) and then measuring their melting curves over a graduated series of temperatures (Fig. 9.27). The melting of heteroduplexes is compared with the melting of homoduplex control molecules (both strands from one species), to derive the percentage of DNA that melted at each temperature. Specifically, comparisons are made by reference to the temperature at which half of the hybrid or of the control DNA melted (**median melting temperature,** or $T_{50}H$). The lower the $T_{50}H$ of a DNA heteroduplex, the poorer the match between the two DNA strands, and the more distantly related the species.

Charles Sibley and Jon Ahlquist have used this method to determine phylogenetic relationships among 168 of the 171 living families of birds, from about 1600 of the 9000 known extant species, and also among the hominoid primate lineages. We will focus on the bird phylogenies first because they are the most extensive studies available for organisms with a poor fossil record. The first requirement for the analysis was to see if DNA differences accumulated at the same rate in all the lineages, so the data could serve to indicate genetic distance among them. By the use of the relative rate test for all possible trios of species (two ingroup and one outgroup), the molecular clock was found to tick at the same average rate in all the bird lineages. Another important feature of these studies was to remove all repetitive DNA and analyze only unique-copy DNA, so spurious results ("noise") were minimized. Huge quantities of DNA were obtained from red blood cells, which remain nucleated at maturity in birds, and the unique-copy sequences in the 2 billion base pairs of a bird genome could then be analyzed.

To convert $T_{50}H$ from an indirect to a direct measure of divergence times, the $T_{50}H$ values were correlated with a dated geologic event that caused an ancestral species to be divided into separate lineages. For example, the common ancestor of the African ostriches and South American rheas ranged across the single landmass of Gondwana before the Atlantic Ocean began to open, when continental drift separated the African and South American continents. Geologic evidence indicates that the Atlantic Ocean became a barrier

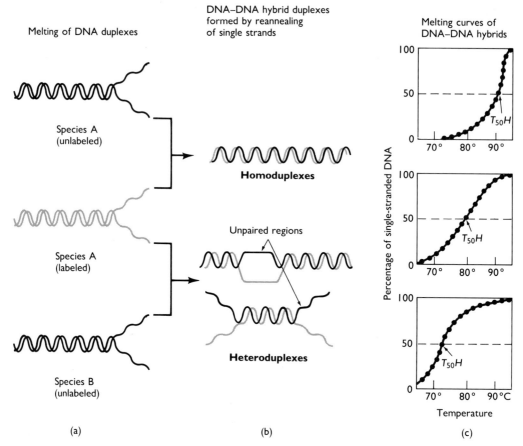

Figure 9.27. Phylogenetic analysis from DNA–DNA hybridization studies. (a) DNA duplexes are melted to single strands, which are (b) reannealed to form homoduplexes and heteroduplexes. A radioactively labeled strand serves as a marker that single strands from different species are present in the molecules to be analyzed. (c) Duplexes with unpaired or mispaired regions will melt to single strands at a lower temperature than is required to melt well-paired double-stranded DNAs. The temperature at which 50 percent of the DNA is single-stranded corresponds to the midpoint melting temperature, or $T_{50}H$. The lower the $T_{50}H$ value, the less homologous are the two strands of a reannealed duplex (less energy required to break fewer hydrogen bonds between strands).

to migration for flightless birds about 80 million years ago; the ostrich and rhea lineages must therefore have diverged at about this time. Dividing 80 million years by the difference in $T_{50}H$ of ostrich/rhea DNA heteroduplexes and $T_{50}H$ of ostrich/ostrich or rhea/rhea homoduplexes yields a calibration constant of millions of years of divergence per 1°C reduction in median melting temperature. From several geologic events and dated divergence between bird lineages, an average calibration constant of 4.5 was obtained. The molecular clock is thus calibrated for chronological time by the equivalence of a median reduction in $T_{50}H$ of 1°C to about 4.5 million years since the two lineages shared an ancestor (Fig. 9.28).

The phylogenetic histories reconstructed from DNA–DNA hybridization data are in agreement in the majority of cases with phylogenies inferred from morphological criteria. In some cases, however, DNA genealogies clearly reveal convergence, not genetic relatedness, as the basis for morphological similarities. For example, the Old World and New World vultures were grouped by some into the same order as other carrion-eating birds of prey, but were placed in separate groups by systematists who emphasized the greater similarities between storks and New World vultures (and condors) than between Old World and New World vultures. Phylogenies reconstructed from DNA comparisons revealed that storks and New World vultures are indeed each other's closest living relatives; convergence was therefore responsible for similarities in habits and appearances of New World and Old World vultures.

In addition to establishing phylogenetic patterns of genetic relatedness, Sib-

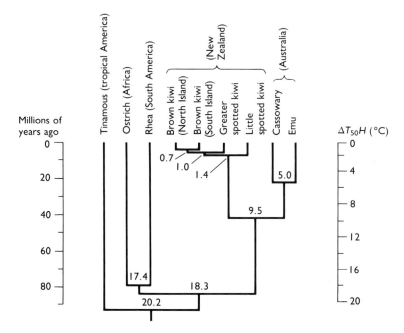

Figure 9.28. Phylogenetic relationships among flightless birds (ratites) from various parts of the world, as determined from DNA–DNA hybridization assays. The number at each point of divergence is the difference in $T_{50}H$ values between homoduplexes and heteroduplexes. The ingroups of ratites and the outgroup, tinamous (flying birds related to the ratites), share an ancestor that lived at least 90 million years ago. The molecular clock is calibrated for chronological time by the equivalence of a median reduction of 1°C in $T_{50}H$ to about 4.5 million years. (Based on studies by C. G. Sibley and J. E. Ahlquist)

ley and Ahlquist provided time frames for divergence that were unknown because of the very poor fossil record for birds. Patterns of migration from one area of residence to another, correlations of climate and changes in the geographical range of species, and other important evolutionary information have been obtained from basic data on genomic changes measured by DNA–DNA hybridization.

Sibley and Ahlquist's analysis of the hominoid primates by the use of DNA–DNA hybridization is one of many studies of phylogenetic relationships in this superfamily of organisms. Their $T_{50}H$ values for heteroduplexes of unique-copy nuclear DNAs indicated the following features for hominoid primate phylogeny:

1. The gibbons (small apes) were the first hominoids to diverge from an ancestor common to all the hominoid primates.
2. The Asian orangutan was the first of the great apes to diverge from an ancestor common to the great apes and human lineage.
3. Humans and chimpanzees are more closely related to each other than either is to the gorilla.
4. The molecular clock appears to tick at a constant rate in hominoid evolution (Fig. 9.29).

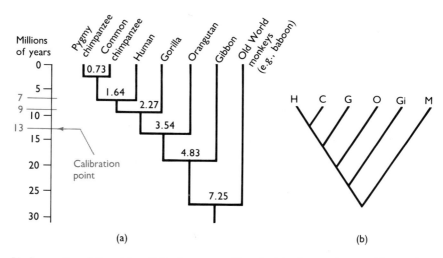

(a) (b)

Figure 9.29. Phylogenetic relationships of the ingroups of hominoid primates (ape and human lineages) and the outgroup of Old World monkeys, as determined from DNA–DNA hybridizations. (a) The sequence of divergence from a common ancestral Old World lineage, showing $\Delta T_{50}H$ value at each divergence point and the times of branching, determined from the molecular clock calibrated by orangutan lineage divergence about 13 million years ago. These studies showed an earlier divergence date of the gorilla lineage (9 million years ago) than of the human and chimpanzee lineages (7 million years ago) from their last common ancestor. (b) Cladogram of relationships based on data in (a), showing that the human and chimpanzee lineages are closer to each other than either is to the gorilla lineage. (Based on studies by C. G. Sibley and J. E. Ahlquist)

Most investigators agree with the first and second points, but opinions differ on the third and fourth. The close similarity among human, chimpanzee, and gorilla molecular and chromosomal features, which indicates their having branched close together in time, makes it difficult to resolve the precise sequence of divergences. The constancy of the molecular clock is at odds with some of the data from other molecular studies, which have shown a slowing rate of hominoid molecular change. We will discuss these problems shortly, after considering other data on molecular phylogenetic analysis.

Restriction Enzyme Analysis of DNAs

Restriction enzymes are endonucleases that break both strands of duplex DNA at specific sites, call *restriction sites,* in sequences of four to six nucleotide pairs that have twofold rotational symmetry (Table 9.4). For analysis, the broken pieces, called *restriction fragments,* are separated according to their lengths by migration in a gel electrophoretic system. Progressively shorter fragments move progressively faster through the gel, and thus the restriction fragments can be isolated and analyzed (Fig. 9.30). *Restriction maps* of a gene or DNA complement are constructed from the electrophoretic data and are compared to determine differences that reflect changes in the nucleotide sequences attacked by each restriction enzyme included in the analysis. If a particular fragment is present in the *Hpa*II digest of one species but not in that of another, we know that the first two nucleotides of the *Hpa*II restriction site are present in the first species but have been modified or are missing in the second species.

Restriction mapping is usually carried out for gene sequences or parts of mitochondrial, chloroplast, or nuclear DNA complements, which may encompass a few thousand nucleotide pairs. By mapping different regions of DNA and combining the individual maps, a large segment or even an entire genome may be reconstructed in some cases (Box 9.2). It is not practical to consider mapping very large DNA complements, because of the time and effort required, but it is conceivable for a genome as small as 16,000 nucleotide pairs, such as occurs in mammalian mitochondria. A nuclear genome consisting of 2 or 3 billion base pairs of DNA, as in birds and mammals, respectively, cannot easily be analyzed in its entirety by any other method than DNA–DNA hybridization.

Restriction mapping of mitochondrial DNA has been carried out by many investigators, and one of the most interesting observations from these studies is the 10-fold higher rate of molecular change in mitochondrial DNA than in nuclear DNA. The rapid pace of mitochondrial DNA evolution makes this genetic material especially suitable for phylogenetic analysis of closely related species, but less useful for species that diverged 80 or 100 million years ago, or earlier. For this reason, close attention has been given to analyzing the phylogenetic relationships of the hominoids by comparing their mitochondrial DNA subjected to restriction enzyme digestions. Unfortunately, contro-

TABLE 9.4
A Selection of Site-Specific Restriction Endonucleases Isolated from Bacteria

Enzyme	Bacterial Source	Restriction site*
EcoRI	*Escherichia coli* RY13	5′—G-A-A-T-T-C—3′ 3′—C-T-T-A-A-G—5′
EcoRII	*Escherichia coli* R245	5′—C-C-T-G-G—3′ 3′—G-G-A-C-C—5′
BamI	*Bacillus amyloliquefaciens*	5′—G-G-A-T-C-C—3′ 3′—C-C-T-A-G-G—5′
HpaII	*Hemophilus parainfluenza*	5′—C-C-G-G—3′ 3′—G-G-C-C—5′
HaeIII	*Hemophilus aegyptius*	5′—G-G-C-C—3′ 3′—C-C-G-G—5′
HindII	*Hemophilus influenza* Rd	5′—G-T-Py-Pu-A-C—3′ 3′—C-A-Pu-Py-T-G—5′
HindIII	*Hemophilus influenza* Rd	5′—A-A-G-C-T-T—3′ 3′—T-T-C-G-A-A—5′
PstI	*Providencia stuartii*	5′—C-T-G-C-A-G—3′ 3′—G-A-C-G-T-C—5′
SmaI	*Serratia marcescens*	5′—C-C-C-G-G-G—3′ 3′—G-G-G-C-C-C—5′
HhaI	*Hemophilus haemolyticus*	5′—G-C-G-C—3′ 3′—C-G-C-G—5′
BglII	*Bacillus globiggi*	5′—A-G-A-T-C-T—3′ 3′—T-C-T-A-G-A—5′

*All known restriction sites are four to six nucleotide pairs long and have a twofold rotational symmetry. Arrows indicate specific sites of cleavage on each strand of the duplex DNA.

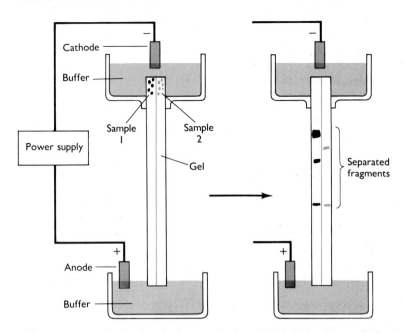

Figure 9.30. Gel electrophoresis of restriction fragments. A sample of restriction fragments is placed at the top of the gel column, and they migrate through the gel when power is applied to produce an electrical field from cathode (−) to anode (+) terminals in buffer. Smaller fragments move faster than larger fragments, and fragments of different size (length) come to rest in different parts of the gel when the power supply is turned off. The separated fragment populations can be analyzed by restriction mapping (see Box 9.2) or base sequencing (see Box 9.3), or can be used for other purposes.

versial interpretations of these data have not resolved the problem of the human/chimpanzee/gorilla cluster (Fig. 9.31).

A particular problem encountered in comparing restriction sites is the relatively widespread occurrence of *restriction fragment length polymorphisms* within each species. In mammals, two individuals picked at random are expected to possess mitochondrial DNAs that differ by 1 to 10 percent in nucleotide sequence. In many studies, one or a very few individuals are analyzed as representatives of a species. Large amounts of mitochondrial DNA may be obtained from fresh or frozen organs or from cloned DNA segments, but the material may still have come from one individual. Errors may be introduced because of such inadequate samples and because proper attention may not be paid to the subspecies or geographical population from which the individual came. Orangutans from Borneo and Sumatra differ by 5 percent in mean nucleotide composition; subspecies of eastern and western common chimpanzees show distinctive restriction patterns of their mitochondrial DNAs; and comparisons between species may thus reflect larger or smaller nucleotide differences, depending on the sources of the individuals analyzed.

Box 9.2   Restriction Mapping of DNA

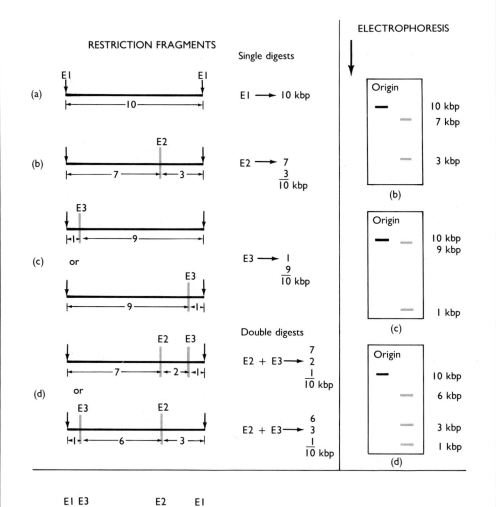

Physical mapping of genomic DNA can be achieved by cutting the DNA with various sequence-specific restriction enzymes and sizing the restriction fragments by gel electrophoresis. DNA molecules or fragments migrate through a gel, such as 1 to 2 percent agarose, under the influence of an applied electrical field. The smaller the molecule or fragment, the faster it migrates in the gel. Larger molecules thus come to rest nearer to the origin (place of application of the DNA sample), and smaller molecules come to rest in the gel at points farther from the origin. Each band of DNA should consist of a homogeneous group of molecules of a particular length in the original sample. These bands can be visualized by appropriate means, such as staining, ultraviolet optics, or radioactivity. The size of molecules or fragments in the set of bands in a gel is determined by reference to fragments or molecules of known size that are run in neighboring tracks or lanes in the gel. Pieces produced by digestion with restriction enzymes may be placed in the order in which they occurred in the original genomic DNA.

A genomic fragment 10 kilobase pairs (kdp) long, produced by restriction enzyme E1(a), is cut by another sequence-specific restriction enzyme (E2). The fragment lengths are determined by gel electrophoresis. As shown in (b), the two fragments measure 7 kbp and 3 kbp, thus showing that enzyme E2 cuts at only one site on the original 10-kbp DNA sequence. All other restriction sites are located with reference to this E2 site. Another restriction enzyme (E3) also cuts the 10-kbp genomic DNA at one specific site, thereby producing fragments that are 9 kbp and 1 kbp long, as seen in (c). In order to determine the E3 site relative to the E2 restriction site, double enzyme digests are prepared and run, with results shown in (d). If the E3 site is to the right of E2, there will be no change in the E2 7-kbp fragment, but the E2 3-kbp piece will be cut by E3 into 2-kbp plus 1-kbp fragments. If, however, the E3 site is to the left of the E2 restriction site, the E2 3-kbp fragment will be unchanged, but the E2 7-kbp fragment will be cut by E3 into 1-kbp plus 6-kbp pieces. Electrophoresis of the double digests will reveal which of the two alternatives is correct. Once the E3 site is located with reference to the E2 site, these restriction sites can be placed in correct order on the physical map of the genomic fragment.

These variations would also be reflected in the calculated divergence times, which are based on sequence divergence between populations or pairs of individuals.

Restriction mapping to determine nucleotide sequence divergence does provide an important shortcut in estimating phylogenetic relationships from DNA data, compared with direct nucleotide sequencing of genes or genomes.

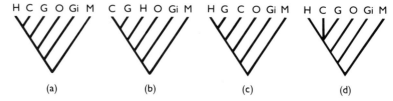

Figure 9.31. Cladograms showing four interpretations of higher primate phylogeny, based on various studies. The interpretations vary only with regard to the human/chimpanzee/gorilla cluster. (a) Human and chimpanzee are more closely related to each other than either is to the gorilla; (b) chimpanzee and gorilla are more closely related to each other than either is to the human; (c) human and gorilla are more closely related to each other than either is to the chimpanzee; and (d) human, chimpanzee, and gorilla lineages are equally related to one another, which implies a three-way split from a last common ancestor as the most recent major divergence event in primate evolution.

But, as with any method, care must be taken to obtain data from adequate samples that reflect the species variability to the greatest degree possible.

Direct Comparisons of Nucleotide Sequences

The most direct and detailed data on nucleotide variation have come from comparisons of individual gene sequences. They can be cloned to obtain large, pure samples for standard gel electrophoretic sequencing, such as the Maxam-Gilbert method (Box 9.3), or sequences from databanks can be compared when available. Direct analysis of gene sequences permits the recognition of substitutions leading to synonymous codons (silent changes) and to codons specifying different amino acids (replacements), and of additions and deletions in sequences. Furthermore, substitutions can be compared in introns and exons and in the transcribed and untranscribed 3' and 5' flanking regions of a gene sequence. Many of these studies are directed toward an understanding of mechanisms of genetic change in evolution, but others concentrate on the phylogenetic implications of sequence divergence.

One of the interesting conclusions from DNA phylogenetic analysis is that the molecular clock ticks more slowly in anthropoid primates (monkeys, apes, and human) than in the more ancient prosimian primates and in other mammals (Fig. 9.32). The reasons for the slowdown in anthropoid molecular evolution are unclear, but various suggestions have been made. Roy Britten and others have proposed that a more efficient DNA-repair system may have evolved in the higher primates. If more nucleotide substitutions are erased by repair of the DNA sequence, fewer mutations would be established in species with more effective repair processes or other genetic modifications.

Another possible reason for the slowdown is that nucleotide substitution rates are slower in species with a longer generation time; that is, the clock

Box 9.3 The Maxam-Gilbert Method for DNA Sequencing

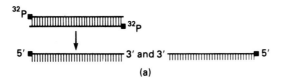

(a)

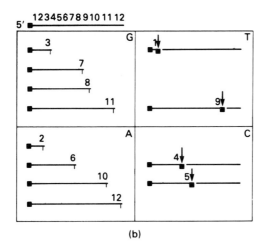

(b)

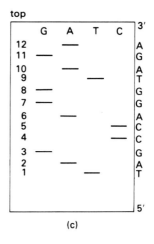

(c)

(a) A pure preparation of DNA, such as a population of a particular restriction fragment 12 base pairs long, is labeled with ^{32}P at the 5′ ends of both strands of the duplexes. Duplexes are separated into two fractions of complementary single strands, and one of these fractions is then sequenced. (b) The fraction of single strands is divided into four approximately equal portions, or aliquots, each of which is treated specifically to identify one of the four bases (G, A, T, or C). Each treatment causes breakage of the DNA strand at the site of the base in question, but the concentration of reagents is adjusted so that each strand has an average of one break. The break may occur by chance at any of the sites occupied by the base involved, and a break will occur at one or another of these sites in the many fragments making up the aliquot of DNA. The different lengths of the broken 12-base-long fragments are shown in panels for G and A; the site of a break in the 12-base-long fragments is indicated by an arrow in the panels for T and C. Each usable fragment retains its original 5′ end, as determined by auto-radiographic detection of the radioactive ^{32}P marker. Fragments lacking the 5′ end are not visible in the autoradiograph, and thus confusion is avoided. (c) Each of the four aliquots migrates in a separate lane in a gel that is subjected to an electrical field in a gel electrophoresis apparatus. Fragments migrate at rates proportional to their length, with the shorter fragments migrating more rapidly toward the bottom of the gel from the origin at the top. By comparing the autoradiographic patterns of the 12 different-sized fragments in the 4 gel lanes, the entire sequence of 12 bases can be read directly. Reading proceeds from the shortest to the longest pieces, going from the 5′ to the 3′ end of the sequence, from the bottom to the top of the gel.

may indeed run on generation time rather than on chronological time, as discussed earlier. In one of a number of their studies on this question, Wen-Hsiung Li and colleagues compared the rates of synonymous nucleotide substitutions in higher primates, rodents, and artiodactyls (even-toed ungulate mammals). The highest average rates were found in rodents (shortest generation time) and the lowest rates in primates (longest generation time). An intermediate rate in artiodactyls coincided with their intermediate generation time (Table 9.5). On the basis of these data, Li suggested that the molecular clock ticks more slowly in the human lineage compared with apes than human compared with Old World monkeys (1.1×10^{-9} versus 2.3×10^{-9}). In the same study, however, detailed comparisons among human, chimpanzee, gorilla, orangutan, Old World monkeys, and New World monkeys showed that the rates are 1.3 to 1.9 times faster in apes and monkeys than in humans. There appeared to be no faster rate in monkeys compared with humans than in apes compared with humans. Both sets of data do show that the human molecular clock has slowed down relative to rates of change in the other anthropoid primates.

Li's maximum parsimony tree for anthropoid primate phylogeny depicts a closer relationship between human and chimpanzee than between human and gorilla lineages. This tree shows the same branching pattern as that proposed by Sibley and Ahlquist for DNA–DNA hybridization data (see Fig. 9.29) and by other investigators based on immunological, protein sequence, and mitochondrial DNA comparisons. Although it might seem reasonable to accept this particular branching pattern based on a number of independent sets of data, the resolution of the divergence sequence in the human/chimpanzee/

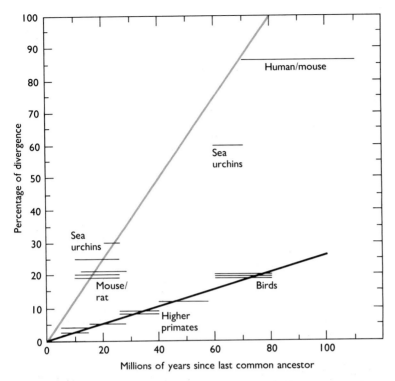

Figure 9.32. Evolutionary rates for various animal groups, based on many molecular phylogenetic studies, show a constant chronological rate of divergence in nucleotide sequences. The rates of change are calculated to be slower for higher primates and birds (black line) than for other animal groups (gray line). The length of the horizontal lines represents the estimated range in time of divergence; for example, 60 to 80 million years ago for pairs of ratite and some other bird species. (Adapted from R. J. Britten, 1986, Rates of DNA sequence evolution differ between taxonomic groups. *Science* 231:1393, Fig. 1)

gorilla cluster has not yet been achieved to everyone's satisfaction. A statistical analysis of the genetic distances between human/chimpanzee and human/gorilla branches, based on a particular mitochondrial DNA analysis, showed that the difference is statistically insignificant (Fig. 9.33). The tree could just as well have shown a three-way split or a closer relationship between human and gorilla than between human and chimpanzee (see Fig. 9.31).

The wealth of known sequences provides data for phylogenetic analysis of many organisms, including viruses, but particular attention has been paid to the higher primates. We are, of course, especially interested in our own evolutionary history, but an increasing number of studies will surely be concerned with many other species in future years. This research may well rely on DNA analysis more than on direct or indirect examination of proteins,

TABLE 9.5
Rates of Synonymous Nucleotide Substitutions Per Site Per Year in Primates, Rodents, and Artiodactyls

Species	Divergence Time (Myr ago)	Gene	Number of Sites	Percentage of Divergence	Rate ($\times 10^{-9}$)
Human/ chimpanzee	7 (5–10)*	$\alpha 1$ globin	108	2.8	2.0 (1.4–2.8)*
		$\alpha 2$ globin	108	0.0	0.0 (0.0–0.0)
		β globin	89	0.0	0.0 (0.0–0.0)
		$^G\gamma$ globin	101	3.0	2.1 (1.5–3.0)
		$^A\gamma$ globin	101	2.0	1.4 (1.0–2.0)
		Total	507	1.6	1.1 (0.8–1.6)
Human/ Old World monkeys	25 (20–30)	$\alpha 1$-antitrypsin	270	11.1	2.2 (1.9–2.8)
		Erythropoietin	145	11.2	2.2 (1.9–2.8)
		δ globin	104	10.4	2.0 (1.7–2.6)
		Insulin	84	18.6	3.7 (3.1–4.7)
		β globin (partial)	72	8.9	1.8 (1.5–2.2)
		Metallothionein II	35	9.1	1.8 (1.5–2.2)
		Total	710	11.6	2.3 (1.9–2.9)
Mouse/rat	15 (10–30)	Aldolase A	184	15.4	5.1 (2.6–7.7)
		Creatine kinase M	251	17.2	5.7 (2.9–8.6)
		Apolipoprotein E	201	17.4	5.8 (2.9–8.7)
		THY-1	116	19.3	6.4 (3.2–9.6)
		Metallothionein I	37	11.5	3.8 (1.9–5.7)
		LDH-A	219	30.9	10.3 (5.2–15.5)
		GPHA	58	30.8	10.3 (5.1–15.4)
		ANF	107	20.4	6.8 (3.4–10.2)
		Growth hormone	124	14.1	4.7 (2.4–7.1)
		Prolactin	127	21.9	7.3 (3.7–11.0)
		α-actin	249	17.9	6.0 (3.0–8.9)
		POMC	154	21.4	7.1 (3.6–10.7)
		Thyrotropin, β	90	25.7	8.6 (4.3–12.8)
		Alpha-fetoprotein	321	37.2	12.4 (6.3–18.9)
		Albumin	274	36.7	12.2 (6.1–18.3)
		Total	2511	23.7	7.9 (4.0–11.9)
Cow/goat	17 (12–25)	β globin	99	13.6	4.0 (2.7–5.7)
		γ globin	105	13.5	4.0 (2.7–5.7)

*Numbers in parentheses refer to the range of divergence time and the rates for these times.
Source: Adapted from W.-H. Li and M. Tanimura, 1987. The molecular clock runs more slowly in man than in apes and monkeys. *Nature* 326:93, Table 1.

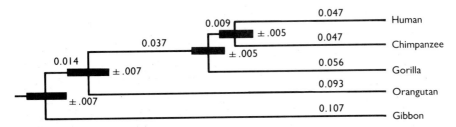

Figure 9.33. A statistical analysis of genetic distances among hominoid primate lineages reveals an insignificant level of difference values between the human/chimpanzee and the human/gorilla lineages. Accordingly, it is not possible to ascertain the actual topology of branching for the human/chimpanzee/gorilla cluster of lineages (see also Fig. 9.31). The evolutionary tree was based on mitochondrial DNA sequences. The numbers indicated on the branches represent the number of nucleotide substitutions per site, and are proportional to branch length. The black bars and the numbers next to them represent 1 standard error on each side of the mean branch length (± 1 S.E). (From M. Nei et al., 1985, Method for computing the standard errors of branching points in an evolutionary tree and their application to molecular data from humans and apes. *Mol. Biol. Evol.* 2:66, Fig. 3)

because DNA methods not only are widely available and fast, but also allow more detailed and accurate observations.

Summary

The reconstruction of phylogenies and the determination of patterns of evolutionary change are based on comparative studies of living species and on geologic and paleontological data. Different criteria to assess relationships are used by the three major schools of systematics: phylogenetics, phenetics, and cladistics. Different evolutionary trees may be constructed for the same groups of organisms, depending on the chosen criteria for relatedness. It is essential to distinguish between ancestral ("primitive") and derived ("advanced") character states, and between homologous features modified during divergence and analogous features produced by convergence. A more thorough knowledge of developmental biology and its genetic basis is important in assessing ancestral and derived character states for phylogenetic analysis.

An evolutionary tree represents a hypothesis of evolutionary history and includes both clades (shared abilities and organization) and grades (portions of a phylogeny with new abilities and organization). A large number of possible evolutionary trees can be constructed from a matrix of character states and combinations, and the "best" tree is selected by the methods of maximum parsimony, compatibility, or distance matrix. No one method is accepted by all evolutionary biologists. Molecular phylogenies produce measures of genetic distances from allozyme polymorphisms, immunological distances from microcomplement fixation or radioimmune assay data, and mutation distances from amino acid sequence variations. These measures, like morphological traits, can be used to construct evolutionary trees. Molecular clocks for particular genes or proteins vary, but are useful only if they "tick" at a constant rate. Constancy can be judged by the calibration-dependent test, in which nucleotide substitutions are plotted against estimated divergence times, or by the relative rate test, in which ingroups are measured against an outgroup. Even if it is constant, the rate of change may vary at different times in a group's history; the reasons for such variation are uncertain.

Phylogenetic analysis based on studies of DNA provide more direct and detailed information than can be obtained from proteins. Various methods are used in measuring variations in genic, genomic, or organelle DNAs. DNA–DNA hybridizations may permit analysis of whole genomes, and these data are used to construct evolutionary trees and determine molecular clocks of nucleotide changes. Restriction enzyme data can be used to produce restriction maps of polymorphic genes, gene clusters, or organelle DNAs, and such maps can be compared and be used for tree construction. The fast pace of change in mitochondrial DNA and the small size of the organelle genome make this a favorite material for study. Direct analysis of nucleotide sequences provides the most detailed information on nucleotide modifications, additions and deletions, and substitutions producing synonymous codons, new restriction enzyme cut sites, or codons specifying new amino acids. All these DNA studies produce evolutionary trees that compare favorably with trees constructed from morphological or protein data and that are generally compatible with the fossil record of evolutionary history.

References and Additional Readings

Avise, J. C., and C. F. Aquadro. 1982. A comparative summary of genetic distances in the vertebrates. *Evol. Biol.* 15:151.

Avise, J. C., and F. J. Ayala. 1975. Genetic differentiation in speciose versus depauperate phylads: evidence from the California minnows. *Evolution* 29:411.

Ayala, F. J., ed. 1976. *Molecular Evolution*. Sunderland, Mass.: Sinauer.

Ayala, F. J. 1986. On the virtues and pitfalls of the molecular evolutionary clock. *Jour. Hered.* 77:226.

Beverley, S. M., and A. C. Wilson. 1985. Ancient origin for Hawaiian Drosophilinae inferred from protein comparisons. *Proc. Natl. Acad. Sci. U.S.* 82:4753.

Britten, R. J. 1986. Rates of DNA sequence evolution differ between taxonomic groups. *Science* 231:1393.

Brown, W. M. 1983. Evolution of animal mitochondrial DNA. In *Evolution of Genes and Proteins*, ed. M. Nei and R. K. Koehn, p. 62. Sunderland, Mass.: Sinauer.

Brown, W. M., E. M. Prager, and A. C. Wilson. 1982. Mitochondrial DNA sequences of primates: tempo and mode of evolution. *Jour. Mol. Evol.* 18:225.

Buth, D. G. 1984. The application of electrophoretic data in systematic studies. *Ann. Rev. Ecol. Syst.* 15:501.

Cracraft, J. 1983. The significance of phylogenetic classifications for systematic and evolutionary biology. In *Numerical Taxonomy,* ed. J. Felsenstein, p. 1. Berlin: Springer-Verlag.

Dickerson, R. E. 1980. Evolution and gene transfer in purple photosynthetic bacteria. *Nature* 283:210.

Doolittle, R. F. 1981. Similar amino acid sequences: chance or common ancestry. *Science* 214:149.

Doolittle, R. F. 1985. The genealogy of some recently evolved vertebrate proteins. *Trends Biochem. Sci.* 10:233.

Farris, J. S. 1983. The logical basis of phylogenetic analysis. In *Advances in Cladistics,* vol. 2, p. 7. New York: Columbia University Press.

Felsenstein, J. 1982. Numerical methods for inferring evolutionary trees. *Quart. Rev. Biol.* 57:379.

Felsenstein, J. 1985. Phylogenies and the comparative method. *Amer. Nat.* 125:1.

Ferris, S. D., A. C. Wilson, and W. M. Brown. 1981. Evolutionary trees for apes and humans based on cleavage maps of mitochondrial DNA. *Proc. Natl. Acad. Sci. U.S.* 78:2432.

Field, K. G., et al. 1988. Molecular phylogeny of the animal kingdom. *Science* 239: 748.

Fitch, W. M. 1982. The challenges to Darwinism since the last centennial and the impact of molecular studies. *Evolution* 36:1133.

Fitch, W. M., and E. Margoliash. 1967. Construction of phylogenetic trees. *Science* 155:279.

Ghiselin, M. T. 1984. Narrow approaches to phylogeny: a review of nine books on cladism. In *Oxford Surveys in Evolutionary Biology,* ed. R. Dawkins and M. Ridley, vol. 1, p. 209. New York: Oxford University Press.

Gilbert, W. 1981. DNA sequencing and gene structure [Nobel lecture]. *Science* 214:1305.

Gojobori, T., W.-H. Li, and D. Graur. 1982. Patterns of nucleotide substitution in pseudogenes and functional genes. *Jour. Mol. Evol.* 18:360.

Goldman, D., P. R. Giri, and S. J. O'Brien. 1987. A molecular phylogeny of the hominoid primates as indicated by two-dimensional protein electrophoresis. *Proc. Natl. Acad. Sci. U.S.* 84:3307.

Goodman, M. 1976. Protein sequences in phylogeny. In *Molecular Evolution,* ed. F. J. Ayala, p. 141. Sunderland, Mass.: Sinauer.

Goodman, M. et al. 1983. Evidence on human origins from haemoglobins of African apes. *Nature* 303:546.

Goodman, M. et al. 1987. Globins: a case study in molecular phylogeny. *Cold Spring Harbor Sympos. Quant. Biol.* 52:875.

Gould, S. J. 1977. *Ontogeny and Phylogeny.* Cambridge, Mass.: Harvard University Press.

Gunderson, J. H. et al. 1987. Phylogenetic relationships between chlorophytes, chrysophytes, and oomycetes. *Proc. Natl. Acad. Sci. U.S.* 84:5823.

Hall, B. D., L. Haarr, and K. Kleppe. 1980. Development of the nitrocellulose filter technique for RNA–DNA hybridization. *Trends Biochem. Sci.* 5:254.

Hall, B. D., and S. Spiegelman. 1961. Sequence complementarity of T2-DNA and T2-specific RNA. *Proc. Natl. Acad. Sci. U.S.* 47:137.

Hennig, W. 1965. Phylogenetic systematics. *Ann. Rev. Entomol.* 10:97.

Hennig, W. 1979. *Phylogenetic Systematics.* Urbana: University of Illinois Press.

Higuchi, R. et al. 1984. DNA sequences from the quagga, an extinct member of the horse family. *Nature* 312:282.

Hori, H. et al. 1985. Evolution of green plants as deduced from 5S rRNA sequences. *Proc. Natl. Acad. Sci. U.S.* 82:820.

Jeffreys, A. J. 1982. Evolution of globin genes. In *Genome Evolution*, ed. G. A. Dover and R. B. Flavell, p. 157. New York: Academic Press.

Jukes, T. H. 1980. Silent nucleotide substitutions and the molecular evolutionary clock. *Science* 210:973.

Jukes, T. H., and J. L. King. 1979. Evolutionary nucleotide replacements in DNA. *Nature* 281:605.

Kemp, T. S. 1985. Models of diversity and phylogenetic reconstruction. In *Oxford Surveys in Evolutionary Biology*, ed. R. Dawkins and M. Ridley, vol. 2, p. 135. New York: Oxford University Press.

Kimura, M. 1980. A simple method for estimating evolutionary rates of base substitution through comparative studies of nucleotide sequences. *Jour. Mol. Evol.* 16:111.

Kimura, M. 1981. Estimation of evolutionary distances between homologous nucleotide sequences. *Proc. Natl. Acad. Sci. U.S.* 78:454.

King, J. L., and T. H. Jukes. 1969. Non-Darwinian evolution. *Science* 164:788.

Kurtén, B. 1969. Continental drift and evolution. *Sci. Amer.* 220(3):54.

Langley, C. H., and W. M. Fitch. 1974. An examination of the constancy of the rate of molecular evolution. *Jour. Mol. Evol.* 3:161.

Lewontin, R. C. 1978. Adaptation. *Sci. Amer.* 239(3):212.

Li, W.-H. 1983. Evolution of duplicate genes and pseudogenes. In *Evolution of Genes and Proteins*, ed. M. Nei and R. K. Koehn, p. 14. Sunderland, Mass.: Sinauer.

Li, W.-H., and M. Tanimura. 1987. The molecular clock runs more slowly in man than in apes and monkeys. *Nature* 326:93.

Li, W.-H., and C.-I. Wu. 1987. Rates of nucleotide substitution are evidently higher in rodents than in man. *Mol. Biol. Evol.* 4:74.

Li, W.-H., C.-I. Wu, and C.-C. Luo. 1985. A new method for estimating synonymous and nonsynonymous rates of nucleotide substitution considering the relative likelihood of nucleotide and codon changes. *Mol. Biol. Evol.* 2:150.

Lowenstein, J. M. 1985. Molecular approaches to the identification of species. *Amer. Sci.* 73:541.

Lowenstein, J. M, V. M. Sarich, and B. J. Richardson. 1981. Albumin systematics of the extinct mammoth and Tasmanian wolf. *Nature* 291:409.

Maxam, A. M., and W. Gilbert. 1977. A new method for sequencing DNA. *Proc. Natl. Acad, Sci. U.S.* 74:560.

Mayr, E. 1981. Biological classification: toward a synthesis of opposing methodologies. *Science* 214:510.

Nathans, D. 1979. Restriction endonucleases, simian virus 40, and the new genetics [Nobel lecture]. *Science* 206:903.

Nei, M. 1972. Genetic distance between populations. *Amer. Nat.* 106:283.

Nei, M. 1987. *Molecular Evolutionary Genetics*. New York: Columbia University Press.

Nei, M., J. C. Stephens, and N. Saitou. 1985. Method for computing the standard errors of branching points in an evolutionary tree and their application to molecular data from humans and apes. *Mol. Biol. Evol.* 2:66.

Nei, M., and F. Tajima. 1985. Evolutionary change of restriction cleavage sites and phylogenetic inference for man and apes. *Mol. Biol. Evol.* 2:189.

O'Brien, S. J. 1987. The ancestry of the giant panda. *Sci. Amer.* 257(5):102.

Pace, N. R. et al. 1986. Ribosomal RNA phylogeny and the primary line of evolutionary descent. *Cell* 45:325.

Patterson, C., ed. 1987. *Molecules and Morphology in Evolution.* New York: Cambridge University Press.

Prober, J. M. 1987. A system for rapid DNA sequencing with fluorescent chain-terminating dideoxyribonucleotides. *Science* 238:336.

Reeck, G. R. et al. 1987. Homology in proteins and nucleic acids: a terminology muddle and a way out of it. *Cell* 50: 667.

Romer, A. S. 1959. *The Vertebrate Body,* 4th ed. Philadelphia: Saunders.

Sakoyama, Y. et al. 1987. Nucleotide sequences of immunoglobulin ε genes of chimpanzee and orangutan: DNA molecular clock and hominoid evolution. *Proc. Natl. Acad. Sci. U.S.* 84:1080.

Sanger, F. 1981. Determination of nucleotide sequences in DNA [Nobel lecture]. *Science* 214:1205.

Sarich, V. M., and A. C. Wilson. 1966. Quantitative immunochemistry and the evolution of primate albumins: microcomplement fixation. *Science* 154:1563.

Sarich, V. M., and A. C. Wilson. 1967. Immunological time scale for hominid evolution. *Science* 158:1200.

Selander, R. K. 1982. Phylogeny. In *Perspectives on Evolution,* ed. R. Milkman, p. 32. Sunderland, Mass.: Sinauer.

Sibley, C. G., and J. E. Ahlquist. 1984. The phylogeny of the hominoid primates, as indicated by DNA–DNA hybridization. *Jour. Mol. Evol.* 20:2.

Sibley, C. G., and J. E. Ahlquist. 1986. Reconstructing bird phylogeny by comparing DNA's. *Sci. Amer.* 254(2):82.

Simpson, G. G. 1951. *Horses.* New York: Oxford University Press.

Smith, H. O. 1979. Nucleotide sequence specificity of restriction endonucleases [Nobel lecture]. *Science* 205:455.

Sogin, M. L., H. J. Elwood, and J. H. Gunderson. 1986. Evolutionary diversity of eukaryotic small-subunit rRNA genes. *Proc. Natl. Acad. Sci. U.S.* 83:1383.

Spiegelman, S. 1964. Hybrid nucleic acids. *Sci. Amer.* 210(5):48.

Steiner, D. F. et al. 1985. Structure and evolution of the insulin gene. *Ann. Rev. Genet.* 19:463.

Steward, C.-B., J. W. Schilling, and A. C. Wilson. 1987. Adaptive evolution in the stomach lysozymes of foregut fermenters. *Nature* 330:401.

Templeton, A. R. 1983. Phylogenetic inference from restriction endonuclease cleavage site maps with particular reference to the evolution of humans and the apes. *Evolution* 37:221.

Templeton, A. R. 1985. The phylogeny of the hominoid primates: a statistical analysis of the DNA–DNA hybridization data. *Mol. Biol. Evol.* 2:420.

Vawter, L., and W. M. Brown. 1986. Nuclear and mitochondrial DNA comparisons reveal extreme rate variations in the molecular clock. *Science* 234:194.

Villaneuva, E. et al. 1985. Phylogenetic origins of the plant mitochondrion based on a comparative analysis of 5S ribosomal RNA. *Jour. Mol. Evol.* 22:46.

Wake, D. B. 1966. Comparative osteology and evolution of the lungless salamanders, family Plethodontidae. *Mem. So. Calif. Acad. Sci.* 4:1.

Wilson, A. C. 1985. The molecular basis of evolution. *Sci. Amer.* 253(4):164.

Wilson, A. C., H. Ochman, and E. M. Prager. 1987. Molecular time scale for evolution. *Trends Genet.* 3:241.

Wu, C.-I., and W.-H. Li. 1985. Evidence for higher rates of nucleotide substitution in rodents than in man. *Proc. Natl. Acad. Sci. U.S.* 82:1741.

Yalow, R. S. 1978. Radioimmuneassay: a probe for the fine structure of biologic systems [Nobel lecture]. *Science* 200:1236.

Zuckerkandl, E., and L. Pauling. 1962. In *Horizons in Biochemistry*, ed. M. Kasha and B. Pullman, p. 189. New York: Academic Press.

Zuckerkandl, E., and L. Pauling. 1965. Evolutionary divergence and convergence in proteins. In *Evolving Genes and Proteins*, ed. V. Bryson and H. J. Vogel, p. 97. New York: Academic Press.

CHAPTER 10

The Timetable of Biological History

The geologic and fossil records provide irrefutable evidence for the time dimension of biological history and for continued diversification of life forms over billions of years. Even if there were no traces of former life, we would feel confident that biological evolution had occurred because the evidence from comparative studies of molecular biology, genetics, biochemistry, physiology, anatomy, and morphology clearly show varied degrees of relationship within and among living groups of organisms. These relationships are more likely to be due to ancestry and descent than to random combinations of characters. In accord with undeniable evidence of genetic relatedness among living species, we would expect fossil life forms to show similar patterns of descent with modification and fossil and living forms to be bound by genetic ties.

The Fossil Record

The fossil record of past life and the geologic record of past events in the history of our planet are intertwined. Fossils are preserved in the rocks, and any proper interpretation of ancient life depends on dating these rock formations and understanding the effects of geologic events on organisms while they lived and after they were deposited. With a timetable of biological and geologic occurrences, we can piece together the overall history of life and discern patterns and trends of evolutionary change from an expanded data base.

Geologic Divisions of Time

The geologic timetable is divided into three major intervals called **eons,** which are large blocks of time separated from each other by major geologic and biological events and discontinuities. The **Archean Eon** and **Proterozoic Eon** encompass the first 4000 million years (Myr), with the boundary between them set as 2500 Myr ago. The **Phanerozoic Eon** began 590 Myr ago, and it is sharply distinguished from the two earlier eons by the presence of many different multicellular life forms, including representatives of all the modern phyla of animals.

When the early geologists examined rocks of progressively younger age, they discovered the discontinuity between rock formations with conspicuous and diverse fossils and older rocks without these remains. The oldest fossiliferous rocks were designated as Cambrian in age, and all the preceding non-fossiliferous rocks became known as **Precambrian** formations. The Precambrian interval is formally divided into the Archean and Proterozoic eons, but the informal designation of Precambrian time is very convenient and widely used, especially by biologists.

There is little detailed geologic or fossil information for the Precambrian eons to lead geologists to subdivide them into smaller intervals, with one exception. The recent discovery of an unusual fauna of multicellular organisms in deposits of the last 100 Myr of Precambrian time is of such significance that the interval has been designed by some as the **Ediacaran Period,** in recognition of the Ediacara location in southern Australia where these fossils were first found. The Phanerozoic Eon, however, is rich in evidence for geologic and biological discontinuities and major events, and it is subdivided on the basis of these features into three primary intervals called **eras.** The **Paleozoic Era** is the interval of "old life"; the **Mesozoic Era,** of "middle life"; and the **Cenozoic Era,** of "modern life," in which we live. All three eras include subdivisions of time designated as **periods,** and the extensive deposits and formations of the Cenozoic period permit their further subdivision into **epochs** (Fig. 10.1).

The three eras of the Phanerozoic are separated by gaps following major mountain-building episodes, but primarily by significant changes in life forms (Table 10.1). The Paleozoic Era ended with upthrusts of the Appalachian Mountains in the United States and the Ural Mountains at the Eurasian border in the Soviet Union, and with a major biological extinction event that included more than 90 percent of the Paleozoic species. The reptiles became the dominant terrestrial animals in the Mesozoic Era, the so-called **Age of Reptiles.** The era ended with the pole-to-pole building of the Rocky Mountains in North America and the rising of the Andes Mountains in South America, as well as the sudden extinction of the dinosaurs and many other groups of animals. The uplift of the Alps and the Himalayas and last upthrusts of the Rocky Mountains took place just before the start of the Pleistocene Epoch of the Cenozoic Era, but the era is particularly marked by the dominance on land

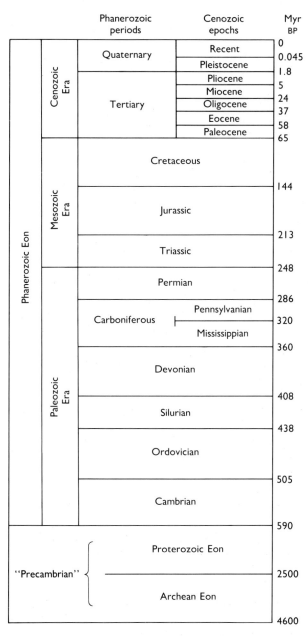

Figure 10.1. The geologic intervals of eons, eras, periods, and epochs and the approximate time each interval began (and the previous one ended). Only the periods of the Cenozoic Era are subdivided into epochs.

TABLE 10.1
Distinctive Biological Features of the Phanerozoic Eon

Millions of Years Ago*	Geologic Interval			Feature
	Era	Period	Epoch	
0.045			Recent	Modern humans worldwide in distribution
1.8		Quaternary	Pleistocene	First hominines (*Homo* species)
5			Pliocene	Origin of human family
24			Miocene	Abundant grazing mammals; primate radiation
37			Oligocene	First anthropoid primates; many birds
58			Eocene	Abundant modern mammals and angiosperms
65	Cenozoic	Tertiary	Paleocene	Placental mammals diversify
144		Cretaceous		Archaic mammals; first modern birds; first modern fishes; angiosperms appear and become dominant; climax of dinosaurs, and extinction at close of the period
213		Jurassic		First mammals; first birds; dinosaurs dominant; therapsids extinct; abundant bony fishes; gymosperm forests
248	Mesozoic	Triassic		Reptiles diversify; first dinosaurs; many insect types; bony fishes diversify; abundant cycads and conifers; mass extinction event
286		Permian		Insects diversify; reptiles diversify; first therapsids; cycads and conifers expand; mass extinction event
360		Carboniferous		First reptiles; amphibians diversify; first conifers; abundant seed ferns; major coal deposits
408		Devonian		First amphibians; terrestrial life diversifies; fishes diversify; first true bony fishes; first sharks; mass extinction event
438		Silurian		Jawless fishes diversify; first jawed fishes; first terrestrial plants and animals
505		Ordovician		Aquatic life only; armored ostracoderm fishes diversify; invertebrates diverse and dominant; mass extinction event
590	Paleozoic	Cambrian		Aquatic life only; all modern animal phyla; first vertebrates; abundant trilobites

*Time that interval began.

Figure I0.2. Cast of a Cambrian fossil trilobite, one of a group of marine invertebrates that serve as index markers for Paleozoic times. (Photograph courtesy of the American Museum of Natural History, New York)

of the mammals and flowering plants, which replaced the dominant reptiles and gymnosperm plants of the Mesozoic Era.

Determinations of absolute time are made by radiometric dating when possible, and by correlations between geologic and fossil deposits in various parts of the world. In some cases, a particular fossil group may serve as a marker of some interval of time, even when radiometric dating is not feasible. The marine **trilobites** were a highly diversified arthropod group that existed only during the Paleozoic Era (Fig. 10.2). Particular species of trilobites serve as markers of the beginning and end of the era, and as index fossils for the Cambrian and other intervals of Paleozoic time. They are useful because of the relative ease of identification and because they were widespread and numerous, leaving ample numbers of fossils in every stratum of Paleozoic age.

It is clear that biologists' primary focus on the Phanerozoic Eon is due to its wealth of plant and animal fossils, from which they can trace evolutionary trends and patterns leading to modern lineages and species. The Precambrian time is obviously very important for evolutionary study, but most of the life forms were microscopic and morphologically limited. There is fossil evidence going back at least 3 500 Myr, which was discussed in earlier chapters in some detail. Major evolutionary episodes took place then to produce prokaryotes and eukaryotes, photosynthetic organisms, sexually reproducing groups, and other significant innovative forms. In this chapter, the main focus will be on the plant and animal fossil records of the past 600 Myr. This focus is due in part to more detailed records, and in part because we wish to trace our own origin back to ancient multicellular lineages.

Fossilization and the Nature of the Fossil Record

A **fossil** is any evidence of former life and is usually found in sediments or rock. An impression made by an organism, such as a footprint or a burrow, is as much a fossil as are actual remains of the body or any of its parts, because it is evidence of former life. Of the three basic groups of rocks—igneous, metamorphic, and sedimentary—virtually all fossils are parts of sedimentary rock deposits.

Igneous rocks are formed by volcanic action in the crust and mantle of the planet. Familiar examples of igneous rocks include silicon-rich, quartz-containing *granite,* which makes up most of the continental crust, and dark *basalt,* which has no quartz and little silicon, but is rich in magnesium and iron. Basalts and other rocks of this type compose most of the oceanic crust, and most of these dense rocks are formed from cooled molten material, or *magma,* which flows up from the mantle and may be extruded through a vent as *lava.* Some igneous rocks are formed from volcanic ash and dust released during eruptions, such as that of Mount St. Helens in Washington State in 1980 and of Mount Vesuvius in A.D. 79, when the cities of Pompeii and Herculaneum were buried. Ancient crustal and volcanic rocks are unlikely to contain fossils because of their age or nature. One exception to this rule is the igneous rock called *tuff,* which is composed of loose volcanic debris that settled to form rock, just as deposits consolidate into sedimentary rock. Very important information concerning fossils can be obtained from tuffs, particularly dating information that can be related to the age of associated fossiliferous deposits.

Metamorphic rocks form at very high temperatures and pressures, producing rocks of different crystal structure and texture from the igneous, sedimentary, or metamorphic starting materials. Even if the parent rocks did contain fossils, the extreme conditions for metamorphism result in the obliteration or distortion of the fossils such as to render them unidentifiable. Some of the commoner metamorphic rocks include *gneiss, schist, slate,* and *marble.* These and other metamorphic rocks are classified according to their mode of formation, the nature of the parent rock material, and other features.

Sedimentary rocks form from accumulations of rock and organic particles cemented together in quiet waters, producing material that is different from the parent substances. *Sandstones* are formed from granite particles cemented together by iron- and magnesium-rich minerals, giving them a rusty color; *shale* is lithified mud; *chalks* are produced from the calcium-rich remains of countless microscopic organisms belonging mainly to the foraminifer group. The breathtaking sandstone cliffs of the Grand Canyon and other parts of the southwestern United States, and the chalk deposits that make up the white cliffs of Dover on the eastern English coast, are spectacular examples of fossil-rich sedimentary rock formations.

Fossilization is a rare event, but it is more likely to happen if one or more particular conditions are met. If there is *rapid burial* of an organism, the remains are less likely to be destroyed by scavengers or predators, and less likely to decay because of the limited oxygen available for bacterial and fungal activities. In addition, buried remains are less subject to mechanical forces of destruction by wind and water or accidents. The chances for fossilization are also enhanced by the presence of *hard parts* that resist destruction. Structures such as the bones or shells of animals and the tough woody tissues of trees and shrubs make up a very large proportion of fossil remains in Phanerozoic deposits.

Fossilization usually begins when organisms or their parts are buried in sediments under water (Fig. 10.3). The organic remains settle to the bottom of the watery graves, where little predation or scavenging may occur and where decay processes are slow or incomplete. The hard parts become mineralized as the soft parts are destroyed or carried away, and fossilized traces or remains become part of sedimentary rocks formed under water. Sediments continue to accumulate if appropriate conditions persist, and lower strata

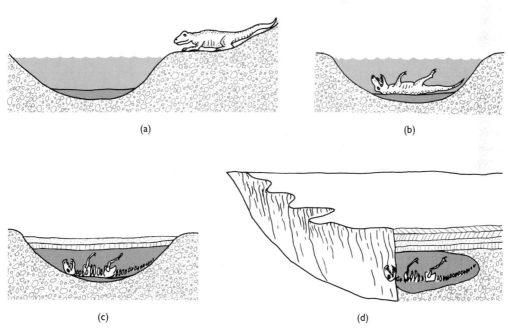

(a)   (b)

(c)   (d)

Figure 10.3. Fossilization. (a) An animal or other organism (b) is buried under water. If not destroyed by predators, scavengers, decay, or other means, (c) the organism is buried under accumulating layers of sediment, and its hard parts are preserved. Layers of sediments solidify, with newer strata added to older ones below. (d) Eventually, erosion or some other physical event may expose all or part of a fossil that had been buried deep within a stratified sedimentary rock formation.

solidify while additional deposits pile up and form new and younger sedimentary rock layers. Layer on layer of sedimentary rock remain long after an aquatic zone has dried up, which is plainly evident in the stratified sandstone making up the walls of steep canyons and rocky extrusions in the American Southwest and many other places in the world.

When sedimentary strata are exposed to air after the covering waters have disappeared, the abrasive forces of wind and water erode the surfaces of the rock formations. Fossils that had been buried deep within these formations may be exposed at the surface or be released entirely from the rock. In many cases, fossils that had been preserved for millions or billions of years are destroyed by natural forces after they have lost their protective rocky shrouds, or been mangled by metamorphism or cataclysmic geologic events.

All these factors help explain the rarity of fossils and the gaps and distortions in the fossil record virtually everywhere in the world. The fossil record is a disproportionate sampling of the variety of life forms because of many other factors as well (Table 10.2). Aquatic forms are more likely to be preserved than terrestrial life because of the requirement for quiet waters in sedimentary rock formation. Organisms lacking hard parts are less likely to be preserved, as stated earlier, and there probably is a richer representation of fossils of species from large populations than small ones, and from stable, enduring populations than transient ones. The low probability of fossilization in general is rendered even lower for small populations of fleeting existence.

TABLE 10.2
Conditions that Generally Favor Fossilization

Favorable Condition	Rationale
Aquatic versus terrestrial habitat	Higher probability for burial under water
Presence versus absence of hard parts in organism	Greater resistance to destruction by decay or by scavengers
Member of large versus small population	Higher probability for sampling of the population
Member of enduring versus transient population	Higher probability for sampling over time of population existence
Rapid versus slow burial of organism	Minimum probability for destruction by decay or by scavengers
Warm and humid versus cool and dry climate	Extensive shallow seas available for burial under water and for sedimentary rock formation
Continental submergence versus uplift	Extensive shallow seas available for burial under water and for sedimentary rock formation

Richer fossil deposits are more likely to occur in times and places of mild, warm climates, which foster the formation of extensive shallow seas over previously exposed land areas; thus more sites become available for fossilization and sedimentary rock formation. There may be fewer fossils deposited when cooler, drier climates prevail and more land area is exposed, and these deposits would include many fewer kinds of organisms. The more land available, the greater the diversity of habitats, with varied selection pressures operating and more speciation in a mosaic of topographically isolated locations. The times that are best for species diversification are thus likely to be those when conditions for fossilization are poorest, particularly for terrestrial organisms.

These considerations would suggest widely different degrees of representation of past life in the fossil record. Among fossils of unicellular organisms we found many diatoms whose walls contained much silica and calcareous foraminifers, but few ameboid or other wall-less groups. Well-represented invertebrate animals include corals (coelenterates), brachiopods (lampshells), certain mollusks (clams, oysters, snails), marine arthropods such as trilobites, and echinoderms (starfish and allies). In contrast, many kinds of wormlike and annelid organisms are missing almost entirely, and relatively few fossil insects are known. There are fewer fossil birds than amphibians and reptiles, and many kinds of mammals from different groups. Among the plants, a fair number of species of woody vascular plants are found, but the soft parts, such as flowers, are rarely preserved; identification is extremely difficult because of this problem. There are relatively few specimens of nonvascular plants, such as algae, mosses, and liverworts, and fungi are poorly represented in the fossil record. These observations confirm the predictions of a distorted and fragmentary fossil record.

Under certain conditions, fossil organisms may be beautifully preserved, soft parts and all, rather than being a handful of woody or bony fragments that have to be painstakingly sorted out during the reconstruction of the original structure or individual. Excellent, intact insects have been found in *amber,* which is the hardened fossilized form of resin exuded from trees such as pines and other conifers. The extraordinary preservation of detail in these insects allows their precise comparisons with living species, and, in many cases, it is clear that little or no change has taken place for as long as 25 million years. Excellent conditions for preservation are also offered by sticky tar pools formed by the evaporation of volatile oils from petroleum springs; the tars are later converted to viscous asphalt. A variety of predator and prey animals have been recovered from such sites, and the La Brea tar pits in Los Angeles, California, are famous for the wealth of birds and mammals from the Pleistocene and Recent epochs of the Cenozoic that they have yielded.

Excellent specimens are preserved as a result of burial under hot volcanic dust and ash, because of rapid desiccation and the absence of destructive processes after burial. Fragile insects are wonderfully preserved, often after falling into water as they became covered with dust and ash and then being fossilized under conventional conditions for sedimentary rock formation. The ruins and biological remains at Pompeii and Herculaneum demonstrate, in astonishing detail, how much can be preserved after quick burial under hot volcanic ash and dust. Bread just out of the oven still sits on tables, although now inedible, and people fleeing the disaster were preserved in positions clearly revealing the agony and surprise of their last moments.

Another unusual situation involves the quick-freezing of tundra-dwelling animals, which may be so well preserved that tissue specimens reveal great cellular detail. Wild claims were made by some Soviets that frozen fossil lizards had been thawed out after thousands of years in the Siberian ice and restored to life. These reports were never substantiated independently by other scientists, and they are not taken seriously. Frozen fossil lizard did appear on the luncheon menu of the Explorers Club some years ago, and the unanimous gourmet verdict was that it tasted terrible.

It used to be assumed that huge herds of Pleistocene and Recent mammoths roaming the frozen tundra had been wiped out by some catastrophe, leaving a great number of well-preserved frozen fossils of these hairy creatures. However, a careful study revealed that only 38 frozen mammoths had been dug up by 1960, but at least 50,000 must have existed at the time these animals died. It hardly seemed likely that catastrophe had been responsible for the frozen specimens or for the extinction of the mammoths, for that matter. When autopsies were performed on the 38 frozen fossils, 34 of them showed considerable damage from scavengers or predators. The other 4 mammoths were well preserved, but all showed clinical symptoms indicating that they had died of suffocation. Each of the 4 specimens had eaten well before its death, as judged by stomach contents, and had been in good health. The fact that 34 animals had been damaged after their death and that the other 4 also showed some signs of decay pointed to slow rather than very rapid burial, as would have occurred after a natural disaster.

Various species of large animals, including woolly rhinoceroses, are found along with woolly mammoths as the major types of frozen fossils. The preponderance of large animals, instead of a random sample of tundra species, among the fossils is significant. It indicates that the fossils probably were unfortunates that were somehow trapped when they strayed near a precipice or another dangerous feature of their landscape. The fossils in the La Brea tar pits provide a similar image of a collection of strays that accidentally blundered into the sticky pools and were trapped there. Predators such as the saber-toothed cat were also trapped, and the reward for their hunger was the same deadly fate that befell their intended prey.

Despite its inadequacies and distortions, the fossil record fits the major predictions based on comparative studies of living organisms. The plant and animal life that has evolved and diversified during the past 600 Myr can be better understood today within the context of large-scale changes in the physical and biological environments of the world. The continual redistribution of landmasses through continental drift has markedly influenced the levels of biological variety evolving in isolation and the composition of fauna and flora freely migrating or prevented from migrating between one landmass and another.

Continental Drift

Until fairly recently, our vision of the Earth was one of large-scale stasis, except for fluctuations in climate, the formation and disappearance of mountains on the continents, and changes in sea level because of glaciation and other geologic and climatic phenomena. The continents and oceans themselves were believed to be basically the same now as they had been billions of years ago; that is, the large-scale topographic features of the Earth's crust were considered to be relatively unchanging. The obvious resemblance of the existing continents to neatly fitting pieces of a jigsaw puzzle had been noted over the centuries, but any idea that this was due to migrations of landmasses was not accepted.

The formal proposal for land migrations, or **continental drift,** based on fossil and geologic information, was made in 1912 by Alfred Wegener, a German meteorologist, and expanded by him in the 1920s and 1930s as new data were obtained. He proposed the existence in Paleozoic times of one large supercontinent, which he called **Pangaea** (Fig. 10.4). He suggested that Pan-

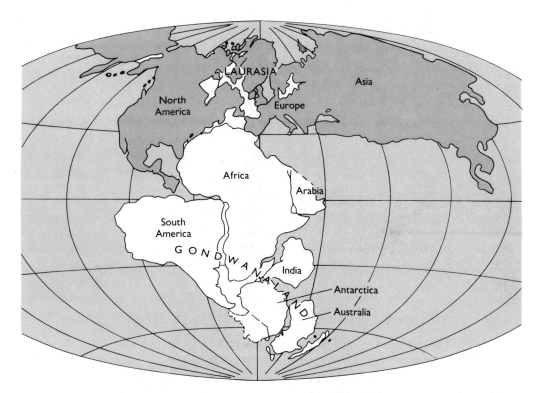

Figure 10.4. The supercontinent Pangaea was proposed by Alfred Wegener as a single pre-Mesozoic landmass that later ruptured to produce Laurasia in the north (gray) and Gondwanaland in the south. The fragmentation of these two regions eventually produced the present-day continents.

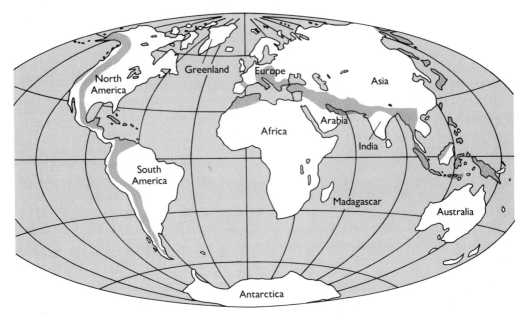

Figure 10.5. Global belts of current mountain-building activity (dark gray) are evident in the great mountain ranges of North and South America, extending almost from pole to pole, and in the mountain chains reaching from western Europe and North Africa to Indonesia and New Guinea in the western Pacific. The volcanically active islands of Japan essentially connect with the active Alaskan regions.

gaea later ruptured to produce two continental landmasses, which he named **Gondwanaland,** in the Southern Hemisphere, and **Laurasia,** in the Northern Hemisphere. These two regions eventually fragmented to form the present-day continents.

Wegener's theory was revived and dismissed at various times until the 1960s, when extensive and independent lines of evidence were presented in support of wandering continents. These and later sets of data form the basis for the modern theories of **plate tectonics,** which is the study of processes and events involved in changing the topography of the Earth's crust.

There was no dispute about the occurrence of substantial topographic changes on the Earth's crust, as was evident in periodic rejuvenation producing new mountain chains, mixtures of strata of different ages due to crustal warping and folding by which younger strata intruded into older formations, and global belts of mountain-building activities (Fig. 10.5). These upheavals and topographic fluxes certainly indicated a pliable crust influenced by volcanism and other geologic processes initiated deep in the mantle of the Earth. They did not by themselves necessarily imply a mobile crust.

Major lines of evidence against a rigid, fixed crust and in support of continental drift came from various lines of study, all of which showed that a supercontinent had existed more than 200 Myr ago and that it had fragmented to produce the modern global map. Patterns of magnetism preserved in iron-bearing rocks of different ages and from different places were used to reconstruct and relate the changing crustal topographies. Computer-programmed fits were found for regions and shorelines now separated by great distances and were reinforced by a variety of geochronological data (Fig. 10.6). The distribution of index fossils on different continents today helped to trace events back to times when the separated lands must have been united. The topography, age, and thickness of the ocean floors proved to be among the

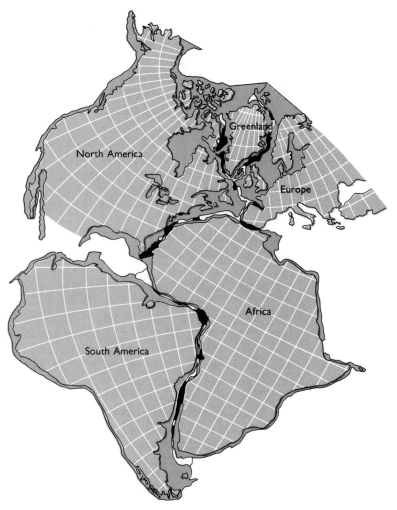

Figure 10.6. Computer-generated best fit of the continents now on opposite sides of the Atlantic Ocean. The fit was made along the 500-fathom line of each continental slope. Overlaps of continental margins are in black, and gaps between margins in pale gray. (From P. M. Hurley, "The Confirmation of Continental Drift." Copyright © 1968 by Scientific American, Inc. All rights reserved)

most persuasive pieces of evidence for drift and for tectonic processes causing drift.

Geologists had believed that the ocean floors would be immensely thick because of sedimentary deposits accumulating for billions of years, relatively secure from the erosive forces of violent winds and waters. Instead, it was found that the crust of the ocean floors is quite thin, and the oldest sediments are only of Mesozoic age. These data formed the basis for the concept of sea-floor spreading, which is a pivotal feature in the overall framework of plate tectonic processes leading to continental drift.

The current view of tectonic processes is based on the properties of the Earth's crust and mantle. The crust, or lithosphere, extends into the interior for about 100 km (62 miles), but the depth varies somewhat from one place to another. The continents and ocean basins of the lithosphere are parts of rigid **tectonic plates** abutting one another and completely covering the Earth's surface (Fig. 10.7). The mantle consists of molten or semimolten rock, on which the plates *float* in various directions relative to one another. There are now eight large tectonic plates and several small ones, but their number and size have changed at different times in the past.

Plate boundaries are of three types: (1) an actively **spreading ridge;** (2) a **subduction zone,** or trench; and (3) a **transform fault.** Plates move away from ridges, where they form, toward subduction zones, where they are carried into the mantle and are destroyed (Fig. 10.8). As two plates move, they may

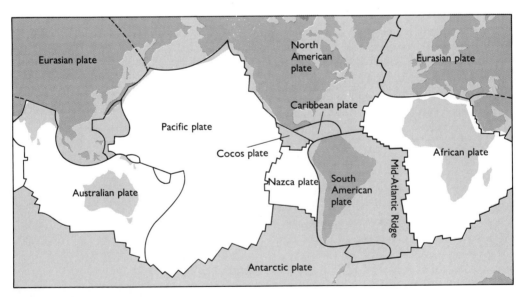

Figure 10.7. The current distribution of tectonic plates of the Earth's crust. Eight large plates and several smaller ones (Cocos plate, Caribbean plate, Nazca plate) have been mapped. The adjoining Pacific, Nazca, and Cocos plates carry no continental landmasses, but all the others do.

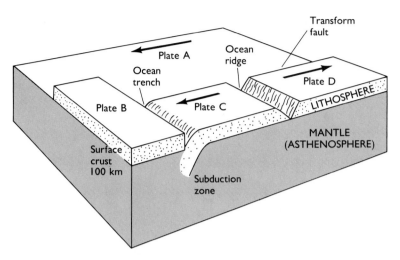

Figure 10.8. The three types of plate boundaries are an actively spreading ridge, between plates C and D; a subduction zone, or trench, between plates B and C; and a transform fault, between plates A and D. Plates form and move away from ridges toward subduction zones, where they are carried into the asthenosphere portion of the mantle and are destroyed. A transform fault is the boundary between two plates that scrape past each other during their movement.

scrape past each other along a transform fault. The San Andreas fault in southern California is such a transform boundary, and the movement of the North American and Pacific plates is responsible for the many earthquakes in that region.

The continuous addition of mantle rock to lithospheric plates is particularly evident at mid-ocean ridges of the ocean floors. **Sea-floor spreading** refers to the addition of material to the ocean floor as molten mantle rock rises up and is extruded at the ridge. The lava moves outward and away from the ridge, pushing away older sediments and forming new lithospheric floor on which sediments are deposited. The progressive change from younger sediments at an ocean ridge to older and older sediments increasingly farther away from it is unambiguous evidence for tectonic processes in the mantle causing changes in the topography of the Earth's crust. Sea-floor spreading is responsible for continental (plate) drift.

If an actively spreading ridge produces a rising zone under a continental landmass, *rifting* occurs and the fragments of the plate move apart (Fig. 10.9). Rifting is evident today in some parts of the world, including the Afar triangle of Ethiopia and the Great Rift Valley, which runs from southern Africa along eastern Africa and up into Turkey. The Afar triangle is the site of a junction of three spreading zones, and is tectonically very active. Interestingly, both the Afar region and the Great Rift Valley were the sites of the origin and most of the evolutionary history of the human family over the past 5 Myr.

Subduction zones are regions where the forward edge of a moving plate

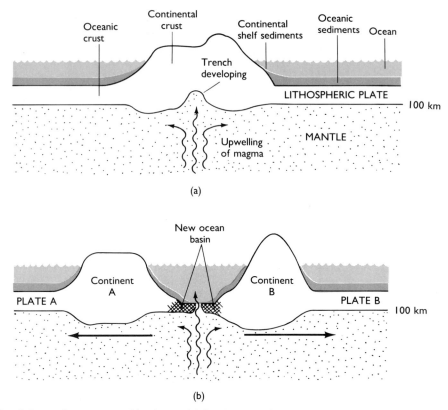

Figure 10.9. Rifting of a continental landmass. (a) Production of a rising zone of upwelling magma from the mantle leads to trench development if it occurs under a continental landmass. (b) Continued upwelling leads to rifting of the landmass as the plate fragments separate and move apart. An ocean opens between plates A and B (fragments of the original plate) in a new ocean basin created from the mantle magma. Sediments are laid down on the magma base of the widening basin.

sinks into the mantle and is destroyed. When an ocean floor occurs at the leading edge of a plate and moves toward a subduction zone located in an oceanic region, a deep trench forms and is bordered by volcanoes or volcanic island chains (Fig. 10.10). Japan and the Philippines are such island chains. When an ocean floor moves toward a continent, it passes under the continental border if the movement is rapid; if it is slower, the oceanic and continental edges pile up together in a major uplift. When the Australian plate, carrying peninsular India, collided with the southern edge of the Eurasian plate during the Cenozoic Era, it thrust itself under the Eurasian plate. The uplift of the Himalayas resulted from the merging of the two plate borders, which doubled the thickness of the crust at the suture line. Similar mountain-building events are now taking place along the western coast of South America, causing the continuing uplift of the Andes Mountains, and were responsible for mountain building in various places in past times. The Ural Mountains rose up when the European and Asian plates collided, and the contorted chain of mountain ranges from the Atlas Mountains of northwestern Africa through the Mediterranean region, the western Alps, the Caucasus, and on to the recently upthrust Himalayas rose at various times when drifting continents collided and caused the buckling of sediments in the ancient seas and pileups at plate sutures.

Various kinds of evidence permit the determination of *relative* rates and directions of plate movements, but until recently there was no way of directly ascertaining the *absolute* values involved. Absolute time can be learned by reference to relatively immobile (fixed) topographic features, which consist primarily of hot spots. **Hot spots** are sites in the mantle, from which magma rises and causes igneous activity within the overlying crust. Volcanoes often form at the surface over these thermal plumes, if the periodic upventing of magma from the mantle leads to volcanic rock deposits that rise from the sea or emerge on land. The volcanic Hawaiian Islands and Galápagos Archipelago, as well as geysers and other evidence of recent volcanic activity in Yellowstone National Park, are associated with hot spots.

By marking the successive positions of a plate over a hot spot, by means of a chain of volcanoes, the absolute direction of a plate's movement and its rate of movement past the hot spot can be determined. The Hawaiian Islands, for example, have been forming one by one in a southeasterly direction for millions of years as the Pacific plate has moved northwestward over the hot spot (Fig. 10.11). Kauai, at the northwestern end of the chain, is 5.6 Myr old, and the islands are progressively younger moving southeastward toward Hawaii, which is less than 1 Myr old. Seamounts exist to the northwest of Kauai,

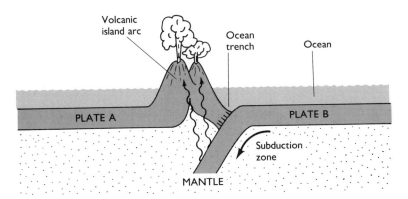

Figure 10.10. When an ocean floor is at the forward edge of a moving plate (B), and it moves toward a subduction zone located in an oceanic region, a deep trench forms and is bordered by volcanic island chains or volcanoes. The leading edge of the moving plate sinks into the mantle at the subduction zone and is destroyed.

representing volcanic islands that formed earlier than Kauai but have been eroded to stumps and now lie under the ocean surface (see Box 9.1). From data for some of the 100 hot spots located around the globe, but mostly at mid-ocean ridges, it has been found that many plates are moving 5 cm or more per year.

Continental landmasses may split up into discrete entities as the result of rifting, as previously mentioned, but they may also be built up by *accretion* from bits and pieces of oceanic sediments and microcontinents, which may also be added to an existing landmass to produce a larger continent (Fig. 10.12). The changing topography of the Earth is the outcome of continental accretion and suturing, which forms larger landmasses, and of rifting, which breaks up larger landmasses. As far as geologists can tell, the plate tectonic

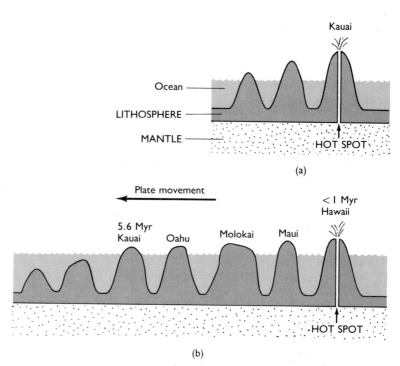

Figure 10.11. Formation of a volcanic island chain as a tectonic plate moves over a hot spot. (a) As the Pacific plate moves northwestward, the upventing of hot plumes of mantle magma from a hot spot leads to volcanic deposits. The island of Kauai formed about 5.6 Myr ago, but older volcanic islands lie to its northwest. (b) Progressive northwestward movement of the plate has led to the creation of additional islands, one by one, with Hawaii being the newest member of the Hawaiian island chain. Direction and rate of plate movement past the hot spot can be determined from the position of islands and seamounts. Recent studies suggest that hot spots are not absolutely fixed in place; they also move, but at a very slow rate compared with the rate of plate movement.

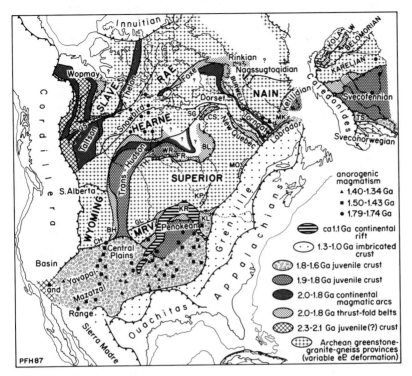

Figure 10.12. North America was assembled from blocks and scraps of various land areas by plate tectonic processes operating for at least 2000 Myr. The pieces illustrated include a core of ancient continental blocks (cratons), the collision-induced mountains (orogens) between cratons, and accreted volcanic island arcs ("juvenile crust"). (P. F. Hoffman, Geological Survey of Canada)

processes in action today have been in action for more than 2000 Myr. The Archean world may have been different, however, from the world in Proterozoic and Phanerozoic times.

The information on Archean times is fragmentary, because very few ancient formations remain in unmetamorphosed condition. From the information available, however, geologists infer that very few large continental landmasses existed in Archean times, more than 2500 Myr ago. Many small protocontinents may have existed, but the Archean crust is considered to have been so fragile and mobile that many ocean basins were produced around and between relatively weak protocontinental areas. Many crustal elements may have begun to consolidate in the Late Archean, but evidence for such consolidation is solid for only the Early Proterozoic, perhaps 2000 to 2500 Myr ago. The tectonic history of the Proterozoic and Phanerozoic eons, unlike the Archean, included episodes of continent formation and breakup, which is the next topic.

The Formation and Breakup of Pangaea

During the Proterozoic Eon, the lands of the Northern Hemisphere included *Laurentia*, a continent that consisted of what is now North America, Greenland, northern Great Britain, and possibly Siberia and Scandinavia; the continent of *Baltica*, which is now northern Europe; the continent of China; and at least two smaller landmasses, Siberia and Kazakhstania. In the Southern Hemisphere, by at least 800 Myr ago but perhaps even earlier, the large mass of Gondwanaland included what is now South America, Africa, Antarctica, Australia, and peninsular India.

The assembly of Pangaea took place in stages during the Paleozoic Era (Fig. 10.13). By the Early Devonian, a broad sea separated Gondwana from the northern continents. In the Late Carboniferous, the northward movement of Gondwana led to its suturing to North America; slightly later, Europe became attached to Asia, with the consequent uplift of the Ural Mountains at the suture line. The eastern part of North America was in contact with Europe and Africa by the Late Permian, and by the end of the Paleozoic Era, Pangaea was fully assembled in the midst of a single vast ocean.

The exact duration of Pangaea's existence is uncertain, but during the Mesozoic Era, Pangaea began to break up. In the Late Triassic, rifting began to separate what is now southern Europe from Africa, opening the Tethys Sea between them. At this time, reptiles had become predominant over the earlier amphibians, and forests of ferns and lycophytes, which earlier fossilized to form major coal deposits, had already diminished and been replaced largely by gymnosperms. Rifting propagated westward during the Jurassic Period, ultimately separating North America and South America, and North America also began to break away from Africa. Pangaea had now been torn in two by rifting, producing the partially separated landmasses of Laurasia in the north and Gondwana in the south (Fig. 10.14).

Major breakup events took place during the Cretaceous Period. Gondwana had remained intact and barely attached to the northern lands until this time, but by the end of the Cretaceous, South America, Africa, and peninsular India had become separate landmasses. Only Australia and Antarctica remained attached as a fragment of the original eastern Gondwana landmass. By the Late Cretaceous, continental rifting had split the huge Laurasian landmass into North America, Greenland, and Eurasia. During these major Cretaceous events, the flowering plants became predominant over the earlier gymnosperms, and the reptiles experienced a sharp decline and were soon to be superseded by mammalian fauna. Mammals and birds had made their appearance in the Jurassic, but remained relatively minor components of Mesozoic life. Until their decline near the end of the Cretaceous, the reptiles had flourished through all the surface upheavals for nearly 150 million years.

In the Cenozoic Era, India ended the northward journey it had begun after splitting from Gondwana and collided with the Asian landmass, uplifting the Himalayas (Fig. 10.15). Australia finally broke away from Antarctica in the

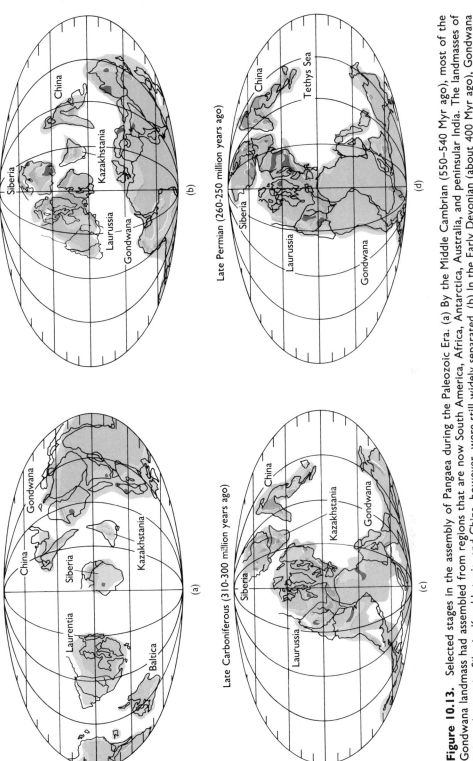

Figure 10.13. Selected stages in the assembly of Pangaea during the Paleozoic Era. (a) By the Middle Cambrian (550–540 Myr ago), most of the Gondwana landmass had assembled from regions that are now South America, Africa, Antarctica, Australia, and peninsular India. The landmasses of Gondwana Baltica, Laurentia, Siberia, Kazakhstania, and China, however, were still widely separated. (b) In the Early Devonian (about 400 Myr ago), Gondwana was separated by a broad sea from the northern continents. (c) By the Late Carboniferous (310–300 Myr ago), Gondwana had moved northward and been sutured to North America (part of Laurussia). Shortly afterward, Europe became attached to western Asia. (d) Near the end of the Paleozoic, in the Late Permian (260–250 Myr ago), the assembly included Siberia and was followed soon after by the joining of China to produce the supercontinent Pangaea. (From R. K. Bambach et al., 1980, Before Pangaea: the geographies of the Paleozoic world. *Amer. Sci.* 68:26, Figs. 5, 10, 12, 14)

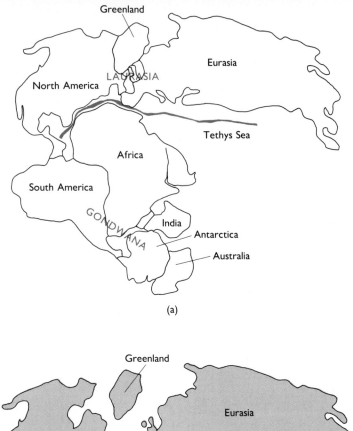

(a)

(b)

Figure 10.14. The breakup of Pangaea during the Mesozoic Era. (a) In the Late Triassic, rifting began to separate the northern region of Laurasia from southern Gondwana, and the Tethys Sea began to open between the two areas. During the Jurassic Period, Pangaea was torn in two by rifting, as North America separated from South America and began to break way from Africa. (b) In the Late Cretaceous, the northern Laurasian landmass had fragmented to what are now North America, Greenland, and Eurasia. By the end of the Cretaceous, the southern landmass of Gondwana had fragmented into the continents of South America, Africa, and peninsular India, but Antarctica and Australia remained attached as a single fragment of the original East Gondwana region (compare Fig. 10.4).

Figure 10.15. Peninsular India moved northward, beginning about 80 Myr BP, and collided with the Asian landmass about 10 Myr BP, leading to the uplift of the Himalayas. Mount Everest, at 29,000 feet, is the tallest mountain in this most recently formed of the world's mountain chains. Numbers show the times (in Myr BP) when the subcontinent reached various positions during its travel. (Adapted from C. M. Powell and B. D. Johnson, 1980. *Tectonophysics* 63:91)

Eocene, and the plate carrying Australia, New Zealand, and New Guinea moved northward. The Australian plate collided with the Eurasian plate in the Miocene, about 15 Myr ago, making contact in the Malay Archipelago and providing a land connection over which animals migrated. After Australia broke away from Antarctica, Antarctica was left isolated over the South Pole. At this time in the Eocene Epoch, changes also occurred in the Northern Hemisphere, particularly the greater separation of North America and Europe from Greenland. North America and Eurasia were connected in the late Eocene by the Bering land bridge between present-day Alaska and Siberia, which served as a corridor for the migration of animals between the two continents.

In the middle Miocene, between 18 and 14 Myr ago, a land corridor was established between Africa and Eurasia following the collision of the plates carrying these continents. During the Pliocene Epoch, between 5 and 1.8 Myr ago, the Isthmus of Panama arose by tectonic activity and separated the

Caribbean Sea from the Pacific Ocean. The Panamanian land bridge was present at least 3.5 Myr ago, providing a migration route between North and South America. The African continent, which had been relatively stable for hundreds of millions of years, began to fragment along spreading zones called *rift valleys*. If these events continue, Africa may split into two or more continents.

Continental drift clearly is still in progress, but some landmasses have experienced greater dislocations than others. Whereas peninsular India, Australia, North America, and other continents shifted substantially during the Cenozoic Era, Eurasia and Antarctica remained relatively stationary, except for rotation during the past 65 Myr. Two of the major ocean trenches have remained relatively fixed, but they now lie east of the Americas instead of west, because of continental drift of the North American and South American plates.

Most of the story of Pangaea has been derived from geologic and geochronological information, but strong supporting evidence has also come from the fossil record. Biogeographical distributions of both living and fossil species can be traced and understood by reference to the rifting and joining of land and ocean regions as the continents wander from place to place on their floating lithospheric rafts. Some of this fossil evidence also serves to show that certain ancient landmasses indeed existed at specified times in the past.

One of the important trace fossils for reconstructing the position of ancient landmasses is the seed fern *Glossopteris,* which formed a large part of the forests of Late Carboniferous and Permian times. Fossils dated as Permian have been found throughout the regions that were components of Gondwanaland (Fig. 10.16). From what is known of the size and nature of *Glossopteris* seeds, it is highly unlikely that they could have been carried to all the present-day southern continents across many thousands of miles of open oceans. It is far more likely that *Glossopteris* grew on a single landmass, which was Gondwanaland. On a smaller scale, but also indicative of the union of South America and Africa in the Permian and Triassic periods, are three reptile species. The Permian fresh-water *Mesosaurus* and the Triassic therapsid *Cynognathus* occur in deposits in southern Africa and parts of South America. The Triassic fossil *Lystrosaurus*, a therapsid herbivore, is found in deposits of the same age and composition in Antarctica, India, and southern Africa. It is inconceivable that any or all of these reptiles were carried or swam over great ocean distances to the separate locations in which their remains have been found. It is also inconceivable that each species evolved identically and by chance into the separated populations in the different regions. Clearly, their present disparate locations are the consequence of the rifting of Gondwanaland and the subsequent drift of continents, carrying their fossil histories with them as they floated off in various directions.

In the remainder of this chapter, we will take a more detailed look at the Phanerozoic fossil record. We should now be in a better position to understand this evolutionary history, which was clearly influenced by changes in

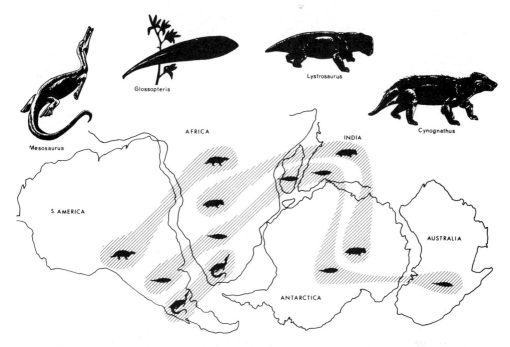

Figure 10.16. Evidence for the reconstruction of Gondwana by the use of four trace fossils, whose records now exist in widely separated parts of the world. The presence of the Permian seed fern *Glossopteris* in deposits from South America, Africa, Madagascar, India, Antarctica, and Australia indicate that these areas were parts of a single entity in the Late Paleozoic. Supporting evidence for ancient Gondwana is provided by the current locations of fossil specimens of the Permian reptile *Mesosaurus* and the Triassic reptiles *Cynognathus* and *Lystrosaurus*. (From E. H. Colbert, 1973, *Wandering Lands and Animals*. New York: Dutton)

the Earth's surface, and by the changing climates and environments that resulted from these and other critical geologic events.

The Procession of Phanerozoic Life

The lineages of living plants and animals can be traced back in time by reference to the fossil record of the past 590 Myr of Phanerozoic history. The three eras are set apart by significant differences in the life forms of the times and by major radiation and extinction episodes that heralded a new dominant biota after the demise of a previous suite of animals. Major changes in plant life also occurred, but none of them coincides with era discontinuities. As so often is the case in biological studies, major emphases are given to animals rather than to plants. Also, the record of animal life is much better documented than that of plant life and provided most of the landmarks used in past and present studies.

Life in the Paleozoic Era

The appearance of abundant and diverse invertebrate animals marks the beginning of the **Cambrian Period,** about 590 Myr before the present time (590 Myr BP). Paleontologists know very little about the ancestry of Cambrian organisms, but representatives of all the living animal phyla have been found in Cambrian deposits (590–505 Myr BP). Fossil evidence is available for Late Precambrian marine invertebrates, but the dating and phylogenetic relationships of these organisms are poorly understood. Traces of soft-bodied Precambrian relatives of modern coelenterates, annelids, and arthropods have been identified in formations less than 1000 Myr old, particularly in the unusual Ediacara fauna discovered in southern Australia. Many Ediacara animals were unique or unusual organisms that left no descendants at the end of the Precambrian. Even those recognizable enough to be associated with living animal phyla remain enigmatic, and cannot be designated as specific ancestors of later lineages.

One of the great evolutionary puzzles is the ancestral link among all the Cambrian animal phyla, which are very distinct from one another and for which no intermediate forms have been found. Their relatively sudden and virtually simultaneous appearance in the Cambrian may have been due to a momentous episode of adaptive radiation, perhaps following a period of 100 Myr when the Ediacara fauna developed, or a much longer interval for which little trace remains. Most of the phylogenetic relationships among animal phyla have thus been based on comparative anatomy and embryology of living groups.

A remarkable slice of Cambrian life was discovered in the **Burgess Shale** of British Columbia, from about 530 Myr BP, which included exceptionally well-preserved soft-bodied animals. At least 10 extinct phyla are present among the Burgess Shale suite of fossils, plus cnidarians, annelids, and the earliest known chordate *(Pikaia),* which resembles *Amphioxus* in some respects. Apart from this unusual assemblage, the major types of Cambrian fossils are those with hard parts, including arthropods, brachiopods, mollusks, and echinoderms, as well as sponges (Porifera). The earliest vertebrate fossils are known only from fragments of the bony external armor typical of *ostracoderms,* and dated very near the end of the Cambrian as 510 Myr old. Ostracoderms were jawless, finless fishes (Agnatha) whose bodies were covered with bony plates rather than scales, as in the later fishes (Fig. 10.17).

In the **Ordovician Period** (505–438 Myr BP), ostracoderms were well established in the seas, and many of the animal phyla underwent adaptive radiations that produced a large number of new classes and orders. Much of this diverse life became extinct at the end of the Ordovician, the earliest known episode of a major event that is properly called a mass extinction. A

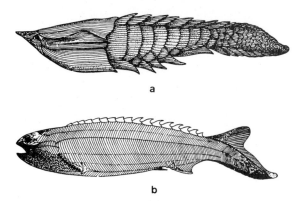

a

b

Figure 10.17. Fossil ostracoderms, such as (a) *Anglaspia* and (b) *Pterolepsis*, were members of the Agnatha, the earliest known group of vertebrates. These fishes were jawless and finless, and their bodies were covered with bony plates rather than scales, as in the later fishes. (From A. S. Romer, 1970, *The Vertebrate Body*, 4th ed. Philadelphia: Saunders)

mass extinction is characterized by the relatively sudden disappearance of many unrelated families of organisms that had been relatively successful and widely distributed.

The **Silurian Period** (438–408 Myr BP) saw a new increase in diversity and highly significant evolutionary grade changes among plant and animal life. The first vascular land plants appeared (Fig. 10.18), with *xylem* tissues for water transport and *phloem* tissue for food transport throughout the plant. This efficient system replaced the slower and inefficient transport by diffusion between neighboring cells and made terrestrial life possible for larger forms.

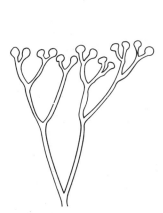

Figure 10.18. *Cooksonia*, the first acknowledged vascular land plant, was rootless and leafless, but had upright, branched stems terminating in sporangia. Specimens of this member of the rhyniophyte group (see Fig. 10.38) have been found in deposits of Middle Silurian to Early Devonian age. (Photograph courtesy of H. P. Banks)

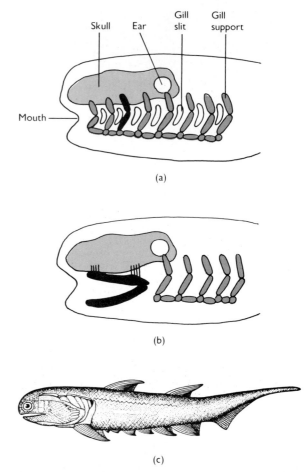

Figure 10.19. The evolution of the vertebrate jaws. (a) Jawless agnathans were ancestral to placoderms, (b) whose jaws evolved from a pair of gill supports (black) in the ancestral agnathan. (c) Reconstruction of *Climatius,* a Devonian fossil placoderm with large fin spines and accessory fins between the pairs of pectoral and pelvic fins.

The leafless, rootless rhyniophytes also provided both habitats and food for new terrestrial animal life, which included scorpion and millipede types. Agnathans diversified in the seas and gave rise to a new grade of fishes with fins and jaws, the order Placodermi. These grade innovations permitted more efficient movement in the water and a more effective feeding apparatus for a greater variety of foods. Placoderm jaws, which evolved opportunistically from a pair of gill supports in the agnathan ancestor, represent one of the significant innovations underlying later vertebrate history (Fig. 10.19).

    The **Devonian Period** (408–360 Myr BP) is referred to as the *Age of Fishes* because the agnathans and placoderms reached their highest level of diversity, and the class of true bony fishes (Osteichthyes) and, later, cartilaginous fishes

(Chondrichthyes) made their appearance. Both the true bony fishes, which include the ray-finned and lobe-finned groups, and the cartilaginous fishes, which include the sharks, rays, and skates, may have evolved from placoderm ancestors, but their origins are uncertain. The two groups of fishes possess paired fins, a grade improvement that allows greater control over the direction and overall efficiency of swimming. The sturgeons, our major source of caviar, are living representatives of one ancient ray-finned Devonian lineage, and only three lungfish and one crossopterygian (the coelacanth) species continue the lobe-finned lineage today (see Fig. 8.16).

Although called the Age of Fishes, the Devonian was the period during which terrestrial life became abundant and diverse. Among the plants, the early vascular forms were joined by their descendant lycophytes (club mosses and horsetails), ferns, and first seed plants (Fig. 10.20). The nonvascular bryophytes (mosses and liverworts) made their appearance in the Devonian, but their ancestry is unknown. Among the land animals that emerged in the Devonian were springtails, an insectlike arthropod, and the first amphibians, which resembled their crossopterygian ancestor in many features, except that they had four limbs instead of paired, lobed fins (Fig. 10.21). The earliest known amphibian fossil, *Ichthyostega,* was found in eastern Greenland, and is a representative of the early labyrinthodont group.

Living on the land requires the extraction of oxygen from the air instead

Figure 10.20. An artist's depiction of a swamp forest of the Carboniferous Period, consisting mainly of lycophytes (club mosses, horsetails), ferns, and seed ferns. Giant club mosses (left) had thick trunks and very small (microphyllous) leaves; giant horsetails (right) had microphyllous leaves arranged in whorls on branches. Ferns and seed ferns (right and center) had large (macrophyllous) leaves, as do living ferns. Living lycophytes are microphyllous, as were their fossil relatives. (Painting by Charles Knight, from the Field Museum of Natural History, Chicago. Negative number 75400, reproduced with permission)

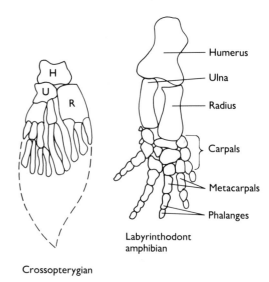

Figure 10.21. Homologies are evident in the skeletal elements of the fleshy fin of a Devonian cros-sopterygian fish and the forelimb of a Permian fossil amphibian. Humerus, ulna, and radius are similar enough to be identified with some certainty, but the smaller bones of the amphibian wrist and digits are less clearly related to the smaller bones of the fish fin. These and other resemblances suggest crossopterygian ancestry for the tetrapod (land) vertebrates (see Fig. 10.22).

of the water, by means of external nostrils and an air passage connected to the lungs. Lungs first evolved in ancient bony fishes, which also possessed gills. During its evolution, the original pair of lungs was reduced to one lung, which is still functional in a few modern forms but has been modified to a swim bladder in modern bony fishes, or teleosts, such as carp, tuna, and other familiar forms. The teleost swim bladder has a buoyancy function, but it sometimes acts as a producer or receptor of sounds. A lung persisted in the lobe-finned fishes, but serves in modern lungfishes and the coelacanth as an organ of fat storage, which provides buoyancy for these marine species. Lungs presumably served a respiratory function in ancient fresh-water cross-opterygians, and continued to do so in their amphibian descendants.

Only three orders of amphibians survive today: the tailed newts and sala-manders; the limbless wormlike forms; and the frogs and toads, which are tailless as adults. Except for a few recently derived forms that are entirely terrestrial, amphibians require an aquatic medium in order to reproduce. Eggs and sperm are shed into the water, where fertilization takes place and where the larval stage metamorphoses to the adult after hatching.

The end of the Devonian was marked by another mass extinction, when many groups of marine invertebrates disappeared, and most of the placoderm and agnathan fishes expired. There is no fossil record for agnathans later than the Devonian, but living agnathans include the lampreys and hagfishes, which have no fossilizable bony skeleton. The **Carboniferous Period** (360–286

Myr BP), which in North America is divided into a Mississippian (360–320) and a Pennsylvanian (320–286) interval, is characterized by major coal deposits. These coal beds resulted from the fossilization of giant lycophytes, pteridosperms (seed ferns), and ferns, huge forests of which covered the tropical and subtropical landscapes of an assembling Pangaea. Many new orders of insects buzzed through these lush forests, including ancient representatives of the grasshoppers, mayflies, cicadas, and roaches.

The amphibians underwent an adaptive radiation in the Carboniferous, producing some species that were more than 20 feet long but very different from modern forms. The first reptiles are represented by *Hylonomus*, a 1-foot-long animal found in Early Pennsylvanian deposits in Nova Scotia, which strongly resembles living amphibians in many major structural features (Fig. 10.22). The reptiles evolved a new grade of organization that made them the first truly land-adapted vertebrates, completely free from the requirement for water in their reproductive cycle. The egg is retained in the female's body and is fertilized there by sperm delivered directly into her body; internal fertilization replaced the amphibian mode of external fertilization. The egg is organized on the **amniote** plan, with the embryo being provided with a nutritious yolk and two sacs. One sac, the amnion, encloses the embryo; the other collects waste products. Once the egg is laid by the mother, the tough outer covering of the egg protects the developing embryo and prevents its desiccation. By the time it hatches, the young reptile has developed into a miniature version of the adult, rather than a larval form that metamorphoses to the adult stage in a watery environment (Fig. 10.23).

Reptilian adaptations to a terrestrial life included many other features, such as desiccation-resistant skin, improved circulatory and respiratory systems, and a more efficient skeletal framework and articulations suited to greater mobility and speed on the land. Later reptiles evolved more effective jaw and tooth structures, which allowed them to apply heavy pressure to foods and slice them with bladelike teeth. Previously, they had to swallow their food whole, as the living amphibians still do.

By the **Permian Period** (286–248 Myr BP), the reptiles had diversified considerably, and were beginning to replace the amphibians as the dominant land fauna. The pelycosaurs, which include finback reptiles, such as *Dimetrodon* and others, were a successful and varied group composed predominantly of carnivores that had appeared in the Pennsylvanian and continued into the Permian. Pelycosaurs are particularly important because they, in turn, were the ancestors of the **therapsids**, or mammal-like reptiles, which gave rise to the mammals in the Mesozoic Era (Fig. 10.24). The therapsids may have been endothermic (warm-blooded), although this remains uncertain, but the pelycosaurs and most or all other reptiles were ectothermic (cold-blooded).

In addition to an abundant flora of lycophytes, seed ferns, and spore-producing ferns, gymnosperm evolution produced the first cycads and a diversity of conifer-type seed plants, which later became the predominant forest trees of a fragmenting Pangaea. Many order of insects appeared in the Permian, including beetles (Coleoptera), true flies (Diptera), and true bugs (Hemi-

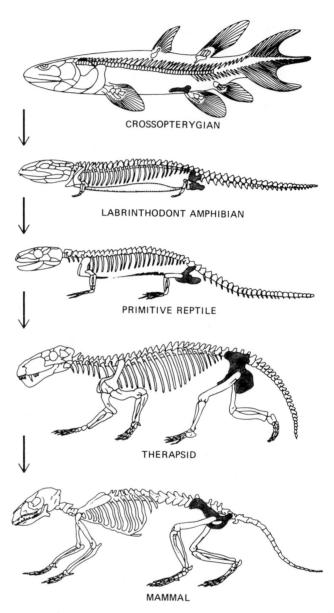

Figure 10.22. Evolutionary changes in skeletal features of the ancestral crossopterygian fish and the tetrapod vertebrate descendant lineages. Note particularly the increasingly sturdy support and articulation of the pelvis, backbone, and upper leg bone, as well as anterior structures. The earliest reptiles, such as *Hylonomus*, had a sprawling posture, whereas limbs of the therapsid supported the body high off the ground, as do those of their mammalian descendants.

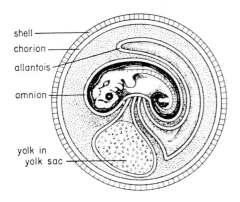

shell
chorion
allantois
amnion
yolk in
yolk sac

Figure 10.23. The amniote egg of land-adapted vertebrates has an outer covering, which may or may not be a rigid shell, that encases the developing embryo situated in a watery cavity bounded by the amnion membrane. Both the amnion and the chorion, another membrane, grow out from the embryo's digestive tract. The yolk sac in the eggs of birds and reptiles is filled with a large mass of nutritive yolk. The allantois grows out from the back end of the embryo's gut, and may function as an embryonic bladder or lung. In a typical mammal, the chorion and surrounding maternal tissues form the placenta, by which the embryo receives nutriments from the maternal bloodstream and eliminates wastes. The blood vessels of the allantois carry nutriments from placenta to embryo (see Fig. 10.30).

ptera). In the seas, the ammonites flourished, as did many other marine invertebrates, and a new grade of bony fishes (Holostei) made their appearance. These fishes include the ancestors of the teleost fishes, which emerged in the Mesozoic Era and today constitute the major group of fishes all over the world.

The end of the Permian was the time of the greatest mass extinction ever known to have occurred, principally affecting marine invertebrate life but also touching vertebrate groups. About 75 percent of the amphibian families and about 80 percent of the reptilian families went extinct, although representatives of the main groups (orders) survived the cataclysm. In contrast with the lobe-finned fishes, which were greatly diminished in the Carboniferous as well as the Permian, the ray-fins survived and were to diversify significantly in later eras. Some estimates place the species kill in the Permian extinction as high as 96 percent. The Paleozoic Era ended as well, but Pangaea had come into being and was to influence Mesozoic and Cenozoic life profoundly when it slowly split up during the next 200 Myr.

Mesozoic Plants and Animals

In the 342 Myr of the Paleozoic Era (590–248 Myr BP), almost all the major living groups of plants and animals had appeared; the exceptions were the birds, mammals, and flowering plants. They evolved in the Mesozoic Era (248–65 Myr BP), but the reptiles remained the predominant land animals throughout the era, and the flowering plants overtook the gymnosperms only in the Late Mesozoic.

In the **Triassic Period** (248–213 Myr BP), the gymnosperms, seed ferns, and spore-producing ferns predominated over the earlier lycophytes. The

a

b

Figure 10.24. Permian reptiles. (a) The pelycosaurs included finback types, such as *Dimetrodon*, and more lizardlike types (at left). Pelycosaurs were ancestral to (b) therapsids, such as *Lycaenops*, which, in turn, were ancestral to the mammals. (Painting by Charles Knight, from the Field Museum of Natural History, Chicago. Negative number 73837, reproduced with permission)

grade innovation of seed production in gymnosperms and seed ferns is a terrestrial adaptation analogous to the evolution of the amniote egg in land vertebrates. Seeds are produced on the mother plant after internal fertilization of the egg, and the embryo develops within a nutrient-containing environment covered by a tough, desiccation-resistant seed coat. In the spore-producing ferns, a tiny green gametophyte plant develops from a germinating spore shed by the large, familiar sporophyte (see Fig. 5.33). Eggs and sperm produced in the gametophyte must have access to water so that the sperm can swim to the eggs, and the fertilized egg leads a precarious existence until it grows into the new spore-producing fern plant. Seed plants can inhabit a wide variety of land habitats, whereas many ferns are tropical or subtropical plants that must grow in a moist environment.

Many diverse invertebrates appeared and flourished, and new amphibians replaced the Paleozoic forms. The first frogs evolved in the Triassic, and fossil salamander bones have been found in Late Triassic deposits. Reptiles proceeded through a series of adaptive radiations that populated the land with cotylosaurs, therapsids, archosaurs, and other groups (Fig. 10.25). Reptiles also reentered the seas in the form of ichthyosaurs, which were fishlike, and of nothosaurs and plesiosaurs, which had paddlelike limbs resembling those of modern seals. Turtles appeared at the end of the Triassic, and the tortoises arose later, in the Jurassic Period. Therapsids continued their evolution toward a mammalian grade, and went extinct in the Jurassic at about the time the first mammals appeared. The transitional nature of many preserved skeletal and dental features often makes it difficult to determine whether a particular fossil should be classified as a therapsid reptile or as a mammal.

The archosaurs diverged during the Triassic into a variety of lineages, all characterized by the insertion of teeth into jaw sockets and by other features. The predatory fresh-water phytosaurs were similar to crocodiles and related to them. There also appeared in the Late Triassic two groups of dinosaurs.

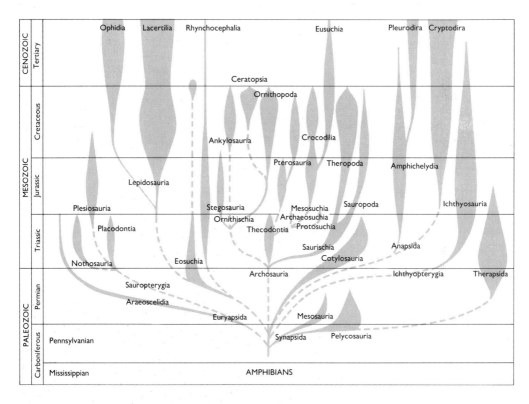

Figure 10.25. Suggested evolutionary history of the reptiles from their origin in the Carboniferous Period, through their greatest diversification in the Mesozoic Era, to the remaining living orders. Note particularly the extinctions at the end of the Triassic and end of the Cretaceous periods. The width of a line or band indicates the relative size of the group at the corresponding geologic interval. Dashed lines indicate uncertainties about relationship or time of origin. (From J. W. Valentine, "The Evolution of Multicellular Plants and Animals." Copyright © 1978 by Scientific American, Inc. All rights reserved)

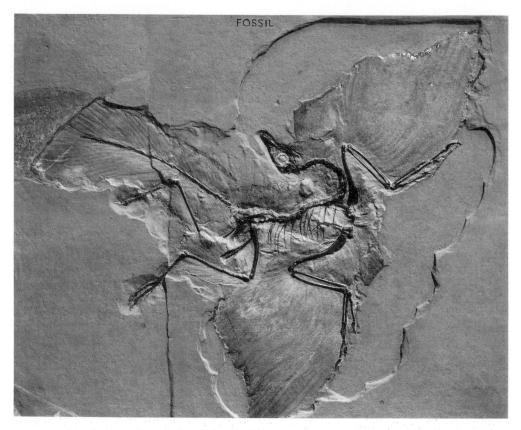

Figure 10.26. Cast of the Jurassic fossil *Archaeopteryx*, the oldest known bird. It closely resembled its carnivorous dinosaur ancestors, except that feathers rather than scales covered the body. The relatively long tail was characteristic of these primitive toothed birds. (Photograph courtesy of the American Museum of Natural History, New York)

The **saurischian dinosaurs** included both herbivores and carnivores and had a three-pronged pelvis, as do other reptiles. The **ornithischian dinosaurs** were herbivores and had a characteristic four-pronged pelvis, as do birds. The names of the two orders, Saurischia and Ornithischia, refer to their pelvic forms. These orders diverged from a dinosaurian stock of the Middle Triassic. In the Late Triassic Period, the vertebrates took to the air for the first time, as pterosaurs. These flying reptiles had large wings made of skin, a tail, and hollow pneumatic bones that facilitated flight.

The Jurassic fossil *Archaeopteryx* is the oldest known fossil bird. It resembled small carnivorous dinosaurs in many respects, but clearly possessed feathers (Fig. 10.26). Its ancestry had long been assigned to primitive archosaurs (thecodonts), but it is now considered to be descended from the small theropod (carnivorous) saurischian dinosaurs of the Jurassic. Birds are thus thought to be descendants of the dinosaurs. Some biologists consider birds to be living dinosaurs, and classify them as such. Although classifications may differ, the consensus is that birds evolved from a dinosaurian lineage.

Reptiles continued to diversify and proliferate in the **Jurassic Period** (213–144 Myr BP), having come through a mass extinction at the end of the Triassic, which claimed many groups of marine invertebrates and amphibians. Many of the dinosaurs evolved to gigantic size, including sauropod (herbivorous) saurischians such as *Brachiosaurus*, about 80 feet long; *Apatosaurus (Brontosaurus)*, nearly 70 feet in length; and *Diplodocus*, about 90 feet from head to tail. Even larger forms have been identified recently. Among the Jurassic ornithischian dinosaurs was *Stegesaurus*, whose rows of dorsal armor plates may have provided a system for cooling the animal, although this suggestion remains speculative.

Archaic mammals appeared about 180 to 190 Myr BP, very near the time that their therapsid ancestors went extinct. Fossil evidence indicates that at least three groups of early mammals evolved in transitional stages from small carnivorous therapsids similar to *Probainognathus*, of 190 to 195 Myr BP. These mammals were small shrewlike nocturnal creatures assigned to the amphilestid, morganucodont, and kuehneotherid groups (see Fig. 10.37). The amphilestids left no known descendants, but the morganucodonts are considered to be related to the living monotremes, and the kuehneotherids may be related to living marsupials and placentals (Fig. 10.27). New fossil discoveries may clarify these suggested relationships.

The transition from therapsid to mammal is among the best-documented examples of gradual divergence by means of anagenetic evolutionary changes. Early in their history, therapsids progressed from a sprawling posture to a more erect carriage, with all four limbs sturdily and flexibly articulated below the body (see Fig. 10.22). Changes in the skull, particularly in the jaw, provide the best fossil evidence for the transition to mammalian features. The reptilian lower jaw is a composite of a small tooth-bearing (dentary) bone and a number of other bones toward the rear of the jaw and at the joint where it is attached to the skull (dentary–squamosal joint). The mammalian lower jaw is primarily a single large dentary bone. The articular and quadrate bones, which form the jaw joint in reptiles, have moved out of the jaw in mammals to become the *malleus* and *incus* of the mammalian middle ear; the *stapes* of the middle ear occur in all tetrapods (Fig. 10.28).

The malleus and incus bones, along with the stapes of the middle ear, provide mammals with a more actue sense of hearing over a much greater range of frequencies than is enjoyed by other vertebrates. The mammalian jaw joint is stronger and, together with new muscles and changing surfaces of the teeth, allows for more complex chewing than in the other living vertebrates. The lower jaw can move forward and backward and from side to side as well as up and down. The larger nasal passages in mammals (and an expanded olfactory lobe) underlie an enhanced sense of smell, which permits more scope in

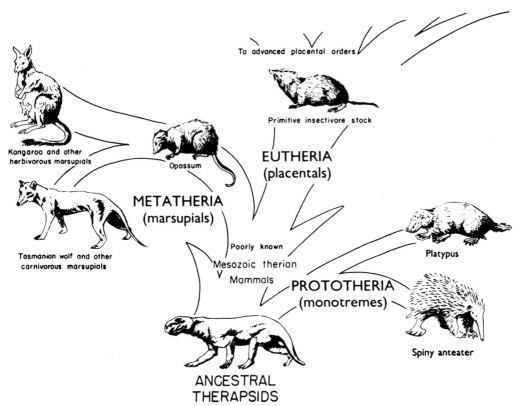

Figure 10.27. Generalized family tree of the mammals. The subclass Prototheria (monotremes) are egg-laying species, now represented by only the Australian platypus and spiny anteater (echidna). Members of the subclass Theria give birth to live young, but important differences distinguish the infraclasses of Metatheria (marsupials) and Eutheria (placentals). In marsupials, the young are born in an early embryonic stage and continue their development inside a pouch (marsupium) formed by a flap of abdominal skin on the mother's belly. In placental mammals, the young develop over a longer gestational time in the mother's body, and are nourished there through a placenta composed of embryonic and maternal tissues. Once hatched or born, all mammalian young suckle on mother's milk before they become capable of obtaining food from their surroundings. (From A. S. Romer, 1970, *The Vertebrate Body*, 4th ed. Philadelphia: Saunders)

finding food or avoiding predators. All these changes in the lower jaw, the jaw joint, the middle ear, the nasal passages, and other skull features began to take place early in therapsid evolution, and ultimately emerged in their mammalian form in the first archaic mammalian species to cross the line of transition to the new grade.

Therapsid evolution apparently did not include a relatively large brain, which distinguishes the mammals, or endothermy. Comparable features of the

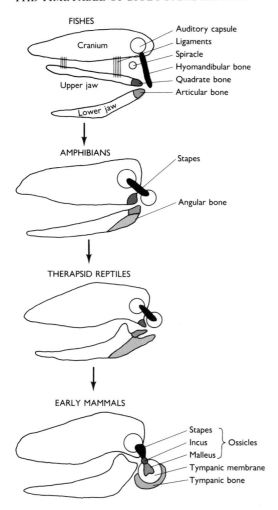

FISHES
Auditory capsule
Cranium
Ligaments
Spiracle
Hyomandibular bone
Upper jaw
Quadrate bone
Articular bone
Lower jaw

AMPHIBIANS
Stapes

Angular bone

THERAPSID REPTILES

EARLY MAMMALS
Stapes
Incus ⎱ Ossicles
Malleus ⎰
Tympanic membrane
Tympanic bone

Figure 10.28. The evolution of the vertebrate ear ossicles and dentary–squamosal joint. The hyomandibular, which supports the jaw in fishes, was modified and incorporated as stapes into the amphibian middle-ear cavity. In therapsid evolution, the stapes was joined by the articular and quadrate bones from the back of the jaws. These bones become the malleus and incus, respectively, of the mammalian middle ear. In addition, the upper jaw fused to the base of the cranium in tetrapod evolution, and the new articulation between the lower jaw and cranium began in the therapsids and advanced in mammals. The mammalian dentary–squamosal joint is much stronger than the joint in other living vertebrates.

vascular structure of long bones in mammals and therapsids, however, suggest similar high metabolic levels in the two groups. The prevailing theory is that at least one later therapsid lineage, perhaps like *Probainognathus*, evolved into creatures that could exist successfully as nocturnal species in the environments dominated by the diurnal dinosaurs and other reptiles. With more acute hearing and smell, more effective chewing apparatus, and body warmth achieved as a result of higher metabolic rate, the mammal-like therapsids may have crossed the threshold into the mammalian grade. The long history of increasingly mammal-like features and the considerable similarities between the small, nocturnal, carnivorous therapsids of 190 to 195 Myr ago and the small, nocturnal, insectivorous mammals of 180 to 190 Myr ago document the transition from therapsid to mammalian grade.

The **Cretaceous Period** (144–65 Myr BP) saw the appearance of modern sharks and rays; the dominance of the teleost fishes; the first snakes and a greater diversity of lizards; giant *Triceratops* and other ornithischian dinosaurs; fierce *Tyrannosaurus rex,* whose 6-inch-long bladelike teeth could tear apart the toughest prey; and more modern birds with a keel-shaped sternum, no teeth, a reduced tail, and fused wing digits, among other features. Birds and mammals have a high metabolic rate and internal control over body temperature, but their endothermic features are the result of independent changes in different lineages during convergent evolution. Both classes of vertebrates also independently evolved a highly developed central nervous system, which permits great control over motor activities, well-developed sight and equilibrium, and complex behavior patterns.

Flowering plants, or **angiosperms,** appeared in the Early Cretaceous and quickly diversified to become the dominant component of the Mesozoic flora. They have maintained this dominance ever since. Little is known about the origin of the flowering plants, but the earliest forms were insect-pollinated and had flowers with a large but variable number of petals and reproductive parts. According to comparative anatomical studies, it is quite possible that various structural similarities in the reproductive structures of seed ferns (pteridosperms) and angiosperms represent homologies indicative of an ancestor–descendant relationship. It has been suggested that a *Glossopteris*-like pteridosperm may have diverged to produce the grade of angiosperms, but the data are inconclusive. The diversity of Cretaceous insects, such as moths and butterflies (Lepidoptera), bees and their allies (Hymenoptera), and true flies (Diptera)—all of which are major pollinators—obviously is correlated with the rapid expansion and diversification of flowering plants in this period.

Many mammals of the Middle Cretaceous were *therian,* producing live young after a period of gestation within the mother, but they had not yet evolved the distinctive marsupial (Metatheria) or placental (Eutheria) suites of characteristics. By the Late Cretaceous, however, members of the modern opossum family and other marsupials were present, as were an assortment of placental mammals somewhat similar to the Insectivora, the order of shrews, moles, and hedgehogs. Other placental types included members of the primate order, to which we belong; rodentlike multituberculates, named for the many bumps (tubercles) on their teeth; and the condylarths, which are now extinct but included lineages ancestral to the living carnivores, ungulates (Perissodactyla, Artiodactyla), and elephants (Proboscidea). All these types were very similar to the insectivores, or have been classified as members of that order, throughout the Cretaceous Period.

The end of the Cretaceous, 65 Myr BP, was notable for the second-"worst" mass extinction. All terrestrial animals larger than about 50 pounds (20–25 kg) went extinct, including the remaining dinosaurs, but few of the fishes, amphibians, or plant groups appear to have been affected. This dramatic termination of the Mesozoic Era has recently been hypothesized as the result of the catastrophic impact on the Earth of a 10-km-wide asteroid, meteorite, or comet from some outer region of the solar system. The clouds of dust spewed into the atmosphere from the enormous crater left after the impact are supposed to have obscured the sun for months, or perhaps years, causing darkness and cold not unlike the conditions described today for a possible "nuclear winter." This topic will be discussed later in the chapter.

The Age of Mammals: The Cenozoic Era

The abundance of fossils in deposits of the **Tertiary Period** (65–1.8 Myr BP) and **Quarternary Period** (1.8 Myr BP-present) permit paleontologists to distinguish significant events in the different epochs within each period. Monotremes and marsupials were established in the Mesozoic Era, but although present in the Cretaceous, placental mammals began to differentiate significantly only in the **Paleocene Epoch** (65–58 Myr BP). By the early **Eocene Epoch** (58–37 Myr BP), most of the living placental orders appeared in a major episode of adaptive radiation, probably from insectivore ancestors (Fig. 10.29). Living members of the placental order Insectivora (shrews, moles, hedgehogs) are characterized more by ancestral than by derived character states. During this time, widely diverse flowering plants and modern orders of insects and other organisms also came into existence.

The class **Mammalia** is subdivided into two subclasses, the **Prototheria** (egg-laying monotremes) and **Theria** (produce live young). The therians are grouped into two infraclasses, the **Metatheria** (marsupials) and **Eutheria** (placentals). The young marsupial is born in an undeveloped embryonic state after a brief period of gestation. It crawls from the uterus up to a pouch (marsupium), formed by a flap of abdominal skin on the mother's belly. Once inside the pouch, the embryo continues its development while nurtured by milk pumped by maternal efforts to the teats, to which the embryo becomes firmly attached. In placental mammals, the unborn young proceed through a longer gestation, during which most of the major organ systems become relatively well developed. The embryo, and later the fetus, is nutured through the **placenta,** an organ composed of embryonic and maternal tissues (Fig. 10.30). Nutrients are delivered by a continuous circulation from the maternal bloodstream through the placenta to the umbilical cord and into the fetus, and wastes are removed from the fetus to the maternal bloodstream through the same pathway. After birth, the youngster is suckled on mother's milk as it continues to develop to a stage of relative independence. Whether egg-laying or therian, all mammals are characterized by hair (fur) covering the skin to a

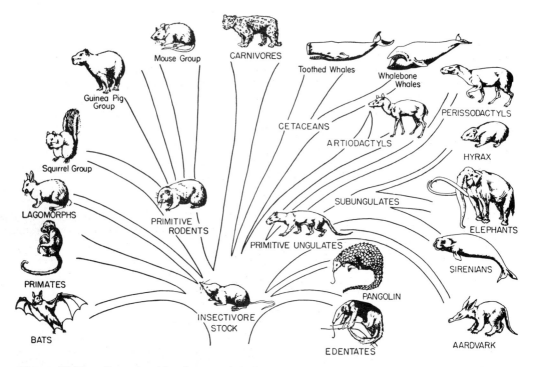

Figure 10.29. Generalized family tree of the living orders of placental mammals, all of which trace their ancestry to an insectivore stock from which they evolved in a major episode of adaptive radiation during the Early Cenozoic Era. (From A. S. Romer, 1970, *The Vertebrate Body*, 4th ed. Philadelphia: Saunders)

greater or lesser degree, and by being suckled on milk produced by maternal mammary glands after being hatched or born.

Australia probably became isolated from the West Gondwana regions before placental mammals had evolved or reached East Gondwana, and its monotreme and marsupial populations flourished in the absence of competing and more advanced placental species (Fig. 10.31). South America separated after early placentals had joined its marsupial fauna, and marsupials fared better there than elsewhere in the world where more advanced placentals evolved during the Cenozoic Era. The marsupial, primitive placental, and unique placental groups of South America continued to evolve in isolation until the Late Cenozoic, when South America was joined to North America by the Isthmus of Panama.

Many Eocene placental groups became quite diverse, but few of them were able to compete successfully with their more advanced descendants in the **Oligocene** (37–24 Myr BP) and **Miocene** (24–5 Myr BP) radiations. Among the primates, the anthropoids (monkeys, apes, hominids) displaced the primitive prosimians (lemurs, lorises, tarsiers) during the Oligocene and Miocene epochs, although the hominid family may not have appeared until 5 Myr BP, at the start of the **Pliocene Epoch** (5–1.8 Myr BP). Primate and human evolution will be discussed in Chapters 11 and 12.

The Quaternary period began with the **Pleistocene Epoch** (1.8–0.045 Myr BP), during which there existed many genera and species similar or identical to those of the **Recent,** or **Holocene, Epoch** of the past 45,000 years. The Pleistocene was a time of great fluctuations in climate, including at least 4 major and up to 20 minor glaciations. The cool and dry Pleistocene climate led to the migration of many plants and animals to more tropical southern regions in glacial periods and farther north during the warm interglacial periods. The last glacial period (the Wisconsin) ended about 10,000 years ago, and we are now in an interglacial phase. Sea levels dropped by several hundred feet or more when water became tied up in massive ice sheets, and rose again when ice melted as the climate warmed. In spite of these extremes of sea level and climate, rates of speciation and extinction appear to have remained essentially unchanged, except for a few groups in North and South America.

Many of the giant mammals, such as mastodons and mammoths, and species of giant bison, elk, beaver, sloth, cats, and others, went extinct between

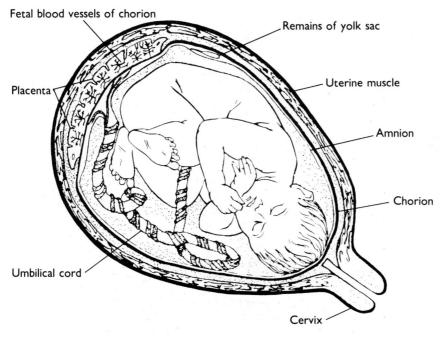

Figure 10.30. A cutaway view through the human uterus, showing the placenta and the fetus shortly before birth (compare Fig. 10.23).

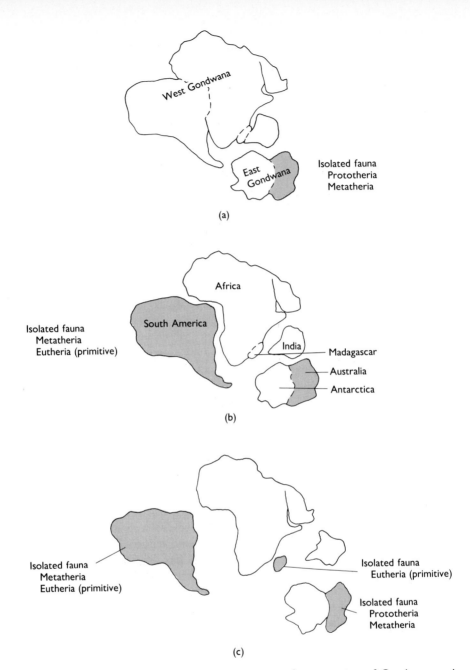

Figure 10.31. The suggested relationship between progressive fragmentation of Gondwana and possible isolated faunas in Australia, South America, and Madagascar by the end of the Cretaceous Period. (a) Prototherians and metatherians may have been isolated in Australia when East Gondwana and West Gondwana became separated. Evolution continued to occur elsewhere and produced primitive eutherians, which were absent from Australia because of its early isolation from the rest of the world. (b) South America split from Africa in the Cretaceous, and its fauna was isolated before advanced eutherians had evolved or reached the South American landmass. (c) By the end of the Cretaceous, Madagascar had separated from the African mainland, carrying its isolated fauna of primitive eutherians. The presence and absence of various mammalian groups in these regions is thus inferred to be due to the time at which landmasses separated from one another in relation to times of evolutionary changes in worldwide mammalian groups.

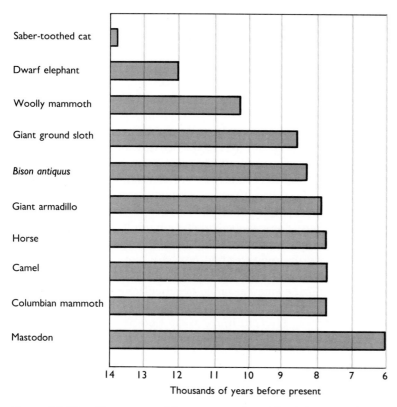

Figure 10.32. The pattern of disappearance from North America of representative groups of mammals between 14,000 and 6000 years ago. The Wisconsin glaciation ended in North America about 10,000 years ago, but extinctions have continued into the current interglacial period.

14,000 and 6000 years ago (Fig. 10.32). Most of these animals disappeared between 10,000 and 6000 years ago, during the interglacial period following the retreat of the Wisconsin glaciers and ice sheets. Other mammals, such as the horses, migrated across the Bering land bridge from North America into Asia and survived there. The horse was reintroduced to its earlier North American home, and to South America, by the Spanish conquistadors in the sixteenth century.

Little information is available with which to explain this most recent episode of wholesale extinction, although some have tied it to human hunters who emigrated from Asia across the Bering land bridge into the Americas at least 12,000 years ago. The "overkill" hypothesis states that human hunting on the scale of mass slaughter was directly responsible for the extinction of many kinds of large mammals in the Americas in the Recent Epoch. The data are ambiguous, however, and archeological and fossil evidence for such gross human activities is virtually nonexistent. Slaughter of animals and decimation of plants certainly go on today at the hands of humans, but whether scattered bands of prehistoric people were as effective killers as the billions of us are today remains to be determined.

The Unpredictability of Evolutionary Direction

Various fossil lineages show a seemingly straight-line progression from an ancestral stock to living descendants. Such apparently continued polarity of change has sometimes been interpreted as evidence that species evolve toward some particular end point of perfection. This viewpoint, called **orthogenesis,** implies purpose in evolution and evokes ideas of vitalism or supernatural plan guiding organisms toward their ultimate destiny. Rather than supporting orthogenesis or its philosophical premises, clear evidence exists in the fossil record showing that most lineages include many more "failures" than "successes." In a familiar metaphor, we find that lineages resemble profusely branched shrubs with many twigs leading nowhere, not a straight-line progression of ancestral and descendant branches.

A particularly-well documented example of reticulate rather than straight-line evolution is the horses of the Cenozoic Era, which evolved from the small, multitoed, browsing *Hyracotherium* of the Eocene Epoch to the large, one-toed, grazing species of *Equus,* the only surviving genus of the 60-million-year-old group. By emphasizing the highlights of the three main features of equine evolutionary change—reduction in digit number, increase in body size, and increase in the height and complexity of grinding teeth—the early paleontologists constructed a progression of forms along a main stem, and deemphasized other forms by showing them as side branches off the main stem.

With more fossil material and a different perception of the framework of change, the evolutionary history of the horses becomes multiply branched and more complex. The peak of diversification occurred in the Miocene and Pliocene epochs, and included a number of successful three-toed, browsing genera, as well as grazers. All became extinct in the Pliocene and early Pleistocene, except for the one-toed, grazing *Pliohippus,* whose only surviving descendant is *Equus* (Fig. 10.33). Rather than being the focus of the entire evolutionary history of the horses, living *Equus* is a side group that happened to survive the wave of extinctions that claimed the rest of the members of the family despite their relative success and diversity in earlier times. In fact, all the odd-toed ungulates (Perissodactyla) are dwindling remnants of this previously dominant order of hoofed mammals. Only the horses, tapirs, and rhinoceroses remain today, and they are represented by a few species. The even-toed ungulates (Artiodactyla), on the contrary, include a variety of deer, antelope, bovines, pigs, and other species.

Evolution is *opportunistic.* Some of the major changes in life forms have originated in structures or systems that either were relatively unimportant or fulfilled functions in ancestral lines different from those they perform in descendant groups. Mammary glands are modified sweat glands, which serve

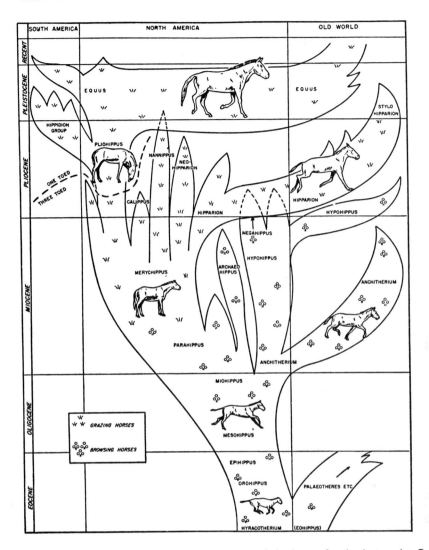

Figure 10.33. General evolutionary history of the horse family during the Cenozoic Era. Major changes in the lineages included increased body size, greater height and complexity of the grinding teeth, and reduction in digit number. Some of these changes are associated with new dietary and habitat features of grazing species from browsing ancestors. *Equus* is a descendant of a side branch of one-toed horses, but its prominence is emphasized by the display according to geographical location rather than some other reference feature. (From *Horses* by G. G. Simpson. Copyright © 1951 by Oxford University Press. Reprinted by permission)

different secretory functions in ancestors and descendants. The jaws of the placoderm fishes and all subsequent vertebrate lines are derived from an anterior pair of skeletal gill supports in the jawless agnathans (see Fig. 10.19), which is the ancestral lineage of all other vertebrates (Fig. 10.34). Jaws and gill supports serve very different functions, and the one evolved from the other opportunistically.

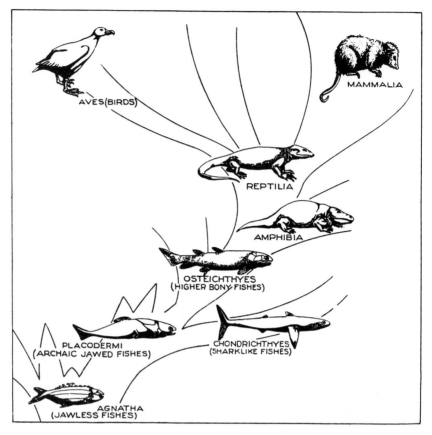

Figure 10.34. Simplified family tree of the eight classes of vertebrate animals. Most agnathans and all placoderms are now extinct. (From A. S. Romer, 1970, *The Vertebrate Body,* 4th ed. Philadelphia: Saunders)

In one sense, evolution is predictable: it is *irreversible*. It is patently clear from the fossil record that a life form never reappears after it has gone extinct, nor would such a reappearance be expected on conventional genetic and evolutionary grounds. Each species evolves from an existing ancestor by mutational and chromosomal changes that are incorporated into an extensive genetic program, which itself has evolved from a collection of earlier events and ancestors. These genetic modifications were fixed or lost in reponse to the actions of genetic drift and natural selection over very long periods of time. Such changes are continuing, and an exact set of reverse changes and selection pressures would be necessary to retrace an earlier sequence from any later derived species. The improbability of mutation and selection pressures being identical to those in past times, and occurring in the correct order in the correct environments, makes extinction an end point from which there is no return.

In every other way, however, the direction and character of evolutionary change are entirely unpredictable. It is only by hindsight that biologists see advantages in some new group of organisms compared with its ancestors. We would never have predicted that mammals would supersede the dinosaurs and other reptiles if we had been observers in the Mesozoic Era and known nothing of the worldwide changes that were to close that era and usher in the Cenozoic. Indeed, we have no idea whether the dominant mammals will be replaced by some new life form in the eras to come, or what that replacement could be.

The Impact of Extinctions on Evolutionary History

Species extinctions during evolutionary history have taken most of the 4000 million kinds of organisms estimated to have existed in the past, leaving only 2 to 3 million species living today. Various proposals have been offered to explain the "normal" or "background" rates of extinction, as well as the more dramatic mass extinctions during the Phanerozoic Eon. The next few sections will focus on these events and proposals and on the influence of extinctions on the rate and diversity of speciation in their aftermath.

Patterns of Extinction

At one time or another, **extinctions** have been blamed on changes in the physical or biological environment, or both. Loss of habitats, predation and competition, disease, and other biological causes, as well as change in climate, glaciation, rise and fall of sea levels, formation or removal of inland waters, and other physical causes have been implicated in species extinction. Although supported by data in some cases, one or another of these factors has not been shown unambiguously to cause most extinctions.

On theoretical grounds, genetically flexible species are expected to be more resistant to physical and biotic changes in their environments, and thus to be more resistant to extinction, than their more closely adapted relatives and neighbors. Comparisons of more generalized and less generalized species have depended largely on morphological criteria applied to fossil forms that endured for different lengths of time. The presumption is that morphologically more advanced species are more "specialized" than their relatives with modest morphological advances and variety; the latter are considered more likely to accommodate genetically to changes around them. The comparative data have shown, however, that morphologically more complex species may have either greater or lesser longevity than their simpler cohorts. These comparisons would be meaningful only if there were information about their different genetic and life histories.

The idea developed from presumptions such as that just described that "overspecialized" species are doomed to early extinction. All species are well adapted to their environments in some feature or another, so it is an arbitrary decision as to whether a species has crossed the threshold and become specialized to a point of lower fitness in its environment. One of the favorite examples of an "overspecialized" species is the extinct giant Irish elk *(Megaloceros giganteus)*, whose huge antlers were regarded as an evolutionary end point of antler increase in the cervine deer. Its massive antlers were considered to have been so burdensome that the Irish elk became unfit, relative to its ancestors, and went extinct as a direct consequence of its headdress. If the relationship between body size and antler size is plotted for this and other species of cervine deer, however, it is clear that the two features are proportional (Fig. 10.35). The Irish elk was no more specialized than its ancestors, whose body size and antler size also increased proportionately during their evolution.

Comparisons of longevities of families, genera, or species *within* a taxonomic group indicate the same average survival time for all the members of that group. Leigh Van Valen has shown that within a group, the probability of extinction is independent of the age of the taxon, which indicates that older taxa are not necessarily more fit than younger taxa (see Fig. 6.21). This contradicts the expectation that long-lived taxa would have a lower probability of extinction than younger taxa, based on the assumption that longer survival time would result from selection over a longer time for greater resistance to environmental change. These data have been interpreted in various ways, none of which is accepted by everyone. One interpretation holds that the probability of extinction may reflect selection causing poorer as well as better levels of adaptation, not just better or more improved adaptive features; these levels would tend to average out to constancy of extinction rate. Another interpretation suggests that each of many kinds of environmental changes bears a small risk of extinction, and because these variations occur continually, they would produce an average extinction rate that appears to be constant. According to these views, therefore, selection causes fluctuations in extinction rates that appear to be constant, but actually are not.

The probability of extinction may be similar within a group, but different rates characterize different groups of organisms. Among marine invertebrates, which make up the bulk of the fossil record, the average longevity of a genus has been calculated to be 78 Myr in bivalves, 16 to 20 Myr in brachiopods, and about 7 Myr in ammonites. Duration times vary from one geologic interval to another, however, so Jurassic bivalves lasted only about 20 Myr, not 78 Myr, and the Cretaceous ammonites winked out after about 500,000 to 700,000 years, compared with their much slower extinction rate in earlier times.

These and similar calculated extinction rates occur throughout intervals of time and are considered to be the "normal" or "background" rates. In contrast with these ongoing occurrences are extinctions that are clustered in relatively brief periods of time. Such brief episodes of extinction may lead to mass

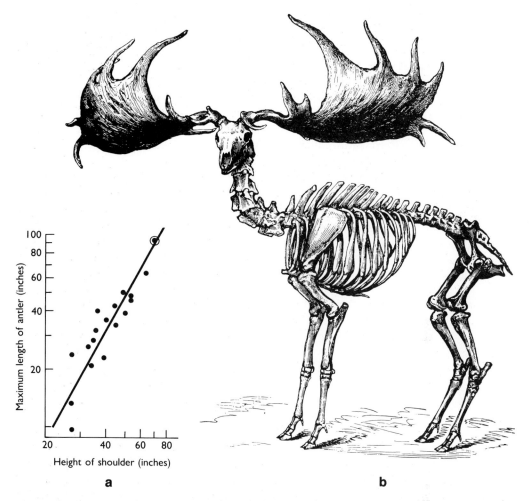

Figure 10.35. Relation between body size and antler size in cervine deer. (a) In general, the antlers of cervine deer increased in size proportionately with body size. (b) The largest known antlers belonged to the Pleistocene Irish "elk" *Megaloceros.* Its antlers were quite large, but were actually exactly the size (circled dot) that could be predicted from the observed trend (straight line in graph). ([a] after S. J. Gould, 1973. *Nature* 244:375. Copyright 1973 by Macmillan Journals Limited. [b] from *Understanding Evolution* by Earl D. Hanson. Copyright © 1981 by Oxford University Press. Reprinted by permission)

extinctions, which affect many different families or higher taxonomic categories in a relatively short interval of geologic time. The high drama of these major events and their reasonably good documentation in the fossil record have made then a focus for many studies. Other incidents of clustered extinctions have occurred on a smaller scale or affected fewer species. Among these are extinctions linked to human activities in prehistoric and historic times.

Apart from the uncertain link between human hunters and the extinction of ice age mammals in the Americas between 14,000 and 6000 years ago, more recent waves of human migrations and colonizations were associated with the disappearance of native species. In some cases, biologists cannot be certain about cause and effect, but in other cases, it is clear that humans were responsible for the eradication of many species native to a newly settled or colonized region. Archeological excavations, radiocarbon datings, and reports written by eighteenth- and nineteenth-century Europeans provide explicit evidence of the extermination of species by human colonizers and by the pest animals and diseases they brought with them to a new land.

Overhunting by humans and depredations by introduced rats, cats, pigs, goats, and other pests were responsible for the extinction of many native species in Hawaii and Madagascar, which were settled by at least 1500 years ago, and in New Zealand by the Maoris, who migrated there from Polynesia at least 1000 years ago. Losses were particularly severe among giant and large species, which were hunted for food, but smaller forms also suffered extinction. Times were particularly hard for the larger species because they had low reproductive potentials, lived in low population densities, and were virtually defenseless, preferred prey of hunters. New Zealand lost its 19 largest flightless birds, including the 9-foot-tall giant moas. Madagascar lost 6 to 12 species of giant elephant birds, 2 species of giant tortoises, and 14 of its largest lemurs, which are prosimian primates found nowhere else in the world. Polynesians decimated nearly 40 bird species in Hawaii, and the Europeans who arrived much later wiped out at least 15 of the remaining 49 native bird species.

In addition to the activities of human hunters and their pests, habitat destruction by human inhabitants led to plant and animal extinctions, as it does now. The clearing and draining of forests, woodlands, and other habitats to make way for agriculture were especially destructive for native species. Today, the economic exploitation of timberlands and the increased need for farmland by a growing human population are responsible for wanton destruction of tropical rain forests and many other regions around the world. The large number of endangered species, the continued destruction of depleted populations of whales and other species, and the daily extinction of one species or another in our own time are a tragic testimonial to the low priority we give to the other species with which we share the planet.

Mass Extinctions During the Phanerozoic

Commonly used measures to assess the intensity of extinction include (1) total taxa going extinct in a particular unit of geologic time, (2) extinctions in a time unit expressed as a percentage of existing diversity, (3) number of extinctions per Myr, and (4) percentage of extinctions per Myr. Despite biases, sampling problems, and difficulties in selecting the most appropriate measure, five incidents stand out consistently as the largest and most profound episodes that are accepted as mass extinctions (Fig. 10.36). They occurred at the end of the Ordovician period (~440 Myr ago), in the Late Devonian Period (~365 Myr ago), in the late or terminal Permian Period (~250 Myr ago), at or near the end of the Triassic Period (~215 Myr ago), and at the end of the Cretaceous Period (65 Myr ago). The end of the Permian marked the end of the Paleozoic Era, and the close of the Cretaceous terminated the Mesozoic Era. The Permian mass extinction was the largest, with estimates of species kill as high as 96 percent.

The occurrence of mass extinctions at the end of geologic periods and eras is due entirely to the fact that geologists and paleontologists separated time intervals during the Phanerozoic according to notable biological events as well as geologic changes. Mass extinction or the disappearance of important index fossils was the most easily detected and dramatic event and so served

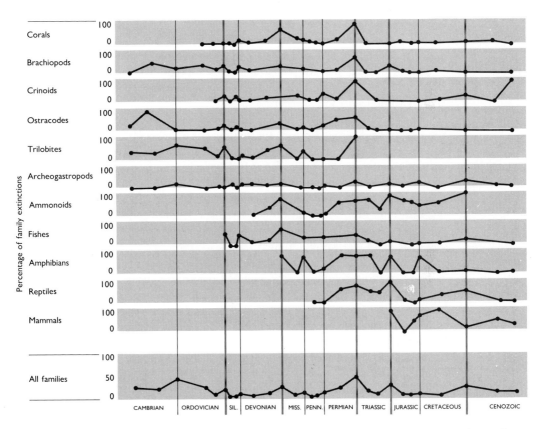

Figure 10.36. The history of extinctions during the Phanerozoic Eon, indicating repeated episodes, of which five conform to the dimensions of mass extinctions (thick vertical lines). The relatively high percentage of family extinctions at the end of the Cambrian was due more to extinction of a few diverse groups, such as families of trilobites, than to extinction of many families of different kinds of animals existing at the time. (From N. D. Newell, "Crises in the History of Life." Copyright © 1963 by Scientific American, Inc. All rights reserved)

to set boundaries for geologic time intervals. For example, the history of the trilobites provides key reference points for Paleozoic time slots. About 60 trilobite families were present during the Cambrian Period, whose end is marked by the extinction of about two-thirds of these families. The last of the trilobites perished during the Permian Period, so these arthropods serve as index fossils for Paleozoic times.

Mass extinction at the end of the Ordovician took many kinds of marine invertebrates, including families of the brachiopod phylum (lampshells), echinoderm phylum (crinoids), and arthropod phylum (trilobites, ostracods). The end of the Devonian saw the disappearance of more families of trilobites, ostracods, corals, ammonoids, and other marine invertebrates, as well as families of fishes, which had appeared as jawless forms in the Late Cambrian. At or near the end of the Permian Period, which closed the Paleozoic Era, at least 50 percent of the known animal families went extinct throughout the world. About 75 percent of the amphibian families and 80 percent of the reptiles died out at this time, but representatives of the main orders (groups of families) survived into the Mesozoic Era, as can be seen in Triassic fossil deposits. Many families of lobe-finned fishes declined or were lost by the end of the Paleozoic, but the ray-finned fishes expanded and gave rise in the Mesozoic Era to the modern bony fishes (teleosts), which predominated then and continue to do so.

The Mesozoic Era was ushered in by many fewer primitive amphibians and reptiles in the Triassic Period, but the few surviving reptile groups diversified and took their place as the predominant vertebrates on land. During the mass extinction at the end of the Triassic Period, a number of the early reptiles and other groups disappeared. The remaining reptiles, however, were the progenitors of the very diverse forms that distinguish the Mesozoic Era, including the dinosaurs and both marine and flying reptiles (see Fig. 10.25). By the end of the Cretaceous Period, which closed the Mesozoic Era, at least one-quarter of the known animal families disappeared, including dinosaurs, marine and flying reptiles, and the last of the molluscan ammonoids.

The surviving reptile groups included the snakes, lizards, turtles, and tortoises, plus remnants of the crocodilians and the rhyncocephalians, whose only remaining species is the tuatara. Except for large species, the birds and mammals came through the extinction reasonably well, and went on to proliferate and diversify in the next 65 Myr. In fact, within the 10 to 12 Myr of the dawn of the Cenozoic Era, virtually all the orders of placental mammals evolved in bursts of adaptive radiations. Whales and bats and primates were well differentiated, along with other placental orders, all having arisen and diversified extremely rapidly from the common shrewlike ancestor of all placental mammals. Placental mammals replaced the reptiles as the predominant terrestrial vertebrates in the Cenozoic Era, or **Age of Mammals** (Fig. 10.37).

The mass extinctions of animals were not accompanied by wholesale disappearance of land plants in any of the five major episodes. The early land flora of the Silurian and Devonian periods was replaced by successful groups of lycophytes and a variety of ferns and seed ferns during the Devonian and afterward. Gymnosperms became the predominant flora near the end of the Paleozoic and continued to flourish until the flowering plants appeared in the Cretaceous Period. Flowering plants became predominant in the Cretaceous, and were relatively unaffected by conditions that wiped out many animal families at the end of the Mesozoic Era (Fig. 10.38).

Different theories have been proposed over the years for the occurrence of a catastrophe that caused the rapid disappearance of many unrelated families of animals in mass extinction episodes. Natural disasters such as profound and widespread changes in climate or in sea level have been favored explanations for a long time. Other ideas, such as the global spread of a devastating disease or the significant reduction in atmospheric oxygen, have little or no support. Various proposals have suggested some extraterrestrial rather than terrestrial cause for mass extinctions, including excessive cosmic radiation from passing stars or nearby supernovas and impacts by large asteroids or meteorites or comets, which had worldwide effects. With little supporting evidence, these ideas of an extraterrestrial trigger for global catastrophe were not taken very seriously. The cause or causes of mass extinctions remained a deep mystery.

In 1980, Luis Alvarez, Walter Alvarez, and their colleagues revived the **impact theory** and presented evidence for an extraterrestrial cause of the mass extinction that occurred 65 Myr ago, at the end of the Cretaceous Period. They found a significant excess of **iridium,** a metal in the platinum group, precisely at the Cretaceous–Tertiary (K–T) boundary but not above or below this narrow line (Table 10.3). Iridium is much more abundant in chondritic meteorites and average solar system material than it is in the Earth's upper mantle or crust, presumably because the platinum group of elements has been depleted in upper layers and concentrated in the Earth's core by physical processes. In their first study, the Alvarezes found up to 120 times more iridium at the K–T boundary in a few locations, but they have since discovered the iridium anomaly in nearly all the 50 complete K–T boundary sites around the world. From the iridium data, the Alvarezes estimated its abundance to be the result of the impact on the Earth of a meteorite or an asteroid measuring about 10 km in diameter.

The impact theory proposes that an asteroid or a similar body from the solar system crashed to Earth and injected about 60 times the object's mass into the atmosphere as pulverized rock, forming a huge crater. A fraction of this dust would have stayed in the stratosphere for several weeks, or perhaps several years, and been distributed worldwide, obscuring the sun completely for months, or perhaps years, according to calculations made by the Alvarezes. The resulting darkness would have caused temperatures to plummet and would have suppressed plant photosynthesis, the ultimate base of the food chain. Such a "nuclear winter"–like change would have led to the extinction of many different life forms in the oceans and on land. Similar but very small-scale pollution of the stratosphere occurred during the eruptions of Mount St. Helens in Washington State in 1980, as clouds of volcanic dust entered the stratosphere and circulated for a long period of time. The gigantic eruption of the volcano on the Indonesian island of Krakatoa, which blew up

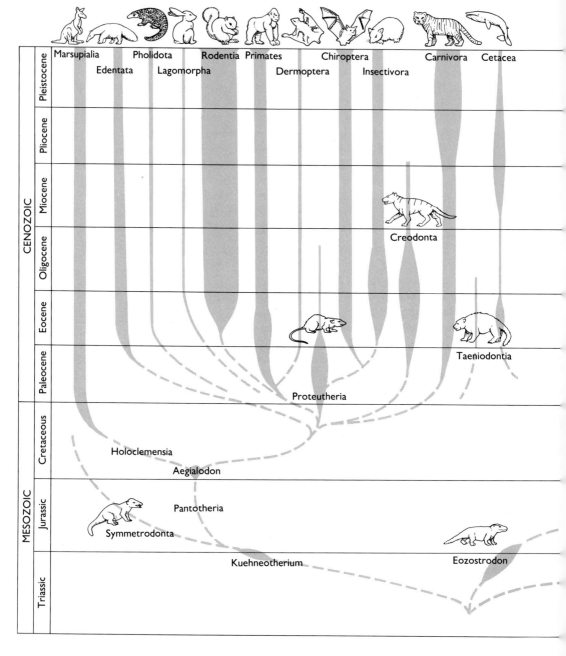

Figure 10.37. Suggested evolutionary history of the mammals from their origin in the Jurassic Period to the diverse living orders. The majority are placental mammals, but remnants of the marsupials and monotremes still exist. The width of a line or band indicates the relative size of the group at the corresponding geologic interval. Dashed lines indicate uncertainties about relationship or time of origin. (From J. W. Valentine, "The Evolution of Multicellular Plants and Animals." Copyright © 1978 by Scientific American, Inc. All rights reserved)

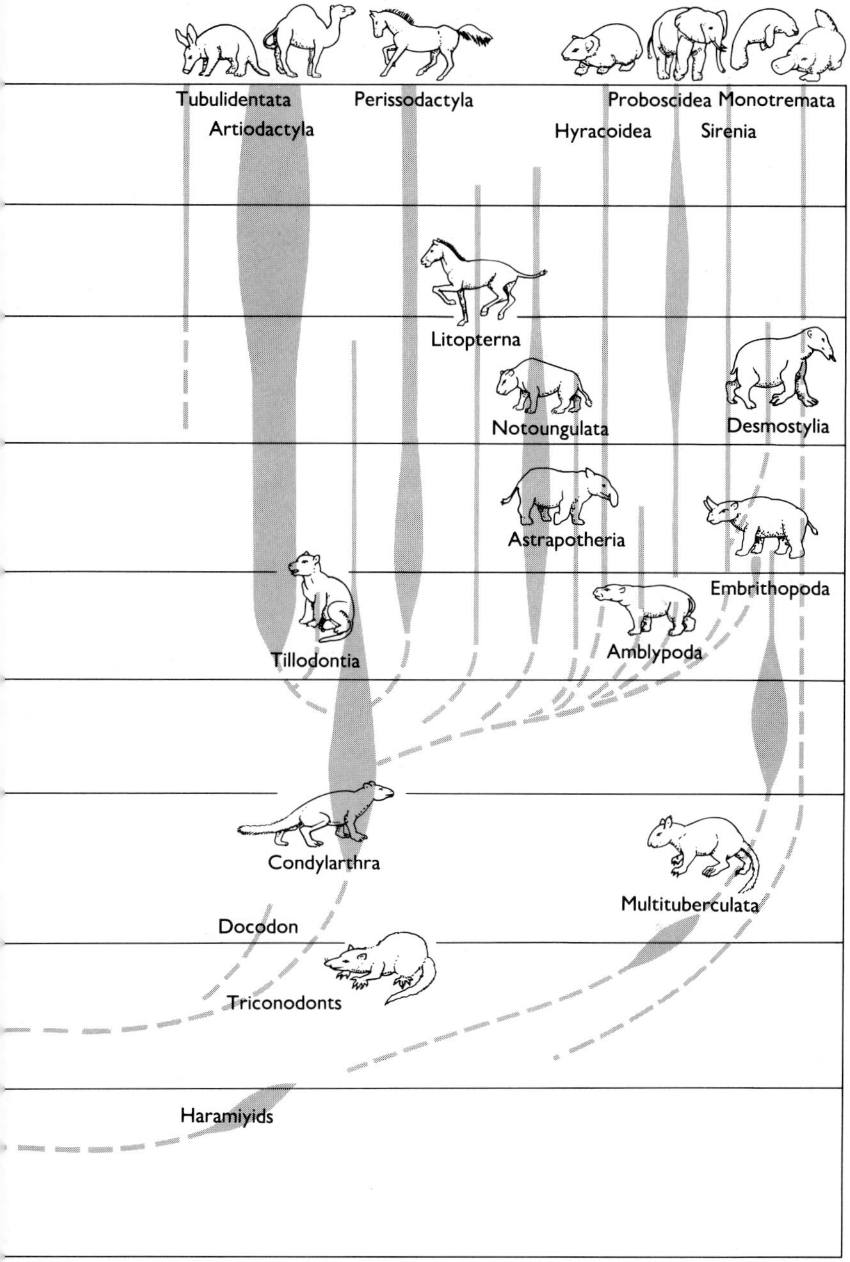

Tubulidentata Perissodactyla Proboscidea Monotremata
    Artiodactyla   Hyracoidea   Sirenia

Litopterna

Notoungulata   Desmostylia

Astrapotheria   Embrithopoda

Tillodontia   Amblypoda

Condylarthra

Docodon   Multituberculata

Triconodonts

Haramiyids

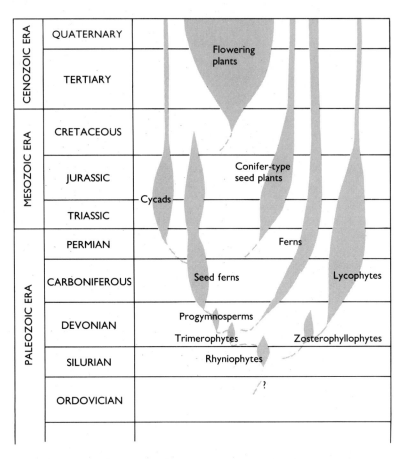

Figure 10.38. Suggested evolutionary history of the vascular plants from the Silurian Period to the present. Seed- and pollen-producing plants joined the assemblage of primitive spore-producing lycophytes (club mosses, horsetails) and ferns in the Devonian. Gymnosperms (naked-seeded plants) appeared by the Permian Period, probably from seed fern ancestry, and were the dominant components of the Mesozoic flora. Flowering plants, whose origins remain uncertain, appeared and diversified in the Cretaceous, and soon became the dominant land plants in the world. This record reveals no correspondence between extinction of plant groups and mass extinctions of animal groups at the end of the Permian, Triassic, and Cretaceous periods. The width of a line or band indicates the relative size of the group at the corresponding geologic interval. (Based on P. G. Gensel and H. N. Andrews, 1987, The evolution of early land plants. *Amer. Sci.* 75:478)

almost the entire island over 100 years ago, provides another analogous situation. The immense quantity of volcanic dust thrown up into the stratosphere caused deep-red sunsets for more than a year afterward.

To test a prediction of their theory, the Alvarezes studied the fossil records of four groups of marine invertebrates (ammonites, cheilostomate bryozoans, brachiopods, and bivalves) known to have been devastated at the end of the

Cretaceous. For each of these groups, they found a pattern showing gradual decline in species for 1 to 10 Myr before, but a very sharp and abrupt disappearance precisely at the K–T boundary (Fig. 10.39). Although they did not include a control group for comparison, the Alvarezes suggested in their report, published in 1984, that an analysis of the fossil record for gastropods, echinoderms, corals, and other marine invertebrates believed to have experienced little extinction at the time would show a very different pattern of species disappearance. One criticism of their study is that the sampling was highly selective; another is that the species already were in a state of decline and merely continued to the point of total extinction by the end of the Cenozoic. This suggestion has also been made for the dinosaurs, which may have consisted of only about 20 genera near the end of the Mesozoic, and simply continued to decline until none was left when the era ended. These criticisms do not imply that a mass extinction did not occur 65 Myr ago or that it was not caused by impact. They do imply that the event would be better understood if a more suitable sample of animal groups was studied, particularly ones that survived the event and others that were very successful and widespread until their abrupt termination at the K–T boundary.

Most plant and animal groups, land forms as well as marine forms, sailed through the end of the Cretaceous, suffering little or no damage. Surviving plants could have held on during the darkness and cold thanks to their resistant spores, seeds, and root systems. New growth or regrowth could then have taken place after the dust settled and the sun reappeared through the clouds. Land animals smaller than 20 to 25 kg came through the episode reasonably well, but virtually all the larger animals succumbed. The mammals of this time consisted of small, nocturnal creatures, which could have weathered

TABLE 10.3
Abundance of Iridium in Acid-insoluble Residues in the Cretaceous–Tertiary Boundary Section at Stevns Klint, Denmark

Distance Above (+) or Below (−) the Boundary (meters)	Abundance of Iridium (parts per billion)	Abundance of Acid-insoluble Residues (%)
+2.7	<0.3	3.27
+1.2	<0.3	1.08
+0.7	0.36 ± 0.06	0.836
Boundary	41.6 ± 1.8	44.5
−0.5	0.73 ± 0.08	0.654
−2.2	0.25 ± 0.08	0.621
−5.4	0.30 ± 0.16	0.774

Source: Adapted from L. W. Alvarez et al., 1980. Extraterrestrial cause for the Cretaceous–Tertiary extinction. *Science* 208:1095, Table 1.

the crisis by subsisting on seeds and roots. Smaller predators also could have survived on the plant-eating prey species. It is very important to learn how long the darkness and cold prevailed, but most of the current information is ambiguous. Despite its incompleteness, the impact theory for the Cretaceous mass extinction remains a testable proposal.

Is There a Periodicity to Mass Extinction Episodes?

The impact theory for the Cretaceous mass extinction revived interest in extraterrestrial causes for other episodes, and perhaps for all mass extinctions. The idea gained momentum from the suggestion made by David Raup and John Sepkoski that mass extinctions have occurred in periodic episodes. In a detailed paper, published in 1984, they claimed that all the mass extinctions, both large-scale and small-scale, had taken place at intervals of about 26 Myr during the past 250 Myr of Phanerozoic time (Fig. 10.40). Raup and Sepkoski's data inspired a number of astronomers to search for possible extraterrestrial periodicities that correspond with mass extinction cycles, and therefore, perhaps have been direct causes of extinction.

Some astrophysicists proposed the occurrence of extraterrestrial phenomena with a periodicity of about 26 to 32 Myr, in accord with the timetable set up by Raup and Sepkoski. One set of proposals suggested a 30-Myr periodicity of the sun's oscillations around the plane of the galaxy, which could perturb solar radiation, cosmic ray intensity, or other galatic phenomena. These,

a

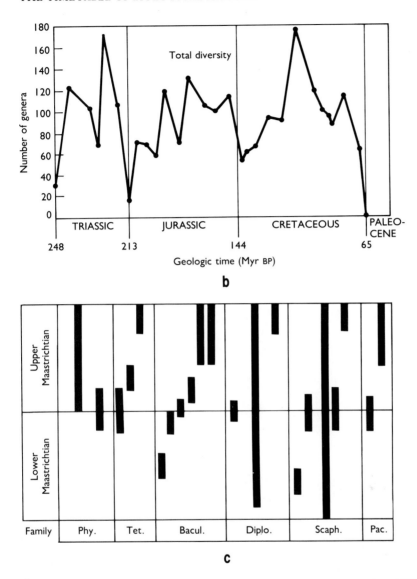

Figure 10.39. Ammonite extinction. (a) The ammonite mollusks were related to living squids and octupuses, but possessed a coiled externai shell. (b) Ammonites experienced waves of generic diversification and extinction throughout the Mesozoic Era, but disappeared entirely at the end of the Cretaceous. (Modified from W. J. Kennedy, in *Patterns of Evolution*, A. Hallam (ed.), pp. 251–304. Elsevier: Amsterdam) (c) The fossil record of six families of ammonites from Danish deposits, plotted according to the Maastrichtian stratigraphic stage at the end of the Cretaceous Period. Gradual declines occurred throughout this 10-Myr interval, but all the remaining families disappeared abruptly at the Cretaceous–Tertiary (K–T) boundary. High concentrations of iridium also occur precisely at this boundary in Denmark. Abbreviations are as follows: Phy., Phylloceratidae; Tet., Tetragonitidae; Bacul., Baculitidae; Diplo., Diplomoceratidae; Scaph., Scaphitidae; and Pac., Pachydiscidae. (Modified from W. Alvarez et al., 1984, Impact theory of mass extinctions and the invertebrate fossil record. *Science* 223:1135, Fig. 2)

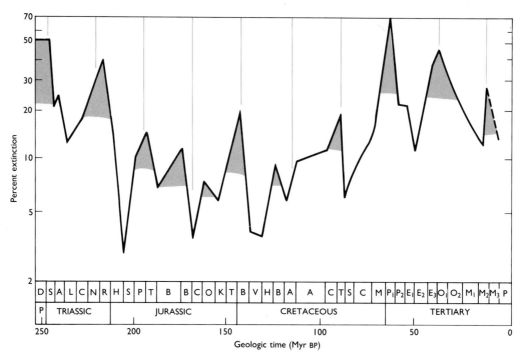

Figure 10.40. The proposed 26-Myr cycle of extinctions (vertical lines) is superimposed on the record of extinctions analyzed in 1984 by David Raup and John Sepkoski for the Permian (P), Triassic, Jurassic, and Cretaceous periods of the Mesozoic Era, and for the Tertiary Period of the Cenozoic Era. The letters refer to stratigraphic stages (for example, M stands for the Maastrichtian ending at the K–T boundary). The relative heights of more recent peaks are exaggerated by this analysis.

in turn, might be responsible for such profound changes in climate and the biosphere that a mass extinction would be triggered on a regular basis.

Another set of proposals postulated a periodicity for comet showers induced by the passage of an unseen solar companion star as it presumably passes through the Oort cloud of comets, at the boundary of the solar system. This postulated but undetected star is presumed to pass through the comet cloud during its eccentric orbit around the sun every 26 Myr. It then perturbs the orbits of the Oort comets, sending about 1 billion of them into paths that reach the inner parts of the solar system. On average, several of these comets might hit the Earth after traveling for about 1 million years from the outer to the inner reaches of the solar system. On the comet's impact with the Earth, clouds of dust would be shot out into the stratosphere, causing global changes that initiate a mass extinction episode. The "death star," referred to as Nemesis, is estimated by its sponsors to be at its maximum of about 2.4 light-years from the sun today and to present no danger to the Earth until 15 Myr from now. These theoretical and unsupported calculations rely on the propositions made by Raup and Sepkoski, including their statement that the last mass extinction took place 11.3 Myr ago. The next event is thus expected to be 26 − 11 Myr, or 15 Myr, from now.

Although proposals of "death stars" and other cosmic nemeses capture our imagination, little or no evidence has been provided in support of the particular extraterrestrial phenomena actually occurring, much less that they are episodic. If periodicity indeed characterizes mass extinctions, which remains to be shown, a cosmic explanation would be the most reasonable basis for the terrestrial effects. Our sun does proceed through a sunspot cycle every 11 Myr, but this is the wrong timing for it to correspond to a mass extinction every 26 Myr.

One study by Walter Alvarez and Richard Muller, reported in 1984, did provide preliminary evidence for periodic impacts on Earth. They analyzed a number of large craters that had been formed between 5 and 250 Myr ago, and determined that they had been formed with a periodicity of about 28 Myr during that span of time (Fig. 10.41). The particular range of time was selected to correspond with Raup and Sepkoski's suggested 26 to 32 Myr periodicity and with their analysis of the fossil record between 5 and 250 Myr ago. From the two sets of data, Alvarez and Muller concluded that impacts on the Earth of extraterrestrial bodies are the most likely cause of the mass extinctions in this time frame. However, iridium anomalies have been found only sporadically and in isolated locations for all presumed extinction times other than the well-documented K–T boundary, which shows a consistent and global distribution of iridium enrichment.

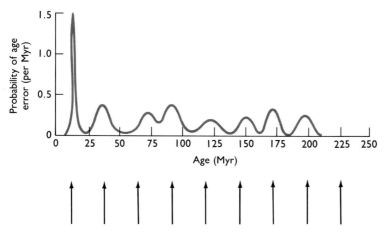

Figure 10.41. A statistical representation of the history of impacts on Earth during the past 250 Myr, for craters having a diameter of 10 km or more. Each part of the curve represents the number of craters analyzed and the relative uncertainty about the age of these craters. The arrows are placed at 28-Myr intervals. A reasonably good fit is evident for periodicity of crater age in most cases, but not all. (Modified from W. Alvarez and R. A. Muller, 1984, Evidence from crater ages for periodic impacts on Earth. *Nature* 308:718, Fig. 1)

In 1986, Raup and Sepkoski published an expanded analysis of the fossil record for 2160 families of marine animals and 11,800 genera that included 9250 extinct forms. They applied statistical methods to analyze all these forms between the Middle Permian and the present, localizing the organisms in detailed stratigraphic stages of the geologic time scale (Table 10.4). *Stratigraphic stages* are time–strata divisions of the geologic time scale, and 39 stages are recognized by international convention for the 590 Myr since the beginning of the Paleozoic Era. This convention permits much more precise timing and reference level than the standard time scale of periods and epochs.

Raup and Sepkoski found eight statistically significant peaks of extinction events for the marine families and genera, and they concluded that these peaks represent eight mass extinctions. They ruled out other times of mass extinction between 268 and 5 Myr ago, based on statistical insignificance when the fossil data were analyzed. The eight proposed mass extinctions were dated from times at or near the end of the Permian up to the middle of the Miocene Epoch (248 or 253, 213 or 219, 194, 144, 91, 65, 38, and 11.3 Myr ago). Of these eight events, only three are recognized to be of such major proportions

TABLE 10.4
Ages of Extinction Events with Dates from Two Time Scales

Period	Stratigraphic Stage or Epoch	End of Interval (Myr BP)		Extinction Event	
		Harland et al.*	Odin†	Families	Genera
Tertiary	Pliocene	2.0	1.9		Possible
Tertiary	Middle Miocene	11.3	11	Significant	Present
Tertiary	Late Eocene	38	34	Significant	Present
Cretaceous	Maastrichtian	65	65	Significant	Present
Cretaceous	Cenomanian	91	91	Significant	Present
Cretaceous	Aptian	113	107		Possible
Cretaceous	Hauterivian	125	114	Doubtful	
Jurassic	Tithonian	144	130	Significant	Present
Jurassic	Callovian	163	150	Doubtful	
Jurassic	Bajocian	175	170	Doubtful	
Jurassic	Pliensbachian	194	189	Significant	Present
Triassic	"Rhaetian"	213	204	Significant	Present
Triassic	Norian	219	209		
Triassic	Carnian	225	220		Doubtful
Triassic	Olenekian	243	239	Doubtful	
Permian	Dzulfian	248	245	Significant	
Permian	Guadalupian	253	252		Present

*W. B. Harland et al., 1980. *A Geologic Time Scale.* Cambridge: Cambridge University Press.
†G. S. Odin, ed., 1982. *Numerical Dating in Stratigraphy.* New York: Wiley.
Source: Adapted from D. M. Raup and J. J. Sepkoski, 1986. Periodic extinction of families and genera. *Science* 231:833, Table 1.

that they can unambiguously be described as mass extinctions (end of the Permian, end of the Triassic, and end of the Cretaceous). The two earlier mass extinctions (end of the Ordovician and end of the Devonian) were not included in the study.

The four extinctions showing a statistically significant periodicity of 26 Myr are the most recent (91, 65, 38, and 11.3 Myr ago), only one of which is widely agreed to be a mass extinction (65 Myr ago). The two other unambiguous mass extinctions (Permian and Triassic) showed a statistically significant 26-Myr periodicity for only two of the four possible combinations of dates: 253 to 219 and 248 to 219, but not 253 to 213 or 248 to 213. Since paleontologists cannot be sure if the events happened near the end or at the end of the Permian and Triassic, there is no basis to determine whether they fit the theoretical 26-Myr interval.

The possibility of periodic mass extinctions caused by periodic extraterrestrial factors remains open and deserving of continued study. All sorts of detailed information are needed, however, before much can be said about either set of periodicities. Part of the problem, of course, is the incomplete nature and disproportionate representation of organisms in the fossil record. The problem is compounded for mass extinctions because they often coincide with profound geologic upheavals that destroy or erase the record of the time. Geologists cannot be sure whether a period or an era ended abruptly, as shown in the geologic record that remains, or if millions of years passed but were erased from the Earth's crust and the transitions only seem to be abrupt.

There is little agreement about the classification of intermediate-level extinctions, between background and mass extinctions. Some of them have been regarded as mass extinctions and others have not, depending on individual interpretations. Was the extinction of many different and successful groups of large mammals in North America between 14,000 and 6000 years ago a mass extinction because of the abruptness and the diversity of vanished life, or was it too localized an event to be classified in the same category as the worldwide disaster at the end of the Permian or Triassic or Cretaceous? We would be in a better position to evaluate general extinction conditions and mechanisms if there were a distinguishable pattern of the whole graded series of events between one extreme and the other. A great deal remains to be learned about basic details of the fossil record and environmental and biotic features of past times. If is often difficult to determine whether the replacement of one group of organisms by another is the result of species turnover due to selection and adaptation phenomena or of the opening of niches following an extinction episode. In either case, some groups of organisms disappear, and other groups take over.

The burst of new multicellular organisms at the beginning of the Cambrian is widely inferred to have been the result of new adaptations and life styles. The adaptive radiation of mammals following the mass extinction of reptiles at the end of the Cretaceous, however, almost certainly depended in large measure on the availability of niches vacated by reptiles and made accessible to mammals. Mammals had lived alongside reptiles for more than 100 million

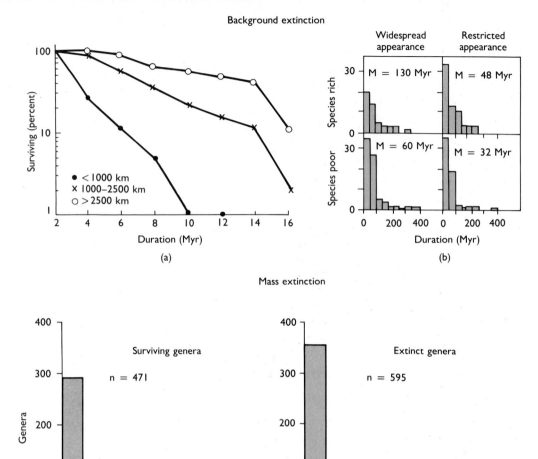

Figure 10.42. The effect of geographical range on survival of Late Cretaceous bivalves and gastropods during background and mass extinctions. (a) During background times, the extinction rate is significantly higher for species having narrow ranges (<1000 km) than for species having intermediate (1000–2500 km) or broad (>2500 km) ranges. (b) Also during background times, species richness and widespread range interact synergistically to influence survivorship. Species-rich groups with predominantly widespread ranges are significantly longer lived than species-poor groups with predominantly restricted ranges. During mass extinctions, the frequency distribution of geographical ranges within genera are not significantly different for (c) survivors and (d) victims. Whether the ranges are widespread or restricted, the probability of extinction in a cataclysmic episode is about the same for both groups of species. (Adapted from D. Jablonski, 1986, Background and mass extinctions: the alternation of macroevolutionary regimes. *Science* 231:129, Fig. 2. Copyright © 1986 by the American Association for the Advancement of Science)

years (180–65 Myr ago), without displacing them in any living zone. With the disappearance of the dominant, competing reptiles, the mammals took over within about 10 million years and had diversified greatly during that time.

Whether or not victims and survivors are a random sample of preextinction species is not clear from the fossil record, except for mass extinctions. It seems that large population size, wide geographical distribution, widespread and efficient dispersion of progeny, richness of species, and other features that protect organisms against background and local extinctions are not effective deterrents in cataclysmic extinction episodes (Fig. 10.42). Much remains to be learned about selectivity in extinctions and its relationship to the kinds of stresses that demolish some organisms and not others in the same times and places. Biologists need a solid understanding of the causes and consequences of extinctions in order to comprehend more fully the nature of speciation in the evolutionary history of life.

Summary

The geologic and fossil records provide a time frame for evolution and evidence for the ancestors and relatives of living species. The 4600 Myr of Earth history are subdivided into eons, eras, periods, and epochs of geologic time. The most recent 590 Myr, or the Phanerozoic Eon, is the time of the appearance of a diversity of multicellular life that included representatives of all the modern animal phyla before the end of the Cambrian Period. The fossil record in sedimentary rocks is only a small sample of former life and is disproportionately represented by organisms with hard parts, which are more likely than those with soft parts to be preserved. From plate tectonics, geologists know that lithospheric plates carrying continental landmasses and ocean basins drift from place to place, producing changes in the topography of the Earth's crust that affect biological evolution. Pangaea was fully formed by the end of the Paleozoic Era and began to break up in the Early Mesozoic, about 200 Myr BP. Speciation and convergence increased as landmasses became isolated as a result of their drifting, and such phenomena were particularly characteristic of mammalian evolution in the Cenozoic Era.

During the Paleozoic Era (590–248 Myr BP), plants and animals invaded the land and evolved to the level of gymnosperms and of reptiles, respectively. The suites of adaptations to terrestrial life sharply separated these forms from their aquatic ancestors. Three major episodes of mass extinction (end of Ordovician, end of Devonian, and end of Permian) reduced the diversity of life forms, but subsequent adaptive radiations replenished the biota and produced more advanced types. In the Mesozoic Era (248–65 Myr BP), reptiles and gymnosperms predominated. Mammals and birds diverged from reptilian lineages in the Jurassic. Flowering plants became the predominant flora during the Cretaceous Period. Mass extinctions occurred at the end of the Triassic and the end of the Cretaceous, the latter event leading to the disappearance of the dinosaurs and many other animals. The Cretaceous mass extinction may have been triggered by the impact of an extraterrestrial body on the Earth, raising clouds of dust that could have blotted out the sun and produced a "nuclear winter," lasting for months or, perhaps, years. Mammals superseded the reptiles as the predominant animal life in the Cenozoic Era (65 Myr–present), and placental mammals that evolved from insectivore ancestors replaced monotremes and marsupials in bursts of adaptive radiations, except on the isolated continent of Australia. All these extinctions and adaptive radiations indicate that evolution is unpredictable and irreversible.

In contrast to a "normal" rate of species extinctions caused by climatic changes, competition for resources, and other factors, mass extinctions have often been attributed to global catastrophes. If there indeed is a 26 to 32 Myr periodicity of mass extinctions, it is most likely that some recurrent extraterrestrial event triggers the catastrophes. The Cretaceous mass extinction, however, is the only one for which an extraterrestrial cause has been suggested by evidence, including the presence of excess iridium at the K–T boundary. Whatever their cause and whether or not they occur periodically, mass extinctions result in the demise of a broad spectrum of organisms enjoying various degrees of successful existence at the time of the catastrophe. Biologists need more information about the causes and consequences of extinctions in order to understand speciation in its broadest aspects.

References and Additional Readings

Archibald, J. D., and W. A. Clemens. 1982. Late Cretaceous extinctions. *Amer. Sci.* 70:377.

Alvarez, L. W. et al. 1980. Extraterrestrial cause for the Cretaceous–Tertiary extinction. *Science* 208:1095.

Alvarez, W., and R. A. Muller, 1984. Evidence from crater ages for periodic impacts on Earth. *Nature* 308:718.

Alvarez, W. et al. 1984. Impact theory of mass extinctions and the invertebrate fossil record. *Science* 223:1135.

Bakker, R. T. 1986. *The Dinosaur Heresies*. New York: Morrow.

Bambach, R. K., C. R. Scotese, and A. M. Zeigler. 1980. Before Pangaea: the geographies of the Paleozoic world. *Amer. Sci.* 68:26.

Banks, H. P. 1968. The early history of land plants. In *Evolution and Environment*, ed. E. T. Drake, p. 73. New Haven, Conn.: Yale University Press.

Ben-Avraham, Z. 1981. The movement of continents. *Amer. Sci.* 69:291.

Boher, B. F., P. J. Modreski, and E. E. Foord. 1987. Shocked quartz in the Cretaceous–Tertiary boundary clays: evidence for a global distribution. *Science* 236:705.

Bonatti, E. 1987. The rifting of continents. *Sci. Amer.* 256(3):96.

Carroll, R. L. 1987. *Vertebrate Paleontology and Evolution*. New York: Freeman.

Cloud, P., and M. F. Glaessner. 1982. The Ediacarian Period and system: Metazoa inherit the Earth. *Science* 218:783.

Colbert, E. H. 1973. *Wandering Lands and Animals*. New York: Dutton.

Conway Morris, S. 1987. The search for the Precambrian–Cambrian boundary. *Amer. Sci.* 75:156.

Conway Morris, S., and H. B. Whittington. 1979. The animals of the Burgess Shale. *Sci. Amer.* 241(1):122.

Cook, P. J., and J. H. Shergold, 1984. Phosphorus, phosphorites and skeletal evolution at the Precambrian–Cambrian boundary. *Nature* 308:231.

Diamond, J. M. 1982. Man the exterminator. *Nature* 298:787.

Dietz, R. S., and J. C. Holden. 1970. The breakup of Pangaea. *Sci. Amer.* 223(4):30.

Edwards, D., and E. C. W. Davies. 1979. Oldest recorded *in situ* tracheids. *Nature* 263:494.

Elliott, D. K., ed. 1986. *Dynamics of Extinction*. New York: Wiley.

Elliott, D. K. 1987. A reassessment of *Astraspis desiderata,* the oldest North American vertebrate. *Science* 237:190.

Fooden, J. 1972. Breakup of Pangaea and isolation of relict mammals in Australia, South America, and Madagascar. *Science* 175:894.

Francis, P., and S. Self. 1983. The eruption of Krakatau. *Sci. Amer.* 249(5):172.

Friis, E. M., W. G. Chaloner, and P. R. Crane, eds. 1987. *The Origins of Angiosperms and Their Biological Consequences*. New York: Cambridge University Press.

Gensel, P. G., and H. N. Andrews. 1987. The evolution of early land plants. *Amer. Sci.* 75:478.

Gillespie, W. H., G. W. Rothwell, and S. E. Scheckler. 1981. The earliest seeds. *Nature* 293:462.

Glaessner, M. F. 1985. *The Dawn of Animal Life*. New York: Cambridge University Press.

Gould, S. J. 1974. The evolutionary significance of "bizarre" structures: antler size and skull size in the "Irish elk," *Megaloceros giganteus. Evolution* 28:191.

Gould, S. J. 1984. The Ediacaran experiment. *Nat. Hist.* 93(2):14.

Gould, S. J., N. L. Gilinsky, and R. Z. German. 1987. Asymmetry of lineages and the direction of evolutionary time. *Science* 236:1437.

Graham, L. 1985. The origin of the life cycle of land plants. *Amer. Sci.* 73:178.

Grayson, D. K. 1987. Death by natural causes. *Nat. Hist.* 96(5):8.

Hallam, A., ed. 1977. *Patterns of Evolution as Illustrated by the Fossil Record*. Amsterdam: Elsevier.

Harland, W. B. et al. 1980. *A Geologic Time Scale*. New York: Cambridge University Press.

Horner, J. R. 1984. The nesting behavior of dinosaurs. *Sci. Amer.* 250(4):130.

Hurley, P. M. 1968. The confirmation of continental drift. *Sci. Amer.* 218(4):52.

Jablonski, D. 1986. Background and mass extinctions: the alternation of macroevolutionary regimes. *Science* 231:129.

James, H. F. et al. 1987. Radiocarbon dates on bones of extinct birds from Hawaii. *Proc. Natl. Acad. Sci. U.S.* 84:2350.

Jardine, N., and D. McKenzie. 1972. Continental drift and the dispersal and evolution of organisms. *Nature* 235:20.

Kerr, R. A. 1985. Plate tectonics goes back 2 billion years. *Science* 230:1364.

Knoll, A. H. et al. 1984. Character diversification and patterns of evolution in early vascular plants. *Paleobiology* 10:34.

Kurtén, B. 1969. Continental drift and evolution. *Sci. Amer.* 220(3):54.

Macdougall, J. D. 1988. Seawater strontium isotopes, acid rain, and the Cretaceous–Teriary boundary. *Science* 239:485.

Martin, P. S., and R. G. Klein, eds. 1984. *Quaternary Extinctions: A Prehistoric Revolution*. Tucson: University of Arizona Press.

McMenamim, M. A. S. 1987. The emergence of animals. *Sci. Amer.* 256(4):94.

Molnar, P. 1986. The geologic history and structure of the Himalaya. *Amer. Sci.* 74:144.

Mossman, D. J., and W. A. S. Sarjeant. 1983. The footprints of extinct animals. *Sci. Amer.* 248(1):74.

Moulton, M. P. and S. L. Pimm. 1983. The introduced Hawaiian avifauna: biogeographic evidence for competition. *Amer. Nat.* 121:669.

Mutter, J. C. 1986. Seismic images of plate boundaries. *Sci. Amer.* 254(2):66.

Newell, N. D. 1963. Crises in the history of life. *Sci. Amer.* 208(2):76.

Nichols, D. J. et al. 1986. Palynological and iridium anomalies at Cretaceous–Tertiary boundary, south-central Saskatchewan. *Science* 231:714.

Norstog, K. 1987. Cycads and the origin of insect pollination. *Amer. Sci.* 75:270.

Odin, G. S., ed. 1982. *Numerical Dating in Stratigraphy*. New York: Wiley.

Ostrum, J. H. 1987. Romancing the dinosaurs. *The Sciences*, May–June, p. 56.

Padian, K., ed. 1987. *The Beginning of the Age of Dinosaurs*. New York: Cambridge University Press.

Quinn, J. F. 1983. Mass extinctions in the fossil record. *Science* 219:1239.

Raup, D. M. 1986. Biological extinction in Earth history. *Science* 231:1528.

Raup, D. M. 1986. *The Nemesis Affair: A Story of the Death of Dinosaurs and the Ways of Science*. New York: Norton.

Raup, D. M., and D. Jablonski, eds. 1986. *Patterns and Processes in the History of Life*. Berlin: Springer-Verlag.

Raup, D. M., and J. J. Sepkoski. 1984. Periodicity of extinctions in the geologic past. *Proc. Natl. Acad. Sci. U.S.* 81:801.

Raup, D. M., and J. J. Sepkoski. 1986. Periodic extinction of families and genera. *Science* 231:833.

Romer, A. S. 1970. *The Vertebrate Body*, 4th ed. Philadelphia: Saunders.

Russell, D. A. 1982. The mass extinctions of the Late Mesozoic. *Sci. Amer.* 246(1):58.

Schopf, J. M., and B. M. Packer. 1987. Early Archean (3.3-billion to 3.5-billion-year-old) microfossils from Warrawoona Group, Australia. *Science* 237:70.

Simpson, G. G. 1944. *Tempo and Mode in Evolution*. New York: Columbia University Press.

Simpson, G. G. 1951. *Horses*. New York: Oxford University Press.

Simpson, G. G. 1953. *The Major Features of Evolution*. New York: Columbia University Press.

Simpson, G. G. 1983. *Fossils and the History of Life*. New York: Freeman.

Sloan, R. E. et al. 1986. Gradual dinosaur extinction and simultaneous ungulate radiation in Hell Creek formation. *Science* 232:629.

Smith, R. B., and R. L. Christiansen. 1980. Yellowstone Park as a window on the Earth's interior. *Sci. Amer.* 242(2):104.

Stanley, S. M. 1984. Mass extinctions in the ocean. *Sci. Amer.* 250(6):64.

Stanley, S. M. 1986. *Earth and Life Through Time*. New York: Freeman.

Stanley, S. M. ed. 1987. *Extinction*. New York: Freeman.

Tims, J. D., and T. C. Chambers. 1984. Rhyniophytina and Trimerophytina from the early land flora of Victoria, Australia. *Palaeontology* 27:265.

Valentine, J. W. 1978. The evolution of multicellular plants and animals. *Sci. Amer.* 239(3):140.

Valentine, J. W. 1985. The evolution of complex animals. In *What Darwin Began*, ed. L. R. Godfrey, p. 258. Boston: Allyn and Bacon.

Van Valen, L. M. 1973. A new evolutionary law. *Evol. Theory* 1:1.

Van Valen, L. M. 1984. Catastrophes, expectations, and the evidence. *Paleobiology* 10:121.

Vink, G. E., W. J. Morgan, and P. R. Vogt. 1985. The Earth's hot spots. *Sci. Amer.* 252(4):50.

Ward, P. D., and P. W. Signor III. 1983. Evolutionary tempo in Jurassic and Cretaceous ammonites. *Paleobiology* 9:183.

Wegener, A. 1924. *The Origin of Continents and Oceans*. London: Methuen; reprint, New York: Dover, 1966.

Whitmire, D. P., and A. A. Jackson IV. 1984. Are periodic mass extinctions driven by a distant solar companion? *Nature* 308:713.

CHAPTER 11

The Origin and Evolution of the Primates

The constellation of adaptive features that contributed to the success of the mammals in the Cenozoic Era also underlies the evolutionary history of the Primates, the order to which we belong. Our ancestral legacies, together with unique derived character states, account for the success of the human species in the Recent Epoch. In this chapter, we will survey some of the highlights of mammalian and primate evolutionary trends, which are carried over into human evolutionary history.

Ancestral Legacies

Basic mammalian innovations that continue to provide adaptive advantages for the primates include endothermy; the birth of live young, which are nourished prenatally and postnatally by the mother; and enlargement of the cerebral portion of the brain, which monitors a broad spectrum of behaviors as well as enhanced intelligence. The Primates are particularly noted for continued enlargement of the cerebrum, enchancement of visual over olfactory sensory apparatus, and grasping hands and feet, whose motor control is highly coordinated with vision. These ancestral legacies provide a basic framework for our evolutionary history.

The Mammalian Legacy in Primate Evolution

Birds and mammals differ from other vertebrates in many respects, one of the most fundamental of which is their internal regulation of constant body temperature, or **endothermy.** The maintenance of constant body temperature requires a high metabolic rate during periods of rest as well as activity and a regulation mechanism. The complex regulation mechanism is under the control of the *hypothalamus*, at the base of the brain, which acts as a heat receptor and thermostat, bringing into action the components of temperature regulation when blood temperature increases or decreases from the set point for the animal. A high metabolic rate generates the heat for constant body temperature, as well as the energy for activity, and is achieved by the predominance on a sustained basis of mitochondrial aerobic oxidations over anaerobic oxidative processes (Table 11.1). Ectothermic reptiles and other animals switch to anaerobic oxidations after relatively brief bursts of activity, and are soon exhausted. They often require hours of recovery time, whereas endotherms recover relatively quickly and can resume a high level of activity shortly after exhaustion. Although exceptions are known, the typical endotherm has greater stamina than the typical ectotherm.

Ectothermic animals can also regulate their body temperature, but they do so mainly by external controls, such as moving into or out of warmer and cooler, sunny and shady places. Because of their reliance on external control of body temperature, ectotherms successfully inhabit warmer climates and can manage colder environments only if they have some means for protection from the cold, such as burrows in which to overwinter. Endothermic animals can exploit a wide range of climates and environments, as well as adopting either a nocturnal or a diurnal life style. Ectotherms are generally restricted to daytime activity, however, because most of them require the warmth of the sun to elevate their body temperature to a level at which they can best function.

TABLE 11.1
Production of ATP during 5 Minutes of Maximal Activity in Reptiles and Rodents

Animal	Mass (g)	Production of ATP (μmole/g)		
		Aerobic	Anaerobic	Total
Montane vole (*Microtus montanus*)	25	40	13	53
Merriam's kangaroo rat (*Dipodomys merriami*)	35	98	9	107
Western fence lizard (*Sceloporus occidentalis*)	13	24	22	46
Coachwhip and racer snakes (*Coluber constrictor* and *Masticophis flagellus*)	262	23	29	52

Source: A. F. Bennett and J. A. Ruben, 1979. Endothermy and activity in vertebrates. *Science* 206:649, Table 2. Copyright © 1979 by the American Association for the Advancement of Science.

Endothermic birds and mammals have additional systems that aid in the regulation of body temperature, including an insulated body covering of feathers or hair and fur. Raising the fur or puffing out the feathers forms a thick insulating layer that aids in heat retention in cold weather. Flattening the fur or feathers against the skin reduces their thickness and permits greater heat loss during hot weather or times of great exertion. Sweat glands, which are uniquely mammalian skin elements, provide another means for body temperature regulation by producing water on the skin, which vaporizes and thereby lowers skin temperature; blood temperature is lowered in turn.

Avian and mammalian endothermy evolved independently because each class is descended from a different reptilian lineage, which have been evolving separately since the Carboniferous Period. Irrespective of their independent evolutionary pathways, endothermic birds and mammals share many of the adaptive features characteristic of internal thermoregulation, among which are sustained high rates of oxidative aerobic metabolism, an insulated body covering, efficient circulatory components for transport and storage of oxygen, and a four-chambered heart that acts as a dual pump that separately directs oxygen-rich blood from the lungs to the body and oxygen-depleted blood back to the lungs from the body (Fig. 11.1).

Endotherms require a considerable amount of food to fuel the aerobic processes by which heat and energy are generated. The broader exploitation of

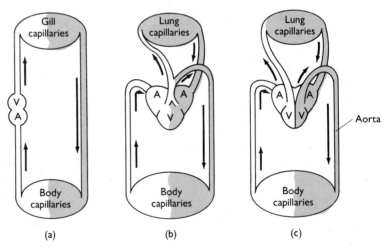

Figure 11.1. Evolution of the vertebrate heart. (a) In fishes, the heart consists of two chambers (ventricle, V; atrium, A), which act together only as a single pump to return oxygen-depleted blood to the gills. (b) In amphibians, the three-chambered heart acts as a partial separation structure as well as a pump. Oxygen-rich (gray) and oxygen-poor blood enter their own atria, but may mix in the single ventricle. (c) In birds and mammals, the left atrium and ventricle act as a separate pump from the right atrium and ventricle. The left side of the heart accepts oxygen-rich blood from the lungs and pumps it to the body. Oxygen-depleted blood from the body enters the right atrium and is pumped back to the lungs by the left ventricle. Oxygen-depleted and oxygen-rich blood supplies do not mix, because the two ventricles have no passageway between them.

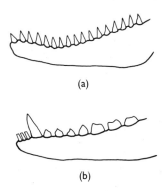

(a)

(b)

Figure 11.2. (a) Most reptiles have one kind of tooth, regardless of their number in a species. (b) Mammals have cutting incisors, tearing canines, and premolars and molars suited to crushing and grinding tough foods. The actual number of different kinds of teeth varies among mammalian groups.

various food sources by mammals compared with living reptiles is the result of changes in the chewing apparatus particularly, and in other systems, such as locomotor adaptations, during mammalian evolution. Mammals have a set of functionally diversified teeth, whereas most living reptiles have a single kind of tooth, and the mammalian jaw and jaw joint are much stronger than the reptilian structures (Fig. 11.2). Food must be digested in the intestinal tract, and nutrients can be extracted and utilized only if they are digested. Living reptiles use their teeth to grasp foods and perhaps slice or crush them to a limited extent before swallowing. In mammals, ingested foods are more thoroughly cut, torn, or ground by the different kinds of teeth and are partially digested by salivary enzyme action during chewing. Thus foods are partially processed before they reach the intestinal tract, and nutrients are thus extracted more efficiently and completely in mammals than in reptiles. In modern birds, which are toothless, the gizzard performs the function of breaking down tough foods and extracting nutrients prior to subsequent digestive tract activities.

The reptiles evolved internal fertilization by copulation, and the amniote egg provides the growing embryo with its own little pond, which is protected by a tough, desiccation-resistant outer covering. However, reptiles continued to produce large numbers of eggs, often left to hatch alone, and the young generally have to fend for themselves after hatching. The therian innovation was the production of live young, which evolved independently in a few other kinds of vertebrates. The placental mammals, unlike the marsupials, established a long period of gestation, and the growing embryo is nourished by the mother while it attains an advanced prenatal state of development. In addition, in egg-laying as well as therian mammals, a period of postnatal parental care allows for the nourishment, protection, and education of the young before they become independent.

The relatively few young born to mammals, compared with the number of eggs produced by reptiles, are an economical reproductive measure made possible by a greatly reduced mortality rate due to predators. During the postnatal interval, an intimate relationship develops between generations as the young suckle and are cared for by parents. This, in turn, makes possible the transmission of learned behaviors by observation and imitation, and minimizes learning by trial and error. The complexity of their behaviors permits mammals to exploit familiar environments and to develop a set of effective potential responses to sudden or extensive changes in their living conditions. These behaviors reflect a substantial number of anatomical and physiological modifications and the overwhelming influence of the brain on almost every activity in a lifetime.

The vertebrate brain is marked out into three regions during embryonic development: the hindbrain, midbrain, and forebrain (Fig. 11.3). The **hindbrain** structures are involved in the regulation of breathing, sleep, swallowing, heartbeat, and blood pressure. By means of the cerebellum, the hindbrain acts as a coordination center for the regulation of posture, equilibrium, and voluntary movements such as locomotion. The **midbrain** includes visual and auditory reflex centers in its upper region, and in its basal region is a relay station and an integration center for various kinds of sensory and motor activities based on information carried to and from different regions of the central nervous system. The **forebrain** has enlarged and expanded considerably during mammalian evolution, and the cerebrum has developed into the most prominent portion (Fig. 11.4). The inner "old" cortex and its overlying layer of "new" cortex (**neocortex**, or gray matter) regulate the conscious, emotional, and intellectual capabilities of all mammals, but have reached their

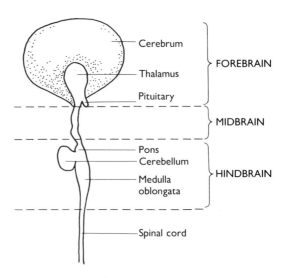

Figure 11.3. The vertebrate brain is the enlarged anterior end of the spinal cord, and is subdivided during embryonic development into three functionally differentiated regions: hindbrain, midbrain, and forebrain.

FOREBRAIN MIDBRAIN HINDBRAIN

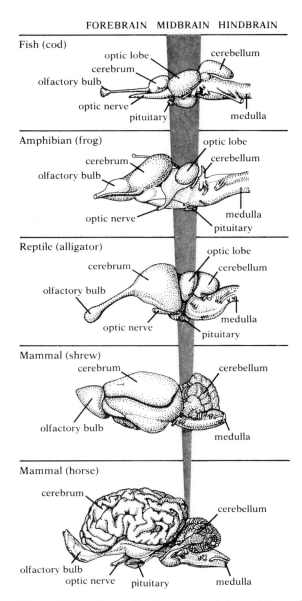

Fish (cod)

optic lobe
cerebrum
olfactory bulb
optic nerve
pituitary
cerebellum
medulla

Amphibian (frog)

cerebrum
olfactory bulb
optic nerve
optic lobe
cerebellum
medulla
pituitary

Reptile (alligator)

cerebrum
olfactory bulb
optic nerve
optic lobe
cerebellum
medulla
pituitary

Mammal (shrew)

cerebrum
olfactory bulb
cerebellum
medulla

Mammal (horse)

cerebrum
olfactory bulb
optic nerve
pituitary
cerebellum
medulla

Figure 11.4. Generalized summary of the evolution of the vertebrate brain. The midbrain portion has been reduced, and the forebrain and hindbrain portions enlarged. Substantial increase in the size and complexity of the cerebrum is particularly characteristic of mammalian evolution. (Reproduced from W. T. Keeton, 1980, *Biological Science,* 3rd ed., Fig. 10.49, with permission of W. W. Norton, New York)

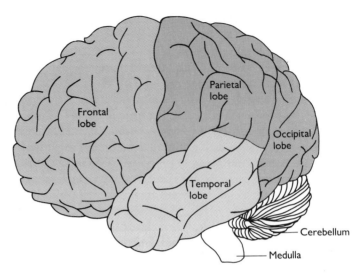

Figure 11.5. The human neocortex surrounds and overlies the older portions of the brain, except for part of the hindbrain. Among other activities, the frontal lobe of the neocortex is associated with motor-related functions; the parietal lobe, with sensory-related functions; the temporal lobe, with auditory functions; and the occipital lobe, with vision. The parietal lobe is relatively larger and the occipital lobe relatively smaller in humans compared with chimpanzees, but brain organization is very similar in the two species.

highest level of influence in Primates, particularly in human beings. The old cortex, hypothalamus, and associated forebrain structures making up the *limbic system* govern basic emotional behaviors. The neocortex, however, exercises a restraining effect on the functioning of the limbic system and on its mediation of basic emotions.

The limbic system evolved earlier than, and now lies underneath, the more highly evolved neocortex. The expansion of the neocortex during primate evolution has resulted in a system that can handle a great amount and quality of information (Fig. 11.5). The neocortex first enlarged its memory store, and in subsequent evolutionary pathways underwent an expansion of its thinking and cognitive regions. The cerebrum executes programs involved in initiating activities, based on sensory inputs and on the accumulated experiences stored as memories. The thinking brain can thus initiate novel patterns of adaptive behavior by recalling experiences and making predictions about the likely course of external events.

Primate Origins and Adaptive Complexes

The order **Primates** is divided differently by various taxonomists, but the classification in Table 11.2 represents a fair consensus of opinion. The suborder of *prosimians* dates from the Paleocene Epoch of the Cenozoic Era, and evolved from insectivore ancestors. The *anthropoid* suborder diverged from a prosimian lineage by at least the late Eocene (or perhaps the early Oligocene), about 40 to 45 Myr ago, and thus did not exist during the 20 to 25 Myr of the Paleocene and earlier Eocene, when prosimians were abundant and diversified.

All the Primates are characterized by adaptations suited to an arboreal existence, including trends leading to a relatively large brain in proportion to body size, with increasing cerebral dominance over older parts of the forebrain (Fig. 11.6). The visual system has evolved toward preeminance over the olfactory sensory system, through modifications leading to more acute, binocular (stereoscopic), color vision. The hands and feet usually possess opposable first digits, which work with long, curving fingers and toes and with modified limbs that permit grasping movements and allow the animal to move freely through the branches of trees. Improved sensory–motor coordination is achieved by expansion of the visual and auditory areas of the cerebrum and integration of sensory inputs with controls over locomotor and other motor activity. The claws have been modified to flattened nails, which exposes more surface area of the digit tip-pads (Fig. 11.7). The tips, in turn, are well supplied with nerves, providing great sensitivity of touch. Variations exist in all these features, either because primate species evolved to different levels or because of subsequent modifications to a derived character state.

The **prosimians** are generally divided into three infraorders, which are further subdivided into superfamilies and families. The lorisoids comprise two relict families, the lorises of Southeast Asia and the galago of central Africa;

TABLE 11.2
Classification of the Living Members of the Order Primates

Suborder	Infraorder	Superfamily	Family	Common Name
Prosimii	Lorisiformes	Lorisoidea	Lorisidae	Loris
			Galagidae	Galago (bush baby)
	Lemuriformes	Lemuroidea	Lemuridae	Lemur
			Indriidae	Indri
			Daubentoniidae	Aye-Aye
	Tarsiiformes		Tarsiidae	Tarsier
Anthropoidea	Platyrrhini	Ceboidea	Callitrichidae	Marmoset, tamarin ⎤ New World
			Cebidae	Cebid monkey ⎦ monkey
	Catarrhini	Cercopithecoidea	Cercopithecidae	Cercopithecid monkey ⎤ Old World
			Colobidae	Colobid monkey ⎦ monkey
		Hominoidea	Hylobatidae	Lesser ape (gibbon, siamang)
			Pongidae	Great ape (orangutan, gorilla, chimpanzee)
			Hominidae	Human

Figure 11.6. The slender loris, an Asian prosimian, shows some of the major evolutionary adaptations of the order Primates, among which are a relatively large brain in proportion to body size, forward-directed eyes, and opposable first digits. The large eyes are typical of nocturnal mammals.

the lemuroids of Madagascar include the lemur, indri, and aye-aye families; and the tarsioids are represented by the single living genus *Tarsius,* now restricted to parts of Sumatra, the southern Philippines, and a few isolated places in other parts of Southeast Asia. The tarsiers are sometimes classified as anthropoids rather than prosimians, because they share with anthropoids a high level of stereoscopic vision and bony ear structures. At one time or another in the past, the tree shrews (Tupaioidea) of Southeast Asia and Indonesia have been classified as members of either the Primates or the Insectivora. They are currently considered to be insectivores, even though their brain has a larger visual area and reduced olfactory area and certain of their skeletal features are lemuroid in nature. In every other way, tree shrews clearly belong with the Insectivora.

The lorisoids are small, thickly furred animals, about the size of a squirrel. Both the lorises of Southeast Asia and the galago of Africa are nocturnal and solitary. Lorises move very slowly in the trees, the tail is rudimentary or absent, and nails are present on all digits except the second, which is clawed. The galago, or bush baby, lives in the trees and moves from branch to branch by agile leaps. It subsists on fruits, leaves, and a variety of other foods.

The lemuroids diversified considerably in Madagascar following its separation and isolation from the African mainland. In the absence of predators and competition from the anthropoid primates, which did not evolve in Madagascar, the lemuroids flourished and diversified. The situation is not unlike conditions fostering marsupial evolution in the absence of placentals in Australia, while the continent was isolated from the rest of the world. The lemuroids posses many unusual adaptive features, compared with other prosimians, and are therefore less useful in furnishing information about the origin of the Primates.

The larger lemur species were annihilated when humans settled in Madagascar about 1500 years ago. One of the giant lemurs, *Megaloadapis*, may still have been in existence in 1658, when a French explorer described seeing an animal about the size of a 2-year-old cow, with a rounded head, fore- and hindfeet like a monkey's, and a face and ears like a human's. The species has been found only as fossils since that time. The remaining lemur species, all of which are endangered as forests are cut down, generally have a skeleton adapted to climbing in trees, claws on one or two digits but nails on the other digits of the hands and feet, a well-developed sense of smell, and a comparatively small brain (Fig. 11.8). Some lemur species are diurnal, whereas all other prosimians are nocturnal; and some lemurs exist in social groups, whereas all other prosimians are solitary creatures. Their eyes are directed forward, but are not entirely enclosed in bony sockets, which is typical of all prosimians. Bony eye sockets developed later only in anthropoid primates (see Fig. 11.10).

The aye-aye *(Daubentonia)* is very unusual in possessing only 18 instead of 36 permanent teeth, as do other prosimians, and in having contantly growing, chisel-like incisor teeth more like those of a rodent than a primate. Its middle finger is greatly elongated and terminates in a claw, which is used to probe for insects and grubs in decaying logs and trees. Its activity in this repect is analogous to woodpecker behavior.

Like most organisms, tarsiers have a mixture of primitive and advanced features; for example, their tarsal bone in the hindlimb is greatly elongated, allowing them to leap as much as 6 feet with great accuracy, and they alone among the prosimians possess nearly perfect binocular vision. Tarsiers are arboreal, nocturnal, solitary creatures, about the size of a young kitten, and

Figure 11.7. A nail covers less of the digit tip-pad area than is covered by a claw.

Figure 11.8. The ring-tailed lemur *(Lemur cattus)* is a social, diurnal prosimian member of the lemuroids endemic to Madagascar. (Photograph courtesy of Ian Tattersall)

are very active. Fossil tarsioid forms from Paleocene and early Eocene deposits in western Europe and North America may well represent a transitional link between lemuroid prosimians and the anthropoid primates, although the matter remains unsettled. In fact, it is not at all clear which prosimian lineage diverged to produce the first anthropoids. A large part of the problem is due to the typically poor fossil records for forest species.

Although most of the primate species are arboreal, there is some difference of opinion about the order of occurrence of the earliest suites of adaptive features. The simplest prosimians are nocturnal and insectivorous, as well as arboreal, and they depend heavily on the sense of smell to locate foods and orient themselves to their surroundings. It has therefore been suggested that the early prosimians first evolved adaptations to a nocturnal life based on insect predation, and later took to the trees as arboreal species. The alternative view is that they inherited a nocturnal, insect-eating life style from their insectivore ancestors, and the earliest primate adaptations therefore were associated with climbing and moving through the trees by clinging to branches and leaping from branch to branch or from tree to tree. These locomotor behaviors are distinct from those of the anthropoid primates, which display quadrupedalism, brachiation, or bipedalism, all of which are derived from the clinging and leaping movements typical of prosimians.

The leaping ability of prosimians is made possible by well-developed hindlimb musculoskeletal structures, which provide the propulsive force for this type of locomotion. In the **quadrupedal** monkeys, which also may leap from place to place, walking on all fours on the ground or in the trees is accompanied by more equal development and propulsive force of both the upper and the lower parts of the body. The gorilla and common chimpanzee in the great ape family are also quadrupedal, and now spend most of their waking time on the ground and use the trees mainly as a place to sleep. **Brachiation** is a specialized locomotion, involving swinging from branch to branch by the use of the forelimbs and upper body; the hindlimbs and lower body are not used for propulsion. The lesser apes (gibbons and siamangs) and the orangutan, alone among the great apes, are habitual brachiators, and spend almost all their time high up in the trees. **Bipedalism,** or walking exclusively on two

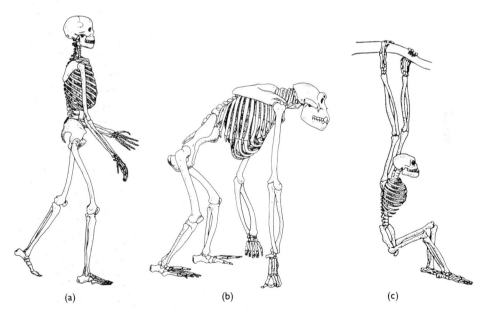

(a) (b) (c)

Figure 11.9. Different limb proportions are associated with different locomotor modes in Primates. (a) Bipedal humans have relatively shorter forelimbs than hindlimbs. (b) Quadrupedal apes and monkeys have forelimbs of about the same or slightly greater length than hindlimbs. (c) Habitual brachiators, such as gibbons and orangutan, have much longer forelimbs than hindlimbs.

Box 11.1   Geographical Distribution of Living Nonhuman Primates

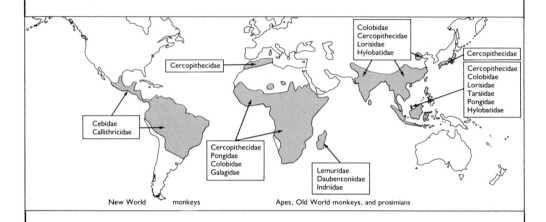

New World monkeys (Callitrichidae and Cebidae) live in tropical and subtropical regions of Central and South America, but all the other nonhuman primates are residents of Africa and Asia. Among the prosimians, some lorisoids are African (Galagidae) and others, Asian (Lorisidae); tarsiers (Tarsiidae) are found only in Indonesia and the Philippines; and the three families of lemuroids (Lemuridae, Indriidae, and Daubentoniidae) are endemic to the island nation of Madagascar, off the southeast coast of Africa. Old World monkeys (Cercopithecidae and Colobidae) are both African and Asian, and some species inhabit cold regions in the Himalayas and in Japan, whereas most are tropical and subtropical. Gibbons and siamangs (Hylobatidae) live only in Asia, but the great apes (Pongidae) exist as remnant populations in equatorial Africa (chimpanzees and gorillas) and Asia (orangutan of Borneo and Sumatra).

legs, is characteristic of only the human family (Hominidae) and is a hallmark of all its species. The relative lengths of forelimbs and hindlimbs provide clues to locomotor modes in fossil as well as in living species. Brachiators have extremely long forelimbs and very short hindlimbs; quadrupedal primates have forelimbs and hindlimbs of approximately equal length; and bipedal hominine primates have relatively short forelimbs and long hindlimbs (Fig. 11.9).

Primates evolved in the tropical and subtropical forests and, except for pre-human and human species in the past few hundred thousand years, have generally remained in warm, humid areas (Box 11.1). Fossils found in Europe and North America are the remains of primates that lived in these regions when the climate was much warmer and subtropical zones extended well into parts of the world that are now temperate. Even in Pliocene and some Pleistocene times, southern and western Europe during the interglacial periods had subtropical or very mild climates and were suitable habitats for primates. As the climate cooled, many of the primates withdrew to warmer regions.

The appearance of the anthropoids in the early Oligocene is correlated with the decline of the prosimians in the late Eocene, and their subsequent restriction to the few outlying pockets they now occupy. Whatever factors caused their decline, the prosimians had already decreased noticably before the anthropoids appeared. In fact, the late Eocene was a time when many animals went extinct, and it has even been labeled a minor mass extinction episode (see Fig. 10.40).

The Evolution of the Anthropoid Primates

The suborder **Anthropoidea** includes all the simians: Old World and New World monkeys, lesser apes, great apes, and the human family. The two groups of monkeys are classified separately in their own superfamilies, and all apes and the human family are grouped together in the superfamily Hominoidea. All the anthropoids are descended from a prosimian lineage and diverged at successive times during the past 40 to 45 Myr.

There are scanty fossil remains for the anthropoids (as well as for prosimians) because their arboreal life style in tropical and subtropical forests provides few conditions suitable for fossilization. Decay occurs rapidly in hot, humid habitats, and even teeth decompose in the corrosive soil of tropical forests, although they are the principal remains for most species. From the ever-growing collection of fossils and from comparative studies of anatomical and molecular features of living species, biologists have at least some of the broad outlines of anthropoid evolutionary history.

New World Monkeys and Old World Monkeys

The **New World monkeys** are strictly arboreal species of tropical and subtropical South America and a few regions of southern Central America, whereas the **Old World monkeys** live in Africa and Asia as both arboreal and primarily terrestrial species. All the monkeys and other anthropoids share a number of features that distinguish them from the prosimians, among which are a highly developed brain with numerous cerebral convolutions, a well-developed visual area in the neocortex, and reduced olfactory components; a bony plate behind the eye socket that separates the forward-directed orbital cavities from regions behind them; and well-developed acute, color, stereoscopic vision (Fig. 11.10). The primarily nocturnal prosimians lack *cone* photoreceptors in the retina, and can see shades of gray in dim light by means of *rod* photoreceptors. The cones in diurnal anthropoids provide the sensory components to see colors in bright light, and anthropoids, like prosimians, function in dim light by use of their retinal rod photoreceptors.

Except for the South American owl (night) monkey *(Aotus),* which presumably evolved its nocturnal habit as a derived character state from diurnal ancestry, all the monkeys and other anthropoids are *diurnal.* They are also *social,* in contrast with the primarily solitary prosimian species. The life style is thus quite different in anthropoid and prosimian primates, and represents a major change in primate evolution. The enhancement of vision and reduction of olfaction in the anthropoids can be discerned in fossils, both from the reduced size of the bony apparatus of the nose and from the progressive recession and retraction of the face to a position below rather than in front of the cranium proper (Fig. 11.11). As we will see later, other changes in the skull, reflecting dietary and postural modifications of importance in life style, occurred during anthropoid evolution.

The anthropoid suborder is conventionally subdivided into two infraorders, the Platyrrhini and Catarrhini. The **platyrrhines,** which include all the New World monkeys, have nostrils that are separated by a wide cartilaginous septum and open laterally. The **catarrhines** include all the Old World monkeys, apes, and human family species, and their nostrils are close together and open down instead of across. It is often convenient to use these designations when discussing the combined evolutionary histories of all the Old World anthropoids or the divergence of the platyrrhine and catarrhine lineages.

The platyrrhines consist today of two families of New World monkeys: the morphologically more primitive marmosets and tamarins of the Callitrichidae, and the morphologically more advanced, familiar monkeys of the Cebidae. Marmosets and tamarins are squirrel-like in some respects, being the smallest of all the monkeys and able to scamper quickly up and down tree trunks by the use of claws, a talent not possessed by the cebid monkeys. Marmosets and tamarins have flattened nails only on each great toe, but the claws on all the remaining digits are evolutionarily derived from nails, according to their histological features.

The cebid monkeys have nails on all 20 digits, but, as in the marmosets and tamarins, the thumb is not opposable to the other digits and is often greatly reduced in size. The cebids have a better developed brain than the marmosets and tamarins, and many genera have a long prehensile tail that can be used as a "fifth limb"; they often cling to a branch only by the prehensile tail (Fig. 11.12). Cebids are quadrupedal in their arboreal living zone, but many of them can leap great distances between trees in the forest canopy. None of the platyrrhine monkeys is terrestrial, and it is inferred that their entire evolutionary history has been arboreal. They rarely descend to the ground, although a few species may do so to drink or forage on the forest floor. Among the more

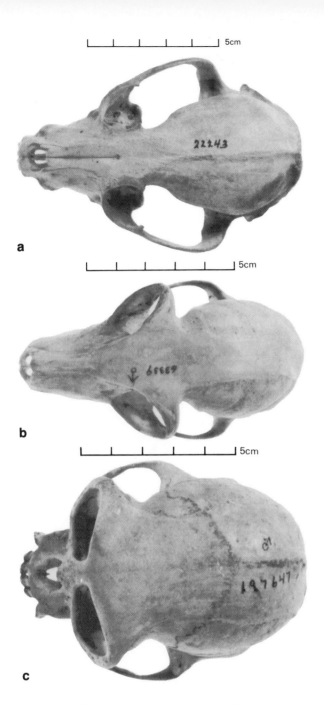

Figure 11.10. Modification of the eye orbits, as seen from above in skulls of (a) raccoon, (b) lemur, a prosimian primate, and (c) Old World monkey, an anthropoid primate. Note particularly the post-orbital bar, which defines the eye sockets in the lemur but not the raccoon, and complete postorbital enclosure by a bony plate behind each eye socket in the monkey. Also note in the monkey, compared with the lemur, the progressive movement of the eyes to the front of the face, the reduced projection of the muzzle, and the large cranium. (From *Primates and Their Adaptations,* Oxford Biology Reader 28, by J. R. Napier. Copyright © 1972 by Oxford University Press. Reprinted by permission)

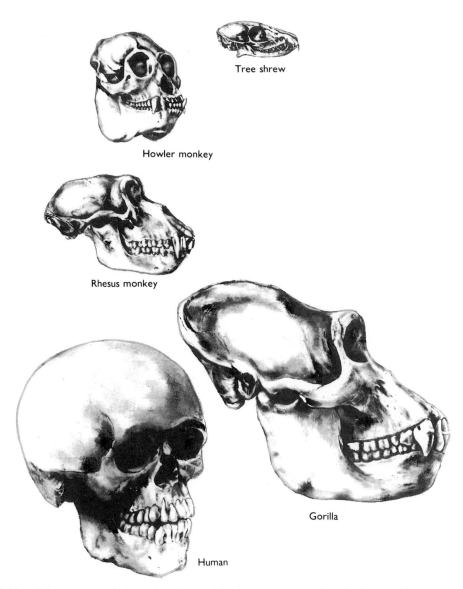

Figure 11.11. Enhancement of vision in anthropoid primates compared with the tree shrew, an insectivore closely related to the ancestral primate stock. The face has receded to a retracted position below rather than in front of the eyes in New World monkeys (howler), Old World monkeys (Rhesus), and apes (gorilla). The face is entirely flattened and nonprojecting in living humans. (Adapted from R. J. Harrison and W. Montagna, 1973, *Man,* 2nd ed., Fig. 1–3. Englewood Cliffs, N.J.: Prentice-Hall)

Figure 11.12. Like many other New World monkeys, the howler *(Alouatta)* is able to cling to a branch by its prehensile tail alone.

familiar cebid monkeys are the howler monkey *(Alouetta)*, spider monkey *(Ateles)*, squirrel monkey *(Saimiri)*, and capuchin monkey *(Cebus)*.

The Old World monkeys are members of the superfamily **Cercopithecoidea**, whose extant species are separated into two families, the Cercopithecidae and the Colobidae. The cercopithecids include omnivorous species, such as the baboons and macaques, and many other common monkeys of Africa; only some of the macaque species *(Macaca)* are Asian, living throughout India and in parts of northern China and Japan. The colobids are African and Asian, and include the colobid monkey *(Colobus)*, proboscis monkey *(Nasalis)*, langurs *(Presbytis)*, and many others, all of which are arboreal and subsist largely on leaves. The colobids are restricted in habitat to forests with nondeciduous trees, because of their diet, and their unusual digestive system, teeth, and chewing muscles are associated with their leafy diet.

Cercopithecids are more diverse than colobids in their habitats, locomotion, and diet. They live in forests, savannas, and high mountain regions; are quadrupedal, but may be arboreal or terrestrial; and eat a large variety of foods, which can be stored in buccal sacs (cheek sacs). The Japanese macaque, which has been intensively studied by behaviorists, is the northernmost cercopithecid and manages reasonably well in a climate that includes cold, snowy winters. Another favorite species in biological research is the Rhesus monkey, a terrestrial macaque of India.

The Old World monkeys are the most diverse group of living anthropoids. Of about 150 species in 36 anthropoid genera, about 75 species in 15 genera are Old World monkeys. Most of the remaining species are New World monkeys, and only 10 species of apes and 1 species of human still exist. This diversity of living species is just the opposite of the diversity known from the fossil record, which contains few monkeys and many more kinds of apes and prehuman species. Whether this difference is due to the better representation in the fossil record of nonforest-dwelling ape and human family members, competition for similar resources, or other factors remains uncertain.

The evolutionary history of the anthropoids has been reconstructed from the spotty fossil record and from molecular and anatomical studies of living species. All these lines of information concur in showing a sequence of divergences in the anthropoids that began with the separation of New World and Old World monkeys (Fig. 11.13). The hominoids diverged later from an Old World catarrhine lineage. Molecular analysis of DNA and proteins indicates the divergence of platyrrhine and catarrhine monkeys about 35 to 37 Myr ago, presumably following an initial divergence of anthropoids and prosimians perhaps 40 to 45 Myr ago. The earliest fossil evidence for the catarrhines, however, is only about 30 to 31 Myr old, and comes from the rich deposits of mid-Oligocene formations in the Fayum region of Egypt.

The Fayum fossils include a fine suite of specimens classified as species of *Aegyptopithecus, Apidium, Parapithecus,* and *Propliopithecus* (Fig. 11.14). All these and other forms appear to be generalized arboreal quadrupeds that had reached monkey status, although Elwyn Simons considers *Aegyptopithecus* and *Propliopithecus,* at least, to be hominoid because of certain apelike rather than monkeylike features. Other paleontologists, such as David Pilbeam and Peter Andrews, suggest that all these mid-Oligocene forms were too primitive and different from modern primates to be identified as monkeys or as early hominoids. Unfortunately, there is a gap of about 10 Myr in the African fossil record, between 31 or 32 Myr (mid-Oligocene) and about 22 Myr (early Miocene) ago. Hominoids already were present in the early Miocene, so the fossil record fails to provide the critical history for the time of divergence of hominoids from an ancestral catarrhine lineage that had produced the monkeys. On the basis of the fossil record, therefore, paleontologists cannot assign the Fayum fossils unambiguously to the monkeys or to the hominoids. This point

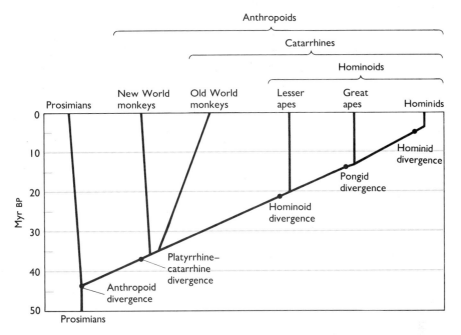

Figure 11.13. Simplified evolutionary history of major divergences in the order Primates, as deduced from various lines of evidence. The anthropoids diverged from a prosimian stock about 40 to 45 Myr BP, and the anthropoid lineage diverged about 35 to 37 Myr BP into platyrrhine and catarrhine branches. The Old World monkeys were the first catarrhines to diverge from an ancestral stock that later evolved to hominoid status. The lesser apes were the first hominoid family to evolve from the ancestral hominoid stock. Later divergences from this hominoid stock produced the great ape lineage, about 14 Myr BP, and the hominid lineage, about 5 to 7 Myr BP.

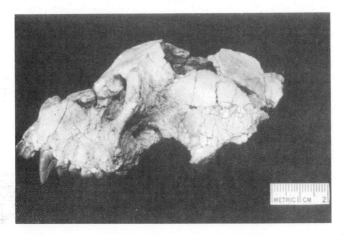

Figure 11.14. *Aegyptopithecus zeuxis*, from the Fayum deposits in Egypt, was a generalized tree-living, fruit-eating quadruped of anthropoid status. Whether this species was a monkey or an early hominoid, however, is a controversial matter. (Photograph by E. L. Simons)

is reinforced by the molecular clocks showing a divergence time of 35 to 37 Myr for the Old World monkeys. By 31 to 32 Myr ago, it is possible that hominoids had branched off, but it is not known whether or not they did.

Fossils with clear relationships to modern Old World monkeys include *Prohylobates* and *Victoriapithecus* from early Miocene deposits only in Africa (Egypt, Libya, Kenya, Uganda). Old World monkeys did not appear in Europe and Asia until the middle Miocene, and were widespread in Europe, Asia, and Africa in the late Miocene. This temporal distribution of fossils indicates that the Old World monkeys originated in Africa and later migrated into Eurasia. The Tethys Sea opened and closed at various times after 17 Myr BP, providing land bridges by which migrations could occur. The fossil record and the geologic record therefore reinforce the premise of an African origin and subsequent spread of the Old World monkeys during the Miocene.

Although the New World monkeys probably were on a separate evolutionary pathway at least by the beginning of the Oligocene, few Oligocene fossils have been found. The fossil *Branisella* from Bolivia is a platyrrhine of Oligocene age, but quite different from modern New World monkey groups. The earliest South American fossils with clear affinities to living forms have come from early Miocene deposits in Argentina (Patagonia). The few other New World monkey fossils have been found in Colombian and Bolivian deposits of middle Miocene age.

Divergence of the Hominoids

The catarrhine superfamily **Hominoidea** is traditionally subdivided into three families: Hylobatidae, lesser apes; Pongidae, great apes; and Hominidae, human family (Table 11.3). All **hominoids** have a larger and more complex brain than the monkeys, no tail, various musculoskeletal features associated with different primary locomotor behaviors, and great flexibility of limb and digit structures. As in the Old World monkeys, the thumb and great toe are opposable in most species, but the degree of opposability varies.

The half-dozen living **hylobatid** species are members of the gibbon genus, *Hylobates*, or the siamang genus, *Symphalangus*, of Malaysia and the Indonesian islands of Borneo and Sumatra. They have been put together by some authorities into the single genus *Hylobates*. They are strictly arboreal and brachiators, swinging freely from branch to branch by exclusive use of their forelimbs and upper body (Fig. 11.15). Typically for brachiators, the thumb is very short and attached far back on the palm. This adaptation permits the animal to grasp and release each branch very efficiently, without having a thumb that must be bent under to hold and then unbent to release the hold. The two hylobatid genera differ in some minor traits; for example, siamangs have a membrane between the first joints of the second and third digits, and an extensible sac in the throat (laryngeal sac) that allows them to produce reverberating, throaty calls heard throughout the area.

TABLE 11.3
Living Members of the Superfamily Hominoidea

Family	Genus	Common Name	Distribution	Locomotion
Hylobatidae	*Hylobates*	Gibbon	Central Asia,	Arboreal brachiator
	Symphalangus	Siamang	Malaysia, Indonesia	Arboreal brachiator
Pongidae	*Pongo*	Orangutan	Borneo, Sumatra	Arboreal brachiator
	Gorilla	Gorilla	Central Africa	Partly terrestrial quadruped
	Pan	Chimpanzee	Central Africa	Partly terrestrial quadruped
Hominidae	*Homo*	Human	Worldwide	Upright biped

The living **pongids** are the single species of orangutan *(Pongo pygmaeus),* which is generally subdivided into two subspecies; the single species of gorilla *(Gorilla gorilla),* which includes the lowland and mountain populations or subspecies, and possibly a third subspecies; and two species of chimpanzee *(Pan):* the common chimpanzee *(P. troglodytes),* and the pygmy chimpanzee *(P. paniscus),* each subdivided into subspecies. The gorilla and chimpanzee are African, and they spend most of their time on the ground. Both these apes are quadrupedal, but they walk on the knuckles of their hands and the soles of their feet (Fig. 11.16). The Asian orangutan is a brachiator, like the Asian hylobatids, and when it walks on the ground, it cups its hands into rounded balls and does not walk on its knuckles. The orangutan's thumb is just a short stump, as expected.

Figure 11.15. Gibbons are habitually arboreal brachiators of the Hylobatidae, the hominoid family of lesser (small) apes. Note the relatively long forelimbs and short hindlimbs, which are typical of habitural brachiators. (From *The Life of Vertebrates,* 3rd ed., by John Zachary Young. Copyright © 1981 by Oxford University Press. Reprinted by permission)

Figure 11.16. The gorilla walks or stands on the knuckles of its hands and the soles of its feet, as does the chimpanzee. Knuckle-walking is considered to be a recent locomotor adaptation in these African pongids. (From *The Life of Vertebrates*, 3rd ed., by John Zachary Young. Copyright © 1981 by Oxford University Press. Reprinted by permission)

All the lesser and great apes are social. The orangutan is the least social of them all when in the forest environment of its native Borneo and Sumatra. The reduced sociality of the orangutan may be due to the large territory required to fulfill its food needs, permitting only a female and her dependent offspring, or a solitary male, to successfully inhabit an area. Both the gorilla and the chimpanzee live in relatively small extended family groups. Hylobatids, on the contrary, live in nuclear family groups composed of the two parents and their young. Sexual dimorphism is typical of all the great apes, but not of the lesser apes. Great ape males are much larger than the females, have larger and more pronounced canine teeth, and exhibit a number of other distinguishing secondary sex characteristics. The correlation between enhanced sexual dimorphism and polygamy, as opposed to reduced or absent sexual dimorphism and monogamy, provides one basis for determining family structure in fossil species. It has been an important element in trying to reconstruct fossil hominid social structure and evolution.

The early evolutionary history of the hominoids is restricted entirely to Africa, between 22 and 17 Myr of the early Miocene. One of the earliest known fossils that can be considered a hominoid is *Proconsul africanus*, discovered in Kenya. Its mixture of features led to its being classified at one time or another as a monkey or as a hominoid, and even as an ancestor of the chimpanzee. Recent discoveries of skeletal structures that were unlabeled or mislabeled in museum collections and proved to belong to *Proconsul* have allowed the most complete reconstruction of an early Miocene hominoid. *Proconsul* appears to have been a very primitive and generalized arboreal quadruped (Fig. 11.17). Along with fragments of other early Miocene hominoids, such as *Limnopithecus, Dendropithecus,* and *Micropithecus,* there is evidence of a diverse group of primarily arboreal, forest-dwelling species that were present at least 22 Myr ago. These forms are not closely allied with any of the modern hominoid families, and probably represent clades that diverged early in the Miocene or late in the Oligocene from an ancestral hominoid lineage.

Apart from these early primitive hominoids, all the other known hominoid fossils are related to the great ape and the human lineages; there is no known fossil record for the hylobatids. On the basis of molecular, anatomical, and paleontological data, it is clear that the gibbon/siamang group was the first of the modern hominoid families to diverge from an ancestral stock that later produced the great ape and human lineages (see Fig. 11.13). The molecular

Figure 11.17. The early Miocene hominoid *Proconsul,* whose remains were found in Kenya, is inferred to have been a primitive and generalized arboreal quadruped. This reconstruction, by Alan Walker and Martin Pickford, includes a few parts initially discovered in 1951 (stippled) and many more parts recently found in museum collections by Walker and Pickford (black). Parts that have not been found include the pelvis and most of the spinal column, as well as various portions of the limbs, digits, skull, and lower jaw. (Courtesy of Alan Walker)

clocks are based on DNA and protein phylogenies and on reference points from the fossil record to calibrate divergence time. The hylobatid divergence has been estimated as sometime between 17 and 25 Myr ago. This wide span of time is due to differences in fossil calibration points and to different sets of molecular data.

Charles Sibley and Jon Ahlquist estimated 18 to 22 Myr ago for gibbon divergence on the basis of DNA–DNA hybridizations, whereas Stephen O'Brien estimated 20 to 25 Myr ago on the basis of an extensive electrophoretic analysis of nearly 400 polypeptides. If the clock is calibrated from the fossil record of the Asian great apes, a different divergence time for gibbons would be obtained, depending on the precise branchpoint of these great apes between 13 and 17 Myr BP. Peter Andrews has compared the fossil and molecular data and concluded that the gibbon/siamang lineage diverged 17 to 20 Myr ago, with 20 Myr as the more likely estimate. The existence of fossil relatives of the great apes that are 16 to 17 Myr old suggests that the lesser apes had diverged before the middle Miocene.

The African fossil record for hominoids is almost blank for the middle and late Miocene Epoch (17–5 Myr BP). European and Asian hominoid fossils are known from this time period. The existence of the earliest hominoids exclusively in Africa during the early Miocene pinpoints Africa as the site of hominoid origins. The linkup of the African plate with the Eurasian mainland by 17 Myr BP provided migration routes for hominoids and marks the separation of the Asian and African ape lineages. Most of the hominoid fossil record from Europe and Asia has been reevaluated in recent years and, together with molecular analysis, has led to a significant revision of earlier ideas about ancestry and relationships of the Asian and African apes to the human family and its evolutionary history.

Ramapithecines and the Pongid–Hominid Divergence

Long after molecular phylogenies suggested that hominids diverged 5 to 7 Myr ago from a common ancestor that had diverged (7–10 Myr BP) to produce the African pongid lineages, many paleontologists continued to insist on much earlier divergence dates. This viewpoint was based on interpretations of characters shared by *Ramapithecus* and hominids, and by *Dryopithecus* and pongids. *Ramapithecus* and other **ramapithecines** lived in open woodlands in Africa and Asia, and possibly in Europe, during the middle and late Miocene. Their teeth were hominidlike in that the canines and incisors were relatively reduced and the premolars and molars were enlarged and coated with a thick layer of enamel. **Dryopithecines,** on the contrary, were forest-dwelling, primitive apes that lived in Africa and Europe during the middle and late Miocene. Their large canines and incisors and thin-enameled, reduced premolars and molars resembled the dentition of living pongids. Based on these sets of shared characteristics, it was assumed that the hominid–pongid split was represented by the ramapithecines and dryopithecines, and had occurred 15 to 20 Myr BP.

Considerable emphasis is given to teeth in assigning relationships, not so much because they are the most reliable indicators as because they often are the only fossil remains from which evolutionary histories can be reconstructed. In addition to shared dental features, dryopithecines lived in forests, as modern pongids do, and they had the same U-shaped dental arcade (palate) as pongids. *Ramapithecus* inhabitied open areas, as hominids presumably did, and the reconstruction of their dental arcade appeared to show a humanlike parabolic shape. An abundance of recently discovered ramapithecine fossils led to the reevaluation of their hominid affinities in teeth and jaw features.

The dental arcade in *Ramapithecus* is neither parabolic nor U-shaped, but some variation of a V-shape. In fact, all the ramapithecines, dryopithecines, and australopithecine hominids show some variation of a V-shaped dental arcade (Fig. 11.18). Furthermore, arcade shape is no longer considered an important diagnostic character for any of these fossil groups. As for teeth, evidence now shows that small, thin-enameled cheek teeth and large canines represent correlated adaptations to the slicing and chewing of soft plant foods, which were the principal diet of the forest-dwelling dryopithecines, as they are of modern pongids. Large, thick-enameled molars and reduced canines are associated with the crushing and grinding of tough plant foods, which typically grow in nonforest habitats where ramapithecine and hominid species lived.

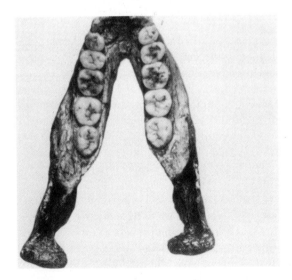

Figure 11.18. The upper jaw of fossil ramapithecines and australopithecines shows some variation of a V-shaped dental arcade, when viewed from below, compared with the U-shaped arcade of apes and the parabolic shape of the human dental arcade (see Fig. 11.26). (Photograph courtesy of Ian Tattersall)

The question that arises is whether similarities between fossil and extant groups represent retention of an ancestral character state, which is a less useful criterion to determine *close* relationships, or of a shared derived character state, which is considered a more useful criterion to assess homology. The situation is more complex than this, however, because derived character states may be the result of either parallel or convergent evolution. As discussed in Chapter 9, this distinction is sometimes difficult to make and may lead to controversial interpretations of the same fossil data.

It is widely acknowledged that the ramapithecine dentition exhibits a number of ancestral character states that have been retained in the hominid lineage, but have evolved to a derived state in the modern pongids. Selection pressures of a similar nature in similar habitats with similar available foods could explain the adaptive basis for the retention of ancestral hominoid features in the evolution of ramapithecines and hominids, independently of each other. It is therefore very unlikely that the similarities in the set of related adaptations mark ramapithecines as prehominids, much less as hominids themselves. Furthermore, ramapithecine dentition varied considerably, and was apelike in some genera and hominidlike in others.

Some ramapithecines had large or intermediate-sized, projecting, apelike canine teeth, whereas *Ramapithecus* and the extinct Asian ape *Gigantopithecus* had small, nonprojecting canines. Even in those ramapithecines with larger canines, however, heavy tooth wear reduced the crowns, so the canines rarely projected beyond the other teeth. Heavy tooth wear and reduced canines apparently go together with large, thick-enameled molars as a character complex associated with the grinding and crushing of tough plant foods.

Although not acknowledged as a hominid or hominid ancestor, *Ramapithecus* and *Sivapithecus* have been classified in the orangutan clade, on the basis of a number of derived characters that they share with the modern Asian pongid. All the fossil specimens assigned with certainty to these two ramapithecine genera (some have put them together in the single genus *Sivapithecus*) have come from Miocene deposits in India, Pakistan, and Turkey, and are dated from 12 to 8 Myr ago. This would put the branchpoint of the orangutan clade at least to 12 Myr ago, but whether the clade originated in Africa or in Asia is uncertain. Many specimens recently found in China, Greece, and Hungary would extend the range of these ramapithecine genera; but paleontologists have no reliable dates for them, and their exact identification has been questioned.

A large number of hominoid teeth from Pasalar in Turkey have been identified by Peter Andrews and others as *Sivapithecus*. These particular fossils have been estimated to be 13 to 14 Myr old, on the basis of associated fauna in the same deposits, not on radiometric data. If this age and identification are correct, it would set the branchpoint of the Asian pongid clade at or before 14 Myr ago. The date is particularly important because it is often used to calibrate hominoid molecular clocks.

Two of the remaining ramapithecines have also been reinterpreted in recent studies. *Gigantopithecus* was once considered a hominid or hominid ancestor because of its hominidlike dentition. But as discussed earlier, the similarities are now regarded as retained hominoid character states and not as an indication of close affinity or ancestry to the hominids. *Gigantopithecus* was restricted to Asian woodlands, and went extinct in the late Miocene without having produced a descendant lineage. *Kenyapithecus*, the only African rama-pithecine known to date, was discovered by Louis Leakey in 1962. The pre-vailing opinion at that time, however, was that the fossil's teeth were so sim-ilar to those of *Ramapithecus* and *Sivapithecus* that it belonged in one or the other of these genera. Recent skeletal remains, together with the new inter-pretations of ramapithecine dentition, have shown that the African species is quite different from the Asian ramapithecines. Accordingly, the species has been reassigned its original name of *Kenyapithecus* and been put into a sepa-rate, but sister, clade from the *Ramapithecus/Sivapithecus/*orangutan clade (Fig. 11.19).

It now is evident that the Asian ramapithecines were a highly varied group, ranging in size from about 20 kg for *Ramapithecus* to about 70 kg for *Gigan-topithecus*. Their postcranial skeletal remains make it clear that the ramapithe-cines, such as *Sivapithecus* and *Gigantopithecus*, were quadrupedal and spent a great deal of their time of the ground. Almost nothing is known about its skeleton, but *Ramapithecus* probably was a terrestrial quadruped, too. All the ramapithecines were woodland inhabitants subsisting on tough plant foods. The entire group disappeared about 8 Myr BP, having diversified considerably in the preceding 4 to 6 Myr of the middle and late Miocene.

Except for *Kenyapithecus*, all the known ramapithecines were Asian and possibly ranged into Europe as well. The very poor fossil record for the mid-

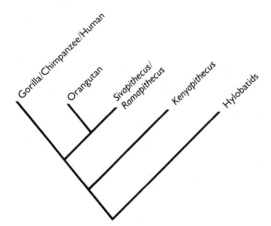

Figure 11.19. Cladogram showing suggested relationships of African and Asian ramapithecines to other hominoid lineages. The living orangutan is inferred to be most closely related to Asian fossil species of *Ramapithecus* and *Sivapithecus*. The African ramapithecine *Kenyapithecus* and the African gorilla/chimpanzee/human lineages are suggested as sister clades of the Asian group.

dle and late Miocene in Africa makes it impossible to know if the ramapithecines originated in Africa or in Asia. There was a land bridge between Africa and Asia by 17 Myr ago, however, and faunal exchanges occurred between the two linked landmasses. It is possible that an earlier African hominoid emigrated to Asia, and there diverged to produce the ramapithecines; *Kenyapithecus* might have reentered Africa after the ramapithecine divergence. It is also possible that the ramapithecines diverged from a hominoid lineage in Africa, and one branch emigrated to Asia, where it diversified, while the other branch remained in Africa. It would help to know where the orangutan clade originated, because an African origin would indicate that its ramapithecine ancestors probably also were African. But we have no information about the place of origin of the orangutan clade, even though all the fossils have been found outside Africa.

The relatively close relationship between the orangutan and the African pongid/hominid clade, based on molecular and anatomical criteria, points to a common ancestor of all the pongids and the hominid lineage. Since it is likely that the ramapithecines were ancestral to the orangutan lineage, it is most probable that some kind of ramapithecine was the last common ancestor before divergence to the first of the pongid clades between 13 and 16 or 17 Myr BP. Sometime during orangutan clade evolution, the lineage evolved a series of derived character states. They included transition from quadrupedal to habitual brachiator locomotion, entry into a forest habitat from the open woodlands of its ramapithecine ancestors, and associated changes in dentition and diet related to its altered life style.

Although the place of origin of the ramapithecine/orangutan clade is uncertain, paleoanthropologists firmly accept an African location for the divergence and origin of the gorilla, chimpanzee, and hominid lineages. This is the most likely possibility, because Africa is the only continent common to all three living groups; the three groups diverged fairly recently, according to molecular and fossil evidence; the hominids clearly originated in Africa; and the three groups are one another's closest living relatives. Unfortunately, virtually no fossil record exists for the African pongids, except for a gorillalike upper jaw from the Samburu Hills in the Great Rift Valley in Kenya. This specimen is 7 to 9 Myr old, which fits nicely with the estimated time of divergence of the gorilla lineage, based on molecular clock analysis. This fossil does not help in deciding whether the gorilla lineage split first, or whether gorilla, chimpanzee, and hominid lineages split three ways from their last common ancestor (see Fig. 9.31).

If there was a three-way split, perhaps 7 to 8 Myr BP, the last common ancestor would most likely have been a ramapithecine. Its hominid descendants presumably retained the ancestral character states associated with a diet of tough plant foods in open places and woodlands. Its pongid descendants may later have evolved the derived character states associated with a diet of soft plant foods in forest habitats, and knuckle-walking as a modification of the retained quadrupedalism. Whether there was an immediate dryopithecinelike state from which the modern pongids evolved, or whether they evolved their derived features in parallel or convergently, remains unknown. If the last common ancestor first diverged to produce a gorilla lineage and later diverged to produce the chimpanzee and hominid lineages, as some molecular data seem to show, an intermediate dryopithecine evolutionary stage is less likely to have occurred in pongid evolution.

From this discussion, it is obvious that there is very little information from the fossil record for the critical times between ramapithecine disappearance about 8 Myr BP and the first fossil hominids from 4 to 5 Myr BP. This kind of fossil evidence is needed to fill in the blanks on hominid evolution and ancestry and to resolve the problem of a two-way or three-way split for pongids and hominids. The question of the origin of the hominids thus remains unanswered. But there is a growing amount of fossil information concerning the evolution of the hominids, once they made their appearance in Africa.

Australopithecines: The First Hominids

The unraveling of human evolutionary history has been a tortuous process, in part because of the scanty fossil record and in part because we are often unable to be objective about our ancestry. Many more fossil specimens are now available, although not nearly enough, and there is now a bit more objectivity than in the past. The fact that our ancestors had a relatively small brain and were apelike in some respects is now accepted, so another hoax like the Piltdown skull would be rejected instantly by modern paleontologists. But some are reluctant to see our ancestors as scavengers rather than as noble hunters, so there is still a tendency toward wishful thinking in human evolution.

Divergence and Evolution of the Hominids

All the hominids share the derived character of **upright bipedalism**—that is, consistent upright posture and the exclusive use of two legs for locomotion. Subdivision into the two subfamilies, Australopithecinae and Homininae, is based on two major features. The **hominines** have a relatively large (and complex) brain in proportion to body size, and complex cultural activity. The **australopithecines** had a small brain in proportion to body size, resembling the apes in this respect, and little or no cultural traditions associated with tool making. These and other features, together with the temporal sequence of the fossil record, clearly point to the australopithecines as the earliest branch of the human family.

Apart from fragmentary remains that are about 5 Myr old and appear to be

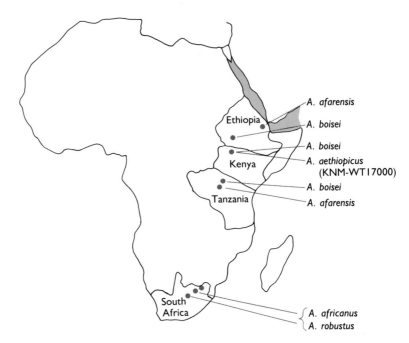

Figure 11.20. East and South African locations in which australopithecine fossils have been found. The gracile A. africanus and robust A. robustus appear to have been exclusively South African, whereas gracile A. afarensis and robust A. boisei occurred only in East Africa. The recently discovered fossil KNM-WT 17000 is a robust form that some paleontologists have named *Australopithecus aethiopicus*. Australopithecines lived at various times between nearly 4 Myr and 1 Myr BP.

hominid in nature, the earliest undeniably australopithecine fossils are nearly 4 Myr old. The australopithecines are found exclusively in South and East Africa, their place of origin and entire evolutionary history until their extinction at about 1 Myr BP (Fig. 11.20). At least four *Australopithecus* species are now known, but a robust australopithecine skull, referred to as specimen KNM-WT 17000 (Kenya National Museum-West Turkana, accession number 17000), that was found in northern Kenya in the region west of Lake Turkana in 1986 may represent a new species *(A. aethiopicus).*

The australopithecines are of two general types: the finer boned, gracile species represented by *Australopithecus africanus* and *A. afarensis;* and the more heavily built, robust species *A. robustus* and *A. boisei.* The majority view among paleontologists is that all these species belong in the single genus *Australopithecus.* Some paleontologists maintain that the gracile and robust species are sufficiently distinct to separate them into a genus of gracile species, *Australopithecus,* and a genus of robust species, *Paranthropus.* We will adhere to the prevailing majority view of a single genus.

The first reported australopithecine fossil was the skull of a young child that had been found in 1924 in a cave in Taung, South Africa, and given to Raymond Dart for study. Dart proposed that the fossil represented an extinct ape intermediate between humans and chimpanzees, and named the species *Australopithecus africanus* (Fig. 11.21). Although the brain was somewhat enlarged compared with an ape's, the immaturity of the fossil made it difficult to know whether it would have become more apelike in adulthood, and the dating was very uncertain. The more hominid quality of the fossil was evident from the small size of the front teeth and from the flattened face, even if the size of the brain was in question. Except for Robert Broom, who joined Dart in a search for more of these fossils and considered the Taung skull to have belonged to a direct human ancestor, the whole idea of hominid status for *Australopithecus* was totally rejected by most of the scientific community.

The vigorous rejection of *A. africanus* as a hominid was based largely on the prevailing view that the first evolutionary change separating man from the beasts was enlargement of the brain, and that upright bipedalism and cultural traditions appeared later. It was widely believed that these "bright apes" had evolved from tree dwellers into terrestrial upright bipeds and had used their freed hands to make tools by which they could defend themselves and hunt for food. The principal human quality, however, is a large brain, and the small-brained *Australopithecus* was considered to be an ape whose humanoid front teeth were the product of evolutionary convergence. In addition, almost everyone accepted *Ramapithecus* as a hominid or hominid ancestor. They failed to see how an intermediate form could have been present 2 to 3 Myr

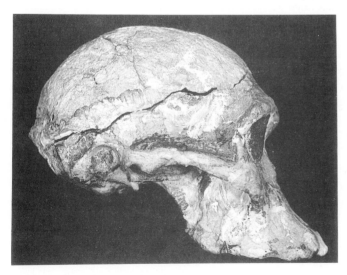

Figure 11.21. *Australopithecus africanus* from South Africa. The skull of an adult from Sterkfontein (specimen Sts 5), found in 1950, dated as about 2.5 Myr old. (Photograph courtesy of the American Museum of Natural History, New York)

BP, a probable age for the Taung skull, when *Ramapithecus* was nearly 15 Myr old and marked the hominid beginnings. The fraudulent Piltdown skull, with a human braincase attached to an orangutan jaw, supported the thesis that brain increase had occurred in an ape and marked the evolutionary change that set humanity apart from the apes (see Box 1.1). *A. africanus* was too ape-like to be on the road to humanity.

In the early 1950s, two particular discoveries changed the viewpoint on hominid evolution. The Piltdown skull was shown to be a hoax, and Broom discovered the hip bones of *Australopithecus*, clearly revealing that they were more like our own and not at all like an ape's. *Australopithecus* was an upright biped, and finally was welcomed into the hominid family. The sequence of change in hominid evolution clearly began with changes in the postcranial limbs and trunk, such that posture and locomotion were modified from ape-like to humanlike. Only afterward did brain size increase, and only then did cultural traditions come into being. These last two features never character-ized the australopithecines during their 3 Myr of evolutionary history, but did mark the later emergence of hominines.

Until 1974, the only australopithecines known were *A. africanus* and the robust type, which many paleontologists simply lumped into the single species *A. robustus* (sometimes referred to as *A. robustus-boisei*). *A. africanus* preceded the robust species by about 500,000 years, and was considered a possible ancestor of both the hominines and the robust australopithecines after a diver-gence about 2 Myr BP. The fossil australopithecines found in the 1970s by Donald Johanson's group at Hadar in the Afar region of Ethiopia proved to be about 3 Myr old, and similar material discovered by Mary Leakey's group at Laetoli, near Olduvai, in Tanzania was reliably dated as 3.75 Myr old. The variety of specimens showed the existence at that time in East Africa of a small-brained form with distinctly apelike face and teeth, but essentially upright and bipedal (Fig. 11.22). The footprints preserved in the volcanic ash at Olduvai leave no doubt that the creatures walked on two feet, and the post-cranial skeletal features of Lucy and other specimens from Afar have the shape of pelvis, angle of femur, and other characteristics of basically upright bipeds.

The discoveries from Ethiopia and Tanzania were named *Australopithecus afarensis* by Johanson and Tim White, and the species was proposed as the common ancestor of the two branches of the hominid family tree. Particu-larly, they proposed that *A. afarensis* diverged to produce a gracile species, *A. africanus*, which gave rise to a robust branch that culminated in *A. boisei*. *A. afarensis* also was suggested as the ancestor of the first hominines, which appeared in East Africa between 2 and 1.8 Myr ago. This phylogeny was appealing to many investigators, but was vigorously challenged by Mary Leakey and her son Richard Leakey, who still prefer their alternative theory of a hypothetical ancestor that diverged earlier than 4 Myr ago to produce the separate branches of australopithecines and hominines. In their view, there-fore, australopithecines are not ancestral to the hominines (Fig. 11.23). The Leakeys and their supporters infer that all the presumed *A. afarensis* fossils actually may represent several species rather than one, and the fossil reconstructions from bits and pieces of material may include important judgmental errors. The controversy continues to the present day.

Yet another discovery of an australopithecine skull by Richard Leakey and Alan Walker, the now famous KNM-WT 17000, reopened the whole question of hominid phylogeny (Fig. 11.24). This fossil is 2.5 Myr old, which puts the robust lineage back to an earlier time than had been known, making it contemporaneous with *A. africanus*. The exaggerated robustness of the skull, with its huge teeth and massive cranial crest and flaring cheek bones, plus a very small brain (410 cc), links it most closely to *A. boisei*. KNM-WT 17000 (*A. aethiopicus*) is unlikely to have been a descendant of *A. africanus*, its contemporary, or of the known *A. robustus* lineage, which came much later. It is also about 500,000 years younger than the known *A. afarensis* fossils, and may or may not have descended from that species. One of several suggestions of its place in hominid evolution is that KNM-WT 17000 represents one branch of a three-way split from *A. afarensis*, and the other two branches are the *A. africanus-robustus* lineage and the hominine lineage (Fig.11.25). The phylogeny remains unclear in the absence of a more complete fossil record.

From the fossil record and molecular data, the origin of the hominids can be traced back to at least 5 Myr BP. There is no information about the ancestral form, but David Pilbeam and others suggest that it was probably a ramapithecine with a relatively small brain and thick-enameled teeth. Whether this hypothetical ancestor produced the australopithecine lineage, like *A. afarensis*, or diverged to produce an australopithecine and a separate hominine branch is not known. The earliest hominines are about 1.8 Myr old, but perhaps go back to nearly 2 Myr. In either case, there is a gap of 1 Myr between the earliest known australopithecine and the earliest known hominine, *Homo habilis*. By the time *H. habilis* appeared in East Africa, 1.8 to 2 Myr BP, both *A. afarensis* and *A. africanus* were gone, but *A. robustus* and *A. boisei* were present. These robust australopithecines coexisted with early hominines for nearly 1 million years, which is evident from the time frame and from fossil deposits in East and South Africa containing remains of both kinds of hominids.

Unfortunately, paleoanthropologists know as little about the causes of extinction of the first hominids as about their origins. There are no living species resembling these early forms, and we can only speculate about their degrees of relationship to later hominids, including ourselves. Molecular data are not readily available from fossils, so studies rely heavily on morphology from the fossil record to help answer some of the important questions about the larger picture of hominid origins and phylogeny. It is quite possible that new fossil discoveries will blur the picture rather than clarify it, just as KNM-WT 17000 has raised more questions than it has answered. As we will see in Chapter 12, new specimens of *H. habilis* uncovered at Olduvai in Tanzania have also raised new questions about the hominine phase of the family history. Before proceeding to the hominines, however, it is pertinent at this point to consider the origins and adaptive properties of the major hominid innovation, upright bipedalism.

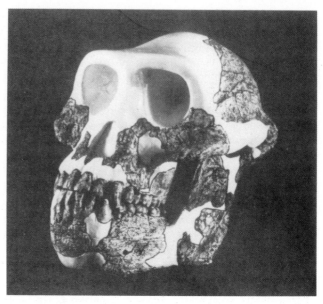

a

Figure 11.22. *Australopithecus afarensis* from Ethiopia. (a) A reconstructed skull, showing the ape-like features of a small cranium, large brow ridges, and massive face and teeth. Note the diastema (space) between the long canine and adjacent incisor teeth in the upper jaw, a typical feature in apes. The diastema accommodates the long, projecting canines on each side of the incisor teeth in the ape lower jaw. (b) Assembled skeleton of Lucy, which is 40 percent complete. The shortened, broadened pelvis and inwardly angled femur, among other features, indicate that she was an upright bipedal hominid. (Photographs courtesy of the Cleveland Museum of Natural History)

The Origin of Erect Posture and Bipedal Locomotion

It is a simple matter to compare the anatomy of living apes and humans to determine the suites of skeletal and muscular changes required for the evolution of an upright biped from a quadrupedal apelike ancestor. It is a far more difficult problem to determine the sequences of changes and the adaptive basis or selection pressures leading to these changes. The problem is compounded by the absence of intermediate forms among living species and among the known fossils.

The location of the **foramen magnum,** the opening in the base of the skull through which the spinal cord passes, differs in quadrupedal apes and bipedal hominids. The opening is located at the far back of the skull in quadrupeds, but is more centrally located in bipeds (Fig. 11.26). The forward shift of the foramen magnum is accompanied by changes in the shape of the occipital bone, which forms the back of the skull, and by a forward shift of the pair of occipital condyles protruding from this bone. These cranial features in fossils

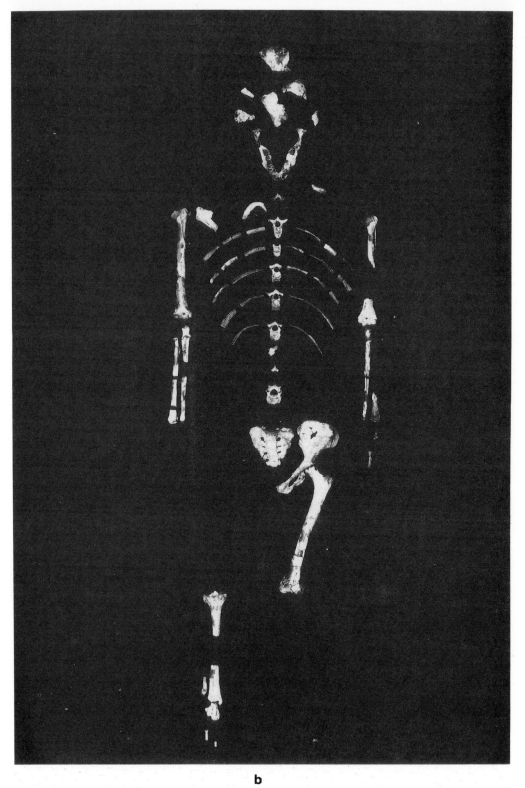

b

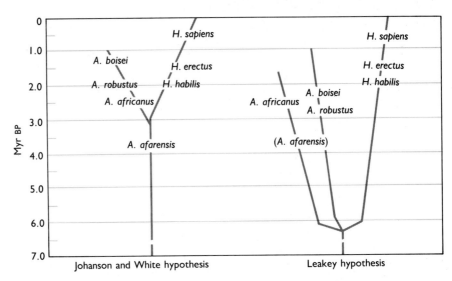

Figure 11.23. Comparison of hominid phylogenies suggested by Johanson and White and by the Leakeys. Donald Johanson and Tim White proposed *A. afarensis* as the common ancestor of an australopithecine and a hominine lineage, which diverged between 2.5 and 3 Myr BP. The Leakeys, however, proposed divergences of australopithecine and hominine lineages from a last common ancestor that was not *A. afarensis*. The nature of this hypothetical common ancestor is unknown.

provide markers of erect posture, even in the absence of postcranial skeletal structures.

Several major modifications in the postcranial skeleton underlie hominid posture and locomotion, including the shape and length of the vertebral column, size and shape of the pelvis, relative proportion of forelimbs to hindlimbs, angle of the femur (upper thigh bone), and construction of the knee (Fig. 11.27). The hominid femur describes an acute angle between the hip and knee joints, which permits a balanced, forward-directed striding walk. The less acute angle of the femur in apes and monkeys leads to an off-balance rocking from side to side as the animal proceeds short distances on two legs. The reshaped hominid pelvis is short, broad, and bowl-shaped, whereas the ape pelvis is elongated and relatively narrow. Important changes in the functions of the gluteal and femoral muscles of pelvis and legs are associated with skeletal modifications for bipedalism.

The gluteus medius and gluteus minimus have changed from extensors to abductors in hominids, and the function of extending the trunk has been assumed by the gluteus maximus. The gluteus maximus is attached to the posterior region of the pelvis, which is greatly broadened in hominids, and the muscle in hominids thus originates well behind the joint of hip and femur (Fig. 11.28). In this location, the power provided by the gluteus maximus, together with a normally extended thigh and straightened knee in bipeds, is in a line of action that causes extension rather than abduction when the muscle

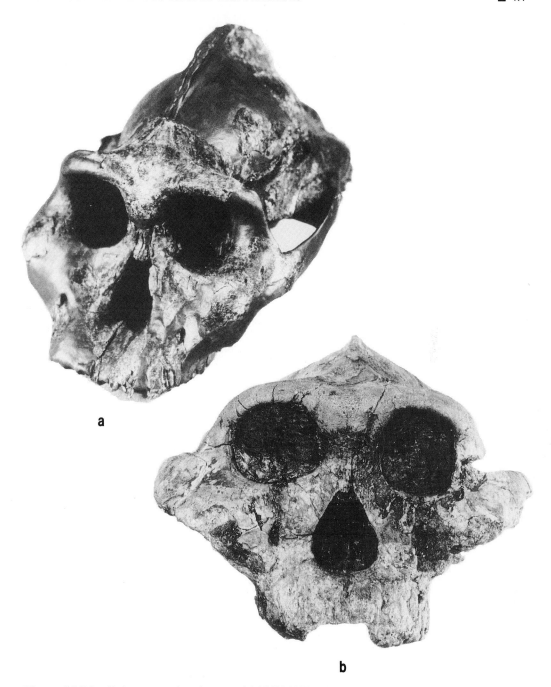

Figure 11.24. Robust australopithecines. (a) KNM-WT 17000, from Kenya, is 2.5 Myr old, making it the oldest known robust form. Its massive cranial crest, flaring cheek bones, huge teeth, and smaller cranium set this fossil apart from (b) *A. boisei* of East Africa (and *A. robustus* of South Africa). (Photograph [a] courtesy of Alan Walker; photograph [b] courtesy of the National Museums of Kenya, Nairobi)

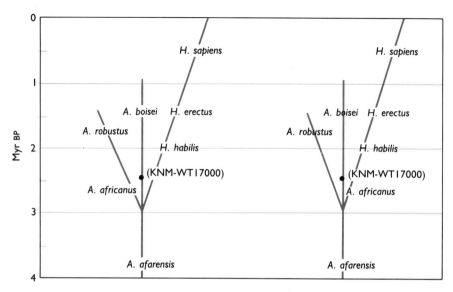

Figure 11.25. Two of a number of hominid phylogenies that attempt to place KNM-WT 17000 (black dot) in an appropriate branch of the family tree. These schemes suggest a three-way split from *A. afarensis*, with KNM-WT 17000 on the same branch as *A. boisei*, because both are East African extreme robust forms. Whether *A. africanus* should be on an australopithecine branch or on the hominine branch as an ancestral form is also a controversial matter at present.

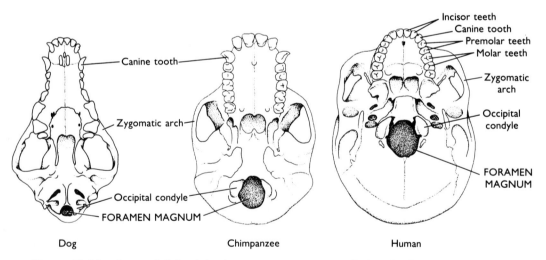

Figure 11.26. Forward shift of the foramen magnum in upright bipeds. Quadrupedal nonprimate mammals, such as the dog, have the foramen magnum at the far back of the skull. A slightly forward shift is evident in apes, such as the chimpanzee, but the foramen magnum is centrally located in humans. Note also in humans the forward shift of the occipital condyles on either side of the opening in the base of the skull. (Reprinted by permission of the University of Chicago Press from *History of the Primates* by W. E. L. Clark, 1965, Fig. 3, p. 22, 5th edition. Copyright © 1965 by the Trustees of the British Museum (Natural History). All rights reserved)

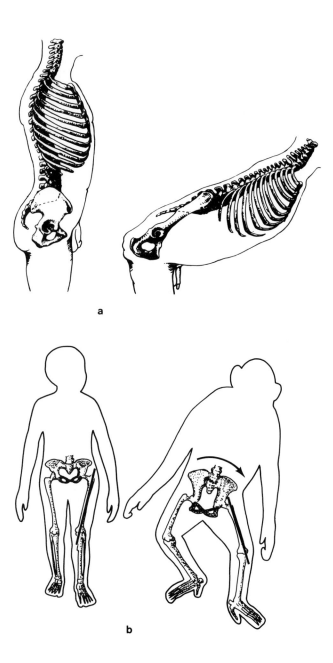

Figure 11.27. Modifications in the postcranial skeleton during evolution of hominid upright bipeds from quadruped ancestry. (a) The pelvis is shorter and broader, and the vertebral column is longer and more curved in humans than in apes. (b) The acute angle described by the femur between pelvis and knee contributes to a balanced, forward-striding walk in humans. In apes, however, the femur describes a wider angle and leads to an off-balance, side-to-side rocking when the animal walks on two legs.

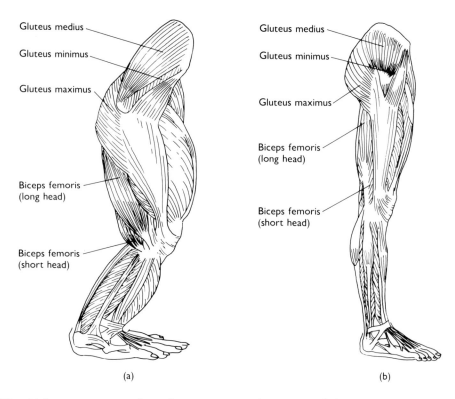

(a)   (b)

Figure 11.28. (a) In apes, two sets of muscles act as principal extensors of the hip in quadrupedal posture and walking. One is the gluteal group (the gluteus medius and gluteus minimus, in particular), which connects the pelvis to the upper part of the femur. The other set is the hamstring group (biceps femoris), which connects the femur to the lower leg bones. (b) The pelvic and femoral muscles work in the opposite way in upright bipedal humans. Gluteus medius and gluteus minimus have changed from extensors to abductors, and the function of extending the trunk has been assumed by the gluteus maximus. The hamstring muscles in humans act mainly as stabilizers and extensors of the hip.

contracts. The posterior location of the upper portion of the gluteus maximus accounts for the more generous proportion of the buttocks in humans than in apes or monkeys.

The human foot has been remodeled into a sturdier support structure associated with bipedalism. The big toe is no longer opposable, and the heel bone (calculus) is massive in proportion to the rest of the foot bones (Fig. 11.29). During walking, the pressure points on the sole of the foot are distributed in a pattern that leads to a certain amount of wear and tear on the tip joint of the big toe. Fossil specimens have been identified as bipeds on the basis of this feature alone.

From the few useful specimens of australopithecine postcranial remains, it appears that these species had an imperfect bipedal gait. The foot has rather long, curved toes; forelimbs are proportionately longer than in hominines; the foramen magnum is not quite central; the pelvis is not as short or as broad as ours; and curved fingers and other features point to the possibility that australopithecines may have been more adept at climbing trees and less adept at walking than the hominines. Notwithstanding these characteristics, australopithecines were upright bipeds and thus possessed the hallmark trait of the hominid family.

What possible adaptive advantages can there have been for the evolution of upright posture and bipedal locomotion? Earlier ideas on this question were based on the mistaken thesis that brain increase was the first hominid change, and that the early hominids were intelligent enough to understand that they could improve their life style by making tools. Once out of the trees and on the ground, bipedal hominids would be at a considerable disadvantage with regard to predators, such as lions and leopards. The hominids had greatly reduced canine teeth, had no claws, and were not as strong or as swift as their quadrupedal predators. But by using their freed hands to make tools and weapons, hominids could hunt for food and defend themselves against attackers. A whole new strategy for living on the grassy African plains would thus have become possible by the hominids' intelligent use of their resources and environment. This scenario is unlikely because australopithecines made no stone tools, did not hunt, and possessed a brain that was unlikely to have been large or complex enough to underlie the high level of intelligent activities attributed to them.

Several theories have been proposed in the past 20 years that concentrate

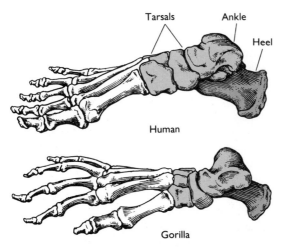

Figure 11.29. The human foot possesses more modifications suited to consistent upright bipedalism than does the foot of predominantly terrestrial apes, which may have much greater body size and weight. The human heel bone (calcaneus), ankle bone (talus), and five tarsal bones, in particular, are larger and sturdier; the big toe is larger and no longer opposable; and the other digits are reduced in size and straight rather than curved. (Adapted from *The Life of Vertebrates*, 3rd ed., by John Zachary Young. Copyright © 1981 by Oxford University Press. Reprinted by permission)

on the adaptive shift as an evolutionary response to a new diet on the open plains. The reduced canines and incisors, the larger molars and premolars, and the anatomy of the jaw and chewing apparatus have led to suggestions that the hominids subsisted principally on tough seeds or nuts. This new diet was available on the plains, and their freed hands allowed hominids to obtain quantities of these foods in an efficient manner. Erect posture also permitted them to observe their surroundings over the top of the grassy vegetation and thus to elude or escape any lurking predators. Although australopithecines apparently were vegetarians, there is no particular reason to associate their plant diet with changes in posture and locomotion. Indeed, baboons and other terrestrial monkeys inhabit the open plains and have remained quadrupedal, and the predominantly terrestrial gorillas and chimpanzees are also quadrupedal. Changes in dentition clearly indicate dietary modifications, but not necessarily in conjunction with locomotor modifications.

Two other theories advanced in recent years have also been met with some skepticism, although each invokes a foraging strategy without the use of stone tools. C. Owen Lovejoy proposed that upright bipedalism was an adaptive response to social behavioral change involving food sharing. He suggested that males became provisioners of food and could carry large amounts of food back to their waiting females and young by the use of hands and arms freed from a locomotor function. The females and young could remain hidden from predators, while the males took all the chances. The preceived sexist overtones of Lovejoy's theory have triggered some alternative views, in which females were the core of the social group and the innovators of new behaviors and other advances. More important, there is no evidence at all of specific social behaviors in hominid or prehominid groups. It certainly is possible to determine whether individuals lived in social groups and whether they were more likely to be monogamous or polygamous. But whether there was a division of labor and responsibility within a social group, cannot be readily ascertained from the fossil record, nor can details of social and family life.

A more recent theory by Pat Shipman suggests that upright bipedalism may represent an adaptation to a scavenging life style in early hominids. Bipedal hominids would have been relatively clumsy walkers, but they would have been able to engage in the long, steady searches for carcasses and to expend energy very efficiently in this ongoing activity. There is evidence of a scavenging life style for early hominines, based on marks lefts by stone tools on the bones of animals that were eaten. There is no similar evidence for australopithecines, who apparently did not use stone tools. In addition, it has been pointed out that modern scavenging animals include those like the hyena, which can tear into a carcass by ripping the tough hide open and exposing the edible inner parts, and others that partake of the feast later because they cannot initiate the opening up of a fresh carcass. Vultures, for example, cannot break through intact animal hide and must wait until the carcass becomes accessible to them, and until the first and fiercer scavengers have had their fill and left the scraps behind. Australopithecines had no stone tools to pierce the hide of a carcass, nor did they have powerful claws or jaws and teeth for such work. If they were scavengers, australopithecines must have waited their turn to eat, when they could get at the carcass and when no strong predators remained. It does not seem to be a likely subsistence strategy when viewed in this perspective.

No theory enjoys a broad consensus of opinion about the origins of the hominids and their locomotor innovation. It is not known whether the primary adaptation was related more to locomotion or to freed hands, or whether one or either of these novel features was tied closely to hominid diet and life style on the grassy savannas of East and South Africa. As Pilbeam and others have remarked, it is entirely possible to deal with the australopithecines quite separately from the hominines. Even though much remains to be learned about hominid origins, a much clearer picture of the significant features underlying hominine evolution has emerged, and in many respects these features appear to be unrelated to australopithecine features. Hominines are characterized by enlargement of the brain, a diet that included meat, manufacture of stone tools, and a perfected bipedalism that is distinct from australopithecine locomotion. We can therefore proceed to the hominine phase of evolutionary history, and examine its significant elements in a separate framework from the australopithecines.

Summary

Basic mammalian innovations inherited by Primates include endothermy, the birth of live young nourished pre- and postnatally by the mother, and a relatively large cerebrum of the brain. Endothermy regulation is under the control of the hypothalamus, and heat plus energy generated by high sustained rates of aerobic oxidations provide stamina and the capacity to exploit various habitats and life styles. A number of anatomical and physiological advances underlie successful endothermy, including a four-chambered heart, efficient oxygen circulation, and an improved chewing apparatus for effective extraction of nutrients in foods. The mammalian reproductive pattern includes an interval of postnatal dependence on parents, which binds parent and young and provides an interval for education. The brain exerts an overwhelming influence on mammalian behaviors, and the expansion of the neocortex of the forebrain permits a growing repertory of intelligent actions.

Primate adaptations of increasing cerebral dominance, visual predominance over olfaction, grasping hands and feet, nails instead of claws, and improved sensory–motor coordination allow free movement in tree canopies of tropical and subtropical forests. Prosimian primates of the Early Cenozoic were mostly nocturnal, solitary, arboreal quadrupeds, and only remnant populations of lemuroids, lorisoids, and tarsiers exist. Anthropoid primates diverged from some prosimian lineage, perhaps 40 to 45 Myr ago, but the earliest known fossils are 30 to 31 Myr old. New World monkeys (platyrrhines) and Old World monkeys (catarrhines) diverged about 35 to 37 Myr BP, and the Old World catarrhine lineage diverged again to produce the hominoid ancestor of the ape and human families, perhaps 31 to 32 Myr ago, or later. Based on molecular data and the scanty fossil record, hylobatids (gibbons and siamangs) diverged about 20 Myr ago; pongids (great apes), about 13 to 17 Myr ago; and hominids, about 5 to 7 Myr BP. The last common ancestor of pongids and hominids may have been one of the ramapithecines, a group that inhabited Africa and Asia between 14 and 8 Myr BP. The orangutan clade diverged first, about 13 to 14 Myr BP, but whether in Asia or Africa is uncertain. The gorilla, chimpanzee, and human lineages diverged in Africa between 7 and 10 Myr BP.

All hominids are consistently upright in posture and bipedal in locomotion. The australopithecine subfamily was relatively small-brained and appears to have lacked tool-making (cultural) traditions. Australopithecines lived exclusively in East and South Africa between 3.75 and about 1 Myr ago, and at least four species are recognized. The robust species disappeared about 1 Myr ago, whereas the two gracile species went extinct much earlier. Various hominid genealogies have been proposed, but one of the favored schemes postulates *Australopithecus afarensis* as the last common ancestor of other australopithecines and of hominines. The cranial and postcranial modifications leading to upright bipedalism are reasonably well understood from comparative studies of living apes and humans. The theories accounting for the shift from quadrupedalism to bipedalism are controversial and are largely undocumented by explicit evidence. Whatever the selection pressures may have been, they were unrelated to an increase in brain size or to tool making, both of which are associated only with hominines at least 2 Myr after upright bipedalism had evolved in the australopithecines.

References and Additional Readings

Andrews, P. 1986. Aspects of hominoid phylogeny. In *Molecules and Morphology in Evolution—Conflict or Compromise,* ed. C. Patterson. New York: Cambridge University Press.

Andrews, P. 1986. Fossil evidence on human origins and dispersal. *Cold Spring Harbor Sympos. Quant. Biol.* 51:419.

Andrews, P. 1986. Molecular evidence for catarrhine evolution. In *Major Topics in Primate and Human Evolution,* ed. B. A. Wood, L. B. Martin, and P. Andrews, p. 107. New York: Cambridge University Press.

Andrews, P., and H. Tobien. 1977. New Miocene locality in Turkey with evidence on the origin of *Ramapithecus* and *Sivapithecus. Nature* 268:699.

Bennett, A. F., and J. A. Ruben. 1979. Endothermy and activity in vertebrates. *Science* 206:649.

Brown, W. M., E. M. Prager, and A. C. Wilson. 1982. Mitochondrial DNA sequences of primates: tempo and mode in evolution. *Jour. Mol. Evol.* 18:225.

Cartmill, M., D. Pilbeam, and G. Issac. 1986. One hundred years of paleoanthropology. *Amer. Sci.* 74:410.

Chiarelli, A. B. 1973. *Evolution of the Primates.* New York: Academic Press.

Ciochon, R. L., and J. G. Fleagle, eds. 1987. *Primate Evolution and Human Origins.* Hawthorne, N.Y.: Aldine de Gruyter.

Crawshaw, L. I. et al. 1981. The early development of vertebrate thermoregulation. *Amer. Sci.* 69:543.

Crompton, A. W., C. R. Taylor, and J. A. Jagger. 1978. Evolution of homeothermy in mammals. *Nature* 272:333.

Cronin, J. E. et al. 1981. Tempo and mode in hominid evolution. *Nature* 292:113.

Dart, R. A. 1925. *Australopithecus africanus,* the man-ape of South Africa. *Nature* 115:195.

Dart, R. A. 1926. Taungs and its significance. *Nat. Hist.* 26:315.

Day, M. H. 1986. *Guide to Fossil Man,* 4th ed. Chicago: University of Chicago Press.

Delson, E., ed. 1985. *Ancestors: The Hard Evidence.* New York: Liss.

Eckhardt, R. B. 1987. Hominoid nasal region polymorphism and its phylogenetic significance. *Nature* 328:333.

Falk, D. 1984. The petrified brain. *Nat. Hist.* 93(9):36.

Goldman, D., P. R. Giri, and S. J. O'Brien. 1987. A molecular phylogeny of the hominoid primates as indicated by two-dimensional protein electrophoresis. *Proc. Natl. Acad. Sci. U.S.* 84:3307.

Hasegawa, M., H. Krishino, and T. Yano. 1985. Dating of the human–ape splitting by a molecular clock of DNA. *Jour. Mol. Evol.* 22:160.

Hay, R. L., and M. D. Leakey. 1982. The fossil footprints of Laetoli. *Sci. Amer.* 246(2):50.

Hill, A. 1985. Early hominid from Baringo, Kenya. *Nature* 315:222.

Holloway, R. L. 1974. The casts of fossil hominid brains. *Sci. Amer.* 231(1):106.

Jacobs, K. H. 1985. Human origins. In *What Darwin Began,* ed. L. R. Godfrey, p. 274. Boston: Allyn and Bacon.

Johanson, D. C., and M. Edey. 1981. *Lucy: The Beginnings of Humankind.* New York: Simon and Schuster.

Johanson, D. C., and M. Taieb. 1976. Plio-Pleistocene hominid discoveries in Hadar, Ethiopia. *Nature* 260:293.

Johanson, D. C. and T. D. White. 1979. A systematic assessment of early African hominids. *Science* 203:321.

Jolly, C. J. 1970. The seed-eaters. A new model of hominid differentiation based on a baboon analogy. *Man* 5:5

Lancaster, J. B. 1978. Carrying and sharing in human evolution. *Human Nat.* 1:82.

Leakey, M. D. et al. 1976. Fossil hominids from the Laetolil beds. *Nature* 262:460.

Leakey, R. E. F. 1976. New hominid fossils from the Koobi Fora formation in northern Kenya. *Nature* 261:574.

Leakey, R. E., and R. Lewin. 1978. *People of the Lake.* New York: Doubleday.

Leakey, R. E., and A. C. Walker. 1985. New higher primates from the early Miocene of Buluk, Kenya. *Nature* 318:173.

Lewin, R. 1987. *Bones of Contention.* New York: Simon and Schuster.

Lovejoy, C. O. 1981. The origin of man. *Science* 211:341.

Lowenstein, J. M., T. Molleson, and S. L. Washburn. 1982. Piltdown jaw confirmed as orang. *Nature* 299:294.

Marks, J. et al. 1986. The primate α-globin gene family: a paradigm of the fluid genome. *Cold Spring Harbor Sympos. Quant. Biol* 51:499.

Martin, L. 1985. Significance of enamel thickness in hominoid evolution. *Nature* 314:260.

Napier, J. 1967. The antiquity of human walking. *Sci. Amer.* 216(4):56.

Patterson, C., ed. 1987. *Molecules and Morphology in Evolution—Conflict or Compromise.* New York: Cambridge University Press.

Pilbeam, D. 1984. The descent of hominoids and hominids. *Sci. Amer.* 250(3):84.

Pilbeam, D. 1986. Hominoid evolution and hominoid origins. *Amer. Anthropol.* 88:295.

Pilbeam, D. et al. 1977. New hominoid primates from the Siwaliks of Pakistan and their bearing on hominoid evolution. *Nature* 270:689.

Pilbeam, D., and S. J. Gould. 1974. Size and scaling in human evolution. *Science* 186:892.

Robinson, T. F., S. M. Factor, and E. H. Sonnenblick. 1986. The heart as a suction pump. *Sci. Amer.* 254(6):84.

Romer, A. S. 1970. *The Vertebrate Body,* 4th ed. Philadelphia: Saunders.

Sarich, V. M., and A. C. Wilson. 1967. Immunological time scale for hominid evolution. *Science* 158:1200.

Schwartz, J. H. 1984. The evolutionary relationships of man and orang-utans. *Nature* 308:501.

Shipman, P. 1986. Baffling limb on the family tree. *Discover,* September, p. 86.

Shipman, P. 1986. Scavenging and hunting in early hominids: theoretical framework and tests. *Amer. Anthropol.* 88:27.

Sibley, C. G., and J. E. Ahlquist. 1984. The phylogeny of the hominoid primates, as indicated by DNA–DNA hybridization. *Jour. Mol. Evol.* 20:2.

Simons, E. L. 1977. *Ramapithecus. Sci. Amer.* 236(5):28.

Smith, F. H., and F. Spencer, eds. 1984. *The Origins of Modern Humans.* New York: Liss.

Steward, C.-B., J. W. Schilling, and A. C. Wilson. 1987. Adaptive evolution in the stomach lysozymes of foregut fermenters. *Nature* 330:401.

Tattersall, I. 1982. *The Primates of Madagascar.* New York: Columbia University Press.

Templeton, A. R. 1985. The phylogeny of the hominoid primates: a statistical analysis of the DNA–DNA hybridization data. *Mol. Biol. Evol.* 2:420.

Terborgh, J. and M. Stern, 1987. The surreptitious life of the saddle-backed tamarin. *Amer. Sci.* 75:260.

Tobias, P. V. 1971. *The Brain in Hominid Evolution.* New York: Columbia University Press.

Tobias, P. V., ed. 1985. *Hominid Evolution: Past, Present, and Future.* New York: Liss.

Tuttle, R. H. 1969. Knuckle-walking and the problem of human origins. *Science* 166:953.

Ueda, S. et al. 1986. Hominoid evolution based on the structures of immunoglobulin epsilon and alpha genes. *Cold Spring Harbor Sympos. Quant. Biol.* 51:429.

Walker, A. C. 1981. Dietary hypotheses and human evolution. *Philos. Trans. Royal Soc. B* 292:57.

Walker, A. C. 1986. 2.5-Myr *Australopithecus boisei* from west of Lake Turkana, Kenya. *Nature* 322:517.

Walker, A. C., and R. E. Leakey. 1978. The hominids of East Turkana. *Sci. Amer.* 239(2):44.

Weiner, J. S., F. P. Oakley, and W. E. LeGros Clark. 1953. The solution of the Pilt-down problem. *Bull. Brit. Mus. (Nat. Hist.) Geol.* 2:141.

Wolpoff, M. H. 1983. *Ramapithecus* and human origins: an anthropologist's perspective of changing interpretations. In *New Interpretations of Ape and Human Ancestry*, ed. R. L. Ciochon and R. S. Corruccini, p. 651. New York: Plenum Press.

Wood, B. A., L. B. Martin, and P. Andrews, eds. 1986. *Major Topics in Primate and Human Evolution*. New York: Cambridge University Press.

CHAPTER 12

Two Million Years of Hominine Evolution

The **hominine** branch of the Hominidae is composed of at least three species of *Homo*, of which two are extinct and only we remain. Hominines originated about 2 Myr ago in Africa, presumably from an australopithecine ancestor such as *Australopithecus afarensis* or *A. africanus*. Beginning at least 1 Myr ago, some founder populations migrated from Africa to Asia and Europe. These first cosmopolitan hominines were members of the species *Homo erectus*. Our own species is probably descended from *H. erectus,* but the overall evolutionary picture is highly controversial. By whatever route we did emerge, the overwhelming influence of cultural evolution has markedly altered the human condition during its past 10,000 years of existence. In this brief span of time, *H. sapiens* developed from isolated bands of nomadic hunter-gatherers into the dominant species on the Earth—the determiners of its fate and of the future of its life forms.

The Early Hominine Species

Europeans were the first paleoanthropologists, and the earliest discoveries of fossil hominines were made in Europe and in isolated Asian outposts by predominantly European scientists. Later discoveries of a larger set of Asian fossils and of African segments of hominine evolutionary history have turned the focus away from Europe to a considerable degree. The ongoing story continues to unfold and be revised as new discoveries are made. But just as with the australopithecines, many of these discoveries have raised new complexities as well as given new insights into hominine evolutionary history.

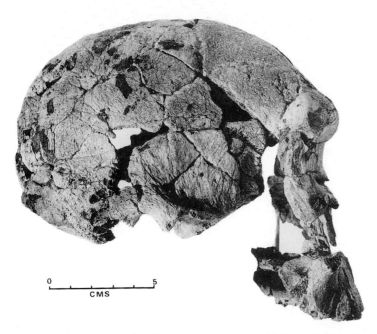

Figure 12.1. KNM-ER 1470 is a representative of *Homo habilis*, the earliest known hominine species. This fossil and others from the East Turkana region in northwestern Kenya are about 2 Myr old, as are the South African examples of this species. (Photograph courtesy of the National Museums of Kenya, Nairobi)

The Enigma of *Homo habilis*

Homo habilis is the oldest known hominine species, although it was discovered long after the fossil record for *H. erectus* and *H. sapiens* was known. Beginning in 1959, Louis Leakey, Mary Leakey, and their colleagues Phillip Tobias and John Napier described a number of specimens from the Olduvai Gorge in northern Tanzania as the fossil remains of a new species, *Homo habilis*. The identification was vigorously disputed for many years, but similar fossils were discovered in Kenya by Richard Leakey (KNM-ER 1470 and others)* and at Sterkfontein in South Africa by Tobias (STw 53) in the 1970s (Fig. 12.1). These and other fossils have strengthened the acceptance of the hominine status of *H. habilis*.

Two particular features provided the basis for *H. habilis* to be considered a hominine: an enlarged braincase, compared with that of australopithecines; and the production of crude stone tools made in the Oldowan tradition and now associated with *H. habilis* culture. The habiline brain size of nearly 700

*Lake Turkana was formerly called Lake Rudolf. Earlier accessions refer to the east or west of Lake Rudolf (ER, WR); later ones, to the east or west of Lake Turkana (ET, WT).

cc is an intermediate value between the maximum of about 500 cc in australopithecines and the average of 1000 cc in *H. erectus*. Their search for such an intermediate was prompted by the Leakeys' prediction that an earlier hominine with a smaller brain than that of *H. erectus* must have existed as an evolutionary link between australopithecines and hominines.

The fragmentary remains of postcranial specimens precluded any substantial reconstruction of the *H. habilis* body, but because it was a hominine there was a general assumption that the species had been more like *H. erectus* than like any australopithecine species. In addition, *H. habilis* skulls and teeth were sometimes found in the same deposits with the remains of robust australopithecines, and the striking differences between the two further emphasized the hominine qualities of *H. habilis*.

A number of studies in the 1980s pointed to various features of *H. habilis*, apart from brain size and tool-making, that were more like those in apes and australopithecines than in hominines. Foot, leg, and hand fossils that had been collected in 1960 by Louis and Mary Leakey at Olduvai were analyzed by Jack Stern and Randall Susman in 1982. Stern and Susman found a mosaic of primitive and advanced features in these structures, including long, curved fingers and toes very similar to those of apes and australopithecines. A new Olduvai Hominid fossil (OH 62), reported in 1987 by Donald Johanson, Tim White, and their colleagues, was reassembled from 302 widely scattered pieces of bone, and was determined to be the remains of *H. habilis*. The nonduplication of bony fragments led Johanson and White to conclude that all the pieces belong to one individual. This reduces the possibility of misinterpretations based on reconstructing an individual from parts of different species or variants of a species. Fossil OH 62 is the most complete example of a *H. habilis* individual known and, importantly, includes postcranial as well as cranial material (Fig. 12.2).

OH 62 closely resembles STw 53 in its face, palate, jaw, and dentition, and KNM-ER 1470 and other habiline skulls as well, which clearly establishes it as *H. habilis*, even though the cranium was too fragmentary to determine its brain size. Forty of the 302 pieces are parts of the right arm and both legs, including the 3 major arm bones (humerus, radius, ulna), the left femur, and the right tibia. The size and character of these limb bones most closely resemble the skeleton of Lucy (Afar specimen AL 288-1), which was found in 1974 in Ethiopia by Johanson and White. Although Lucy is about 3 Myr old and assigned to *Australopithecus afarensis*, and OH 62 is 1.8 Myr old and belongs to *H. habilis*, the two fossils are about the same height (about 3–3.5 feet), and the humerus is nearly as long as the femur in both of them. This proportion of the humerus as 95 percent the length of the femur is very close to the equal length of upper arm and upper leg bones in apes. It is very different from the modern human proportion of the humerus as 70 percent the length of the femur (see Fig. 11.9) These proportions clearly show that *H. habilis*, like australopithecines and apes, had relatively long arms. OH 62 and the hand and foot bones analyzed by Stern and Susman indicate that *H. habilis* was quite

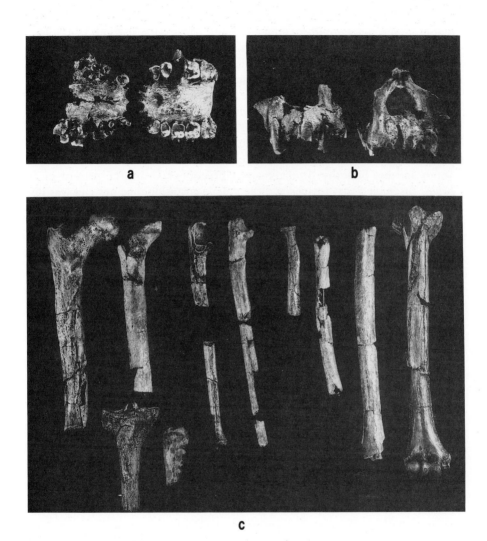

Figure 12.2. Fossil OH 62 is one of several representatives of *H. habilis* found at Olduvai in north-ern Tanzania. Comparisons of (a, b) OH 62 and Stw 53, an accepted habiline fossil from South Africa, show shared features of the palate, face, jaw, and dentition. Bar = 2 cm. (c) Comparison of OH 62 and Lucy (AL 288-1), a much older fossil assigned to *Australopithecus afarensis* and found in Ethiopia, shows strong resemblances of the long bones of the leg and arm. Bar = 4 cm. The Olduvai specimens are to the right in each comparison. (From D. C. Johanson et al., 1987, New partial skeleton of *Homo habilis* from Olduvai Gorge, Tanzania. *Nature* 327:205, Fig. 2. Copyright by Macmillan Journals Limited)

apelike in body characteristics and probably as adept at tree-climbing as are the apes, and perhaps the australopithecines, too.

The dentition of Olduvai habiline fossils was earlier inferred to be more hominine that australopithecine in character, particularly in tooth size. Johanson and White reevaluated habiline tooth size in proportion to body size in OH 62. They determined that its molars and premolars were as large as australopithecine cheek teeth and not nearly as small as hominine teeth were expected to be. Previous conclusions had been based on a presumption of much larger body size for *H. habilis* than the 3 to 3.5 feet height of OH 62. Yet another nonhominine feature of *H. habilis* has been proposed from recent studies of its pattern of dental growth and development. According to comparative studies of apes, australopithecines, and hominines reported by Holly Smith and by Timothy Bromage and Christopher Dean, the pattern in *H. habilis* resembled that in apes and the gracile australopithecines and was distinct from the pattern in *H. erectus* and *H. sapiens*.

The interest in dental growth characteristics is that the pattern reflects the relative duration of the period of maturation, and a longer time until maturation to adulthood permits extended social and intellectual development of the individual. Dentition development in apes and humans differs in the pattern of tooth eruption and in the time over which tooth eruption occurs. In apes, for example, the canine teeth erupt earlier in relation to the first molars than they do in humans, and the three molars in apes emerge at the ages of about 3, 6 or 7, and 10 or 11 years, whereas the ages in humans are 6, 12, and 18 years (Fig. 12.3). By the use of new criteria to establish an absolute time scale for hominid dental development, Smith concluded that the pattern of tooth eruption in gracile australopithecines and *H. habilis* had been distinctly apelike. On the same temporal criteria used by Smith, but by a different method, Bromage and Dean concluded that these species had possessed an apelike timetable for tooth eruption. Based on these studies, it would appear that significant prolongation of development occurred later in hominine evolution. The standards used in these studies and the interpretations based on them, however, have been disputed by Alan Mann and others. Mann considers the habiline patterns of dental growth and development to have been distinctly human, or at least neither apelike nor totally humanlike in all respects. These matters remain to be resolved by further studies.

H. habilis becomes more enigmatic as new and often conflicting information is obtained. This hominine species clearly differed from australopithecines in brain size and in having a primitive stone-tool tradition (Oldowan). In other respects, *H. habilis* retained the suite of apelike features typical of australopithecines, which more and more are being viewed as upright bipedal apes rather than full-fledged hominids. The problem is partly taxonomic, because systematists have separated apes and hominids into different families largely on the criterion of upright bipedalism. The genetic distance between apes and hominids is more like that between genera than between families (Table 12.1), since we share more than 98 percent of our genes with the African pongids. It is not genetically inconsistent to emphasize the apelike traits

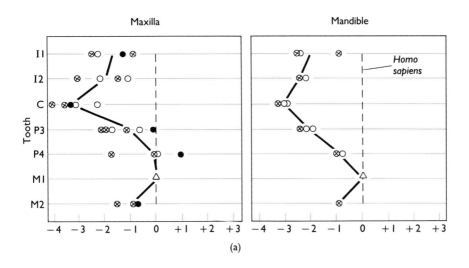

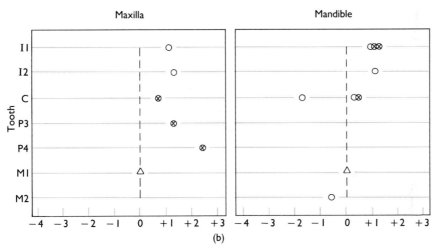

Relative time of tooth development

Figure 12.3. Pattern for 11 fossil hominids of the deviation from human timetables of tooth development in the upper (maxilla) and lower jaw (mandible). Teeth in each half-jaw are identified as the two incisors (I1, I2), canine (C), premolars (P3, P4), and first two molars (M1, M2). Fossil teeth are scored as years of advance (+) or delay (−) relative to tooth M1; dashed line at zero is the position for modern humans. (a) Three gracile species fall into a distinct group, consisting of A. afarensis (open circles), A. africanus (crossed circles), and H. habilis (solid circles). (b) A second distinct group is composed of the robust australopithecines, A. robustus (open circles) and A. boisei (crossed circles). (From B. H. Smith, 1986, Dental development in Australopithecus and early Homo. Nature 323:327, Fig. 3. Copyright by Macmillan Journals Limited)

of hominids and to review our ideas about the scope of evolutionary change during divergence of the African pongid and hominid lineages 5 to 7 Myr ago.

Whatever its taxonomic status may be within the human family, *H. habilis* clearly shared many features with the gracile australopithecines and probably was their descendant. Remains of *H. habilis* have been found in South Africa, Tanzania, Kenya, and Ethiopia, so its geographical distribution overlaps with that of both *A. afarensis,* and *A. africanus* (Fig. 12.4). Its closest relative in time, however, was *A. africanus,* which disappeared at about the time that *H. habilis* appeared, about 2 Myr BP. There is no fossil evidence for *A. afarensis* after about 3 Myr BP, which leaves a gap of about 1 Myr between its fossil record and that of *H. habilis.* On this basis, it is more likely that *A. africanus* was the ancestor of *H. habilis.* It is quite possible, however, that new fossil discoveries in the future, just as now, will cause paleoanthropologists to modify their ideas. After all, information was obtained only in 1974 for the exis-

TABLE 12.1
Estimates of Standard Genetic Distances Among Taxa of Various Ranks

Taxa	Number of Taxa	Number of Loci Analyzed	Genetic Distance (D)
Local races			
human	3	62	.011–.029
mice	4	41	.010–.024
lizards	3	23	.001–.017
fish	4	33	.000–.003
Subspecies			
mice	2	41	.194
lizards	4	23	.335–.351
plants (peppers)	4	26	.02–.07
Species			
monkeys (macaques)	6	30	.02–.10
birds	4	27	.01–.028
lizards	4	27	.12–.27
fish	6	42	.36–.52
Genera			
birds	12	26	.05–.69
fish	3	31	.47–1.3
Families			
human/chimpanzee	2	42	.62

Source: Adapted from M. Nei, 1987. *Molecular Evolutionary Genetics.* New York: Columbia University Press. Table 9.3. With permission from Columbia University Press, New York.

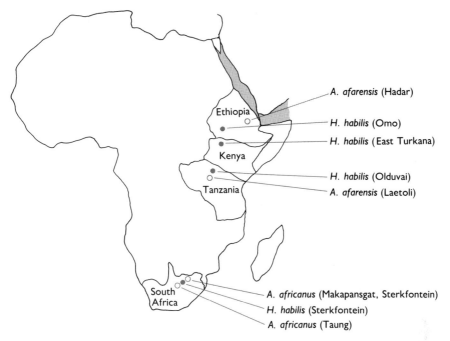

Figure 12.4. The geographic distribution of *H. habilis* fossils overlaps with locations of fossil remains of the gracile australopithecines *A. afarensis* in East Africa and *A. africanus* in South Africa.

tence of *A. afarensis*, and in the 1980s for the considerably apelike features of *H. habilis*.

Homo erectus: An Unambiguous Hominine

The australopithecines and *Homo habilis* originated and evolved in South and East Africa during the Pliocene Epoch (5–1.8 Myr BP). Fossil evidence collected in the past few decades reveals the emergence of *Homo erectus* about 1.6 to 1.7 Myr ago as a Pleistocene evolutionary replacement of the habiline species in Africa. Before the discovery of these ancient African fossils, the oldest known specimens of *H. erectus* were from East Asia and were not more than 700,000 years of age. The East Asian fossils included the so-called Java man *(Pithecanthropus erectus)*, found in Indonesia in 1891, and Peking man *(Sinanthropus pekinensis)*, found in a town south of Peking (now Beijing), China in the early 1920s (Fig. 12.5). In the apparent absence of earlier *H. erectus* specimens, it was widely assumed that the species had evolved in East Asia, which was believed to be the cradle of human evolution. In earlier decades, australopithecines were not accepted as hominids, and *H. habilis* was

Figure 12.5. Reconstruction of the skull of *Homo erectus* from China ("Peking man"). (Photograph courtesy of the American Museum of Natural History, New York)

unknown. The current African fossil record has greatly extended the time frame for *H. erectus*, established its time and place of origin, and indicated that the species became cosmopolitan later in its history.

The Olduvai Hominid skull OH 9, which was found in 1960, was identified by Louis Leakey as a specimen of *H. erectus* over 1 Myr in age. This identification was not generally accepted because there appeared to be no curvature of the cranium base. Whether this was due to distortion during fossilization was widely debated; but in any case, the controversial status of OH 9 diverted interest away from any consideration of an ancient existence for *H. erectus* in Africa. In addition, the wealth of Asian fossils discovered over many decades commanded greater attention than a lone, distorted African skull.

All this changed in the 1970s and 1980s with discoveries in East Africa and elsewhere on that continent of unambiguous specimens of *H. erectus*, which were much older than the known Asian fossils. One of the most remarkable finds was an almost complete skeleton of a 12-year-old boy (KNM-WT 15000), reported by Richard Leakey's group in 1985 from Kenya (Fig. 12.6). This fossil has been reliably dated as about 1.6 Myr old. His height of 5 feet as an adolescent was well above the heights of earlier hominids, and had he lived into his teens, this boy might well have grown taller. From the variety of total *H. erectus* fossil materials, it is clear that *H. erectus* had hands and feet essentially like ours, not the curved digits of earlier hominid species. Its postcranial features, as well as cranium, jaws, and dentition, were more closely allied to human traits, but sufficiently different in relative size, shape, and robustness to distinguish *H. erectus* from modern *H. sapiens*.

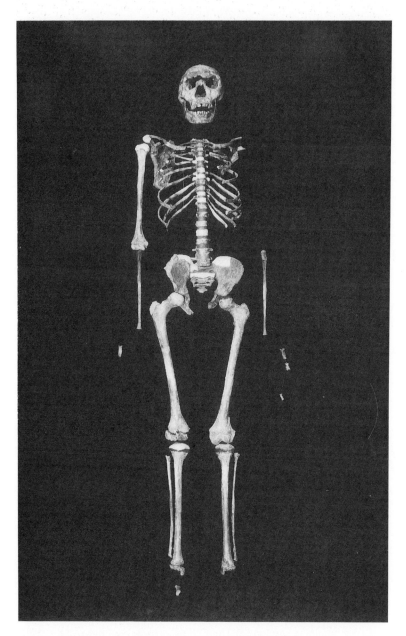

Figure 12.6. Assembled remains of KNM-WT 15000 represent a 12-year-old boy assigned to *H. erectus*. This fossil specimen from northwestern Kenya has been dated as 1.6 Myr of age. (Photograph courtesy of the National Museums of Kenya, Nairobi)

Figure 12.7. Stone tools of two hominine species. (a) The Oldowan tools credited to H. *habilis,* such as these choppers, are considered as cruder than (b) Acheulean tools made by H. *erectus,* exemplified by a hand ax (left) and scrapers (right). (Photographs courtesy of the American Museum of Natural History, New York)

A significant and distinguishing feature of all *H. erectus* populations is a braincase ranging from about 775 to 1225 cc, which is intermediate between the values of nearly 700 cc for *H. habilis* and an average of 1300 cc for modern humans. Behaviorally, *H. erectus* was well advanced beyond the habilines. The *H. erectus* stone-tool tradition (Acheulean) was more varied and sophisticated in its manufacture than the pebblelike Oldowan tools made by *H. habilis* (Fig. 12.7). *H. erectus* used fire by at least 1.4 Myr BP, whereas no evidence exists for the use of fire by habilines. In addition, from the nature of the tools and the condition of large amounts of animal bones associated with later *H. erectus* fossils, it is clear that this species hunted as well as foraged for food. This is the first hominid species known to have hunted and to have regularly included meat in its diet.

H. erectus can be dated reliably to 1.6 to 1.7 Myr BP, but paleoanthropologists are uncertain about the time that founding individuals first emigrated to Asia. It probably was earlier than 700,000 to 800,000 years ago, because the earliest known Asian fossil record for *H. erectus* goes back at least that far. Changes in climate about 1 Myr ago during the Pleistocene ice ages, with the consequent lowering of sea level and emergence of land connections, would have allowed *H. erectus* individuals to migrate to adjoining landmasses. The species inhabited a broad geographical range throughout East Asia, from far northern China to southernmost parts of Indonesia, for 500,000 years or more. Typical *H. erectus* disappeared from East Asia about 200,000 years ago, according to interpretations of the fossil record.

The presence of typical *H. erectus* in East Asia and in Africa is well established, but it is not at all clear regarding Europe, or western Asia for that matter. A diverse and confusing group of fossils between 500,000 and about 200,000 years of age are known from a broad band of localities across Europe and western Asia, but their identification and affinities are highly controversial. These fossils have a larger braincase than typical *H. erectus*, smaller teeth, and some other advanced features, but they retain many primitive traits that distinguish them from modern *H. sapiens*. Similar forms have been found in many parts of Africa and have been dated as middle-Pleistocene fossils (730,000–128,000 years BP). These forms are important to an understanding of human evolution, particularly the time frame and tempo of change leading to modern *H. sapiens*.

The Origin of Modern Humans

All modern humans have a distinctive skull anatomy that includes a vertical forehead without prominent brow ridges, a high rounded cranium of thin bone, a protruding chin, and relatively small teeth set in jaws of reduced size and robustness (Fig. 12.8). From the fossil record, which consists mainly of skulls and teeth, different evolutionary scenarios have been proposed for the origin of modern humans and for the time and place of their appearance.

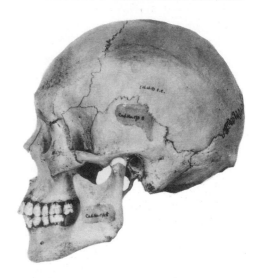

Figure 12.8. Skull of anatomically modern *Homo sapiens*. Note the vertical forehead, absence of prominent brow ridges, high rounded cranium, protruding chin, and relatively small teeth set in jaws of reduced size and robustness compared with those of *H. erectus* (see Fig. 12.5). (Photograph from the American Museum of Natural History, New York)

Recent molecular studies emphasize an African origin for modern *Homo sapiens*, but these studies have been disputed by some authorities. This matter has not yet been resolved.

A Diversity of Pleistocene Fossils

The many fossil specimens of middle Pleistocene age (730,000–128,000 years BP) from Africa, Asia, and Europe present a somewhat bewildering variety that has defied a consensus on taxonomic identification. On the basis of typically fragmentary cranial and dental remains, paleoanthropologists have suggested a series of modifications from a typical *Homo erectus* to typical modern *H. sapiens* during this Pleistocene interval (Fig. 12.9) The major modification is increased size of the brain, but reduced robustness and character of bones are also evident and described in great detail in the literature. Interpretations of this temporal sequence of fossil material range from the identification of all *H. erectus*, *H. sapiens*, and intermediate forms as members of one gradually evolving species *(H. sapiens)*, to separate species identifications for two or more of these types.

The prevailing trend in the recent literature has been to identify three groups of hominines among the fossil collections. One group consists of typical smaller brained (775–1225 cc) *H. erectus*; the second group includes unambiguous larger brained and less robust (anatomically modern) *H. sapiens*; the third group is composed of a variety of less *erectus*-like and more *sapiens*-

TABLE 12.2
Range of Cranial Volumes for Three Groups of Hominines

Hominine Group	Localities	Cranial Volumes (cc)
Homo erectus	Asia, Africa	775–1225
Archaic *H. sapiens*		
non-Neanderthals	Asia, Africa, Europe	1100–1430
Neanderthals	Europe, Middle East	1245–1740
Modern *H. sapiens*	Worldwide	1000–2000

like forms that are referred to as "archaic" *H. sapiens.* Although the bone structure is quite robust, the brain size of archaic *H. sapiens* is greater than in typical *H. erectus* and within the range of modern humans (Table 12.2).

As discussed in the preceding section, *H. erectus* appeared in Africa about 1.6 to 1.7 Myr ago, and had spread to East Asia by at least 700,000 to 800,000 years ago. No unequivocal *H. erectus* fossils have been found in Europe or western Asia. Although typical *H. erectus* appears to have inhabited East Asia until about 200,000 years ago, all the middle Pleistocene fossils from Africa, Europe, and western Asia make up a graded series of archaic *H. sapiens* forms, from older more *erectus*-like to younger more *sapiens*-like specimens (see Fig. 12.9). Dating is extremely difficult for most of these archaic *H. sapiens,* which compounds the problem, but there is a reasonable correlation in many cases

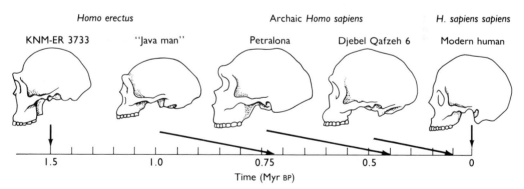

Figure 12.9. Representative hominine fossils of the past 1.5 Myr, tracing an apparently gradual transition in skull features from *H. erectus* through archaic *H. sapiens* to modern *H. sapiens sapiens.* KNM-ER 3733, from Kenya (1.5 Myr), and "Java man," from Indonesia (0.7 Myr), are assigned to *H. erectus.* The Petralona skull, from Greece (about 0.4 Myr), and Djebel Qafzeh skull, from Israel (about 0.1 Myr), are examples of archaic *H. sapiens.* The Djebel Qafzeh fossil is a Neanderthal type of archaic *H. sapiens.*

between more recent dates and the more *sapiens*-like features of the fossil skulls.

One of the best known of the archaic *H. sapiens* populations is the **Neanderthal,** who inhabited Europe and Asia as far east as Afghanistan and Uzbekistan in the Central Asian region of what is now the Soviet Union. Neanderthals were the first fossil hominines to be identified with human evolution, although their human affinities were not understood in 1856 when remains were discovered in a cave in the Neander Valley (*tal* is the German word for "valley") near Düsseldorf in western Germany. The familiar "classic" Neanderthals were a more robust physical type of the late Pleistocene and early Recent epochs, between about 90,000 and 30,000 to 35,000 years ago (Fig. 12.10). Earlier populations of a less robust physical form have been placed as far back as 150,000 to 125,000 years ago, the end of the middle Pleistocene, in localities across Europe. A pre-Neanderthal suite of characters has been recognized by some authorities in fossils such as those from Petralona (Greece) and Arago (France), but the dates are very uncertain—perhaps 200,000 to 700,000 years ago—as are the affinities of these specimens.

Neanderthals have long been treated as a special group of archaic *H. sapiens*, partly because paleoanthropologists know a great deal about them and partly because they have been viewed as most closely resembling modern *H. sapiens*. Classic Neanderthals established a relatively sophisticated stone-tool tradition (Mousterian), which included spear points and hand axes that were used in hunting game animals, and a variety of scrapers and cutting tools for removing

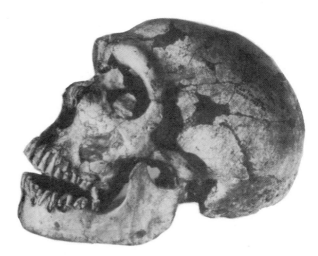

Figure 12.10. Classic Neanderthal skull of an adult male, catalogued as La Ferrassie I, and dated about 40,000 to 50,000 years of age. This fossil was found in a rock shelter in the Dordogne region of south-central France. Numerous Neanderthal remains have been discovered in this part of France, as have assemblies of younger Cro-Magnon fossils, which are assigned to anatomically modern *H. sapiens sapiens.*

meat from bones, preparing hides for clothing, and other uses. They constructed open-air encampments as well as lived in caves, cared for their injured, and buried their dead in graves in which they also placed tools, food, and medicinal plants. Average brain size was slightly larger in Neanderthals (1400 cc) than in modern humans (1300 cc) which may be a reflection of their more robust physical nature and proportionate cranial size. According to some studies of endocasts and other means for analyzing brain organization, Broca's area and Wernicke's area, two speech centers in the human brain, were present in Neanderthals. Neanderthals had a protruding bun at the rear of the braincase, but the arched contour of the cranium base, allowing for a lowered larynx, is another indication of the possibility that they could speak.

Some disagreement exists about the taxonomic status of Neanderthals. The earlier view, still held by some paleoanthropologists, was of a separate species of *Homo neanderthalensis*. The current, wider opinion is that Neanderthals are a subspecies of our own species; they are *H. sapiens neanderthalensis*, while we are *H. sapiens sapiens*. This identification rests in part on the perceived close affinities between them and us, but more on the possibility that certain fossils found in Israel may represent hybrids between Neanderthal and anatomically modern humans who coexisted in that region. There is no agreement at present on the presumed hybrid features of these fossil skulls.

Neanderthals and modern human populations inhabited separate geographical regions until about 40,000 years ago, when Neanderthals began to be supplanted by modern human populations in a westward direction. The first replacements took place in the Near East (about 40,000 years BP), the next in eastern and central Europe (about 35,000 years BP), and last in western Europe (30,000–33,000 years BP). The reasons for the disappearance of Neanderthals remain unknown, but various suggestions have been made, of course. They may have evolved into modern *H. sapiens sapiens*, perhaps in western Asia, and gradually been superseded by their more advanced descendants. Such a scheme implies that modern Europeans, at least, have a separate evolutionary history from the human populations in Africa and the Far East. This idea was popular when little or nothing was known or acknowledged about non-European human populations earlier than about 40,000 years ago. Such a viewpoint has been eroded considerably by new discoveries and reevaluations of African and East Asian human populations that predate the European groups.

Other suggestions concerning the demise of the Neanderthals are premised on the absence of an ancestor–descendant relationship between them and modern humans. It may be that the westward-advancing modern humans wiped out the Neanderthals, who are therefore viewed as an extinct group without living descendants. Some paleoanthropologists consider it more likely, though, that the resident Neanderthal and migrant modern human populations interbred, as subspecies often do, producing a modern human type with some retained Neanderthal alleles. In this case, unlike the previous suggestion, Caucasians are the descendants of populations of *H. sapiens sapiens* who were modified to some undetermined degree by gene flow from Neanderthals. Neither of these suggestions has been ruled out, nor has either been reinforced by clear-cut evidence.

Setting aside the "Neanderthal problem" for the moment, what about the other archaic *H. sapiens* populations of a non-Neanderthal or non-pre-Neanderthal type, and the origin of modern *H. sapiens* in Africa and East Asia? More is known today than in the past few decades about middle and late Pleistocene *H. sapiens* in Africa. Dating has been extremely difficult in many localities, particularly in North Africa and in parts of South Africa, but some estimates can be made on the basis of faunal associations or geologic information, with occasional radiometric dates as well. The distinction between *H. erectus* and archaic *H. sapiens*, grading from *erectus*-like to *sapien*-like, is a bit ambiguous in Africa as elsewhere. However, throughout Africa there appears to have been a collection of archaic *H. sapiens* instead of *H. erectus* between 500,000 to 700,000 years BP and 100,000 to 200,000 years BP. Fossils of modern *H. sapiens* also are distributed throughout Africa in deposits between 50,000 and 100,000 years of age, and perhaps as old as 125,000 years in at least one case.

The oldest known fossils of modern humans have come from two sites in South Africa (Fig. 12.11). The Border Cave remains, which have been dated between 90,000 and 110,000 years BP, include parts of an adult cranium and face that are unmistakably modern in all respects. In addition to these fossils from northeastern South Africa, near Mozambique, several specimens of modern humans have been found in the Klasies River Mouth region, at the southern end of South Africa. Most of the Klasies fossils are at least 70,000 years old, but may be older, and one specimen in a bed beneath all the others may be 100,000 to 125,000 years old.

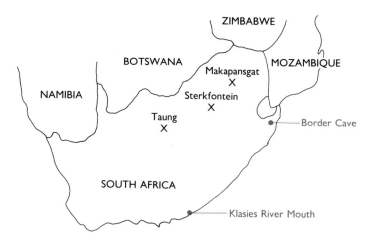

Figure 12.11. The oldest known fossils of anatomically modern humans have been found at Border Cave and at the Klasies River Mouth region in South Africa. Some of these remains may be up to 125,000 years of age.

There now exists a fairly well-documented sequence of *H. erectus,* a gradation of more *erectus*-like to more *sapiens*-like archaic *H. sapiens,* and, since about 100,000 years BP, modern *H. sapiens.* This fossil record has contributed to the increasingly popular viewpoint that all of hominid speciation took place in Africa. Molecular analyses of modern human populations have also been interpreted as showing that Africans are the most ancient group and that all modern human populations trace their ancestry back to an original *H. sapiens* stock from Africa.

The Origin and Divergence of Modern *Homo sapiens*

As just stated, the prevailing opinion is that each hominid species, including our own, originated in Africa. From the fossil evidence alone, the oldest known fossils and artifacts attributed to modern *Homo sapiens* have come from South Africa. Although dating has been uncertain for East Asian human fossils, they appear to be somewhat younger than their African relatives. The fossil record clearly shows, however, that modern humans did not exist in Europe and western Asia until about 40,000 years ago, long after such populations were established in Africa and East Asia.

The fossil evidence from these temporal and geographical parameters can be interpreted in a straightforward sequence. Some members of the ancestral African group remained in Africa, evolving there to the variety of sub-Saharan indigenous populations. Other members of the African stock migrated to East Asia, earlier than 50,000 years ago. This East Asian group evolved in isolation from the ancestral Africans, and fanned out in all directions over the next few tens of thousands of years (Fig. 12.12). Australia and the Papua-New Guinea region received its first migrants about 40,000 years ago. Migrations proceeded westward into central Asia and Europe between 40,000 and 30,000 years ago. The first inhabitants of the Americas were present by 12,000 years BP or earlier, subsequent to migration from Asia eastward into North America. These last Asian migrants were the ancestors of the present-day Aleuts, Inuits (Eskimos), and Native Americans of North, Central, and South America. Within 1000 years after the first known appearance of Asian migrants in North America, some of their members had progressed all the way down to the southernmost tip of South America.

This admittedly simple scenario omits references to repeated colonizations of some areas in ancient times and to gene flow or cultural interactions between populations. Some information concerning cultural interactions can be gained from the fossil record of artifacts, but any data on gene flow or genetic lineages are difficult or impossible to obtain from fossil morphology. Useful information on genetic relationships can be obtained from molecular analysis, however, and such information has been forthcoming since the 1970s. Even more detailed data have come from studies reported in the 1980s. These studies, together with independent evidence from a growing body of fossil materials, have strengthened the case for an African origin of modern

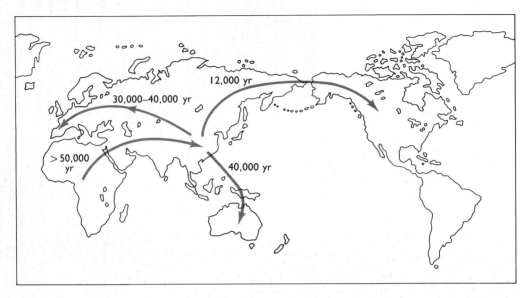

Figure 12.12. Simplified pattern of inferred migrations of modern humans from Africa to East Asia earlier than 50,000 years ago, and from East Asia to other parts of the world at various times between 40,000 and 12,000 years ago. This scheme omits references to reentries into and recolonizations of original sites, and emphasizes only migrations that culminated in the spread and establishment of human populations outside their place of origin.

humans and for a subsequent divergence to produce an ancient Eurasian population.

Studies of differences in gene frequencies, individual gene sequences, and clusters of tightly linked genes reflect the accumulation of mutations over time and the relatedness of nuclear DNAs in samples drawn from different human populations. In one such study, reported in 1986, James Wainscoat and his colleagues analyzed restriction cut-site patterns, referred to as **haplotypes,** in the β-globin gene cluster of individuals from eight populations (Fig. 12.13). The 5 restriction enzymes could produce 32 (2^5) different haplotypes, but only 14 were found; of these 14, only 4 were predominant (Table 12.3). One haplotype ($- - - - +$) was the most common in Africans and absent from almost all the non-African populations, in which three other haplotypes predominated ($+ - - - -,\ - + - + +,\ - + + - +$) but were a minor component in the African populations. Genetic distance analysis of the haplotype data revealed a very large genetic distance between the African population and all the non-Africans (Table 12.4). These data were interpreted as showing a major division of human populations into an African and a Eurasian group. Although the time of divergence and identification of the ancestral group could not be determined from these data, Wainscoat indicated that they at least were consistent with the prevailing idea that a founder population migrated from Africa and subsequently gave rise to all non-African populations.

Variations in nuclear DNA arise by means of recombination as well as mutation, and different patterns may be influenced by selection, random genetic drift, and gene flow, which may obscure relationships and patterns of ancestry and descent. These problems are minimized significantly in studies using mitochondrial DNA, particularly in mammalian species in which little or no recombination is known to occur. The mitochondrial genome is trans-

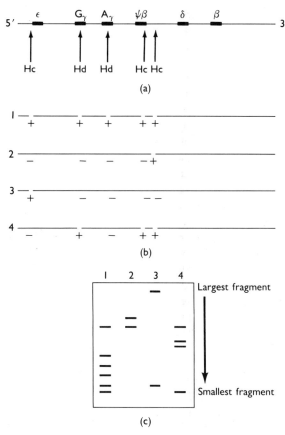

Figure 12.13. Restriction analysis of the β-globin gene cluster in human nuclear DNA. (a) The 50-kbp-long segment includes sequences for functional ε, ^{G}γ, ^{A}γ, δ, and β globins, and the nonfunctional pseudogene ψβ (pseudobeta). The segment can be cut at three sites by restriction enzyme HincII (Hc), and at two sites by HindIII (Hd), as illustrated. (b) A total of 32 (2^5) haplotypes are possible, of which 4 are shown. (c) Each of these four haplotypes (as well as any of the others) can be recognized by the electrophoretic pattern produced as DNA fragments migrate through the gel under the influence of an electrical field. (Based on J. S. Wainscoat et al., 1986, Evolutionary relationships of human populations from an analysis of nuclear DNA polymorphisms. *Nature* 319:491)

TABLE 12.3
Numbers of β-Globin Haplotypes Determined for 601 Chromosomes Analyzed in Eight Human Populations from Different Geographical Regions

Haplotype*	British	Cypriot	Italian	Asian Indian	Thai	Melanesian	Polynesian	African
+−−−−	16	59	32	58	29	117	43	3
−+−++	15	14	15	28	0	29	6	6
−++−+	5	8	2	15	2	3	0	1
+−−++	0	0	0	0	0	11	0	0
−++−−	1	1	1	3	0	0	0	0
−++++	0	0	0	0	0	5	1	0
++−++	0	0	0	3	0	0	0	0
−−−−−	0	0	0	1	0	2	1	0
+++−+	0	0	0	1	0	0	0	0
−−−++	0	0	0	1	1	0	0	0
++−−−	0	0	0	1	0	0	0	0
−+−−−	0	0	0	0	0	0	0	2
−+−−+	0	0	0	0	0	4	4	12
−−−−+	0	0	0	0	0	2	0	37
Totals	37	82	50	111	32	173	55	61

*A polymorphic restriction site is denoted as being present in a particular haplotype by a + symbol and absent by a − symbol (see Fig. 12.12).
Source: Reprinted with permission from J. S. Wainscoat et al., 1986. Evolutionary relationships of human populations from an analysis of nuclear DNA polymorphisms. *Nature* 319:491, Table 1. Copyright 1986 by Macmillan Journals Limited.

mitted from one generation to the next exclusively through the female, in the egg, and can be considered to represent a set of genes modified predominantly or exclusively by mutation. In addition to sequence divergence that is barely influenced, if at all, by processes other than mutation, the rate of mutation in primate mitochondrial DNA is from 2 to 10 times higher than in nuclear DNA. These and other features of mitochondrial DNA have made it an experimental material of choice for many investigators of human evolution. One such study, reported in 1987 by Rebecca Cann and her colleagues, has provided evidence for an African origin of all modern human populations.

Cann and her colleagues analyzed mitochondrial DNA haplotypes produced by 12 restriction enzymes in 147 individuals from 5 geographically distinct human populations (Table 12.5). The average of 370 restriction sites per individual covered about 9 percent of the 16,569 base pairs in the human mitochondrial DNA genome. The 147 individual mitochondrial DNAs were divisible into 134 distinct types, which were arranged in a genealogical tree constructed by the parsimony method (Fig. 12.14). This minimal-length evolutionary tree, which posits the fewest mutations and intercontinental migrations, shows two major branches. One primary branch consists entirely of Africans, and the other includes all five populations. From the genealogical

TABLE 12.4
Genetic Distances Calculated from the β-Globin Haplotype Data for the Eight Human Populations Shown in Table 12.3

	British	Cypriot	Italian	Thai	Indian	Melanesian	Polynesian	African
British	—							
Cypriot	0.6908	—						
Italian	0.5363	0.4012	—					
Thai	1.3064	0.7366	0.9695	—				
Indian	0.6472	0.6533	0.6869	1.0734	—			
Melanesian	1.0239	0.6843	0.6977	0.9533	0.9872	—		
Polynesian	1.1368	0.6650	0.7323	0.8305	1.0360	0.5975	—	
African	2.8658	2.8399	2.8300	2.9619	2.8719	2.7742	2.7288	—

Source: Reprinted with permission from J. S. Wainscoat et al., 1986. Evolutionary relationships of human populations from an analysis of nuclear DNA polymorphisms. *Nature* 319:491, Table 2. Copyright 1986 by Macmillan Journals Limited.

tree, hypothetical African ancestor *a*, whose DNA type characterizes individuals 1 to 7, was interpreted to be the last common ancestor of all the 134 DNA types. The second primary branch is the divergence from hypothetical African ancestor *b* to produce the other African and all the non-African DNA types, characterized by individuals 8 to 134. According to this interpretation, therefore, the African ancestral population is the progenitor of all modern human populations.

Cann and her co-workers determined from sequence variations that Afri-

TABLE 12.5
Sequence Divergence of Mitochondrial DNAs Within and Between Five Human Populations from Different Geographical Regions

Sample Size	Population	Percent Sequence Divergence*				
		African	Asian	Australian	Caucasian	New Guinean
20	African	**0.47**	0.04	0.04	0.05	0.06
34	Asian	0.45	**0.35**	0.01	0.02	0.04
46	Australian	0.40	0.31	**0.25**	0.03	0.04
21	Caucasian	0.40	0.31	0.27	**0.23**	0.05
26	New Guinean	0.42	0.34	0.29	0.29	**0.25**

*Measures of sequence divergence are related to one another by the equation $\delta = \delta_{XY} - 0.5(\delta_X + \delta_Y)$. Values on the diagonal (boldface) represent the mean pairwise divergence between individuals in populations (δ_X). Values below the diagonal are the mean pairwise divergences between individuals from two different populations (δ_{XY}). Values above the diagonal (δ) are divergences between populations.
Source: Modified from R. L. Cann, M. Stoneking, and A. C. Wilson, 1987. Mitochondrial DNA and human evolution. *Nature* 325:31, Table 1. Reprinted with permission; copyright 1987 by Macmillan Journals Limited.

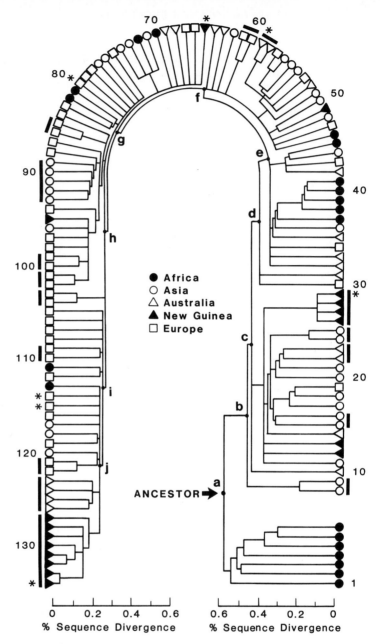

Figure 12.14. The most parsimonious genealogical tree for 5 human populations, showing the 134 types (numbered) of human mitochondrial DNA produced from restriction digests involving 12 restriction enzymes. One primary branch includes mitochondrial DNA types that originated from presumed African ancestor *a,* and is now present in seven African individuals (1–7). The second primary branch represents the other African and all non-African types produced after the initial divergence of presumed ancestor *b.* The lineages in the second primary branch are assigned to subsequent presumed ancestors *c* to *j.* Clusters of mitochondrial DNA types specific to a given geographical region (black bars), and types found in more than one individual (asterisks), are also shown. (From R. L. Cann et al., 1987, Mitochondrial DNA and human evolution. *Nature* 325:31, Fig. 3. Copyright by Macmillan Journals Limited)

cans are the most diverse of all the populations, which is expected for the most ancient population because it has accumulated more mutations in the longest period of time. On the basis of an estimated mutation rate of 2 to 4 percent per million years, and of mutations accumulating at a constant rate, the last common ancestor of all surviving mitochondrial DNA types existed 143,000 to 285,000 years ago, and all other ancestral types (*b* to *j* in Fig. 12.14) may have existed 62,000 to 225,000 years ago (Table 12.6). The mitochondrial DNA results provided no clear information on times of migration of presumptive ancestral African founders. It could be concluded, however, that the earliest migration from Africa was to Asia, on the basis of Asian DNA types being second in diversity only to African and of Asian sequences being the closest to the African ancestral sequence represented by *a* in the genealogical tree.

These results on mitochondrial DNA diversity and percentage divergence are consistent with the scheme of an African ancestry for all human populations. The data were further interpreted as showing that Asia was a less likely site of origin for modern *H. sapiens* and that little or no gene flow took place between migrant Africans and resident Asians in ancient times. The results and interpretations of these studies have been accepted by some but challenged by others, at least on some points. Some investigators remain uncon-

TABLE 12.6
Ancestors, Lineages, and Extents of Sequence Divergence in the Genealogical Tree for 134 Types of Human Mitochondrial DNA*

Ancestor	Total	Number of Descendant Lineages or Clusters Specific to a Region					Percent Divergence	Age (years)†
		Africa	Asia	Australia	Europe	New Guinea		
a	7	1	0	0	0	0	0.57	143,000–285,000
b	2	0	1	0	0	0	0.45	112,000–225,000
c	20	0	7	3	1	3	0.43	108,000–215,000
d	2	0	0	1	1	0	0.39	98,000–195,000
e	14	2	2	4	2	0	0.34	85,000–170,000
f	19	1	7	4	4	1	0.30	75,000–150,000
g	10	2	3	2	2	1	0.28	70,000–140,000
h	30	2	4	0	15	1	0.27	68,000–135,000
i	8	1	0	0	6	0	0.26	65,000–130,000
j	22	1	3	1	5	1	0.25	62,000–125,000
All	134	10	27	15	36	7	—	—

*See Fig. 12.13.
†Assuming that the mitochondrial DNA divergence rate is 2–4% per million years.
Source: Reprinted with permission from R. L. Cann, M. Stoneking, and A. C. Wilson, 1987. Mitochondrial DNA and human evolution. *Nature* 325:31, Table 3. Copyright 1987 by Macmillan Journals Limited.

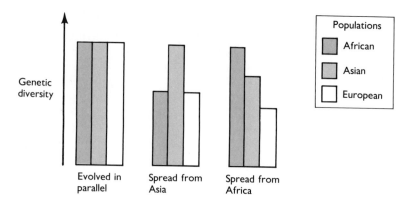

Figure 12.15. Relative levels of genetic diversity predicted for three hypothetical alternatives for the origin and subsequent spread of human populations. If all populations evolved in parallel, they should be equally and substantially variable. If humans originated in Asia and spread to other regions, the Asian population should have accumulated the greatest diversity over its longest time of existence. If Africa was the site of human origin, African populations should have accumulated the greatest diversity. The third alternative is best supported by available evidence.

vinced about the site of origin as Africa rather than Asia. Others dispute the timetable of events inferred from the data by Cann and her colleagues. The timetable depends on an estimate of 2 to 4 percent per million years for mitochondrial DNA modification, a figure that is not accepted by all molecular evolutionists. Other disputes include the conclusion by Cann that all existing human mitochondrial DNAs can be traced back to a single female (Mother Eve), whom we all share as an ancestor. Whatever problems have been raised, however, can be analyzed by additional studies, which are eagerly awaited.

What can we say about human ancestry from the combination of fossil and molecular data thus far available? Even if the common lineage of all mitochondrial DNA diversity in present-day humans existed in Africa 143,000 to 285,000 years ago, it does necessarily imply that the transformation of archaic to modern *H. sapiens* took place during this time, as Cann and others have clearly stated. Mitochondrial DNA does not determine morphological and cultural traits, and there is a high probability that mitochondrial haplotypes from one or more evolutionarily crucial populations are now extinct. The molecular data can help, however, to resolve the alternative viewpoints that the ancestors of modern humans evolved (1) in parallel in different parts of the Old World, (2) in Asia and subsequently spread to Africa and Europe, or (3) in Africa and subsequently spread to Asia and from there to Europe.

If modern humans evolved in parallel in Africa, Asia, and Europe, mitochondrial DNAs of great antiquity and diversity would be expected in all these populations; such results were not obtained by Cann and her co-workers. If Asia was the original site of modern human origin, the greatest antiquity and diversity would be expected to occur in Asian populations. The greatest non-African divergences correspond to times of only 90,000 to 180,000 years, whereas the African mitochondrial DNAs correspond to 143,000 to 285,000 years. In addition, Asian DNA diversity is not as great as African variability. Taken all together, the molecular data are consistent with an African origin of human populations and a later appearance of their descendants in Asia and Europe (Fig. 12.15).

On this basis of an apparently unitary origin of present-day human populations, we must also conclude that most of the different populations of *H. erectus* and of archaic *H. sapiens* in Africa, Asia, and Europe left no descendants. Some *H. erectus* population somewhere, and some archaic *H. sapiens* population somewhere, must have had descendants who emerged as modern *H. sapiens* in Africa. When and where the transformation took place from *H. erectus* to archaic *H. sapiens* remain uncertain on the basis of the available fossil record. When and where archaic *H. sapiens* was transformed to modern *H. sapiens* has not been determined to everyone's satisfaction, but there is now a growing consensus that this event took place in Africa, sometime before 200,000 years ago but no later than 100,000 years ago.

Cultural Components in Human Evolution

Human beings dominate the planet today, largely because of technological innovations produced quite recently in human cultural evolutionary history. All of cultural evolution rests on biological foundations established long before the appearance of modern *H. sapiens* about 100,000 to 200,000 years ago. By far, the most important feature of our biological heritage is a large and complex brain, which gives us the intelligence and foresight to manage our affairs and to seek improvements in the quality of life. In addition, through reorganized and modified vocal apparatus and neocortical speech centers, we can communicate ideas and thoughts to plan for the future, taking into account memories, judgments, and insights gained from past and present experiences.

The Innovation of Language

Every human being is born with the capacity to learn to speak a language, unless there is some anatomical or mental deficiency that precludes speech. A major human anatomical feature that permits the vocalization of articulate and complex sounds is the large size of the **pharynx,** the space between the back of the nasal cavity and the **larynx,** or voice box, in the throat (Fig. 12.16). The pharynx is part of the passageway for foods from the mouth to the esophagus leading to the stomach, but it also is a sound chamber in which fundamental sounds from the vocal cords can be modified. The larger pharynx allows a much greater modification of the sounds emitted from the larynx and vocal cords in humans than in other mammals.

N = Nasal cavity, S = Soft palate,
T = Tongue, P = Pharynx, L = Larynx,
E = Epiglottis, V = Vocal fold

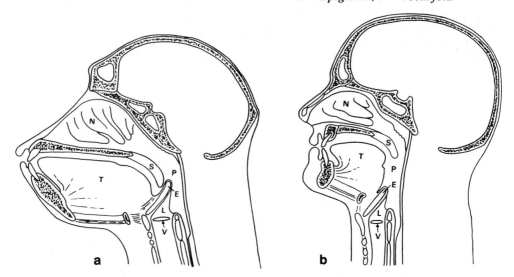

Figure 12.16. Simplified diagram of structures associated with vocalization in (a) apes and (b) humans. The smaller pharynx in apes provides an inadequate sound chamber for articulated language, whereas the much larger human pharynx allows considerable modification of sounds from the larynx and vocal cords. (Adapted from J. T. Laitman, 1984, The anatomy of human speech. *Nat. Hist.* 93(8):20)

Enlargement of the human pharynx is the consequence of the lowering of the larynx in the throat, which takes place sometime between the ages of 1.5 and 2 years. Before the child is 2, the larynx is situated high up in the neck opposite the first to third of the seven cervical vertebrae. By the second year, the larynx begins its descent and ultimately becomes positioned in a region opposite the fourth to nearly the seventh cervical vertebrae. The inarticulate sounds of the human infant are due to the relatively high position of its larynx and to the smaller pharyngeal space at the back of the throat. Articulate speech becomes anatomically possible only after the descent of the larynx. The relatively rapid acquisition of language by the growing child probably reflects a fundamental genetic wiring for rapid learning to communicate. A human child can master language and symbolism long before it has gained a significant level of motor control and coordination.

According to studies by Jeffrey Laitman and a number of his colleagues, it appears that the position of the larynx corresponds to the shape of the bottom of the skull, or **basocranium** (Fig. 12.17). The basic mammalian configuration is one in which the basocranium is fairly flat, and the larynx is positioned high up in the neck. In humans, however, the basocranium is markedly arched, or flexed, and the larynx is situated low in the neck. On the basis of these two different configurations of basocranial shape and larynx position, Laitman and his colleagues examined fossil skulls in order to reconstruct their vocal tracts, using statistical methods that had been developed in their studies of living mammals.

From museum collections of fossil hominids, Laitman and his co-workers were able to analyze a large enough sample of skulls with intact basocrania to derive some important correlations. They found that the australopithecines possessed a flat, nonflexed basocranium, essentially like the chimpanzee's. Reconstruction of their vocal tracts showed the larynx position to be high up in the throat. From these reconstructions, Laitman concluded that australopithecines probably had a very restricted vocal repertoire compared with modern humans, and it would have been impossible for them to produce some of the universal vowel sounds of human speech.

From their studies to date of fossil *Homo erectus* and *H. sapiens* skulls, Laitman and his colleagues believe that the trend toward basocranial flexion may have begun in *H. erectus* but not been completed until the appearance of *H. sapiens*, perhaps 300,000 to 400,000 years ago. These data are preliminary, and any conclusions are beset with problems stemming from the fragmentary nature of fossil materials, accurate dating, and the identification of fossil species and their range of variability. If these tentative conclusions can be substantiated, it would imply that fully articulate speech occurred in *H. sapiens*,

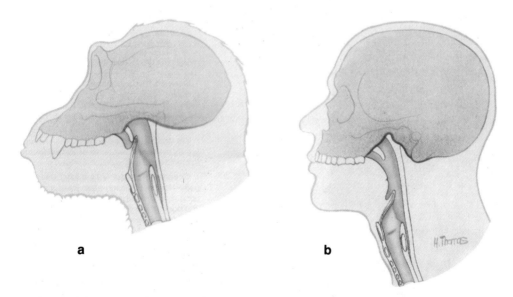

Figure 12.17. Simplified diagram showing the correspondence between (a) a relatively flat basocranium and high position of the larynx in apes, and (b) a markedly arched basocranium and low position of the larynx in humans. Such correspondences have been used to estimate vocal abilities in fossil hominids. (Adapted from J. T. Laitman, 1984, The anatomy of human speech. *Nat. Hist.* 93(8):20)

whereas a more limited degree of sound modification characterized the *H. erectus* phase of hominine evolution. It will be interesting to see what *H. habilis* basocrania are like, and whether they resembled australopithecines in this feature, as in so many other skeletal traits.

Little is known about the genetic modifications underlying descent of the larynx, or what, if any, adaptation this may have represented. Nonhuman mammals (and early hominids) can breathe at the same time they drink or swallow, because the high position of the larynx allows it to lock into the back of the nasal cavity to separate the breathing and swallowing pathways. In adult humans, however, the lowered larynx causes the respiratory and digestive tracts to cross above the larynx. We cannot take air into the lungs and food or water into the digestive tract at the same time without choking. As many of us are aware, it is not uncommon for food to lodge in the entrance to the larynx, cutting off the air supply to the lungs and causing suffocation if the block is not removed or the trachea opened surgically. Human infants do not have this problem because their larynx is high up in the neck and can lock into the back of the nasal cavity, providing an unobstructed passageway in which air can move from the nose to the lungs at the same time that food or water travels in a separate pathway from the mouth to the esophagus and stomach.

The initial adaptation may have had little or nothing to do with the position of the larynx, but may have been concerned with the shape of the basocranium, which is the roof of the upper respiratory tract and an integral component in breathing. It would be pointless to speculate without more information from developmental biology, physiology, and anatomy. Whatever the original adaptation may have been, if indeed there was an adaptive basis for the developmental change, it led to the anatomical basis for human speech.

Speech centers in the brain are crucial components in articulating and in comprehending spoken language (Fig. 12.18). Some tentative data derived from endocasts and other means for topographic studies of the brain show the possible existence of Broca's area and Wernicke's area in Neanderthals. Whether biologists can obtain meaningful information about brain organization and its relationship to articulate speech from fossils remains to be determined. But it is clear from clinical studies that the ability to speak and comprehend speech depends on speech centers and other regions of the brain as well as on a suitable vocal anatomy that includes tongue, palate, lips, larynx, vocal cords, pharynx, jaws, and lungs under the control of the motor cortex of the human brain. Most of these structures, however, are not preserved in fossils.

It is obvious that the capacity for articulate speech is highly advantageous, for speech is one means by which thoughts and ideas can be communicated among individuals. Together with greater intelligence made possible by a larger and more complex brain, foresight and planning for the future have become basic ingredients in human existence. The invention of writing provided a more effective system for communication and for the preservation of information from past experiences. The speed of communication and the

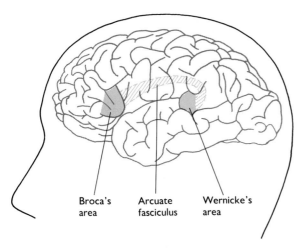

Figure 12.18. Speech centers in the human brain include Broca's area and Wernicke's area, which are connected by a bundle of nerves called the arcuate fasciculus. Broca's area is associated with motor control of the muscles of lips, jaw, tongue, soft palate, and vocal cords during speech. Wernicke's area appears to be involved with the content and comprehension of speech. Both speech centers are usually located as slightly protruding regions on the surface of the left cerebral hemisphere.

transmission of the most complex practical and theoretical information are greatly expanded today by computers and other technological systems. These aids to human intelligent activities are inventions produced by the joint abilities of the brain, dexterous and manipulative hands, and symbolic communication through language, all of which are part of our biological heritage. We change rapidly as a result of cultural evolution, which is based on learning and the transmission of learned information from one generation to the next and among individuals within each generation. None of this would be possible, however, without the appropriate genetic legacies established over millions of years of biological evolution.

Cultural Evolution

Most of the 2 Myr of hominine evolution was characterized by a relatively slow pace and quality of change in tool-making traditions, and by the prevalence of a nomadic life style based at first on scavenging and foraging and later on hunting and foraging for food. By about 30,000 years ago, Paleolithic humans used needles to sew and ornament their clothing, and they produced some of the most striking works of art in the paintings and sculptures that have been found in Europe, the Near East, and elsewhere in the Old World (Fig. 12.19).

Although many human populations continued their nomadic life style until recently, as some groups still do in all parts of the world, by about 12,000

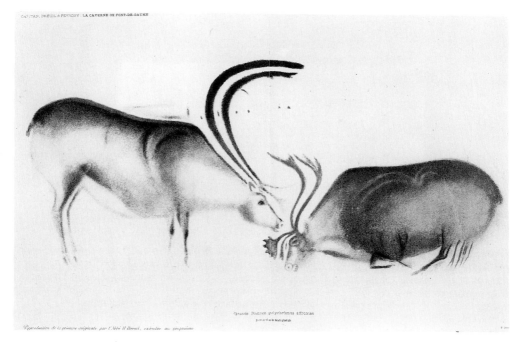

Figure 12.19. Cro-Magnon cave painting of male and female reindeer, from Fonte de Gaume, France. (Photograph courtesy of the American Museum of Natural History, New York)

years ago in the Middle East there occurred a momentous change in subsistence patterns. Permanent villages were constructed in conjunction with a mode of existence that involved gathering the seeds of various grasses, including wild wheat, barley, and millet. According to many archeologists, the first villagers were hunter-gatherers who built settlements to store their collected wild grains, grind them into flour, and prepare them for consumption. Their houses, storage pits, ovens, heavy grinders, and other nonportable investments almost certainly led these people to establish a more permanent home base instead of the temporary campsites more suitable to an unburdened nomadic existence. Many such permanent sites have been uncovered in parts of what are now Israel, Jordan, Syria, Iraq, and other countries in the Middle East.

The archeological evidence indicates that the domestication of plants and animals did not begin until about 2000 years after the first villages had been established. The earliest remains of domesticated sheep, goats, wheat, barley, and oats have been discovered in Iraq, Iran, and nearby regions, in sites that have been dated as about 10,000 years old. Within the next few thousand years, a greater variety of cereal grains was brought under cultivation in the Middle East, and pigs, cows, camels, horses, donkeys, and dogs were domesticated. The Middle East was one of three great centers for the origin of crop plants and domestic animals, and their spread to distant regions. The other major areas were located in the central New World, particularly Central America and western parts of South America, and in China (Fig. 12.20). A few important staples were developed in other parts of the world, but these regions are not considered as major agricultural centers.

New World agriculture is better known from Central America than from South America. The transition from hunting and foraging to growing food crops apparently took place at about the same time in the Americas as in the Middle East and China. In the New World, however, crop cultivation and animal domestication preceded the establishment of permanent villages. Corn and amaranth were the earliest cereal grains to be cultivated, and dogs and turkeys subsequently were domesticated. There were few suitable animals for domestication in the New World as compared with the Old World fauna of 10,000 to 12,000 years ago. By about 4500 to 5000 years ago, however, South American Indians had domesticated llamas, alpacas, and guinea pigs (none of which were indigenous to Central America) and put the potato into cultivation.

Anthropologists do not know why New World peoples postponed the establishment of village life until about 5500 years ago, when they had been growing corn at least 9000 years ago and had been collecting wild grains even earlier. Perhaps they continued a primarily nomadic or seminomadic exis-

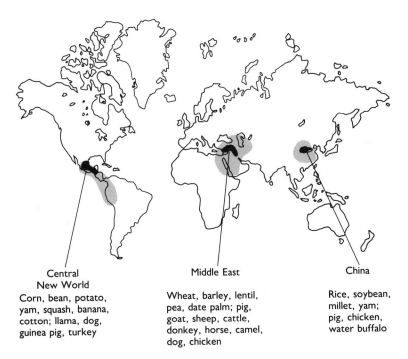

Central New World	Middle East	China
Corn, bean, potato, yam, squash, banana, cotton; llama, dog, guinea pig, turkey	Wheat, barley, lentil, pea, date palm; pig, goat, sheep, cattle, donkey, horse, camel, dog, chicken	Rice, soybean, millet, yam; pig, chicken, water buffalo

Figure 12.20. Three geographical centers are generally recognized as the major sites of the origin and spread to distant regions of domesticated plants and animals. Many of the important agricultural species indigenous to each region are listed.

tence for a longer time because they had an inadequate domesticated meat supply and relied on hunting the variety of wild game to obtain meat, supplementing this diet with collected and cultivated cereal grains. Whatever the reason or reasons may be, the very different pace of technological development in the Old World and the New World almost certainly is related to the different types of agriculture that were possible in the two regions.

Old World agriculture rapidly incorporated a large number of domesticated animals, some of which could be fed with stored grains and be used to haul heavy loads and pull plows. Land transport on sleds, rollers, and, eventually, wheels was efficient, and laid the foundation for a complex technology and economy. The wheel invented by the New World peoples, however, could not be used for transport because no suitable animals were available for hauling loads or for plowing fields. Without a technology based on the wheel, New World inhabitants could not easily develop the many processes that require pulleys, gears, and other elements of advanced technology (Fig. 12.21).

The expansion of agriculture and the establishment of village life in the Middle East generally promoted greater security and a higher standard of living, which probably led to an increase in populations and population size. Populations and their agricultural and other cultural traditions fanned out in all directions in Europe and Asia, and to a more limited degree in Africa and the New World. The earlier stages of population growth are associated principally with the development and improvement of agricultural practices. More recent population growth is based on advances in technology as well as agriculture. It was not until the mid-nineteenth century that the human population reached the 1 billion mark (Fig. 12.22). During the 150 years since then, the number of human beings has increased to over 5 billion, and it continues to grow at an alarming rate. The concern about a rapidly expanding human population is based in part on the depletion of finite world resources, and in part on the fact that billions of people live at a bare subsistence level, with little chance to improve their quality of life. Modern technology and the amenities of life that you and I take for granted are still not available in the impoverished and underdeveloped countries of the world. It seems improbable that the people in these countries can realize an improvement in the quality of life if the populations expand but their resources do not.

Although some complain that advancing technology leads to more impersonal interactions among people, we depend on technological developments for basic needs as well as conveniences. A deeper and more exact understanding of natural phenomena depends on sophisticated research methods and instruments. Advances in medicine and health care have more than doubled the human life expectancy in this century alone, and have improved the quality of life to a remarkable degree. Even if organ transplants remain restricted to a handful of patients, we take for granted the availability of eyeglasses and a host of similar technological aids. We can be immunized against a number of diseases that devastated earlier human populations. Cultural evolution allows human beings to enhance their lives in myriad ways that do

Figure 12.21. Wheels and pulleys are technological aids important to manufacturing on a large scale.

not require genetic changes in biological evolution. The speed of cultural evolutionary change, particularly in recent years, has far outpaced biological change.

Cultural evolution is an outgrowth of our intelligent actions. The ability of human beings to adapt to constant change in every aspect of life is unique among organisms, and depends on our biological heritage of a large and complex brain. Our continued success as a species depends on the adaptability made possible by intelligence, which gives us a level of control over our

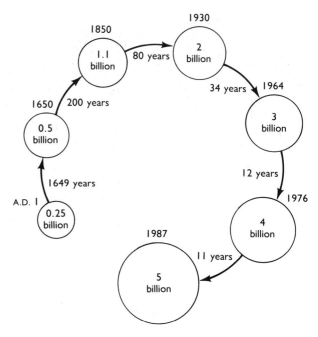

Figure 12.22. Growth of the human population in the past two millennia. The number of people is estimated to have doubled after 1649 years up to 1650, after about 200 years between 1650 and 1850, after about 80 years between 1850 and 1930, but after only 46 years from 1930 to 1976. The 5 billion mark is estimated to have been reached in July 1987, a mere 11 years for an additional 1 billion people in the world.

future. This remarkable and unparalleled feature is the unique quality of the human species among all the life forms of the past and present.

Summary

According to available evidence from the fossil record, all three recognized hominine species originated in Africa. The earliest of these species, *Homo habilis*, seems to have existed briefly (2–1.7 or 1.8 Myr BP) and only in East and South Africa. Habilines had an enlarged brain (about 700 cc), compared with that of australopithecines (400–500 cc), and produced the simple Oldowan stone-tool tradition. From a growing suite of fossil specimens, it now seems that *H. habilis* was distinctly apelike (as were the australopithecines) in its dental and body features, quite unlike the later hominine species. *H. habilis* was replaced in Africa by *H. erectus*, about 1.7 Myr BP, and both species coexisted with australopithecines until about 1 Myr ago.

Homo erectus had an enlarged braincase (775–1225 cc), perfected upright bipedalism, shortened forelimbs, larger body size, and cultural traditions that included a variety of stone tools and the use of fire by 1.4 Myr BP. Founder populations migrated from Africa into East Asia by 800,000 to 1 million years ago, but seem not to have reached western Asia or Europe at this time. By the middle Pleistocene (730,000–128,000 years ago), a diversity of forms intermediate between *H. erectus* and *H. sapiens* existed in Africa, Asia, and Europe. It is not clear whether all these forms belong to "archaic" *H. sapiens*, but such archaic types were present on all three continents by at least 500,000 years ago.

Anatomically modern humans may have originated in Africa from an archaic *Homo sapiens* stock some time between 100,000 and 200,000 years ago, and subsequently migrated to other parts of the world. According to this view, archaic *H. sapiens* populations in East Asia and others in Europe and western Asia (Neanderthals) left no descendants. The oldest known modern human fossils are from South Africa and are possibly 90,000 to 125,000 years old. Independent evidence from molecular studies also points to Africa as the ancient and single place of origin of all existing human populations. These data would rule out an earlier theory that African, Asian, and European human populations evolved independently and in parallel from different archaic progenitors in these separate locations.

The relatively recent success of the human species is due largely to ongoing cultural evolution superimposed on a biological evolutionary heritage. The human ability to transmit varied and complex acquired knowledge within and between generations is an outcome of biological modifications that include a reorganized vocal apparatus that permits articulate speech and a large and complex brain that makes possible complicated learning and reasoning. An important landmark in cultural evolution was the invention of agriculture and its influence on the establishment of settled village life in place of the earlier nomadic life style. Agriculture was initiated in various localized regions, but three major geographical centers are associated with the domestication of plants and animals and the export of these new practices to distant parts of the world. Archeological remains indicate the Middle East, the central New World, and China as the major sites of agriculture between 7,000 and 10,000 years ago.

Technological advances have spread rapidly since that time. Technology has the potential to improve the quality of life for people, regardless of their genetic variations. By actions made possible through cerebral dominance of the brain, human beings exert a level of control over themselves and their surroundings to a degree that is unique in the history of biological evolution.

References and Additional Readings

Andrews, P. 1986. Fossil evidence on human origins and dispersal. *Cold Spring Harbor Sympos. Quant. Biol.* 51:419.

Bellwood, P. S. 1980. The peopling of the Pacific. *Sci. Amer.* 243(5):174.

Bendall, D. S., ed. 1985. *From Molecules to Men*. New York: Cambridge University Press.

Bernardi, G., and G. Bernardi. 1986. The human genome and its evolutionary context. *Cold Spring Harbor Sympos. Quant. Biol.* 51:479.

Bromage, T. G., and M. C. Dean. 1985. Re-evaluation of the age at death of immature fossil hominids. *Nature* 317:525.

Brown, F. et al. 1985. Early *Homo erectus* skeleton from west Lake Turkana, Kenya. *Nature* 316:788.

Cann, R. L. 1987. In search of Eve. *The Sciences,* September–October, p. 30.

Cann, R. L., M. Stoneking, and A. C. Wilson. 1987. Mitochondrial DNA and human evolution. *Nature* 325:31.

Ciochon, R. L., and J. G. Fleagle, eds. 1987. *Primate Evolution and Human Origins.* Hawthorne, N.Y.: Aldine de Gruyter.

Cold Spring Harbor Symposia on Quantitative Biology. 1986. *Molecular Biology of Homo sapiens.* Vol. 51. Cold Spring Harbor, N.Y.: Cold Spring Harbor Laboratory.

Conroy, G. C., and M. W. Vannier. 1987. Dental development of the Taung skull from computerized tomography. *Nature* 329:625.

Cronin, J. E. et al. 1981. Tempo and mode in hominid evolution. *Nature* 292:113.

Davis, R. S., V. A. Ranov, and A. E. Dodonov. 1980. Early man in Soviet Central Asia. *Sci. Amer.* 243(6):130.

Day, M. H. 1986. *Guide to Fossil Man,* 4th ed. Chicago: University of Chicago Press.

Delson, E. 1985. Late Pleistocene human fossils and evolutionary relationships. In *Ancestors: The Hard Evidence,* ed. E. Delson, p. 296. New York: Liss.

Delson, E., ed. 1985. *Ancestors: The Hard Evidence.* New York: Liss.

Dobzhansky, T. 1962. *Mankind Evolving.* New Haven, Conn.: Yale University Press.

Gowlett, J. A. J. et al. 1981. Early archaeological sites, hominid remains and traces of fire from Chesowanja, Kenya. *Nature* 294:125.

Groube, L. et al. 1986. A 40,000 year-old human occupation site at Huon Peninsula, Papua New Guinea. *Nature* 324:453.

Grüsser, O. -J., and L. -R. Weiss. 1985. Quantitative models on phylogenetic growth of the hominid brain. In *Hominid Evolution: Past, Present, and Future,* ed. P. V. Tobias, P. 457. New York: Liss.

Higham, C. F. W. 1984. Prehistoric rice cultivation in southeast Asia. *Sci. Amer.* 250(4):138.

Johanson, D. C. et al. 1987. New partial skeleton of *Homo habilis* from Olduvai Gorge, Tanzania. *Nature* 327:205.

Kalb, J. E. et al. 1984. Early hominid habitation in Ethiopia. *Amer. Sci.* 72:168.

Kaufman, L., and K. Mallory. 1987. *The Last Extinction.* Cambridge, Mass.: MIT Press.

Laitman, J. T. 1984. The anatomy of human speech. *Nat. Hist.* 93(8):20.

Laitman, J. T. 1985. Evolution of the hominid upper respiratory tract: the fossil evidence. In *Hominid Evolution: Past, Present, and Future,* ed. P. V. Tobias, p. 281. New York: Liss.

Latorre, A., A. Moya, and F. J. Ayala. 1986. Evolution of mitochondrial DNA in *Drosophila subobscura. Proc. Natl. Acad. Sci. U.S.* 83:8649.

Leakey, R. E. F. 1976. New hominid fossils from the Koobi Fora formation in northern Kenya. *Nature* 261:574.

Leakey, R. E., and R. Lewin. 1978. *People of the Lake.* New York: Doubleday.

Leroi-Gourhan, A. 1982. The archaeology of Lascaux Cave. *Sci. Amer.* 243(6):130.

Lewin, R. 1987. *Bones of Contention.* New York: Simon and Schuster.

Mangelsdorf, P. C. 1986. The origin of corn. *Sci. Amer.* 255(2):80.

Mann, A. 1975. *Paleodemographic Aspects of the South African Australopithecines.* Philadelphia: University of Pennsylvania Press.

Napier, J. 1962. The evolution of the hand. *Sci. Amer.* 207(6):56.

Nei, M. 1987. *Molecular Evolutionary Genetics.* New York: Columbia University Press.

Patterson, C., ed. 1987. *Molecules and Morphology in Evolution—Conflict or Compromise.* New York: Cambridge University Press.

Pickersgill, B. 1977. Taxonomy and the origin and evolution of cultivated plants in the New World. *Nature* 268:591.

Rightmire, G. P. 1984. *Homo sapiens* in sub-Saharan Africa. In *The Origins of Modern Humans,* ed. F. H. Smith and F. Spencer, p. 295. New York: Liss.

Rukang, W., and L. Shenglong. 1983. Peking man. *Sci. Amer.* 248(6):86.

Savage-Rumbaugh, E. S., D. M. Rumbaugh, and S. Boysen. 1980. Do apes use language? *Amer. Sci.* 68:49.

Shipman, P. 1984. Scavenger hunt. *Nat. Hist.* 93(4):20.

Shipman, P. 1986. Scavenging and hunting in early hominids: theoretical framework and tests. *Amer. Anthropol.* 88:27.

Smith, B. H. 1986. Dental development in *Australopithecus* and early *Homo. Nature.* 323:327.

Smith, F. H., and F. Spencer, eds. 1984. *The Origins of Modern Humans.* New York: Liss.

Stoneking, M., K. Bhatia, and A. C. Wilson. 1986. Rate of sequence divergence estimated from restriction maps of mitochondrial DNAs from Papua New Guinea. *Cold Spring Harbor Sympos. Quant. Biol.* 51:433.

Stringer, C. B. 1984. Fate of the Neanderthal. *Nat. Hist.* 93(12):6.

Stringer, C. B. 1985. Middle Pleistocene hominid variability and the origin of Late Pleistocene humans. In *Ancestors: The Hard Evidence,* ed. E. Delson, p. 289. New York: Liss.

Stringer, C. B., and P. Andrews. 1988. Genetic and fossil evidence for the origin of modern humans. *Science* 239:1263.

Susman, R. L., and J. T. Stern. 1982. Functional morphology of *Homo habilis. Science* 217:931.

Tobias, P. V. 1971. *The Brain in Hominid Evolution.* New York: Columbia University Press.

Tobias, P. V., ed. 1985. *Hominid Evolution: Past, Present, and Future.* New York: Liss.

Trinkaus, E., and W. W. Howells. 1979. The Neanderthals. *Sci. Amer.* 241(6):118.

Trinkaus, E., and F. H. Smith. 1985. The fate of the Neandertals. In *Ancestors: The Hard Evidence,* ed. E. Delson, p. 325. New York: Liss.

Valladas, H. et al. 1988. Thermoluminescence dating of Mousterian "Proto-Cro-Magnon" remains from Israel and the origin of modern man. *Nature* 331:614.

Vandermeersch, B. 1985. The origin of the Neandertals. In *Ancestors: The Hard Evidence,* ed. E. Delson, p. 306. New York: Liss.

Wainscoat, J. S. et al. 1986. Evolutionary relationships of human populations from an analysis of nuclear DNA polymorphisms. *Nature* 319:491.

Walker, A. C., and R. E. Leakey. 1978. The hominids of East Turkana. *Sci. Amer.* 239(2):44.

Wolpoff, M. H. 1985. Human evolution at the peripheries: the pattern at the eastern edge. In *Hominid Evolution: Past, Present, and Future,* ed. P. V. Tobias, p. 355. New York: Liss.

Wolpoff, M. H., W. X. Zhi, and A. G. Thorne. 1984. Modern *Homo sapiens* origins: a general theory of hominid evolution involving the fossil evidence from East Asia. In *The Origins of Modern Humans,* ed. F. H. Smith and F. Spencer, p. 411. New York: Liss.

Wood, B. 1985. Early *Homo* in Kenya, and its systematic relationships. In *Ancestors: The Hard Evidence,* ed. E. Delson, p. 206. New York: Liss.

Wood, B. A., L. B. Martin, and P. Andrews, eds. 1986. *Major Topics in Primate and Human Evolution.* New York: Cambridge University Press.

Yunis, J. J., and O. Prakash. 1982. The origin of man: a chromosomal pictorial legacy. *Science* 215:1525.

Glossary

- **abiotic synthesis** the formation of organic molecules in the absence of life.
- **activation energy** the energy required for a chemical reaction to proceed.
- **adaptability** the capacity of an organism to undergo modification and function more successfully in a given environment, thereby improving its chances for leaving descendants; flexibility.
- **adaptation** the evolutionary process by which organisms undergo modification, allowing them to function more successfully in a given environment; any characteristic of an organism that improves its chances for surviving and for leaving descendants in a given environment.
- **adaptationist program** a conceptual view that presumes that each characteristic has an adaptive value, and focuses on determining what that value is to the organism.
- **adaptedness** the state of being best fit to function in a given environment; fitness.
- **adaptive landscape** a conceptual plot in two or three dimensions that depicts multiple peaks of average fitness for frequencies of two or more genes present in a population, and valleys of lower fitness between peaks.
- **adaptive radiation** the evolution from a generalized ancestral stock of diverse descendant species that come to occupy a variety of living zones in a relatively brief interval of geologic time.
- **aerobic respiration** the major oxygen-dependent processes for oxidation of glucose and other organic fuel molecules, producing ATP by means of the Krebs cycle and oxidative phosphorylation coupled to electron transport toward molecular oxygen.
- **Age of Mammals** the colloquial reference to the Cenozoic Era, when mammals were the dominant land vertebrates.

- **Age of Reptiles** the colloquial reference to the Mesozoic Era, when reptiles were the dominant land vertebrates.
- **allele** one of a series of alternative forms of a given gene.
- **allele frequency** *See* **gene frequency.**
- **allometry** the relation between the rate of growth of a part of an individual and the rate of growth of another part or of the whole, usually evident by change in relative proportions as the total body size increases or decreases.
- **allopatric speciation** the formation of species as the result of differentiation of conspecific populations in geographic isolation. *See also* **parapatric speciation, peripatric speciation, sympatric speciation.**
- **allopatry** the existence of different populations of a species in geographical isolation.
- **allozymes** the alternative electrophoretic forms of an enzyme coded by alternative alleles of a gene.
- **amniote** the pattern of a terrestrial vertebrate egg, characterized by an embryo provided with a yolk and two membranous sacs (amnion and allantois).
- **anaerobic** the absence of oxygen; a characteristic of a process of an organism that does not require oxygen.
- **anagenesis** continuous evolutionary change in a lineage without subdivision or branching into two or more descendant lineages. *See also* **cladogenesis.**
- **analogous** referring to structures or processes that have superficial similarities but are not genetically related. *See also* **convergence.**
- **aneuploid** an individual whose chromosome number is not an exact multiple of the typical haploid set for the species. *See also* **polypolid.**
- **aneuploid series** a group of related species whose chromosome numbers are not exact multiples of the basic haploid set, and usually evolve a lower haploid number than the ancestral set.
- **angiosperms** the flowering plants.
- **Animalia** the animal kingdom of multicellular organisms.
- **Anthropoidea** the suborder of Primates that includes the monkey, ape, and human families, and is distinguished from the suborder of prosimian primates.
- **antibody** a protein produced and secreted by plasma cells (activated B cells) of the lymphatic system that is capable of binding with its specific antigen in an immune response; an immunoglobulin.
- **anticodon** the triplet of nucleotides in a transfer RNA molecule that associates by complementary base pairing with a specific triplet codon in the messenger RNA molecule during its translation at the ribosome.
- **antigen** a foreign substance that stimulates an animal to produce specific neutralizing antibodies in an immune response.
- **archaebacteria** the heterogeneous collection of methanogenic, thermo-acidophilic, and halophilic bacteria that is presumed to be most like the

ancestral cell lineage, from which all prokaryotic and eukaryotic organisms are descended. *See also* **eubacteria.**

· **Archean Eon** the geologic time interval that spans the first 2100 Myr since the Earth was completed, or from 4600 Myr to 2500 Myr BP. *See also* **Phanerozoic Eon, Proterozoic Eon.**

· **assortative mating** sexual reproduction in which partners tend to be more alike or less alike than if selection for mating were random.

· **atmosphere** the gaseous envelope of a planet or moon. *See also* **hydrosphere, lithosphere.**

· **australopithecines** the members of the subfamily of Hominidae that were upright in posture and bipedal in locomotion, but lacked a large brain and cultural traditions; the first known members of the human family.

· **autotroph** an organism that obtains its carbon and energy for macromolecular synthesis from inorganic sources. *See also* **heterotroph.**

· **bacteriochlorophyll** any of a group of light-absorbing porphyrin-based pigments that function in bacterial photosynthesis.

· **balancing selection** the natural selection strategy that favors the heterozygote over the homozygotes and thereby maintains a stable genetic polymorphism in a population; heterozygote advantage.

· **basocranium** the base of the skull, which may be flat or arched.

· **Big Bang theory** the proposal that the universe began with an explosion and will continue to expand until only a void remains. *See also* **Oscillating Universe theory.**

· **bipedalism** the mode of locomotion in which two legs are used consistently to move from place to place.

· **bottleneck** *See* **population bottleneck.**

· **brachiation** the mode of locomotion by which the individual moves from branch to branch by grasping with both hands and exerting propulsive force by the use of the upper body.

· **Burgess Shale** a Cambrian rock formation in British Columbia, Canada, that contains animal fossils with well-preserved soft parts as well as hard parts.

· **calibration-dependent test** a method to analyze the constancy of a molecular clock by comparing rates of divergence from plots of nucleotide substitution against times of species divergence estimated from the fossil record.

· **Cambrian Period** the first geologic time interval of the Paleozoic Era, from 590 to 505 Myr BP, and noted for the first appearance in the fossil record of representatives of all the living animal phyla.

· **Carboniferous Period** the fifth geologic time interval of the Paleozoic Era, from 360 to 286 Myr BP, noted for extensive formations of coal deposits; in North America, it is subdivided into Mississippian and Pennsylvanian times.

- **catarrhine** a member of the infraorder of anthropoid Primates that includes the Old World monkeys, apes, and human family, whose nostrils are close together and open down. *See also* **platyrrhine.**
- **catastrophism** the concept that cataclysmic or catastrophic events have been responsible for the changing topography of the Earth throughout its history. *See also* **uniformitarianism.**
- **C_3 cycle** that part of the dark reactions of photosynthesis in which CO_2 is reduced to 3-phosphoglyceraldehyde, from which hexose sugars are produced and ribulose 1,5-bisphosphate is regenerated; Calvin cycle.
- **Cenozoic Era** the third geologic time interval, from 65 Myr BP to the present, of the Phanerozoic Eon.
- **centric fusion** breakage in the short arms of two acrocentric chromosomes, followed by fusion of the long arms into a single metacentric or submetacentric chromosome, usually with the loss of the two small fragments of the original chromosomes.
- **Cercopithecoidea** the primate superfamily of Old World monkeys, which includes the cercopithecid and colobid monkeys of Asia and Africa and, together with the hominoid superfamily, are members of the catarrhine anthropoids.
- **chain of being** the Aristotelian concept of a graded sequence of organisms that range from the lowest forms to the Supreme Being, making up an indefinite number of links in the chain of past and present life.
- **clade** a taxon consisting of a single species and its descendants, representing a monophyletic branch on an evolutionary tree.
- **cladistics** a system of classification that attempts to reconstruct phylogenies in terms of branched sequences of ancestors and descendants, each branch composed of a monophyletic group whose members share uniquely derived character states. *See also* **phenetics, phylogenetics.**
- **cladogenesis** the branching episodes of speciation in a phylogeny. *See also* **anagenesis.**
- **cladogram** the portrayal of a phylogeny in which each branch represents a monophyletic group of ancestor and descendants, which share uniquely derived character states.
- **ClB method** an experimental design used for detecting X-linked lethal, detrimental, and morphological mutations in *Drosophila melanogaster,* based on observed versus expected proportion of males, or of their phenotypes, in a progeny; the name refers to X chromosome markers: *C,* crossover suppressor (an inversion); *l,* lethal; and *B,* bar eye, a dominant morphological trait.
- **cline** the gradient of gene frequencies or phenotypes along a geographical transect of the range of the species.
- **coacervate** the fluid-filled droplet formed in dilute aqueous solutions of macromolecules and/or colloidal particles, whose boundary develops as a film of bound water molecules; the protobiont model proposed by A. I. Oparin.
- **codon** the triplet of nucleotides in a gene sequence or its messenger RNA

copy, that specifies 1 of the 20 amino acids or 1 of the 3 punctuations (STOP) included in the genetic code.
- **compatibility method** a means of choosing the most likely one of a number of possible evolutionary trees reconstructed from a set of data, as the tree most compatible with the largest clique of characters. *See also* **distance matrix methods, parsimony method.**
- **conjugation** the temporary union of two unicellular organisms, with at least one receiving genetic material from the other; in bacteria, the exchange is unidirectional, from donor to recipient cell, and usually involves the transfer of only a portion of the genome. *See also* **transduction, transformation.**
- **continental drift** the movement of landmasses on lithospheric plates carried over the underlying semimolten mantle by plate tectonic processes; the concept first developed by Alfred Wegener that the world's continents once formed parts of a single landmass (Pangaea) and have drifted to their present positions over a period of about 200 Myr.
- **convergence** the evolution by means of genetically different programs of superficially similar adaptive structures and habits in unrelated species. *See also* **divergence, parallel evolution.**
- **cosmology** the branch of astronomy that deals with the structure and origin of the universe.
- **creation scientist** an individual who denies the occurrence of biological evolution and adheres instead to a literal interpretation of Genesis.
- **Cretaceous Period** the third, and last, geologic time interval of the Mesozoic Era, from 144 to 65 Myr BP, which closed with a major episode of mass extinction.
- **crossing over** the physical exchange of homologous chromosomes or DNA segments, leading to recombination of linked genes.
- **C value** the characteristic amount of DNA in the genome of a species, measured in picograms, daltons, or base pairs of DNA.
- **C value paradox** the reference to the lack of correlation between the size of the genome and the level of morphological or evolutionary complexity of different species.
- **cytochromes** the group of electron transport enzymes containing heme or related prosthetic groups that undergo valency changes of their iron atom in coupled oxidation-reduction reactions.
- **cytoplasm** the protoplasmic contents of the cell, exclusive of the nucleus or nucleoid.

- **deletion** the loss of part of a chromosome or DNA molecule from the genome; one of the four classes of change in chromosome structure; one of the categories of mutations.
- **dendrogram** the portrayal of a branched evolutionary tree showing ancestor and descendant relationships for one or more related lineages.
- **descent with modification** the Darwinian evolutionary concept of gradual change from ancestral to descendant forms over time.

- **deterministic processes**   processes that lead to a particular direction of change with time, such as natural selection. *See also* **stochastic processes.**
- **developmental constraints**   the concept that evolutionary changes in morphology are limited to some degree and are more likely to be modifications of later than of earlier stages in ontogeny because of their different levels of effect on the development of structures and functions of the organism.
- **Devonian Period**   the fourth geologic time interval of the Paleozoic Era, from 408 to 360 Myr BP, noted for the appearance of land vertebrates (amphibians) and for a major episode of mass extinction at its close.
- **differential gene expression**   the transcription and translation of some genes at the same time that other genes in the same cell are inactive, leading to different phenotypes in genetically identical cells or individuals.
- **differential reproduction**   the Darwinian concept that the more fit individuals are more likely to survive and leave descendants than are the less fit individuals in a population, and that by such natural selection, populations change over time, or evolve.
- **diploid**   a cell, an individual, or a species that has two sets of homologous chromosomes in the nucleus; $2n$.
- **directional selection**   the natural selection strategy that results in a shift in the population mean toward one of the homozygous classes having higher adaptive value in a given environment, and in a reduction of the genetic diversity in a population.
- **distance matrix methods**   a number of procedures used to reconstruct molecular phylogenetic trees, by the computation of genetic or evolutionary distance for all pairs of species or populations, and a consideration of all the relationships deduced from these distance values. *See also* **compatibility method, parsimony method.**
- **divergence**   the evolutionary diversification of related species by means of accumulated genetic changes. *See also* **convergence, parallel evolution.**
- **diversifying selection**   the natural selection strategy by which divergent phenotypic extremes in populations are favored, leading to decrease in heterozygotes and increase in one or another of the homozygous classes in given subdivisions of a heterogeneous environment, and to an overall increase in genetic diversity; disruptive selection.
- **DNA–DNA hybridization**   a method of molecular hybridization involving the formation of duplex DNA molecules by the hydrogen bonding of complementary single-stranded molecules or parts of molecules under standard conditions; a test for DNA homology in heteroduplexes formed from single strands of DNA isolated from different species, and a basis for deducing phylogenetic relationships.
- **DNA methylation**   the addition of methyl groups to cytosine residues in DNA to form 5-methylcytosine, which may be associated with inactivation of gene expression in the methylated segments of chromosomes.
- **DNA transformation**   *See* **transformation**
- **drift**   *See* **genetic drift.**
- **dryopithecine**   a member of an extinct group of Miocene apes that inhab-

ited forested regions in Europe and Asia, and used to be considered ancestral to the modern great apes.

· **duplication**   an extra copy of a nucleotide sequence, gene, part of a chromosome, or whole chromosome in a cell or individual.

· **Dust Cloud theory**   the proposal that a solar system originates as a huge cloud of dust and gases in interstellar space that condenses to become a flattened spinning solar nebula, from which a sun and planets ultimately develop.

· **Ediacaran Period**   a somewhat informal designation of the last 100 Myr of the Precambrian, when a variety of enigmatic invertebrate forms were present but left few or no descendants among Phanerozoic animals.

· **effective population number**   the actual number of breeding individuals in a population, rather than the total census of individuals present; symbolized as N_e.

· **electron transport chain**   electron carriers, such as cytochromes, that transfer electrons from donor to receptor along a gradient of decreasing energy and that cause the release of energy at each transfer step.

· **endosymbiosis**   the mutually beneficial association of two or more kinds or organisms, in which one serves as the host and the other(s) reside within the host.

· **endosymbiosis theory**   the proposal by Lynn Margulis that mitochondria and chloroplasts in eukaryotic cells are evolutionary descendants of free-living prokaryotic organisms that originally were endosymbionts in their host.

· **endothermy**   the metabolic regulation of its body temperature by an animal.

· **enhancer**   the regulatory sequence of nucleotides that enhances the transcriptional activity of a gene in eukaryotes and their viruses, and may be at some distance from the gene on the same strand of the DNA.

· **enzyme**   the unique protein catalyst of biological systems, which modulates the rate of chemical reactions but is not consumed by them.

· **Eocene Epoch**   the second geologic time interval of the Tertiary Period of the Cenozoic Era, from 58 to 37 Myr BP, noted for adaptive radiation of placental mammals.

· **eon**   the largest subdivision of geologic time, whose boundaries are set by major mountain-building events (Archean and Proterozoic) or by biological history (Phanerozoic).

· **epigenetic**   referring to the influence on development of prior conditions and events during each step of ontogeny, in addition to the determination of a developmental program through gene action.

· **epoch**   the subdivision of geologic time within the periods of the Cenozoic Era, based on a greater wealth of preserved information in the geologic and fossil records than is typical of older deposits, which have been subjected to longer times of erosion and modification.

· **equilibrium**   the state of unchanging conditions, such as the maintenance of the same allele frequencies in the gene pool of a population through successive generations.

- **era** a subdivision of geologic time within the larger interval of an eon.
- **eubacteria** all the bacteria exclusive of the archaebacteria, from which they differ in such characteristics as the chemical composition of cell walls, ether- versus ester-linkage of branched versus unbranched fatty acid chains, and sensitivity to particular antibiotics. *See also* **archaebacteria.**
- **eukaryote** an organism with a well-defined, membrane-bounded nucleus surrounded by cytoplasm containing one or more membranous organelles; a unicellular or multicellular member of the kingdom Protista, Plantae, Animalia, or Fungi. *See also* **prokaryote.**
- **Eutheria** the placental mammals. *See also* **Metatheria.**
- **evolution** change with time, specifically referred to as biological evolution in the case of living systems as distinct from nonliving physical matter or systems.
- **evolutionarily stable strategy** the proposal by John Maynard Smith that no mutant strategy of doing something can become established in a population under the influence of natural selection if all its members adopt the same strategy in particular situations in which they may find themselves.
- **evolutionary stasis** a pattern evident in the fossil record as a long interval of little significant phenotypic change in a species from the time it appears until it vanishes.
- **exon** a coding segment of a split gene.
- **exon shuffling** the incorporation of one or more exon (coding) segments from one gene sequence into another gene sequence, presumably by intron-mediated recombination, resulting in new or enhanced function or potential in the protein product of the recipient gene.
- **extinction** the termination of an evolutionary lineage.

- **fermentation** the oxidation of carbohydrate in oxygen-independent pathways, yielding only partially oxidized end products, such as pyruvate, rather than CO_2.
- **fitness** the relative ability of an organism to survive and transmit its genes to the next generation; symbolized as w.
- **fluctuation test** a statistical analysis of experimental populations designed to determine whether fit variants are selected spontaneous mutants that were present in a population prior to exposure to a selective agent, or arose in response to the need of the organism to survive in a changed environment (Darwinism versus Lamarckism). *See also* **replica plating test.**
- **foramen magnum** the opening in the base of the skull through which the spinal cord passes and is continuous with the brain inside the skull.
- **forebrain** the anterior of the three divisions marked out by constrictions in the embryonic vertebrate brain, and particularly involved in conscious, emotional, and intelligent activities.
- **fossil** any evidence of former life, including bones, shells, and impressions.

- **fossilization** the preservation of parts of or whole organisms or their traces, usually by mineralization and incorporation into forming sedimentary rock.
- **founder effect** an aspect of the model of peripatric speciation, in which a few individuals in peripherally isolated locations around a main population establish genetically distinct derived entities, which may evolve to new species status; the isolated populations first are influenced by genetic drift, but later are subject to different evolutionary pressures in different environments, leading to a diversity of gene pools that are distinct from one another and from that of the main population.
- **frequency-dependent selection** the natural selection strategy that favors a particular phenotype when it is uncommon, but acts against the same phenotype when it is common; selection of one phenotype over another, depending on the frequency of the phenotype in a given environment.
- **Fungi** the kingdom of eukaryotic organisms that obtain nutrients by absorption, lack photosynthetic structures and processes, have few, if any, complex tissues formed from tubular filaments of cells, and reproduce sexually or asexually by spores.

- **gel electrophoresis** the method for separating and identifying charged molecules or their fragments in solution, according to their movement through a porous, supporting gel made of starch, agar, or polyacrylamide subjected to an electrical field.
- **gene amplification** the repeated replication of some genes, producing many copies, at the same time that other genes in the same cell do not replicate.
- **gene expression** the readout of encoded genetic information in DNA (or RNA) during transcription and translation, which leads to phenotypic development in the cell or individual.
- **gene flow** the exchange of genes among different populations of a species as migrants move from one local population to another and influence allele frequencies in the gene pools of these subpopulations.
- **gene frequency** the percentage of each allele of a given gene in a population; allele frequency.
- **gene pool** the collection of alleles shared by members of a population or species and transmitted randomly from parents to offspring in reproduction.
- **gene rearrangement** the genomic alteration that leads to phenotypic diversity by the synthesis of different proteins from different combinations of a relatively few genes, as in antibody production from immunoglobulin genes.
- **genetic distance** a measure of the number of nucleotide substitutions per locus that have occured during the separate evolution of two populations or species; symbolized as *D. See also* **immunological distance, mutation distance.**
- **genetic drift** the random fluctuations of gene frequencies in populations, the effects of which are most evident in small populations.

- **genetic identity**   a measure of the proportions of alleles that are identical in two populations or species; symbolized as I.
- **genetic load**   the average number of detrimental or lethal alleles per individual in a population; harmful, hidden variation in the individual.
- **genetic recombination**   the production of new genotypic combinations of alleles of linked genes, by the exchange of homologous DNA or chromosome segments through crossing over.
- **genetic revolution**   a speciation pattern characterized by substantial and rapid genetic differentiation of the main popualtion.
- **genetic system**   the organization and mode of transmission of the genetic material for a given species.
- **genotype**   the genetic constitution of a cell or an individual. *See also* **phenotype.**
- **Gondwana(land)**   the name of the southern landmass of the supercontinent of Pangaea, or of the separated region following the breakup of Pangaea; the Mesozoic landmass that included present-day South America, Africa, India, Antarctica, and Australia. *See also* **Laurasia.**
- **grade**   a portion of a phylogeny that has evolved new abilities or organization. *See also* **clade.**
- **group selection**   the concept of group benefit from the altruistic behavior of individuals, which thereby reduce their own reproductive success.

- **haplotype**   a particular pattern of cuts in a DNA segment made by some number of restriction enzymes, and often depicted as a series of plus (cut) and minus (not cut) signs denoting the nature of the relevant DNA sequence; sites that are cut have a nucleotide sequence recognized by the enzymes, and sites that are not cut have an altered nucleotide sequence.
- **Hardy-Weinberg principle**   the statement that genotype frequencies depend on allele frequencies at a given locus, and will remain constant at any values in an infinitely large, randomly mating population in which mutation, migration (gene flow), and selection are negligible or nonexistent; the principle is formulated as $(p_A + q_a)^2 = p^2\ AA + 2pq\ Aa + q^2\ aa = 1$.
- **heavy-chain class switching**   during development of B lymphocytes, the rearrangement of constant-segment *(C)* genes such that one or more upstream genes are deleted and the downstream genes are spliced to an existing *VDJ* sequence to make a new *VDJC* combination, which encodes a different functional class of antibodies (immunoglobulins) due to the presence of a different constant segment in the heavy chain.
- **heme**   the iron-containing porphyrin derivative in hemoglobins and in such enzymes as catalase and cytochromes.
- **heterochrony**   a change during evolution in the timing of ontogenetic development in descendants compared with the ancestor, often resulting in the retention of juvenile (ancestral) features by the adults of a species. *See also* **neoteny, paedomorphosis, progenesis.**
- **heteroduplex**   a double-stranded nucleic acid in which each chain has a

different origin, and the efficiency of base pairing serves as a measure of strand complementarity and homology. *See also* **DNA–DNA hybridization.**

- **heterokaryon**   a somatic cell that contains genetically different nuclei, which divide independently but together influence phenotype development.
- **heterotroph**   a cell or an organism that obtains its carbon and energy from the oxidation of organic molecules. *See also* **autotroph.**
- **heterozygote**   a diploid or polyploid cell or individual that has different alleles at one or more loci.
- **hindbrain**   the posterior of the three divisions marked out by constrictions in the embryonic vertebrate brain, and forming the medulla and cerebellum during development.
- **Holocene Epoch**   the second geologic time interval of the Quaternary Period of the Cenozoic Era, from 45,000 years ago to the present; Recent Epoch.
- **hominid**   an australopithecine or a hominine member of the human family, Hominidae, characterized by consistent upright posture and bipedal locomotion.
- **hominine**   a member of the genus *Homo* in the subfamily Homininae of the human family of species, and characterized by upright bipedalism, an enlarged brain in proportion to body size, and established cultural traditions.
- **hominoid**   a member of the anthropoid superfamily Hominoidea, composed of the families of lesser apes (Hylobatidae), great apes (Pongidae), and prehumans and humans (Hominidae).
- **Hominoidea**   the superfamily of anthropoid catarrhines characterized by the absence of a tail, a larger and more complex brain, and other features that distinguish it from the catarrhine superfamily of Old World monkeys (Cercopithecoidea).
- **homologous**   referring to genes, structures, or processes that are descended from common ancestry and are genetically related, even if they have changed substantially during their separate evolutionary histories after divergence.
- **hot spot**   a location where a column of magma rises as a thermal plume from the semimolten mantle and causes heating and igneous activity evident in the overlying crust as volcanoes, geysers, and similar geologic phenomena.
- **Hubble's law**   the speed of recession of a galaxy is directly proportional to its distance from the observer, so a galaxy twice as far away as another will be moving twice as fast, three times as far away it will be moving three times as fast, and so forth; the proportion is expressed by the formula $v = Hx$, where v is the speed of recession, x is the distance from the observer, and H is the proportionality constant, or Hubble constant.
- **hybrid inviability**   a postmating reproductive isolating mechanism in which hybrids between different populations or species fail to survive to reproductive age.
- **hybrid production, lack of**   a premating reproductive isolating mechanism

in which members of different populations or species cannot interbreed, thereby keeping their gene pools separate and distinct.

· **hybrid sterility** a postmating reproductive isolating mechanism in which hybrids between different populations or species are infertile, and thus cannot further the exchange of genes between the two parental groups.

· **hydrogen burning** the atomic reactions in which protons are consumed in the synthesis of helium and heavier elements in the hot interiors of stars.

· **hydrosphere** the portion of a planet maintained in the liquid state. *See also* **atmosphere, lithosphere.**

· **hylobatid** a gibbon or siamang member of the lesser-ape family Hylobatidae, one of the hominoid families of Primates.

· **hypersensitive site** in chromatin, the place of cleavage by very mild nuclease digestion and an indicator of actively transcribing regions of chromosomes.

· **igneous rock** mineral matter formed by volcanic action from the crust and mantle of the planet, including granite of the continental crust and basalts making up a large proportion of oceanic crust. *See also* **metamorphic rock, sedimentary rock.**

· **immunoglobulin** any of five classes of antibody molecules, each consisting of two identical heavy and two identical light polypeptide chains, that are made in B lymphocytes and secreted when the lymphocytes become plasma cells on activation by antigens; the molecules that neutralize invading antigens in immune responses, abbreviated as IgA, IgD, IgE, IgG, IgM.

· **immunological distance** the measure of relationships between species based on intensities of immunological cross-reactions determined by microcomplement fixation; the formula is $d_1 = 100 \times \log_{10} \text{I.D.}$, where I.D. is the index of dissimilarity between members of a pair of species. *See also* **genetic distance, mutation distance.**

· **impact theory** the proposition that one or more mass extinction episodes were caused by the impact on Earth of a meteorite, an asteroid, or another extraterrestrial object, which led to global conditions that were too harsh to sustain many life forms.

· **inbreeding coefficient** the measure of the proportion of heterozygosity reduced by inbreeding, relative to the heterozygosity in a randomly mating population with the same allele frequencies; symbolized as F.

· **independent assortment** the random distribution to the gametes of members of pairs of alleles at loci on different chromosomes.

· **ingroup** a set of closely related species whose characteristics are compared with those in an outgroup of less closely related species in order to determine whether a charaacter state is ancestral or derived, whether molecular change has occurred at the same rate in the different ingroup species, and other purposes.

· **inheritance of acquired characteristics** the theory of Jean-Baptiste Lamarck that inherited changes occur in the organism in direct response to its needs in a given environment; use and disuse, Lamarckism.

- **intron** a noncoding segment, or intervening sequence, between exons in a split gene.
- **inversion** the structural rearrangement of part of a chromosome such that it is reinserted in reverse order rather than its original orientation before breakage and reunion.
- **iridium** a member of the platinum group of metallic elements that if found in higher abundance in one location than another is inferred to have been deposited following the impact of an extraterrestrial body on the Earth's surface, thus accounting for its excess in such a location.
- **isolating mechanism** a geographical or biological barrier that prevents successful interbreeding between two or more closely related populations or species.

- **Jurassic Period** the second geologic time interval of the Mesozoic Era, from 213 to 144 Myr BP, noted for the appearance of the first birds and mammals.

- **kingdoms** the major taxonomic categories for organisms, including the Monera, Protista, Fungi, Plantae, and Animalia recognized by many, but not all, authorities.
- **kin selection** a restricted aspect of the concept of group selection, in which the altruists are relatives of the group that benefits from their unselfish behavior in taking risks.

- **larynx** the structure in the throat to which vocal folds (vocal cords) are attached, providing a system for vocalization, and that leads into the trachea to the lungs in a set of passageways for air inflow and outflow.
- **Laurasia** the name of the northern landmass of the supercontinent of Pangaea, or of the separated region following the breakup of Pangaea; the Mesozoic landmass that included present-day North America, Europe, Greenland, and most of Asia. *See also* **Gondwana(land).**
- **light-year** the distance traveled in 1 year by light moving at 186,000 miles per second, equivalent to about 6 trillion miles; a measure of cosmic distance.
- **linkage disequilibrium** the occurrence of alleles at two or more loci in higher or lower frequencies when they are present together than would be expected from their individual frequencies; the nonrandom association of alleles of two or more linked or unlinked genes in a population.
- **liposome** a phospholipid-bounded vesicle that can take macromolecules into its interior selectively; a protobiont model.
- **lithosphere** the portion of a planet or moon maintained in the solid state; the rocky crust of the Earth, which exists as a number of mobile tectonic plates. *See also* **atmosphere, hydrosphere.**
- **living fossil** the colloquial term for an extant species that is morphologically very similar to extinct ancestral species of its ancient past, such as the living lungfish and tuatara.

- **lysosome**   the membrane-bounded eukaryotic organelle that contains a variety of acid hydrolases capable of digesting virtually any biologically important organic compound.

- **Mammalia**   the class of tetrapod vertebrates characterized by body hair (fur) and the suckling of young on mother's milk after birth or hatching.
- **mass extinction**   the relatively sudden disappearance of many unrelated families of organisms, usually animal families.
- **maximum parsimony method**   *See* **parsimony method.**
- **median melting temperature**   the temperature at which 50 percent of the DNA in solution remains single-stranded during reannealing and formation of duplex molecules from single strands of one or different species; symbolized as $T_{50}H$, T_m.
- **Mesozoic Era**   the second geologic time interval, from 248 to 65 Myr BP, of the Phanerozoic Eon, noted for the appearance of the dinosaurs, birds, mammals, and flowering plants.
- **metabolism**   the sum total of biochemical synthesis and degradation reactions in biological systems.
- **metamorphic rock**   mineral matter formed by the alteration of other rocks at very high temperature and pressure, and possessing different crystal structure and texture from the parent materials, such as slate from shale or marble from limestone. *See also* **igneous rock, sedimentary rock.**
- **Metatheria**   the marsupial mammals. *See also* **Eutheria.**
- **methylation**   *See* **DNA methylation.**
- **midbrain**   the middle portion of the three divisions marked out by constrictions in the embryonic vertebrate brain, and includes in its basal region an integration center and relay station for various sensory and motor activities based on information traveling to and from the central nervous system.
- **migration coefficient**   the fraction of migrant genes in a population or gene pool; symbolized as *m*.
- **Miocene Epoch**   the fourth geologic time interval of the Tertiary Period of the Cenozoic Era, from 24 to 5 Myr BP, noted for adaptive radiation of the anthropoid primates.
- **Modern Synthesis**   the framework for speciation and other evolutionary processes and concepts developed in the 1930s and 1940s by collaborative efforts of geneticists, systematists, and paleontologists that provided the basis for modern neo-Darwinian theory, which includes all their perspectives.
- **modifier alleles**   the alternative forms of genes that influence or modify the expression of nonallelic genes.
- **molecular constraints**   a concept stating that more essential parts of proteins are less variable than less essential or nonessential parts, which reflects differences in the rate of evolutionary change in portions of the gene coding for different portions of its protein product.
- **molecular evolutionary clock**   the rate at which mutations accumulate in a given gene, which usually is constant for neutral changes but varies for

changes influenced by natural selection; evolutionary clock; molecular clock.
- **Monera** the kingdom of unicellular organisms with a prokaryotic plan of cell organization, including all the forms of bacteria, cyanophytes, and prochlorophytes.
- **monomer** the basic subunit of a larger functional molecule, particle, or cellular entity.
- **monophyletic** referring to an evolutionary lineage that is derived from and often includes the single ancestral form. *See also* **polyphyletic.**
- **monosomy** the condition in which one chromosome of a pair is missing; $2n - 1$.
- **multigene family** a set of genes derived by duplication of an ancestral gene and varying from that gene and from one another in nucleotide sequence, but still encoding functionally similar polypeptides or RNAs.
- **mutation** the process by which a gene is altered in its base composition or sequence or in its organization; a modified gene resulting from mutational processes.
- **mutation distance** a measure of the number of nucleotide substitutions per locus that have occurred during the separate evolution of two populations or species, estimated from alterations in amino acid sequences of the gene product according to the standard genetic code. *See also* **genetic distance, immunological distance.**

- **natural selection** the differential reproduction of genetically diverse individuals in populations, which results in individuals of greater fitness leaving more offspring than those of lesser fitness, thereby leading to changes in gene frequencies over time; Darwinism.
- **Neanderthal** a type of archaic *Homo sapiens* who inhabited Europe and West Asia during the late Pleistocene and early Recent epochs.
- **neocortex** the more recently evolved layer of cerebral cortex, which overlies the older limbic system and exercises a restraining effect on basic emotional behaviors governed by the limbic system.
- **neoteny** the retardation of ontogenetic and phylogenetic development in evolution, leading to the retention of juvenile (ancestral) features in the adults of the species. *See also* **paedomorphosis, progenesis.**
- **neutral evolution theory** the proposal by Motoo Kimura that molecular evolution is guided primarily by the stochastic processes of mutation and genetic drift, rather than primarily by natural selection and other deterministic processes.
- **New World monkeys** the anthropoid superfamily of platyrrhines located exclusively in subtropical and tropical Central and South America, and consisting of the family of cebid monkeys (Cebidae) and the family of marmosets and tamarins (Callitrichidae); the platyrrhine monkeys of the Americas as opposed to the catarrhine monkeys of the Old World regions of Europe, Africa, and Asia.
- **nonoxidizing atmosphere** the gaseous envelope of a planet or moon in

which nitrogen, hydrogen, and other gases—but not molecular oxygen—may be present.

- **normalizing selection** *See* **stabilizing selection.**
- **nucleoid** the localized region of genomic DNA that is not separated by a membrane from the surrounding cytoplasm in prokaryotic cells.
- **nucleus** the membrane-bounded compartment containing the chromosomes in eukaryotic cells, and the principal hallmark of eukaryotic cell organization.

- **Old World monkeys** the anthropoid superfamily of catarrhines of Europe, Africa, and Asia, which is subdivided into two families (Cercopithecidae and Colobidae), and is distinct from the platyrrhine New World monkeys of the Americas.
- **Oligocene Epoch** the third geologic time interval of the Tertiary Period of the Cenozoic Era, from 37 to 24 Myr BP, noted for the appearance of the anthropoid primates.
- **ontogeny** the development of the individual from fertilization to maturity.
- **operator** a nontranscribed regulatory nucleotide sequence that is capable of interacting with a specific repressor protein and thereby controlling the transcription of one or more structural genes adjacent to it in a bacterial operon.
- **operon** the gene cluster consisting of a promoter, an operator, and adjacent structural genes that function coordinately in bacterial gene expression.
- **Ordovician Period** the second geologic time interval of the Paleozoic Era, from 505 to 438 Myr BP, noted for a diversity of ostracoderm fishes and dominant invertebrate life and the first known major mass extinction.
- **ornithischian dinosaurs** the herbivorous reptiles of the Dinosauria, having a four-pronged pelvis, rather than the typical reptilian three-pronged pelvis of the saurischian dinosaurs.
- **orthogenesis** the unsupported concept that related groups of organisms evolve in a straight-line progression over a long period of time from some ancestral form to some predetermined descendant form.
- **orthologous genes** the descendants of an ancestral gene that was present in the last common ancestor of two or more species, such as the β-globin genes in vertebrate animals. *See also* **paralogous genes.**
- **Oscillating Universe theory** the proposal that the universe proceeds through repeated cycles of expansion and contraction and will exist forever. *See also* **Big Bang theory.**
- **outgroup** a set of species less closely related to an ingroup of closely related species that serves as a basis for comparison in order to determine whether an ingroup character state is ancestral or derived, whether molecular change has occurred at the same rate in different ingroup species, and other purposes.
- **oxidative phosphorylation** the syntheses of ATP coupled to respiratory electron transport during aerobic respiration.

- **paedomorphosis** the evolutionary phenomenon of retention of juvenile (ancestral) features in adults of a species, resulting from phylogenetic modifications in ontogenetic timing, or heterochrony. *See also* **neoteny, progenesis.**
- **Paleocene Epoch** the first geologic time interval of the Tertiary Period of the Cenozoic Era, from 65 to 58 Myr BP, noted for adaptive radiation of the mammals and their replacement of the reptiles as the dominant land vertebrates.
- **Paleozoic Era** the first geologic time interval, from 590 to 248 Myr BP, of the Phanerozoic Eon, noted for the first representatives of all the living animal phyla and many other evolutionary events.
- **Pangaea** the Paleozoic supercontinent that fragmented during the Mesozoic Era to produce the present-day continental landmasses. *See also* **Gondwand(land), Laurasia.**
- **panmictic** referring to a population in which mating is completely random.
- **parallel evolution** the occurrence of independent evolutionary modifications of the same or a similar kind in closely related groups that have similar developmental programs inherited from a relatively recent common ancestor, and often confused with convergent evolution.
- **paralogous genes** the descendants of a duplicated gene, such as the α-globin and β-globin genes in vertebrate animals. *See also* **orthologous genes.**
- **parapatric speciation** the formation of reproductively isolated species from conspecific populations in adjacent locations. *See also* **allopatric speciation, peripatric speciation, sympatric speciation.**
- **paraphyletic** referring to a monophyletic lineage that does not include all its descendent groups in a classification scheme, such as the paraphyletic class Reptilia, which does not include its avian or mammalian descendant classes.
- **parsimony method** a means by which the most likely tree is selected from among all the possible evolutionary trees constructed from a matrix of morphological or molecular data; the chosen tree is the most parsimonious one, having the minimum number of evolutionary changes required for the phylogenetic reconstruction from the data set. *See also* **compatibility method, distance matrix methods.**
- **peak shifts** movements from one adaptive peak to another in an adaptive landscape, influenced by selection and other evolutioanry processes.
- **period** a subdivision of geologic time within the larger interval of an era.
- **peripatric speciation** a mode of speciation by which small isolated conspecific populations at the periphery of a main population undergo extensive and diverse modifications of their gene pools in a "genetic revolution," which is set in motion by the founder effect due to the few starting individuals in each peripheral isolate. *See also* **allopatric speciation, parapatric speciation, sympatric speciation.**
- **Permian Period** the sixth, and last, geologic time interval of the Paleozoic

Era, from 286 to 248 Myr BP, noted for the appearance of therapsid reptiles and for the most extensive of the known mass extinctions of the Phanerozoic Eon.

- **Phanerozoic Eon**   the third of the three largest subdivisions of geologic time, covering the past 590 Myr, and noted as the time of greatest biological evolutionary diversification; the current geologic time interval, which is subdivided into the Paleozoic, Mesozoic, and Cenozoic eras. *See also* **Archean Eon, Proterozoic Eon.**
- **pharynx**   the space at the back of the throat, which extends from the nasal cavity down to the esophagus and may act as a sound chamber for vocalization.
- **phenetics**   a system of classification that attempts to reconstruct phylogenies by emphasizing similarities among taxa in terms of numerical scores and quantitative measures incorporated into an index of similarity. *See also* **cladistics, phylogenetics.**
- **phenogram**   the phenetic portrayal of a phylogeny as a branched tree whose taxa are linked by estimates of overall similarity of a sample of characters, with no evaluation of whether these characters are ancestral or derived. *See also* **cladogram, dendrogram.**
- **phenotype**   the observable properties of an organism, produced by the genotype in conjunction with the environment. *See also* **genotype.**
- **photophosphorylation**   the light-dependent syntheses of ATP during photosynthesis, by a cyclic or noncyclic pathway.
- **photosynthesis**   the capture by chlorophylls and other light-sensitive pigments of light energy and its conversion to chemical energy in ATP and NADPH, and the subsequent use of this chemical energy to make sugars from reduction products of CO_2; the processes in green cells and certain bacteria of food manufacture from light energy and CO_2.
- **photosystem I**   the photochemical reaction complex in photosynthetic plants and bacteria through which $NADP^+$ is reduced, but molecular oxygen is not evolved.
- **photosystem II**   the photochemical reaction complex to which water provides electrons, leading to the release of molecular oxygen, and the electrons are passed from reaction-center chlorophylls to an electron transport chain linked to photosystem I; the complex of light-harvesting pigments found exclusively in oxygenic photosynthetic organisms.
- **phylogenetics**   a system of classification that attempts to reconstruct phylogenies from lineages traced back to common ancestral forms, based on a spectrum of traits that have undergone both anagenetic and cladogenetic modifications along shared pathways of evolutionary change, and depicted in a phylogenetic tree. *See also* **cladistics, phenetics.**
- **phylogenetic studies**   the genealogical analysis of living organisms and their ancestors and of their patterns of evolutionary change.
- **phylogeny**   the family history of living organisms and their ancestors during evolution; an evolutionary genealogy.

- **placenta** an organ formed from embryonic and maternal tissues, through which a therian mammal embryo or fetus is nourished before birth.
- **planetesimal** a small preplanetary body that may enlarge and become a planet during solar system formation from a solar nebula or dust cloud.
- **Plantae** the kingdom of eukaryotic organisms that have cells with rigid cell walls, are primarily photosynthetic, are usually anchored to a substratum, and include some groups of algae, nonvascular bryophytes, and a variety of vascular land forms that reproduce by spores or seeds.
- **plasma cell** the B cell, or activated B lymphocyte, which produces and secretes antibodies (immunoglobulins) to invading antigens in an immune response.
- **plasma membrane** the mosaic bilayer of phospholipids and proteins that encloses the living protoplasm of the cell and functions as a dynamic boundary between the cell and its surroundings.
- **plate tectonics** the phenomena and processes involved in changing the topography of the Earth's crust by means of the movement of lithospheric plates over the underlying mantle of semimolten rock.
- **platyrrhine** a member of the anthropoid infraorder of Primates that includes the New World monkeys, whose widely separated nostrils open across the nose. *See also* **catarrhine.**
- **pleiotropy** the phenomenon of a single gene being responsible for a number of distinct and seemingly unrelated traits.
- **Pleistocene Epoch** the first geologic time interval of the Quaternary Period of the Cenozoic Era, from 1.8 to 0.045 Myr BP, noted for the appearance and diversification of the hominines.
- **Pliocene Epoch** the fifth, and last, geologic time interval of the Tertiary Period of the Cenozoic Era, from 5 to 1.8 Myr BP, noted for the appearance of the hominids.
- **polymer** the association of monomers in a larger molecule.
- **polymorphism** the presence of two or more allelically different classes in the same interbreeding population, as distinct from a monomorphic population with only one of the possible number of allelic combinations.
- **polyphyletic** referring to a lineage of species supposedly derived from two or more different ancestral forms. *See also* **monophyletic.**
- **polyploid** a cell or an individual that has more than two sets of chromosomes.
- **pongid** a member of the Pongidae, the family of great apes that includes the orangutan, gorilla, and chimpanzee.
- **population bottleneck** a severe reduction in population size, usually accompanied by reduced diversity of the gene pool
- **population flush** an expansion in population size subsequent to a bottleneck, often leading to altered gene frequencies in the new gene pool.
- **population genetics** the study of the genetic composition of populations by the analysis of gene frequencies and of the processes that lead to changes in these frequencies and their effects on subsequent generations.

- **posttranscriptional control** any regulatory mechanism that influences gene expression by acting on or modifying the RNA transcript, rather than transcription itself.
- **preadaptation** a new feature of an organism that becomes more fit as a consequence of the inherited trait, but usually long after that trait had arisen in an environment for which it had no particular adaptive value.
- **Precambrian** the colloquial term for the first 4000 Myr of geologic time, before the relatively sudden appearance of diverse life forms in the Cambrian Period; the time interval that includes the Archean and Proterozoic eons.
- **Primates** the order of mammals that includes all the prosimian and anthropoid species, which are characterized by features such as an enlarged brain in proportion to body size, increasing cerebral dominance, a well-developed visual system, grasping hands and feet, nails in place of claws, and a primarily arboreal life style, although some exceptions are known for each of these features in one or more living species.
- **progenesis** the acceleration of ontogenetic development during evolution, leading to the retention of juvenile (ancestral) features in the adults of the species. *See also* **neoteny, paedomorphosis.**
- **prokaryote** a cellular organism lacking a membrane-bounded nucleus, typically a bacterial, cyanophyte, or prochlorophyte member of the kingdom Monera. *See also* **eukaryote.**
- **promoter** a specific nucleotide sequence in DNA, to which RNA polymerase binds and initiates transcription; a nontranscribed regulatory sequence upstream of the structural gene and essential for gene expression.
- **prosimian** a member of the suborder of Primates that includes the lemuroids, lorisoids, and tarsioids, and distinct from species in the anthropoid suborder.
- **proteinoid microsphere** the small, spherical structure that forms in water from proteinlike polymers that are produced from heated mixtures of dry amino acids; a protobiont model.
- **Proterozoic Eon** the second of the three largest subdivisions of geologic time, from 2500 Myr to 590 Myr BP, and noted for the absence of multicellular organisms for most of its duration. *See also* **Archean Eon, Phanerozoic Eon.**
- **Protista** the kingdom of nearly or exclusively unicellular eukaryotic organisms.
- **protobiont** the prelife entity that ultimately evolved into a true life form, or eubiont, and has been suggested to be a structure resembling a coacervate, liposome, or proteinoid microsphere.
- **proton–proton fusion cycle** the set of thermonuclear reactions by which helium is produced from protons in the hot interiors of stars; one of the hydrogen-burning processes in stars.
- **protostar** an early stage of star formation in a solar nebula or dust cloud.
- **Prototheria** the subclass of mammals that lay eggs, represented today by the platypus and spiny anteater (echidna) of Australia. *See also* **Theria.**

- **punctuated equilibrium** a proposal by Niles Eldredge and Steven Jay Gould that evolutionary lineages are characterized by brief episodes of rapid speciation separated by long intervals of stasis, or equilibrium, of the species thus produced, in contrast to the alternative of gradual evolution of new taxa.
- **pyruvic acid** a three-carbon organic acid that is the product of glycolysis and a reactant in many biochemical pathways.

- **quadrupedalism** the mode of locomotion in which all four limbs are used to move from place to place.
- **Quaternary Period** the second, and current, geologic time interval of the Cenozoic Era, begun 1.8 Myr BP.

- **ramapithecine** a member of a hominoid lineage of the middle and late Miocene, with a mixture of features resembling pongids and hominids, and inhabitants of open woodlands in Asia, Africa, and possibly Europe; a member of the genera *Ramapithecus, Sivapithecus, Kenyapithecus, Gigantopithecus,* and others related more closely to the living orangutan than to African pongids or to hominids.
- **random genetic drift** *See* **genetic drift.**
- **rDNA** ribosomal DNA; the repeated ribosomal genes clustered at a nucleolar-organizing region of a nucleolar-organizing chromosome in the eukaryotic genome.
- **reading frame** the sequence of codons translated into a polypeptide product of gene expression, beginning at a START codon and ending at a downstream STOP codon.
- **recapitulation** the theory that during ontogeny, the individual passes through stages resembling the adult forms of its successive ancestors, and embodied in Ernst Haeckel's theory that early ontogenetic features represent ancestral character states and later ontogenetic features represent derived character states, or "ontogeny recapitulates phylogeny."
- **Recent Epoch** the second geologic time interval of the Quaternary Period of the Cenozoic Era, from 45,000 years ago to the present; Holocene Epoch.
- **Red Queen hypothesis** the explanation proposed by Leigh Van Valen for the general observation that presumably more fit descendant species are not longer lived than their presumably less fit ancestors; the hypothesis states that natural selection acts more to maintain adaptive states than to improve them in ever-changing environments, and any species is doomed to extinction as a consequence of having finite genetic resources for change.
- **regulatory sequence** a nucleotide sequence, such as a promoter or an enhancer, that acts on or modifies gene expression but does not alter the structure of the gene product.
- **relative rate test** a calibration-independent test for the constancy of the rate of nucleotide substitutions in different lineages during their evolution,

thus determining whether they have the same molecular clock and can be compared reliably in phylogenetic analysis.

· **repetitive DNA**   reiterated sequences of a DNA segment or of genes in a genome, which may be of varied length and occur in hundreds, thousands, or millions of copies.

· **replica plating test**   an experiment designed to determine whether fit variants are selected spontaneous mutants that were present in a population prior to exposure to a selective agent, or arose in response to the need of the organism to survive in a hostile environment, and characterized by identifying such variants in environments lacking the agent throughout the course of the experiment. *See also* **fluctuation test.**

· **repressor**   the protein product of a gene that regulates transcription of a bacterial operon through binding to the operator, thereby preventing RNA polymerase from moving along the template to catalyze transcription of the adjacent structural genes.

· **reproductive isolation**   the absence of interbreeding between members of different species, or the ineffectiveness of such matings because of the production of sterile or inviable hybrid offspring.

· **restriction enzyme**   any of a class of endonucleases that cut both strands of duplex DNA at specific sites of specific sequences composed of four to six base pairs showing rotational symmetry.

· **ribosomal DNA**   *See* **rDNA.**

· **rooted tree**   an evolutionary tree or a phylogeny that includes ancestral and descendant species, thus indicating the direction of evolutionary change. *See also* **unrooted tree.**

· **R plasmid**   an autonomously replicating duplex DNA molecule that consists of a resistance transfer factor (RTF) and one or more transposable *r* determinants, and that can be transferred from one bacterial cell to another, influencing the phenotypic expression of the cells.

· **saurischian dinosaurs**   the herbivorous and carnivorous reptiles of the Dinosauria, having the typically reptilian three-pronged pelvis, rather than the four-pronged pelvis of the ornithischian dinosaurs.

· **sea-floor spreading**   the tectonic process in which new material is added to the ocean floor from molten mantle rock that rises up and is extruded at an ocean ridge, adding mantle rock to existing lithospheric plates on which new sediments are deposited; the process responsible for continental (plate) drift.

· **sedimentary rock**   mineral matter formed from the accumulation of rock and organic particles cemented together in quiet waters, such as sandstone, limestone, shale, and chalks. *See also* **igneous rock, metamorphic rock.**

· **selection coefficient**   the proportion of reduction in fitness of a genotype, or its measure of selective disadvantage; symbolized as *s*.

· **selective disadvantage**   the reduction in fitness of a genotype, or ability to survive and reproduce; symbolized by the selection coefficient *s* in the formula $w = 1 - s$.

- **sibling species**   species that are almost identical morphologically but are reproductively isolated.
- **Silurian Period**   the third geologic time interval of the Paleozoic Era, from 438 to 408 Myr BP, noted for the appearances of jawed fishes and terrestrial plants and animals.
- **Social Darwinism**   the thesis presented by Herbert Spencer, a British social theorist and contemporary of Darwin, proposing that society functions as an organism and is responsive to a deterministic mechanism analogous to natural selection, leading progressively to the natural corollaries of a perfect social order and a superior biological condition.
- **speciation**   the divergence of one population into two or more populations of different organisms that are closely related but usually reproductively isolated; the formation of new species from an ancestral population.
- **species**   a group of individuals that can breed with one another but not with members of other species.
- **splitting**   a proposed mode of speciation characterized by the relatively rapid separation of an ancestral population into two or more descendant populations that are genetically altered and reproductively isolated, rather than by gradual divergence of ancestral to descendant forms over a relatively long period of time.
- **spreading ridge**   one of the three kinds of tectonic plate boundaries; the boundary of a plate at which it forms and from which it moves away. *See also* **subduction zone, transform fault.**
- **stabilizing selection**   the elimination by natural selection of those alleles that produce deviations from the mean population phenotype, leading to the maintenance of a stable but reduced level of genetic diversity in the gene pool; normalizing selection.
- **stochastic processes**   random processes, such as mutation and genetic drift, that lead to no particular direction of change in an evolutioanry lineage. *See also* **deterministic processes.**
- **stromatolite**   the pillarlike mound of layered mats of debris left by the activities of photosynthetic prokaryotic organisms, usually in a shallow body of water.
- **structural gene**   a nucleic acid sequence that codes for RNA or protein, and distinct from regulatory genes or sequences, which control the expression of structural genes.
- **subduction zone**   one of the three kinds of tectonic plate boundaries; the boundary of a plate where it is carried into the mantle and destroyed. *See also* **spreading ridge, transform fault.**
- **subspecies**   subdivisions of a species that are reproductively compatible with one another but which occupy different areas or living zones and thus have less opportunity to interbreed.
- **supergene**   a block of genes transmitted from one generation to another as a single unit.
- **sympatric speciation**   divergence within a panmictic population that may result in reproductively isolated descendant species in the absence of spatial

isolation. *See also* **allopatric speciation, parapatric speciation, peripatric speciation.**

- **sympatry**   the existence of different subpopulations of a species in the same geographical space.

- **taxon**   the general name for a taxonomic group of any rank.
- **tectonic plates**   the subdivisions of the Earth's lithosphere that abut one another and can float from place to place on the underlying semimolten mantle, carrying their continental landmasses and ocean basins to different locations on the planet.
- **Tertiary Period**   the first geologic time interval of the Cenozoic Era, from 65 to 1.8 Myr BP, noted for mammalian diversification and dominance on land.
- **therapsids**   the group of mammal-like reptiles that included the ancestral stock of the mammals, and disappeared in the Early Jurassic Period.
- **Theria**   the subclass of mammals that bear live young, consisting of the marsupial and placental mammalian species. *See also* **Prototheria.**
- **tit for tat**   the theory proposed by Robert Axelrod and William Hamilton for the evolution of socially cooperative behavior of individuals that repeatedly interact.
- **transcription control**   any mechanism that regulates gene expression by influencing transcription (on, off, increase, decrease).
- **transduction**   the unidirectional transfer of genes from one bacterial cell to another through the agency of a virus, which acts as a vector for transfer. *See also* **conjugation, transformation.**
- **transformation**   the heritable modification of the properties of a recipient cell by DNA transferred from a donor cell. *See also* **conjugation, transduction.**
- **transform fault**   one of the three kinds of tectonic place boundaries; the boundary at which two plates scrape past each other during their movement over the mantle of the Earth. *See also* **spreading ridge, subduction zone.**
- **transitional forms**   the series of modified populations produced during the gradual evolution of descendant species from an ancestral species; "missing links" in an evolutionary lineage.
- **translational control**   any mechanism that regulates expression by influencing translation, such as modulation of gene expression in cells having long-lived messenger RNA, or genome rearrangements in immunoglobulin synthesis.
- **translocation**   the structural rearrangement of chromosomes through the transfer of segments between nonhomologous chromosomes.
- **transposable genetic element**   a unique DNA sequence that can move from one site to another in a genome, through insertion and excision, and thereby influence gene expression; a mobile genetic element.
- **transposon**   a transposable genetic element that is over 2000 base pairs long and includes genes for insertion into a chromosome as well as genes unrelated to insertion.

- **Triassic Period**   the first geologic time interval of the Mesozoic Era, from 248 to 213 Myr BP, noted for the first dinosaurs and a major episode of mass extinction at its close.
- **trilobite**   a marine arthropod that existed only in the Paleozoic Era and that serves as an index fossil for at least some of the periods of the era.
- **trisomy**   the condition in which one chromosome is present in three copies instead of the usual two copies in a diploid cell or individual; $2n + 1$.

- **unicellular**   the designation of a single-celled organism, such as a bacterium or protist, in contrast to a multicellular organism.
- **uniformitarianism**   the concept that the geologic phenomena evident on the Earth today are consequences of the same processes that have been operating throughout Earth history, which are the principal forces responsible for the present-day geologic features of the planet. *See also* **catastrophism.**
- **unrooted tree**   an evolutionary tree or network that makes no statement about the direction of evolutionary change, but can be rooted if the group of species can be connected to an ancestral stem and thus indicate direction of evolutionary change.
- **upright bipedalism**   the characteristic of consistent upright posture and locomotion on two legs; the major hominid trait.
- **use and disuse**   *See* **inheritance of acquired characteristics.**

- **wild type**   the most common allele or phenotype, or one designated as "normal."
- **wobble hypothesis**   the proposal by Francis Crick that a transfer RNA may recognize two or more different codons in messenger RNA because of reduced constraint ("wobble") of the 5′ base of the tRNA anticodon in pairing with the 3′ base of the mRNA codon, thus explaining the requirement for only 32 kinds of tRNA in translation of 61 different codons in mRNAs.

Index